Nonlinear Analysis

Hwai-Chiuan Wang

National Tsing Hua University Press
December 2003

Hwai-chiuan Wang
Department of Mathematics
National Tsing Hua University
Hsinchu, Taiwan

2000 Mathematics Subject Classification:
Primary 46B10, 47A10, 58E05, 35J20, 35J25

Hwai-chiuan Wang
Nonlinear Analysis

ISBN 957-01-4476-9

Printed in Taiwan 2003.12

PREFACE

This volume is intended as an essentially self-contained exposition of portions of the theory of nonlinear analysis. It grew out of lecture notes for graduate courses by the author at National Tsing Hua University, Hsinchu, Taiwan.

For many decades, great mathematical interest has focused on problems associated with linear operators and the extension of the results of linear algebra to an infinite-dimensional context. When one drops the assumption of linearity, the associated operator theory and the many concrete problems associated with such a theory represent a frontier of mathematical research. Just as in the linear case, these results were inspired by and are highly relevant to concrete problems in mathematical analysis. We shall take up a variety of analytic and topological techniques for the study of nonlinear problems and their applicability to a variety of concrete problems taken from various fields of mathematical analysis.

This volume contains: function analysis; fixed point theorems in Banach spaces; calculus in Banach spaces; bifurcation theory for operators on Banach spaces; Sobolev spaces in $\mathbb{R}$ and in $\mathbb{R}^N$; $N > 1$; the theory of compact operators, Fredholm operators, and monotone operators; the variational methods for nonlinear equations; the topological degree of the compactness perturbation of the identity map; and the applications of topological degrees to ODEs and PDEs.

Hwai-Chiuan Wang
July 2003

Contents

PREFACE   iii

1 FUNCTIONAL ANALYSIS   1

2 FIXED POINT THEOREMS   19
- 2.1 Banach Fixed Point Theorem 19
- 2.2 Brouwer Fixed Point Theorem 24
- 2.3 Leray-Schauder Fixed Point Theorem 25

3 CALCULUS ON BANACH SPACES   27
- 3.1 Fréchet Derivatives and Gâteaux Derivatives 27
- 3.2 Sum Rule, Chain Rule, and Product Rule 30
- 3.3 Mean Value Theorems 32
- 3.4 Sard Theorem 35
- 3.5 Fundamental Theorem of Calculus 36
- 3.6 Inverse Function Theorem 39
- 3.7 Implicit Function Theorem 42
- 3.8 Taylor's Theorem 47

4 BIFURCATION THEORY   53
- 4.1 Morse Theory 53
- 4.2 Krasnoselskii Theorem and Rabinowitz Theorem 66

5 SOBOLEV SPACES $W^{1,p}(I)$ 77
- 5.1 Motivation 77
- 5.2 The Sobolev Space $W^{1,p}(I)$ 79
- 5.3 Sobolev Spaces $W^{m,p}(I)$ 94
- 5.4 The Space $W_0^{1,p}(I)$ 99
- 5.5 The Dual Space of $W_0^{1,p}(I)$ 103

6 SOBOLEV SPACES $W^{1,p}(\Omega)$ 107
- 6.1 Elementary Properties of Sobolev Spaces $W^{1,p}(\Omega)$ 107
- 6.2 Characterizations 116
- 6.3 Derivatives 119
- 6.4 Extension Operators 125
- 6.5 The space $W^{m,p}(\Omega)$ 134
- 6.6 Sobolev Inequalities 134

6.7   The Space $W_0^{1,p}(\Omega)$ 152
6.8   Potential Estimates 161
6.9   Application of Chain Rules 170
6.10 Lipschitz Functions 179

7   OPERATOR THEORY   **185**
7.1   Compact Operators 185
7.2   Fredholm Operators 194
7.3   Monotone Operators in Hilbert Spaces 200

8   VARIATIONAL METHODS   **211**
8.1   Stampacchia Theorem and Lax-Milgram Theorem 211
8.2   Ljusternik-Schnirelman Constrained
      Theorems 219
8.3   Ljusternik-Schnirelman Theorems 234
8.4   Concentration-Compactness principle 236
8.5   Ekeland Variational Principle 258
8.6   Best Sobolev Constant 263
8.7   Palais-Smale Sequences 269
8.8   Global Compactness 276

9   TOPOLOGICAL DEGREE THEORY   **297**
9.1   Finite Dimensional Degree Theory 297
9.2   Brouwer Fixed Point Theorem 323
9.3   The Borsuk Theorem 334
9.4   Multiplicative Property of Degrees 341
9.5   The Infinite-Dimensional Degree Theory 347
9.6   Leray-Schauder Index Theory 356
9.7   Leray-Schauder Fixed Point Theorems 362

10 TOPOLOGICAL DEGREE APPLICATIONS   **367**
10.1 Regularity Theorems and
      Maximum Principles 367
10.2 Elliptic Quasilinear Equations 370
10.3 Connectedness of Solution Sets 376
10.4 Global Results for Eigenvalue Problems 382
10.5 Sturm-Liouville Nonlinear Problems 391
10.6 Strictly Nonlinear Sturm-Liouville Problems 394
10.7 Applications of Bifurcation Problems 398
10.8 Bifurcation at Infinity 404
10.9 Existence of Pairs of Positive Solutions 410
10.10 Existence of Pairs of Solutions in S_k 412

Contents

BIBLIOGRAPHY 417

INDEX 421

Chapter 1

FUNCTIONAL ANALYSIS

In this chapter we present various theorems of Functional Analysis.

Theorem 1.1 (*Hahn-Banach Theorem*)

(*i*) *Let $p(x)$ be a seminorm defined on a linear space X, M a linear subspace of X, and $f(x)$ a linear functional defined on M with $|f(x)| \leq p(x)$ for $x \in M$. Then f can be extended to a linear functional F on X with $|F(x)| \leq p(x)$ for $x \in X$.*

(*ii*) *Let X be a normed linear space, M a linear subspace of X, and $f : M \to \mathbb{R}$ a continuous linear functional with norm $\|f\|_M = \sup\{f(x)|x \in M, \|x\| \leq 1\}$. Then, there is a continuous linear functional $F : X \to \mathbb{R}$ such that $F|_M = f$, and $\|F\|_X = \|f\|_M$.*

Theorem 1.2 (*Open Mapping Theorem*) *Let X, Y be Banach spaces, and $f : X \to Y$ a continuous linear map with $f(X) = Y$. Then f is open: the image of every open set is open. If f is also injective, then f has a continuous linear inverse.*

Theorem 1.3 (*Banach-Steinhaus Theorem*) *Let X and Y be Banach spaces, $B(X,Y)$ the space of all continuous linear operators of X into Y, and $\{f_\alpha\}_{\alpha \in \Lambda}$ a family in $B(X,Y)$. If, for each $x \in X$, the set $\{f_\alpha(x)\}_{\alpha \in \Lambda}$ is bounded, then $\{f_\alpha\}_{\alpha \in \Lambda}$ is uniformly bounded: there is $c > 0$ such that $\|f_\alpha\| \leq c$ for all $\alpha \in \Lambda$. In particular, if there is $\lim_{n \to \infty} f_n(x)$ for each $x \in X$, then there is $c > 0$ such that $\|f_n\| \leq c$ for $n = 1, 2, ...$, and there is $f \in B(X, Y)$ such that $\lim_{n \to \infty} f_n(x) = f(x)$ for all $x \in X$. (However, f_n does not necessarily converge to f strongly in $B(X,Y)$).*

Theorem 1.4 *A weakly convergent sequence $\{x_n\}$ in a normed linear space X has a unique limit x such that there is a constant $M > 0$ that satisfies $\|x_n\| \leq M$ for each n and $\|x\| \leq \liminf_{n \to \infty} \|x_n\|$.*

Proof. See Brezis [7, p. 35] ∎

Theorem 1.5 (*Banach-Alaoglu Theorem*) *Let X be a Banach space with dual space X^*. If $\mathbb{A} \subset X^*$ is a bounded set, then $\mathbb{A}$ is w^*-compact.*

Lemma 1.6 (*F. Riesz Lemma*) *Let M be a proper closed linear subspace of a normal linear space X. Then for each θ, $0 < \theta < 1$, there is $x_0 \in X$ such that $\|x_0\| = 1$, $d(x_0, M) \geq \theta$.*

Proof.   Choose $x_1 \in X \backslash M$. Because M is closed, $d = d(x_1, M) > 0$. For $\epsilon = \frac{d(1-\theta)}{\theta}$, take $m_0 \in M$ such that

$$\|x_1 - m_0\| < d + \epsilon.$$

Set

$$x_0 = \frac{x_1 - m_0}{\|x_1 - m_0\|}.$$

Then $\|x_0\| = 1$, and for $m \in M$

$$
\begin{aligned}
\|x_0 - m\| &= \|\frac{x_1 - m_0}{\|x_1 - m_0\|} - m\| \\
&= \frac{1}{\|x_1 - m_0\|} \|\|x_1 - m_0\|m - x_1 + m_0\| \\
&= \frac{1}{\|x_1 - m_0\|} \|(\|x_1 - m_0\|m + m_0) - x_1\| \\
&\geq \frac{d}{\|x_1 - m_0\|} > \frac{d}{d + \epsilon} = \theta.
\end{aligned}
$$

Thus, $d(x_0, M) \geq \theta$.   ∎

Theorem 1.7 *Let X be a normed linear space. If the unit sphere $\{x \in X | \|x\| = 1\}$ is compact, then X is finite-dimensional.*

Proof.   Choose $x_1 \in X$, $\|x_1\| = 1$, and set $M_1 = <x_1>$. If $M_1 = X$, then we are done. Otherwise, take $x_2 \in X \backslash M_1$, $\|x_2\| = 1$, $d(x_2, M_1) \geq \frac{1}{2}$, and set $M_2 = <x_1, x_2>$. Similarly, if $M_2 = X$, then we are done. Otherwise we continue this process. Either we stop at a finite process, or we obtain a sequence $\{x_n\}_{n=1}^{\infty}$ in X such that $\|x_n\| = 1$, $\|x_n - x_m\| \geq \frac{1}{2}$ if $n \neq m$. This sequence does not have any convergent subsequence, showing that the unit sphere is not compact. This is a contradiction, so $X = M_n$, a finite- dimensional space, for some finite n.   ∎

Theorem 1.8 (*Banach Theorem*) *A Banach space X is reflexive if and only if the closed unit ball $B_1(0)$ in X is weakly sequentially compact. In particular, bounded sets in a Hilbert space are weakly sequentially compact.*

Lemma 1.9 *Let N be a finite-dimensional subspace of a normed linear space X. Then there is a closed linear subspace M such that $X = N \oplus M$.*

Proof.   Set $N = <x_1, \cdots, x_n>$. Take $f_1, \cdots, f_n$ in X^* such that $f_i(x_j) = \delta_{ij}$ and set $A = <f_1, \cdots, f_n>$. Let $M = A^{\perp}$. Then $X = N \oplus M$. In fact, for $x \in X$, if $x \notin M$ then $f_j(x) \neq 0$ for some j, $1 \leq j \leq n$. Setting $\alpha_i = f_i(x)$ for each i, we obtain

$$f_k(x - \sum_{i=1}^{n} \alpha_i x_i) = f_k(x) - \alpha_k = 0 \text{ for each } k.$$

Thus, $x - \sum_{i=1}^{n} a_i x_i \in M : x \in N + M$. Moreover, if $x \in N \cap M$, then

$$x = \sum_{i=1}^{n} a_i x_i, \quad f_j(x) = 0 \text{ for each } j,$$

then $a_j = 0$ for each j, so $x = 0$. ∎

Lemma 1.10 *Let N be a closed linear subspace of a normed linear space X, of codim $N = \dim N^\perp = n$. Then, there is an n-dimensional subspace M of X such that $X = N \oplus M$.*

Proof.

(i) $N = (N^\perp)^\perp$: If $x \in N$, then $f(x) = 0$ for $f \in N^\perp$. Thus, $x \in (N^\perp)^\perp$. If $x \notin N$, because N is closed, we have $d(x, N) > 0$. Then the Hahn-Banach Theorem 1.1 implies that there is $f \in X^*$ such that

$$f(x) = d(x, N) > 0,$$
$$\|f\| = 1,$$
$$f(N) = 0.$$

Thus, $f \in N^\perp$, $f(x) > 0$, and $x \notin (N^\perp)^\perp$.

(ii) $X = N \oplus M$: Set $N^\perp = \langle f_1, \cdots, f_n \rangle$. Note that $x \in N$ if and only if $f_j(x) = 0$ for each j by (i). Take $x_1, \cdots, x_n \in X$ such that $f_j(x_i) = \delta_{ij}$. If $\sum_i \alpha_i x_i = 0$, then $f_j(\sum_i \alpha_i x_i) = 0$, or $\alpha_j = 0$ for each j. Hence, $x_1, \cdots, x_n$ are linearly independent. Set $M = \langle x_1, \cdots, x_n \rangle$. For $x \in X$, set $w = \sum_k f_k(x) x_k$, then $w \in M$ with $f_j(x - w) = 0$ for each j, or $x - w \in N$. Thus $x \in N + M$. If $x \in N \cap M$ then $x = \sum_i \beta_i x_i$, and $f_j(x) = 0$ for each j. Hence, $\beta_i = 0$ for each i, and we obtain $x = 0$. We conclude that $X = N \oplus M$. ∎

Theorem 1.11 (*Riesz-Fréchet Representation Theorem*) *Let H be a Hilbert space with dual H'. Assume $\varphi \in H'$. Then there is a unique $f \in H$ such that*

$$\varphi(v) = (f, v) \text{ for each } v \in H,$$
$$\|f\|_H = \|\varphi\|_{H'}.$$

Proof. See Theorem 8.1. ∎

Theorem 1.12 *Let E be a real Banach space, and let $C \subseteq E$ be a convex set. Then C is $\sigma(E, E')$-closed if and only if C is strongly closed.*

Proof. By the Hahn-Banach Theorem 1.1. ∎

Theorem 1.13 (*Mazur Theorem*) *Let E be a normed linear space, and $A \subset E$ a weakly relatively compact subset. Then its convex closure conv A is compact. In other words, if $x_n \rightharpoonup x$ in E, then there is a sequence (y_n),*

$$y_n = \sum_{k=n}^{N_n} a_k x_k,$$

where $a_k \geq 0$, $n \leq k \leq N_n$, and $\sum_{k=n}^{N_n} a_k = 1$, such that $y_n \to x$ in E.

Proof. For fixed $n \in N$, let $A = \cup_{k=n}^{\infty} \{x_k\}$, and let conv A be the convex hull of A. By the hypothesis, because x must be in the weak closure of A, then x is in the weak closure of conv A; but the weak closure of conv A is the closure of conv A. Therefore we obtain

$$y_n = \sum_{k=n}^{\overset{*}{N_n}} a_k x_k,$$

where $a_k \geq 0$, $n \leq k \leq N_n$, $\sum_{k=n}^{N_n} a_k = 1$, such that $\|y_n - x\| \leq 1/n$. ∎

Remark 1.14 *The Mazur Theorem 1.13 is false if a Banach space is replaced by a locally convex topological vector space.*

Definition 1.15 *Let E be a Banach space. A map $f : E \to (-\infty,\ \infty]$ is lower semicontinuous if $f(x) \leq \liminf_{n \to \infty} f(x_n)$ whenever $\{x_n\}$ converges to x in E.*

Theorem 1.16 *Let H be a Hilbert space, and A a positive self-adjoint continuous linear operator on H. Define $f(x) = (Ax, x)$, then f is a weakly lower semicontinuous function on H.*

Proof. Let $x_n \rightharpoonup x$ weakly in H. Then

$$0 \leq (A(x_n - x), (x_n - x)) = (Ax_n, x_n) - (Ax_n, x) - (Ax, x_n) + (Ax, x)$$

Thus

$$f(x_n) = (Ax_n, x_n) \geq (Ax_n, x) + (Ax, x_n) - (Ax, x)$$

We have

$$\liminf_{n \to \infty} f(x_n) \geq \lim_{n \to \infty} (x_n, Ax) + \lim_{n \to \infty} (Ax, x_n) - (Ax, x)$$
$$= (x, Ax) + (Ax, x) - (Ax, x)$$
$$= (Ax, x) = f(x). \qquad \blacksquare$$

Theorem 1.17 *Suppose E is a Banach space, and $\varphi : E \to (-\infty,\ \infty]$ is convex and lower semicontinuous with respect to the strong topology. Then φ is weakly lower semicontinuous. In particular, if $x_n \rightharpoonup x$ weakly, then*

$$\varphi(x) \leq \liminf_{n \to \infty} \varphi(x_n).$$

Theorem 1.18 *Let E be a reflexive Banach space, and $K \subseteq E$ any convex, bounded and closed set. Then K is weakly compact.*

Now we come to the useful classical existence result of extremes.

Theorem 1.19 *Let E be a reflexive Banach space, and $A \subseteq E$ a nonempty closed, convex set. Assume that $\varphi : A \to (-\infty, \infty]$ is a convex, lower semicontinuous function, $\varphi \not\equiv \infty$ such that*

$$\lim_{\substack{x \in A \\ \|x\| \to \infty}} \varphi(x) = \infty. \tag{1.1}$$

Then φ attains its minimum on A: there is $x_0 \in A$ such that

$$\varphi(x_0) = \min_{x \in A} \varphi(x).$$

If A is bounded, then the condition (1.1) will be omitted.

Proof.   See Theorem 8.4.   ∎

Theorem 1.20 (*Projection on a closed convex set*) *Let H be a Hilbert space, and let $K \subseteq H$ be a nonempty closed convex set. For $f \in H$ there is a unique $u \in K$ such that*

$$|f - u| = \min_{v \in K} |f - v|$$

and u can be characterized by

$$\begin{cases} u \in K, \\ (f - u, v - u) \leq 0 \ \text{ for } \ v \in K. \end{cases}$$

Moreover, if we denote $u = P_K f$, then

$$|P_K f_1 - P_K f_2| \leq |f_1 - f_2| \ \text{ for } \ f_1, f_2 \in H.$$

Proof.   See Theorem 8.6.   ∎

Corollary 1.21 *Let H be a Hilbert space and $M \subseteq H$ a closed linear subspace. Then P_M is a linear operator. Moreover, $u = P_M f$ is characterized by*

$$\begin{cases} u \in M, \\ (f - u, v) = 0 \ \text{ for } \ v \in M. \end{cases}$$

Proof.   See Corollary 8.7.   ∎

Theorem 1.22 (*Weierstrass Theorem*) *For $a, b \in \mathbb{R}$, $a < b$, the polynomials are dense in $C([a, b])$.*

For the proofs of Theorem 1.23 and Corollary 1.25, see Dieudonne [18].

Theorem 1.23 (*Stone-Weierstrass Theorem*) *Let X be a compact metric space, and let $C(X)$ be the Banach algebra of all real-valued continuous functions on X under the supreme norm. Let A be a subalgebra of $C(X)$ containing the constant functions and separate points, then given any x, $y \in X$, if we can find $f \in A$ with $f(x) \neq f(y)$, then A is dense in $C(X)$.*

Theorem 1.24 (*Tietze-Urysohn Extension Theorem*) *Let E be a metric space, A a closed subset of E, and f a continuous bounded function of A into $\mathbb{R}$. Then there is a continuous function g of E into $\mathbb{R}$ that coincides with f in A and satisfies*

$$\begin{cases} \sup_{x \in E} g(x) = \sup_{x \in A} f(x), \\ \inf_{x \in E} g(x) = \inf_{x \in A} f(x). \end{cases}$$

Corollary 1.25 *Let A, B be two nonempty closed sets in a metric space E such that $A \cap B = \phi$. Then there is a continuous function f in E with values in $[0,1]$, such that $f(x) = 1$ in A, and $f(x) = 0$ in B.*

Theorem 1.26 (*Dugundji Extension Theorem*) *Let X be a metric space and A a closed subset of X. Let E be a locally convex topological vector space over real or complex numbers and C a convex subset of E. Then any continuous map $f : A \to C$ admits a continuous extension $F : X \to C$.*

Proof.   For $x \in X \backslash A$, take a bounded open neighborhood V_x of x such that

$$c \operatorname{dist}(V_x, A) \leq \operatorname{diam} V_x \leq \operatorname{dist}(V_x, A).$$

Then $\{V_x\}$ is an open covering of $X \backslash A$. Recall that a metric space is paracompact, and consequently $X \backslash A$ is paracompact: there is a locally finite refinement $\{U\}$ such that each U is open, $\{U\}$ covers $X \backslash A$, each U is contained in some V_x, and for each $x \in X \backslash A$ there is an open neighborhood O_x of x such that O_x is disjoint from all but a finite number of the $U's$. For $U_0 \in \{U\}$, define for $x \in X \backslash A$,

$$\lambda_{U_0}(x) = \frac{\operatorname{dist}(x, \ X \backslash U_0)}{\sum_U \operatorname{dist} (x, \ X \backslash U)}.$$

Note that dist $(x, X \backslash U) > 0$ if and only if $x \in U$. Because each $x \in X \backslash A$ is contained in some U, we obtain $0 \leq \lambda_{U_0}(x) \leq 1$. Because

$$|\operatorname{dist}(x, X \backslash U) - \operatorname{dist}(y, X \backslash U)| \leq \operatorname{dist}(x, y),$$

dist $(x, X \backslash U)$ is continuous. For $x \in X \backslash A$, $\lambda_{U_0}|_{O_x}$ has the form

$$\frac{\operatorname{dist}(x, \ X \backslash U_0)}{\sum_U \operatorname{dist} (x, \ X \backslash U)},$$

hence, $\lambda_{U_0}|_{O_x}$ is continuous, and consequently λ_{U_0} is continuous on $X\backslash A$, and $\lambda_{U_0}(x) = 0$ if and only if $x \notin U_0$. For each U, choose $a_U \in A$ satisfying $\text{dist}(a_U, U) < 2\text{dist}(A, U)$. Let the extension F of f be given by

$$F(x) = \begin{cases} \sum_U \lambda_U(x)f(a_U) \text{ for each } x \in X\backslash A, \\ f(x) \text{ for each } x \in A. \end{cases}$$

Because for each $x \in X\backslash A$, $\lambda_U(x) = 0$ except for finitely many U's, and $\sum_U \lambda_U(x) = 1$, and $f(a_U) \in C$, we have $F(x) \in C$. If $x \in X\backslash A$, then $F(x)$ is a finite sum of continuous functions. Hence, F is continuous on $X\backslash A$. However, F is continuous in the interior of A because $F = f$ on $\text{int}(A)$. Let $x_0 \in \partial A$, and $W \subset E$ be any convex open set containing 0. Because f is continuous on A, there is $\alpha > 0$ such that $a \in A$ and $\text{dist}(x_0, a) < \alpha$ imply

$$f(a) - f(x_0) \in W.$$

Let $O = \{x \in X \mid \text{dist}(x, x_0) < \frac{\alpha}{6}\}$. We must prove $F(x) - F(x_0) \in W$ for $x \in O$.

(i) Assume $x \in X\backslash A$, $\text{dist}(x, x_0) < \frac{\alpha}{6}$ and $\text{dist}(x, a_U) < \frac{\alpha}{2}$. Then

$$\text{dist}(x_0, a_U) \leq \text{dist}(x_0, x) + \text{dist}(x, a_U)$$
$$< \frac{\alpha}{6} + \frac{\alpha}{2} < \alpha,$$

so $f(a_U) - f(x_0) \in W$.

(ii) Assume $x \in X\backslash A$, $\text{dist}(x, x_0) < \frac{\alpha}{6}$, and $\text{dist}(x, a_U) \geq \frac{\alpha}{2}$. Then

$$\text{dist}(x, a_U) \geq 3\,\text{dist}(x, x_0) \geq 3\,\text{dist}(x, A). \tag{1.2}$$

Suppose $x \in U \subset V_x$,

$$\text{diam} V_x \leq \text{dist}(V_x, U)$$

so

$$\text{diam} U \leq \text{dist}(V_x, A) \leq \text{dist}(U, A).$$

In this case

$$\begin{aligned} \text{dist}(x, a_U) &\leq \text{dist}(a_U, U) + \text{diam } U \\ &< 2\text{dist}(A, U) + \text{diam } U \\ &< 3\text{dist}(U, A) \\ &\leq 3\text{dist}(A, x), \end{aligned} \tag{1.3}$$

and we have a contradiction between (1.2) and (1.3). Therefore, if $\text{dist}\,(x, x_0) < \frac{\alpha}{6}$, $\text{dist}(x, a_U) \geq \frac{\alpha}{2}$, then $x \notin U : \lambda_U(x) = 0$. Finally, for $x \in X\backslash A$, $\text{dist}\,(x, x_0) < \frac{\alpha}{6}$, we have

$$F(x) - F(x_0) = \sum_U \lambda_U(x)f(a_U) - f(x_0)$$
$$= \sum_U \lambda_U(x)(f(a_U) - f(x_0)).$$

From the above arguments, for each U either $\lambda_U(x) = 0$, or $f(a_U) - f(x_0) \in W$. The sum is actually a finite sum and $\sum_U \lambda_U(x) = 1$ with $0 \leq \lambda_U(x) \leq 1$. Hence, $F(x) - F(x_0) \in W$. ∎

Recall that the Hilbert cube I^∞ can be characterized by

$$I^\infty = I \times I \times \cdots = \{x = \{x_n\} \in \ell^2 \mid |x_n| \leq \frac{1}{n}\},$$

and I^∞ is a compact metric space.

Theorem 1.27 (*Urysohn Theorem*) *Any compact metric space admits an embedding in the Hilbert cube I^∞.*

Theorem 1.28 (*Radon Theorem*) *Let X be a uniformly convex Banach space and $\{f_k\}$ a sequence in X that converges weakly to $f \in X$. If $\|f_k\| \to \|f\|$, then $f_k \to f$ strongly : $\|f_k - f\| \to 0$ as $k \to \infty$.*

Proof.   The theorem is trivial to prove for the case $f = 0$. Let $f \neq 0$, and without loss of generality assume that $\|f\| = 1$, and that $\|f_k\| \neq 0$ for all k. Moreover, set $g_k = \|f_k\|^{-1} f_k$, and $h_k = f$ for each k. By the Hahn-Banach Theorem 1.1, there is $u_0 \in X^*$ such that $u_0(f) = 1$, and $\|u_0\| = 1$. Because $f_k \rightharpoonup f$ weakly, and $\|f_k\| \to \|f\|$ as $k \to \infty$, we have

$$u_0(g_k) = \|f_k\|^{-1} u_0(f_k) \to u_0(f) = 1 \ \ as \ \ k \to \infty. \tag{1.4}$$

Using $\|u_0\| = 1$ and $u_0(f) = 1$, we obtain

$$\begin{aligned} |1 + u_0(g_k)| &= |u_0(f) + u_0(g_k)| \\ &\leq \|f + g_k\| \\ &\leq 2. \end{aligned} \tag{1.5}$$

Let $k \to \infty$, in (1.4), (1.5) to have $\|g_k + f\| \to 2$, and consequently, $\|g_k - f\| \to 0$ because X is uniformly convex. Finally,

$$\begin{aligned} \|f - f_k\| &\leq \|f - g_k\| + \|g_k - f_k\| \\ &= \|f - g_k\| + \|\|f_k\|^{-1} f_k - f_k\| \\ &\leq \|f - g_k\| + \|f_k\|\|1 - \|f_k\|^{-1}\| \\ &= \|f - g_k\| + |1 - \|f_k\|| \to 0 \ \ as \ \ k \to \infty \end{aligned}$$

We conclude that $\|f - f_k\| \to 0$ as $k \to \infty$. ∎

Theorem 1.29 (*Clarkson Inequalities*)
Let $u, v \in L^p$ with $1 < p, q < \infty$, $p^{-1} + q^{-1} = 1$.

(i) *For $2 \le p < \infty$, we have*

$$\|u+v\|_{L^p}^p + \|u-v\|_{L^p}^p \le 2^{p-1}(\|u\|_{L^p}^p + \|v\|_{L^p}^p),$$
$$\|u+v\|_{L^p}^q + \|u-v\|_{L^p}^q \ge 2(\|u\|_{L^p}^p + \|v\|_{L^p}^p)^{q-1}.$$

(ii) *For $1 < p \le 2$, we have*

$$\|u+v\|_{L^p}^q + \|u-v\|_{L^p}^q \le 2(\|u\|_{L^p}^p + \|v\|_{L^p}^p)^{q-1},$$
$$\|u+v\|_{L^p}^p + \|u-v\|_{L^p}^p \ge 2^{p-1}(\|u\|_{L^p}^p + \|v\|_{L^p}^p).$$

Recall that a Banach space X is called uniformly convex if any sequences $\{g_k\}$, $\{h_k\}$ in X satisfy $\|g_k\| = \|h_k\| = 1$ for $k = 1, 2, ...,$ and $\|g_k + h_k\| \to 2$ as $k \to \infty$ implies $\|g_k - h_k\| \to 0$, as $k \to \infty$. A uniformly convex Banach space is reflexive, but the converse is not true.

Theorem 1.30 *We have the following properties:*

(i) *For $1 < p < \infty$, the Lebesgue space L^p is uniformly convex.*

(ii) *A Hilbert space is uniformly convex.*

Proof.

(i) Let $\{g_k\}$ and $\{h_k\}$ be two sequences in L^p, with $\|g_k\|_{L^p} = \|h_k\|_{L^p} = 1$, for $k = 1, 2, ...,$ and $\|g_k + h_k\|_{L^p} \to 2$ as $k \to \infty$. By the Clarkson inequalities, if $2 \le p < \infty$

$$\|g_k + h_k\|_{L^p}^p + \|g_k - h_k\|_{L^p}^p \le 2^{p-1}(\|g_k\|_{L^p}^p + \|h_k\|_{L^p}^p) = 2^p,$$

and if $1 < p \le 2$

$$\|g_k + h_k\|_{L^p}^q + \|g_k - h_k\|_{L^p}^q \le 2(\|g_k\|_{L^p}^p + \|h_k\|_{L^p}^p)^{q-1} = 2^q.$$

Let $k \to \infty$ to obtain $\|g_k - h_k\|_{L^p} \to 0$.

(ii) Instead of using the Clarkson inequalities for the L^p space, we apply the parallelogram law

$$\|u+v\|^2 + \|u-v\|^2 = 2(\|u\|^2 + \|v\|^2)$$

to a Hilbert space to obtain the conclusion.   ∎

Theorem 1.31 (*Ascoli Theorem*) *Let K be a compact metric space and F a bounded subset of $C(K)$. Suppose that F is uniformly equicontinuous, that is, for $\epsilon > 0$, there is $\delta > 0$ such that $d(x, y) < \delta$ implies $|f(x) - f(y)| < \epsilon$ for every $f \in F$. Then F is relatively compact in $C(K)$.*

Remark 1.32 *We use the following notation: $w \subset\subset \Omega$ means that $\overline{w}$ is compact and $\overline{w} \subset \Omega$.*

Definition 1.33 *A sequence (ρ_n) of functions is called a regularity sequence if $\rho_n \in C_c^\infty(\mathbb{R}^N)$, supp $\rho_n \subset B(0, 1/n)$, $\int \rho_n = 1$ and $\rho_n \geq 0$ on $\mathbb{R}^N$.*

Remark 1.34 *Let*

$$\rho(x) = \begin{cases} \exp(\dfrac{1}{|x|^2 - 1}) \text{ for } |x| < 1, \\ 0 \text{ for } |x| \geq 1, \end{cases}$$

Then $\rho \in C_c^\infty(\mathbb{R}^N)$, supp $\rho \subset B(0,1)$, and $\rho \geq 0$ on $\mathbb{R}^N$. Let $\rho_n(x) = c\, n^N \rho(nx)$ and $c = (\int \rho)^{-1}$. Then (ρ_n) is a regularity sequence.

Theorem 1.35 *Let (ρ_n) be a regularity sequence. Then:*

(i) *If $f \in C(\mathbb{R}^N)$, then $\rho_n * f \to f$ uniformly on each compact set in $\mathbb{R}^N$.*

(ii) *If $f \in L^p(\mathbb{R}^N)$ with $1 \leq p < \infty$, then $\rho_n * f \to f$ in $L^p(\mathbb{R}^N)$.*

Theorem 1.36 *(Fréchet-Kolmogorov Theorem) Let $\Omega \subset \mathbb{R}^N$ be open and $w \subset\subset \Omega$. Let F be a bounded subset of $L^p(\Omega)$, $1 \leq p < \infty$. Suppose for $\epsilon > 0$ and $\delta > 0$, there is $\delta < dist(w, \Omega^c)$ such that $\|\tau_h f - f\|_{L^p(w)} < \epsilon$ for $h \in \mathbb{R}^N$, $|h| < \delta$, $f \in F$. Then $F|_w$ is relatively compact in $L^p(w)$.*

Proof.   Without loss of generality, we may suppose Ω is bounded. For $f \in F$, denote

$$\widetilde{f}(x) = \begin{cases} f(x) \text{ for } x \in \Omega, \\ 0 \text{ for } x \notin \Omega, \end{cases}$$

$$\widetilde{F} = \{\widetilde{f} \mid f \in F\}.$$

Then $\widetilde{F}$ is bounded in $L^p(\mathbb{R}^N)$ and in $L^1(\mathbb{R}^N)$. Because $w \subset\subset \Omega$, take U such that $w \subset\subset U \subset\subset \Omega$. Let (ρ_n) be a regularity sequence. We claim that

$$\|\rho_n * \widetilde{f} - \widetilde{f}\|_{L^p(w)} < \epsilon \text{ for } \widetilde{f} \in \widetilde{F}, \text{ for } n > \frac{1}{\delta}, \tag{1.6}$$

where δ, ϵ are as hypothesized. In fact,

$$|\rho_n * \widetilde{f}(x) - \widetilde{f}(x)| \leq \int_{\mathbb{R}^N} |\widetilde{f}(x - y) - \widetilde{f}(x)| \rho_n(y) dy$$

$$= \int_{B(0,1/n)} |\widetilde{f}(x - y) - \widetilde{f}(x)| \rho_n(y) dy.$$

By the Minkowski Inequality for Integrals,

$$\|\rho_n * \widetilde{f} - \widetilde{f}\|_{L^p(w)} \leq \int_{B(0,1/n)} \|\widetilde{f}(x - y) - \widetilde{f}(x)\|_{L^p(w)} \rho_n(y) dy.$$

$$< \epsilon \text{ for } n > \frac{1}{\delta}.$$

For fixed n, the family $H_n = (\rho_n * \widetilde{F})|_{\overline{w}}$ satisfies the hypothesis of the Ascoli Theorem: in fact,

$$\|\rho_n * \widetilde{f}\|_{L^\infty(\mathbb{R}^N)} \leq \|\rho_n\|_{L^\infty(\mathbb{R}^N)} \|\widetilde{f}\|_{L^1(\mathbb{R}^N)} \leq c_n \text{ for } \widetilde{f} \in \widetilde{F}.$$

For x_1, $x_2 \in \mathbb{R}^N$,

$$|(\rho_n * \widetilde{f})(x_1) - (\rho_n * \widetilde{f})(x_2)| \leq |x_1 - x_2| \|\rho_n * \widetilde{f}\|_{Lip(\mathbb{R}^N)}$$

$$\leq |x_1 - x_2| \|\rho_n\|_{Lip(\mathbb{R}^N)} \|\widetilde{f}\|_{L^1(\mathbb{R}^N)}$$

$$\leq d_n |x_1 - x_2| \text{ for } \widetilde{f} \in \widetilde{F}.$$

By applying the Ascoli Theorem 1.31, we can show that H_n is relatively compact in $C(\overline{w})$. Note that $L^p(\overline{w})$-open sets are also $C(\overline{w})$- open. In fact, let O be a L^p-open set, $x_0 \in O$. Then there is $r > 0$ such that $\|x - x_0\|_{L^p} < r$ implies $x \in O$. Now, if $\|x - x_0\|_{L^\infty} < r/c$, then $\|x - x_0\|_{L^p} \leq c\|x - x_0\|_{L^\infty} < r$, and consequently, $x \in O$. Therefore, H_n is relatively compact in $L^p(w)$ and there are $k's$ ϵ-balls in $L^p(w)$-norm, which covers H_n. By (1.6), for $\epsilon > 0$, take $n_0 > \frac{1}{\delta}$ such that

$$\|\rho_{n_0} * \widetilde{f} - \widetilde{f}\|_{L^p(w)} < \epsilon \text{ for } \widetilde{f} \in \widetilde{F}. \tag{1.7}$$

By (1.7) there are $k's$ 2ϵ-balls in $L^p(w)$-norm that cover $F|_w$: that is, $\widetilde{F}|_w$ so $F|_w$ is totally bounded. Thus, $F|_w$ is relatively compact in $L^p(w)$. ∎

Theorem 1.37 *Let $\Omega \subset \mathbb{R}^N$ be open and F a bounded set in $L^p(\Omega)$, $1 \leq p < \infty$. Suppose:*

(i) For $\epsilon > 0$, $w \subset\subset \Omega$ and $\delta > 0$, $\delta < dist(w,$ there is $\Omega^c)$ such that

$$\|\tau_h f - f\|_{L^p(w)} < \epsilon \text{ for } h \in \mathbb{R}^N, |h| < \delta, \text{ for } f \in F;$$

(ii) For $\epsilon > 0$, there is $w \subset\subset \Omega$ such that $\|f\|_{L^p(\Omega\backslash w)} < \epsilon$ for $f \in F$.

Then F is relatively compact in $L^p(\Omega)$.

Proof.   Given $\epsilon > 0$, take $w \subset\subset \Omega$ such that $\|f\|_{L^p(\Omega\backslash w)} < \epsilon$ for $f \in F$. By Theorem 1.36, $F|_w$ is relatively compact in $L^p(w)$. Cover $F|_w$ by finite ϵ-balls in L^p-norm: $F|_w \subset \cup_{i=1}^k B(g_i, \epsilon)$ with $g_i \in L^p(w)$. Set

$$\widetilde{g}_i(x) = \begin{cases} g_i(x) \text{ for } x \in w, \\ 0 \text{ for } x \in \Omega\backslash w. \end{cases}$$

Clearly,
$$F \subset \cup_{i=1}^{k} B(\widetilde{g}_i, 2\epsilon).$$
This asserts that F is relatively compact in L^p and the theorem follows. ∎

Theorem 1.38 *Let $\{f_n\}$ be a sequence in $L^p(\Omega)$ and let $f \in L^p(\Omega)$ be such that $\|f_n - f\|_{L^p(\Omega)} \to 0$. Then, there are a subsequence $\{f_{n_k}\}$ and a function h in $L^p(\Omega)$ such that*

$$f_{n_k}(x) \to f(x) \text{ a.e. in } \Omega,$$
$$|f_{n_k}(x)| \le h(x) \text{ a.e. in } \Omega \text{ for } k = 1, 2, \cdots.$$

Proof. Take a subsequence (f_{n_k}) of (f_n) such that

$$\|f_{n_{k+1}} - f_{n_k}\|_{L^p} \le \frac{1}{2^k} \text{ for } k = 1, 2, \cdots.$$

Set

$$g_n(x) = \sum_{k=1}^{n} |f_{n_{k+1}}(x) - f_{n_k}(x)| \text{ for } n = 1, 2, \cdots,$$

$$g = \sum_{k=1}^{\infty} |f_{n_{k+1}} - f_{n_k}|.$$

Then, $\|g_n\|_{L^p(\Omega)} \le 1$, and $\|g\|_{L^p(\Omega)} \le 1$, and $g_n(x) \to g(x)$ a.e. in Ω. For $m > \ell \ge 2$,

$$|f_{n_m}(x) - f_{n_\ell}(x)| \le |f_{n_m}(x) - f_{n_{m-1}}(x)| + \cdots + |f_{n_{\ell+1}}(x) - f_{n_\ell}(x)|$$
$$\le g(x) - g_{\ell-1}(x) \text{ a.e.},$$

and consequently,
$$|f_{n_m}(x) - f_{n_\ell}(x)| \le g(x) \text{ a.e. in } \Omega. \tag{1.8}$$

$(f_{n_m}(x))$ is a Cauchy sequence in $\mathbb{R}$ for each $x \in \Omega$. There is a function u such that $f_{n_m}(x) \to u(x)$ a.e. in Ω. Letting $\ell \to \infty$ in (1.8),

$$|f_{n_m}(x) - u(x)| \le g(x) \text{ for } m = 2, 3, \cdots.$$

Thus $u \in L^p(\Omega)$. Setting $h(x) = g(x) + |u(x)|$ for $x \in \Omega$, then $h \in L^p(\Omega)$ and

$$|f_{n_m}(x)| \le h(x).$$

By the Lebesgue Dominated Convergence Theorem,

$$\|f_{n_m}(x) - u(x)\|_{L^p} \to 0.$$

Hence, $u(x) = f(x)$ a.e. in Ω and $f_{n_m}(x) \to f(x)$ a.e. in Ω.

Theorem 1.39 (*de Beppo Levi Theorem*) *Let $\{f_n\}$ be a sequence in $L^1(\Omega)$ such that $f_1 \leq f_2 \leq \cdots$ and $\sup_{n \in \mathbb{N}} \int f_n < \infty \in L^1(\Omega)$. Then f_n converges a.e. in Ω to a function f. Moreover, $f \in L^1(\Omega)$ and $\|f_n - f\|_{L^1(\Omega)} \to 0$.*

Theorem 1.40 (*Lebesgue Dominated Convergence Theorem*) *Let $\{f_n\}$ be a sequence in $L^1(\Omega)$. Suppose that f_n converges a.e. in Ω to a function f and there is a function $g \in L^1(\Omega)$ such that $|f_n| \leq g$ a.e. in Ω for each n. Then $f \in L^1(\Omega)$ and $\|f_n - f\|_{L^1(\Omega)} \to 0$.*

Theorem 1.41 *Let E be a measure space with $|E| < \infty$. If $f \in L^\infty(E)$, then*

$$\|f\|_{L^\infty(E)} = \lim_{p \to \infty} \|f\|_{L^p(E)}.$$

Proof.   Let $M = \|f\|_{L^\infty(E)}$. For any $M' < M$, the set

$$A = \{x \in E | |f(x)| > M'\}$$

has a positive measure, and

$$\|f\|_{L^p(E)} \geq \left(\int_A |f|^p \right)^{1/p} \geq M'|A|^{1/p}.$$

Therefore,

$$\liminf_{p \to \infty} \|f\|_{L^p(E)} \geq M' \liminf_{p \to \infty} |A|^{1/p} = M',$$

and consequently,

$$\liminf_{p \to \infty} \|f\|_{L^p(E)} \geq M = \|f\|_{L^\infty(E)}. \tag{1.9}$$

Conversely,

$$\|f\|_{L^p(E)} = \left(\int_E |f|^p \right)^{1/p} \leq M|E|^{1/p},$$

so

$$\limsup_{p \to \infty} \|f\|_{L^p(E)} \leq M \limsup_{p \to \infty} |E|^{1/p} = M = \|f\|_{L^\infty(E)}. \tag{1.10}$$

By (1.9), (1.10),

$$\lim_{p \to \infty} \|f\|_{L^p(E)} = \|f\|_{L^\infty(E)}. \qquad \blacksquare$$

Theorem 1.42 *Let E be a measure space with $|E| = \infty$. If $f \in L^p(E)$, $1 \leq p \leq \infty$, then*

$$\|f\|_{L^\infty(E)} = \lim_{p \to \infty} \|f\|_{L^p(E)}.$$

Proof. Without loss of generality, we may assume $f \not\equiv 0$. Let $M = \|f\|_{L^\infty(E)} > 0$. For $M' < M$, consider the set

$$A = \{x \in E \mid |f(x)| \geq M'\}.$$

Then $0 < |A|$. In addition, $|A| < \infty$, otherwise

$$\infty = |A|M' = \int_A M' \leq \int_A |f| \leq \int_E |f|,$$

a contradiction. Now,

$$\|f\|^p_{L^p(E)} = \int_E |f|^p \geq \int_A |f|^p \geq M'^p |A|.$$

Therefore,

$$\liminf_{p \to \infty} \|f\|_{L^p(E)} \geq M',$$

or

$$\liminf_{p \to \infty} \|f\|_{L^p(E)} \geq M = \|f\|_{L^\infty(E)}. \tag{1.11}$$

Because

$$\|f\|^p_{L^p(E)} = \int_E |f|^{p-1}|f| \leq M^{p-1}\|f\|_{L^1(E)},$$

we have

$$\limsup_{p \to \infty} \|f\|_{L^p(E)} \leq \lim_{p \to \infty} (M^{1-1/p})\|f\|^{1/p}_{L^1(E)} = M. \tag{1.12}$$

By (1.11), (1.12)

$$\lim_{p \to \infty} \|f\|_{L^p(E)} = \|f\|_{L^\infty(E)}. \qquad \blacksquare$$

Lemma 1.43 (*Nemytskii Lemma*) *Let Ω be of finite measure and let $g : \bar{\Omega} \times \mathbb{R} \to \mathbb{R}$ be a Caratheodory function: $z \to g(x, z)$ is continuous for a.e. $x \in \Omega$, and $x \to g(x, z)$ is measurable for $z \in \mathbb{R}$. If $\{u_j\} \subset L^1(\Omega)$ such that $u_j \to u$ in measure, then $g(\cdot, u_j) \to g(\cdot, u)$ in measure as $j \to \infty$.*

Proof. Set $\Gamma_{j,\epsilon} = \{x \in \Omega \mid |g(x, u_j(x)) - g(x, u(x))| < \epsilon\}$. We claim that for ϵ_1, $\epsilon_2 > 0$, there is $N \in N$ such that $j \geq N$ implies $|\Gamma_{j,\epsilon_1}| > |\Omega| - \epsilon_2$. For $x \in \Omega$, $\delta(x, \epsilon_1)$, $0 < \delta(x, \epsilon_1) \leq 1$ exists such that $|z - u(x)| < \delta(x, \epsilon_1)$, implying $|g(x, z) - g(x, u(x))| < \epsilon_1$. In fact, let $\Omega_k = \{x \in \Omega \mid \delta(x, \epsilon_1) \geq \frac{1}{k}\}$. Then $\Omega_1 \subset \Omega_2 \subset \cdots$, $\Omega = \cup_{k=1}^\infty \Omega_k$, and $\lim_{k \to \infty} |\Omega_k| = |\Omega|$. Choose $m > 0$ such that $|\Omega_m| > |\Omega| - \frac{\epsilon_2}{2}$ and define $\Lambda_j = \{x \in \Omega \mid |u_j(x) - u(x)| < \frac{1}{m}\}$. Because $u_j \to u$ in measure, there is $N > 0$ such that $j \geq N$ implies $|\Omega_j| > |\Omega| - \frac{\epsilon_2}{2}$. Because $\Lambda_j \cap \Omega_m \subset \Gamma_{j,\epsilon_1}$, we have $\Gamma^c_{j,\epsilon_1} \subset \Lambda^c_j \cup \Omega^c_m$, and consequently,

$$|\Gamma^c_{j,\epsilon_1}| \leq |\Lambda^c_j| + |\Omega^c_m| < \frac{\epsilon_2}{2} + \frac{\epsilon_2}{2} = \epsilon_2.$$

Thus, $|\Gamma_{j,\epsilon_1}| > |\Omega| - \epsilon_2$. $\qquad \blacksquare$

Lemma 1.44 *Let $1 \leq p, q \leq \infty$. Suppose $g : \bar{\Omega} \times \mathbb{R} \to \mathbb{R}$ is Caratheodory, there are $a(x) \in L^q(\Omega)$ and a constant $b > 0$ such that*

$$|g(x, z)| < a(x) + b|z|^{p/q} \text{ for } x \in \Omega, z \in \mathbb{R}.$$

Then $u \to g(\cdot, u)$ of $L^p(\Omega)$ into $L^q(\Omega)$ is continuous.

Proof.

(i) Assume that Ω is of finite measure. For $u \in L^p(\Omega)$, we have

$$\int_\Omega |g(x, u(x))|^q dx \leq 2^{q-1} \int_\Omega (a(x)^q + b^q|u(x)|^p) dx < \infty,$$

and it follows that $g(\cdot, u) \in L^q(\Omega)$. Fix $u_0 \in L^p(\Omega)$, and let

$$h(x, z) = g(x, u_0(x) + z) - g(x, u_0(x)),$$

then

$$|h(x, z)| \leq a(x) + b|u_0(x) + z|^{p/q} + a(x) + b|u_0(x)|^{p/q}$$
$$\leq a_1(x) + b_1|z|^{p/q},$$

where $a_1(x) = 2a(x) + b|u_0(x)|^{p/q} + bc|u_0(x)|^{p/q} \in L^q(\Omega)$, $b_1 = bc > 0$. Note that $h(x, 0) = 0$ for $x \in \Omega$. We claim that $u \to h(\cdot, u)$ is continuous at 0. In fact, let $u_j \to 0$ in $L^p(\Omega)$. Then $u_j \to 0$ in measure, and consequently, $h(\cdot, u_j) \to h(\cdot, 0) = 0$ in measure. Set $A_{j,\epsilon} = \{x \in \Omega | |h(x, u_j(x))| < \epsilon\}$, for ϵ_1, $\epsilon_2 > 0$, there is $N > 0$ such that $j \geq N$ implies $|A_{j,\epsilon_1}| > |\Omega| - \epsilon_2$. Then we have

$$\int_{A_{j,\epsilon_1}} |h(x, u_j(x))|^q dx \leq \epsilon_1^q |A_{j,\epsilon_1}| \leq \epsilon_1^q |\Omega|,$$

$$\int_{\Omega \setminus A_{j,\epsilon_1}} |g(x, u_j(x))|^q dx \leq 2^{q-1}[\int_{\Omega \setminus A_{j,\epsilon_1}} a_1^q(x) dx + b^q \int_{\Omega \setminus A_{j,\epsilon_1}} |u_j(x)|^p dx],$$

for ϵ_1, ϵ_2 small, $j > N$ implies $\int_\Omega |h(x, u_j)|^q dx < \epsilon$.

(ii) Assume that Ω is of infinite measure. Let the sequence $\{u_n\}$ and u be in $L^p(\Omega)$ such that $\|u_n - u\|_{L^p(\Omega)} \to 0$. Applying the Vitali Convergent Theorem, given $\epsilon > 0$, there is a finite measure set $\Omega_\epsilon \subset \Omega$ such that $\int_{\Omega_\epsilon^c} |u|^p \, dx < \epsilon$, $\int_{\Omega_\epsilon^c} |u_n|^p \, dx < \epsilon$ for any $n \in \mathbb{N}$, and $\int_{\Omega_\epsilon^c} a(x)^q dx < \epsilon$. Thus,

$$\int_{\Omega_\epsilon^c} |g(x, u_n) - g(x, u)|^q dx \leq 2^{2q-2} \int_{\Omega_\epsilon^c} [2a(x)^q + b^q (|u_n|^p + |u|^p)] \, dx \quad (1.13)$$

$$< 2^{2q-1}(1 + b^q)\epsilon. \quad (1.14)$$

Because $|\Omega_\epsilon| < \infty$ and $\|u_n - u\|_{L^p(\Omega_\epsilon)} \leq \|u_n - u\|_{L^p(\Omega)}$, by ($i$), we have $g(x, u_n) \to g(x, u)$ in $L^q(\Omega_\epsilon)$. By (1.14), we obtain $g(x, u_n) \to g(x, u)$ in $L^q(\Omega)$. ∎

Lemma 1.45 (*Brezis-Lieb Lemma*) *Suppose, for some p, $0 < p < \infty$, $f_n \to f$ a.e. in Ω and $\|f_n\|_{L^p(\Omega)} \le c < \infty$ for each n. Then $\lim_{n \to \infty}(\|f_n\|_{L^p}^p - \|f_n - f\|_{L^p}^p)$ exists, and*

$$\lim_{n \to \infty}(\|f_n\|_{L^p}^p - \|f_n - f\|_{L^p}^p) = \|f\|_{L^p}^p.$$

Proof.   For $\epsilon > 0$, there is $c_\epsilon > 0$ such that for $0 < p < \infty$ we have

$$\left| |a + b|^p - |a|^p \right| \le \epsilon |a|^p + c_\epsilon |b|^p.$$

In fact, we may assume that $ab \ne 0$ and let $\varphi(t) = t^p$ for $t > 0$, then

$$\varphi'(t) = p\, t^{p-1}.$$

The Mean Value Theorem implies

$$|a + b|^p - |a|^p = \varphi(|a + b|) - \varphi(|a|) = p\, \theta^{p-1}(|a + b| - |a|)$$

where θ lies between $|a|$ and $|a + b|$. Thus, for $p' = \frac{p}{p-1}$,

$$
\begin{aligned}
\left| |a + b|^p - |a|^p \right| &= p\, \theta^{p-1} \big| |a + b| - |a| \big| \\
&\le p\, \theta^{p-1} |b| \\
&= (p^{1/p'} \epsilon^{1/p'} \theta^{p-1})(p^{1/p} \epsilon^{-1/p'} |b|) \\
&\le \frac{p\, \epsilon \theta^{(p-1)p'}}{p'} + \frac{p\, |b|^p}{\epsilon^{p-1}} \\
&= \frac{p}{p'}\, \epsilon\, \theta^p + \frac{p\, |b|^p}{\epsilon^{p-1}}.
\end{aligned}
$$

However,

$$\theta \le |a| + |a + b| \le 2|a| + |b|,$$

or

$$\theta^p \le (2|a| + |b|)^p \le c_p[2^p|a|^p + |b|^p],$$

so

$$\left| |a + b|^p - |a|^p \right| \le \epsilon' |a|^p + c_{\epsilon'} |b|^p.$$

Let $f_n = f + g_n$ for each n. Then $g_n \to 0$ as $n \to \infty$, and $\|g_n\|_{L^p(\Omega)}^p \le c < \infty$ for each n. Set

$$w_{\epsilon,n}(x) = (\, | \, |f_n(x)|^p - |g_n(x)|^p - |f(x)|^p | - \epsilon|g_n(x)|^p)_+$$

where $a_+ = \max(a, 0)$. We have $w_{\epsilon,n}(x) \to 0$ a.e. in Ω as $n \to \infty$, and

$$
\begin{aligned}
\big| |f_n|^p - |g_n|^p - |f|^p \big| &\le \big| |f_n|^p - |g_n|^p \big| + |f|^p \\
&\le \epsilon|g_n|^p + c_\epsilon|f|^p + |f|^p.
\end{aligned}
$$

That is,
$$w_{\epsilon,n} \leq c_\epsilon |f|^p + |f|^p \in L^1.$$

By the Lebesgue Dominated Convergence Theorem,
$$\int w_{\epsilon,n}(x)dx \to 0 \ \ as \ \ n \to \infty.$$

However,
$$||f_n|^p - |g_n|^p - |f|^p| \leq w_{\epsilon,n} + \epsilon |g_n|^p$$

or
$$\int ||f_n|^p - |g_n|^p - |f|^p| \leq \int w_{\epsilon,n} + \epsilon \int |g_n|^p,$$

and we have
$$\overline{\lim}_{n\to\infty} \int ||f_n|^p - |g_n|^p - |f|^p| \leq \epsilon c \ \ \text{for each} \ \ \epsilon > 0$$

or
$$\lim_{n\to\infty} (||f||_{L^p}^p - ||f_n - f||_{L^p}^p) = ||f||_{L^p}^p. \qquad \blacksquare$$

Notes Most theorems in this chapter can be found in Brezis [7]. Some Banach spaces considered in these theorems can be extended to topological vector spaces; see Choquet [13], Larsen [29], and Rudin [41]. For the general theories of Functional Analysis see Brezis [7], Choquet [13], Dieudonne [18], Dunford-Schwartz [22], Larsen [29], Reed-Simon [40], Rudin [41], and Yosida [52].

Chapter 2

FIXED POINT THEOREMS

In this chapter we present various fixed point results.

2.1   Banach Fixed Point Theorem

Theorem 2.1 (*Banach Fixed Point Theorem*) *Suppose that (X, d) is a complete metric space, $M \subset X$ a nonempty closed set, and $T : M \to M$ satisfies*

$$d(Tx, Ty) < kd(x, y)$$

for a fixed k, $0 \leq k < 1$, and for all x, $y \in M$. Then:

(i) *For an arbitrary choice of initial point x_0 in M, let $x_{n+1} = Tx_n$, for $n = 0$, $1, 2, \cdots$, then the sequence (x_n) converges to $a \in M$ such that $Ta = a$;*

(ii) *There is a unique fixed point $a \in M$ such that $Ta = a$;*

(iii) *Error estimates: for each $n = 0, 1, 2, \cdots$*

$$d(x_n, a) \leq k^n (1 - k)^{-1} d(x_0, x_1),$$
$$d(x_{n+1}, a) \leq k(1 - k)^{-1} d(x_n, x_{n+1});$$

(iv) *Rate of convergence: for all $n = 0, 1, 2, \cdots$, we have*

$$d(x_{n+1}, a) \leq kd(x_n, a).$$

Proof.

(i) For an arbitrary choice of initial point x_0 in M, let $x_{n+1} = Tx_n$, for $n = 0, 1, 2, \cdots$, then the sequence (x_n) is Cauchy: this follows from

$$
\begin{aligned}
d(x_n, x_{n+1}) &= d(Tx_{n-1}, Tx_n) \\
&\leq kd(x_{n-1}, x_n) \\
&\leq k^2 d(x_{n-2}, x_{n-1}) \leq \cdots \\
&\leq k^n d(x_0, x_1),
\end{aligned}
$$

19

$$d(x_n, x_{n+m}) \leq d(x_n, x_{n+1}) + d(x_{n+1}, x_{n+2}) + \cdots + d(x_{n+m-1}, x_{n+m})$$
$$\leq (k^n + k^{n+1} + \cdots + k^{n+m-1})d(x_0, x_1)$$
$$\leq k^n(1-k)^{-1}d(x_0, x_1).$$

Because X is complete, there is $a \in M$ such that $x_n \to a$ as $n \to \infty$. Now, $x_{n+1} = Tx_n \to Ta$, so $Ta = a$.

(ii) If $Ta = a$, $Tb = b$, then

$$d(a, b) = d(Ta, Tb) \leq kd(a, b),$$

which forces $d(a, b) = 0$ or $a = b$.

(iii) Letting $m \to \infty$ in $d(x_n, x_{n+m}) \leq k^n(1-k)^{-1}d(x_0, x_1)$, we obtain

$$d(x_n, a) \leq k^n(1-k)^{-1}d(x_0, x_1).$$

Letting $m \to \infty$ in

$$d(x_{n+1}, x_{n+m+1}) \leq d(x_{n+1}, x_{n+2}) + \cdots + d(x_{n+m}, x_{n+m+1})$$
$$\leq (k + k^2 + \cdots + k^m)d(x_n, x_{n+1})$$
$$\leq k(1-k)^{-1}d(x_n, x_{n+1}),$$

we obtain

$$d(x_{n+1}, a) \leq k(1-k)^{-1}d(x_n, x_{n+1}).$$

(iv) $d(x_{n+1}, a) = d(Tx_n, Ta) \leq kd(x_n, a).$ ∎

Theorem 2.2 (*Method of Continuity*) *Let B be a Banach space and V a normed linear space, and let T_0, $T_1 : B \to V$ be two bounded linear operators. For $t \in [0, 1]$, set $T_t = (1-t)T_0 + tT_1$. Assume that there is $c > 0$ such that*

$$\|x\|_B \leq c\|T_t x\|_V \text{ for each } t \in [0, 1], \tag{2.1}$$

then T_0 is surjective if and only if T_1 is surjective.

Proof.   We apply inequality (2.1) to show T_t is injective, for each $t \in [0, 1]$. First we assume that T_s is surjective for some $s \in [0, 1]$, and then T_s is bijective. Consider the inverse map $T_s^{-1} : V \to B$. Then,

$$\|T_s^{-1}\| = \sup_{x \neq 0} \frac{\|T_s^{-1}x\|_B}{\|x\|}$$

$$= \sup_{y \neq 0} \frac{\|y\|_B}{\|T_s y\|} \leq c.$$

For $t \in [0,1]$, and $y \in V$, the equation $T_t x = y$, can be rewritten as

$$
\begin{aligned}
T_s x &= y + (T_s - T_t)x \\
&= y + (1-s)T_0 x + sT_1 x - (1-t)T_0 x - tT_1 x \\
&= y + (t-s)(T_0 - T_1)(x),
\end{aligned}
$$

or

$$
x = T_s^{-1} y + (t-s)T_s^{-1}(T_0 - T_1)x.
$$

Set

$$
Tx = T_s^{-1} y + (t-s)T_s^{-1}(T_0 - T_1)x \text{ for } x \in B.
$$

For $x_1, x_2 \in B$,

$$
\begin{aligned}
\|Tx_1 - Tx_2\|_V &= |t-s|\|T_s^{-1}\|\|T_0 - T_1\|\|x_1 - x_2\|_B \\
&\leq |t-s|c\|T_0 - T_1\|\|x_1 - x_2\|_B.
\end{aligned}
$$

Let $\eta = 1/2[c(\|T_0\| + \|T_1\|)]^{-1}$, if $|t-s| < \eta$, then

$$
\|Tx_1 - Tx_2\|_V < 1/2\|x_1 - x_2\|_B.
$$

By the Banach Fixed Point Theorem 2.1, there is $x \in B$ such that $Tx = x$, and it follows at once that $T_t x = y$. Therefore, T_t is surjective for all t, $|t-s| < \delta$. Divide the interval $[0,1]$ into subintervals of length less than $\eta/2$, and apply the above process to prove that each T_t is surjective for $t \in [0,1]$. In particular, T_0 and T_1 are surjective if either of them is surjective. ∎

Theorem 2.3 (*Continuous Dependence on a Parameter*) *Suppose that (X, d) is a complete metric space, P a metric space, $M \subseteq X$ a nonempty closed set, for each $p \in P$, $T_p : M \to M$ satisfying $d(T_p x, T_p y) \leq kd(x, y)$ for a fixed k independent of p, $0 \leq k < 1$ and for all $x, y \in M$. Assume that given $p_0 \in P$, for each $x \in M$, $\lim_{p \to p_0} T_p x = T_{p_0} x$. Then for each $p \in P$, there is a unique $x_p \in M$ such that $T_p x_p = x_p$, $\lim_{p \to p_0} x_p = x_{p_0}$.*

Proof. By the Banach Fixed Point Theorem 2.1, for each p, there is a unique $x_p \in M$ such that $T_p x_p = x_p$. Then,

$$
\begin{aligned}
d(x_p, x_{p_0}) &= d(T_p x_p, T_{p_0} x_{p_0}) \\
&\leq d(T_p x_p, T_p x_{p_0}) + d(T_p x_{p_0}, T_{p_0} x_{p_0}) \\
&\leq kd(x_p, x_{p_0}) + d(T_p x_{p_0}, T_{p_0} x_{p_0}),
\end{aligned}
$$

and therefore

$$
d(x_p, x_{p_0}) \leq (1-k)^{-1} d(T_p x_{p_0}, T_{p_0} x_{p_0}) \to 0 \text{ as } p \to p_0.
$$

Consider the initial problem

$$\begin{cases} x'(t) = f(t, x(t)) \text{ on } [t_0 - c, t_0 + c], \\ x(t_0) = p. \end{cases} \tag{2.2}$$

If f is continuous near (t_0, p), we can rewrite (2.2) as

$$x(t) = p + \int_{t_0}^{t} f(s, x(s))ds \text{ for each } t \in [t_0 - c, t_0 + c]. \tag{2.3}$$

∎

Theorem 2.4 (*Picard-Lindelof*) *Given a rectangle*

$$Q = \{(t, x) \in \mathbb{R}^2 \mid |t - t_0| \le a, |x - p_0| \le b\}.$$

Suppose that $f : Q \to \mathbb{R}$ is continuous, then

$$|f(t, x) - f(t, y)| \le L|x - y| \text{ for } (t, x), (t, y) \in Q,$$
$$|f(t, x)| \le k \text{ for } (t, x) \in Q,$$

where $L \ge 0$, $k > 0$ are fixed. Then:

(i) *If we set $c = \min(a, \frac{b}{k})$ and $p = p_0$, then (2.2), then consequently (2.3) has exactly one continuous solution $x(t)$ on $[t_0 - c, t_0 + c]$;*

(ii) *Let $x_0(t) \equiv p_0$, and for $n = 0, 1, 2, \cdots$, let*

$$x_{n+1}(t) = p_0 + \int_{t_0}^{t} f(s, x_n(s))ds.$$

Then the sequence $\{x_n\}$ converges uniformly on $[t_0 - c, t_0 + c]$ to the solution $x(t)$ of (2.3);

(iii) *The equation (2.3) has exactly one continuous solution $x_p(t)$ on $[t_0 - d, t_0 + d]$ for each $p \in [p_0 - e, p_0 + e]$ for small d, e;*

(iv) *If $p \to p_0$, then $x_p(t) \to x_{p_0}(t)$ uniformly on $[t_0 - d, t_0 + d]$.*

Proof.

(i) Let $E = C([t_0 - c, t_0 + c])$ with supreme norm $\|\cdot\|$, and $M = \{x \in E | \|x - p_0\| \le b\}$. Consider the norm

$$\|x\|_1 = \max_{t \in [t_0 - c, t_0 + c]} |x(t)| e^{-L|t - t_0|}.$$

Then,
$$e^{-Lc}\|x\| \leq \|x\|_1 \leq \|x\|.$$

Define
$$T_p x(t) = p + \int_{t_0}^t f(s, x(s))ds \text{ for all } t \in [t_0 - c, t_0 + c].$$

(a) M is closed in $(E, \|\cdot\|_1)$: let $(x_n) \subset M$, $x \in E$ such that $\|x_n - p_0\| \leq b$ for n, and $\|x_n - x\|_1 \to 0$ as $n \to \infty$. Because $e^{-Lc}\|y\| \leq \|y\|_1$ for $y \in E$, we have $\| x - x_n\| \to 0$ as $n \to \infty$. Thus, $\| x - p_0\| \leq b : x \in M$.

(b) $T_{p_0} : M \to M$: because if $x \in M$, $\| x - p_0\| \leq b$, for $t \in [t_0 - c, t_0 + c]$
$$\|T_{p_0} x - p_0\| \leq \int_{t_0}^t |f(x, x(s))|ds$$
$$\leq ck \leq b. \cdot$$

Therefore, $T_{p_0} x \in M$ for each $x \in M$.

(c) T_{p_0} is α-contractive on M under the norm $\|\cdot\|_1$: for $x, y \in M$: $\|x - p_0\| \leq b$, $\| y - p_0\| \leq b$. Then,

$$\|T_{p_0} x - T_{p_0} y\|_1 = \max_{t \in [t_0 - c, t_0 + c]} \int_{t_0}^t [f(s, x(s)) - f(s, y(s))]ds e^{-L|t - t_0|}$$

$$\leq \max_{t \in [t_0 - c, t_0 + c]} \int_{t_0}^t L\|x - y\|e^{-L|t - t_0|}ds$$

$$\leq L\|x - y\| \max_{t \in [t_0 - c, t_0 + c]} \left| \int_{t_0}^t e^{L|s - t_0| - L|t - t_0|}ds \right|$$

$$\leq \alpha\|x - y\|_1$$

where $\alpha = 1 - e^{-Lc} < 1$. The integral is computed separately for $t_0 \leq t$ and $t \leq t_0$:

$$\int_{t_0}^t e^{L|s - t_0| - L|t - t_0|}ds = \begin{cases} \int_{t_0}^t e^{L(s-t)}ds = \frac{1}{L}(1 - e^{L(t_0 - t)}) & \text{for } t_0 \leq t \\ \int_{t_0}^t e^{L(t-s)}ds = -\frac{1}{L}(1 - e^{L(t - t_0)}) & \text{for } t \leq t_0 \end{cases}$$

By $(a), (b)$ and (c) and applying Theorem 2.3 to obtain a unique $x \in M$ such that $x = T_{p_0} x$:

$$x(t) = p_0 + \int_{t_0}^t f(s, x(s))ds \text{ for } t \in [t_0 - c, t_0 + c],$$

such $x(t)$ also satisfies (2.2).

(ii) Follows from the Banach Fixed Theorem 2.1 (i).

(iii) Note that (c) of (i) is still true if we replace p_0 by p. For Step (b) of (i), if $x \in M$, $\|x - p_0\| \leq b$, we have

$$\|T_p x - p\| \leq \int_{t_0}^t |f(s, x(s))| ds \leq dk \quad \text{for } t \in [t_0 - d, t_0 + d],$$

and for $p \in [p_0 - e, p_0 + e]$,

$$\|T_p x - p_0\| \leq \|T_p x - p\| + \|p - p_0\| \leq dk + e \leq b,$$

where d and e are small enough. In this case, $T_p : M \to M$. We then apply the Banach Fixed Theorem 2.1 to show that for each p near p_0, there is an $x_p \in M$ such that $x_p = T_p x_p$.

(iv) If $p \to p_0$, then for each $x \in M$,

$$\|T_p x - T_{p_0} x\|_1 = |p - p_0| \to 0.$$

Then (iv) follows from Theorem 2.3.   ∎

2.2   Brouwer Fixed Point Theorem

For proofs of the following theorems, see Chapter 9.

Notation: let $B_1(0)$ be the unit ball with center 0 in $\mathbb{R}^N$ and $S^{N-1} = \partial B_1(0)$.

Theorem 2.5 (*Brouwer Fixed Point Theorem*) *Let* $f : \overline{B_1(0)} \to \overline{B_1(0)}$ *be continuous, then* f *admits a fixed point.*

Definition 2.6 *Let* X, Y *be topological spaces, and* $A \subset X$ *a subspace.*

(i) *If* $f : X \to A$ *is continuous with* $f|_A = I$ *the identity map, then* f *is called a retraction of* X *on* A;

(ii) *If* $f : A \to Y$ *is continuous with no continuous extension* $F : X \to Y$, *then* f *is called essential.*

Theorem 2.7 *The following are equivalent:*

(i) (*Non-retraction of unit ball on its boundary*) *There is no retraction* $f : \overline{B_1(0)} \to S^{N-1}$;

(ii) (*Unit sphere essential on the unit ball*) *The inclusion* $i : S^{N-1} \to \mathbb{R}^N \backslash \{0\}$ *is essential;*

(iii) (*Brouwer Fixed Point Theorem*) *Let* $f : \overline{B_1(0)} \to \overline{B_1(0)}$ *be continuous, then* f *admits at least one fixed point;*

(iv) *Let* $K \subset \mathbb{R}^N$ *be a compact convex set, and* $f : K \to K$ *a continuous function, then* f *admits a fixed point;*

(v) (*KKM Theorem, Knaster-Kuratowski-Mazurkiewicz Theorem*) *Let* E *be a Hausdorff Topological Vector Space, and let* $x_1, \cdots, x_m \in E$, $X_1, \cdots, X_m$ *be closed sets in* E *such that*

$$\mathrm{conv}\{x_{i_1}, \cdots, x_{i_k}\} \subset X_{i_1} \cup \cdots \cup X_{i_k}$$

for $\{i_1, \cdots, i_k\} \subset \{1, 2, \cdots, m\}$. *Then* $\cap_{i=1}^m X_i \neq \emptyset$;

(vi) *Let* E *be a Hausdorff Topological Vector Space,* $X \subset E$ *a subset such that for every* $x \in X$ *there is an associated closed set* $F(x)$ *of* E *and there is at least one* $x_0 \in X$ *such that* $F(x_0)$ *is compact. Suppose for each finite family* $\{x_1, \cdots, x_m\}$ *in* X, *conv* $\{x_1, \cdots, x_m\} \subset \cup_{i=1}^m F(x_i)$. *Then* $\cap_{x \in X} F(x) \neq \emptyset$;

(vii) (*Ky Fan Minimax Inequality*) *Let* E *be a Hausdorff Topological Vector Space and* $K \subset E$ *a compact convex set. Assume there is* $f : K \times K \to \mathbb{R}$ *such that for each fixed* $x \in K$, $y \to f(x, y)$ *is a lower semicontinuous function, and for each fixed* $y \in K$, $x \to f(x, y)$ *is a quasiconcave function. Then,*

$$\min_{y \in K} \max_{x \in K} f(x, y) \leq \sup_{x \in K} f(x, x);$$

($viii$) (*Hartman-Stampacchia Theorem*) *Let* $K \subset \mathbb{R}^N$ *be a compact convex set, and* $A : K \to \mathbb{R}^N$ *continuous. Then there is* $x \in K$ *such that* $(Ax, y - x) \geq 0$ *for* $y \in K$.

2.3   Leray-Schauder Fixed Point Theorem

For proofs of the following fixed point theorems see Chapter 9.

Theorem 2.8 (*Schauder Fixed Point Theorem*) *Let* E *be a Banach space,* $K \subset E$ *a compact convex set,* $T : K \to K$ *a continuous function. Then* T *admits a fixed point in* K.

A variant of theorem 2.8 is as follows.

Theorem 2.9 (*Schauder Fixed Point Theorem*) *Let* E *be a Banach space,* $Q \subset E$ *a convex closed bounded set,* $S : Q \to Q$ *a compact operator. Then* S *admits a fixed point in* Q.

Theorem 2.10 (*Leray-Schauder Fixed Point Theorem*) *Let E be a Banach space, $S : E \to E$ a compact operator. Suppose there is a real number $r > 0$ such that the equality $u = \sigma S(u)$ where $u \in E$, $\sigma \in [0, 1]$, implies that $\|u\| < r$. Then S admits a fixed point in $B_r(0)$.*

Theorem 2.10 has the following two variants.

Theorem 2.11 (*Leray-Schauder Fixed Point Theorem*) *Let E be a Banach space, $B_r(0)$ the ball in E with center at 0, radius r, $S : \overline{B_r(0)} \to E$ a compact operator such that $S(\partial B_r(0)) \subset B_r(0)$. Then S admits a fixed point in $B_r(0)$.*

Theorem 2.12 (*Leray-Schauder Fixed Point Theorem*) *Let E be a Banach space, and $T : [0, 1] \times E \to E$ a compact operator such that $T(0, u) = 0$ for $u \in E$. If there is some $r > 0$ such that the equality $u = T(\sigma, u)$ for $u \in E$, $\sigma \in [0, 1]$ implies that $\|u\| < r$, then for each $\sigma \in [0, 1]$, $T(\sigma, \cdot)$ admits a fixed point in $B_r(0)$.*

Notes Fixed point theorems have many applications in many branches of mathematics, such as applying them to prove the existence of solutions of ordinary differential equations and partial differential equations, and to prove the implicit function theorem. For more details of fixed point theory see Dugundji-Granas [21], Rabinowitz [37], and Zeidler [53].

Chapter 3

CALCULUS ON BANACH SPACES

In this chapter we study the derivatives and the integrals of functions from one Banach space into another Banach space.

3.1 Fréchet Derivatives and Gâteaux Derivatives

We begin with some notation. Let X and Y be Banach spaces, or briefly B-spaces, $U(0) \subset X$ an open neighborhood of 0, and $g : U(0) \to Y$. We will write

$$g(x) = o(\|x\|) \text{ as } x \to 0 \text{ if } \|g(x)\|/\|x\| \to 0 \text{ as } x \to 0,$$
$$g(x) = o(1) \text{ as } x \to 0 \text{ if } \|g(x)\| \to 0 \text{ as } x \to 0.$$

We denote by $L(X, Y)$ all continuous linear operators of X into Y.

Definition 3.1 *Let $U(x)$ be a neighborhood of x in X. Consider the map $f : U(x) \to Y$. f is F-differentiable at x if there is $T \in L(X, Y)$ such that*

$$f(x + h) - f(x) = Th + o(\|h\|) \text{ as } h \to 0. \tag{3.1}$$

Remark 3.2

(i) Such T in Definition 3.1 is unique. In fact, if there are T_1, T_2 satisfying

$$f(x + h) - f(x) = T_i h + o(\|h\|) \text{ as } h \to 0, i = 1, 2,$$

then, for $T = T_1 - T_2$,

$$\overline{\lim}_{h \to 0} \frac{\|Th\|}{\|h\|} \leq \overline{\lim}_{h \to 0} \frac{\|f(x + h) - f(x) - T_1 h\|}{\|h\|}$$
$$+ \overline{\lim}_{h \to 0} \frac{\|f(x + h) - f(x) - T_2 h\|}{\|h\|} = 0.$$

We have $\lim_{h \to 0} \frac{\|Th\|}{\|h\|} = 0$. Hence, for $\epsilon > 0$, there is $\delta > 0$ such that $\|h\| < \delta$ implies $\|Th\| < \epsilon\|h\|$. For any $0 \neq y \in X$, let $h = \frac{\delta y}{2\|y\|}$, then $\|h\| = \frac{\delta}{2}$. Then,

$$\left\| T\left(\frac{\delta y}{2\|y\|}\right) \right\| < \epsilon \left\| \frac{\delta y}{2\|y\|} \right\| = \frac{\delta}{2}\epsilon$$

27

and consequently, $\|Ty\| < \epsilon\|y\|$. Because ϵ is arbitrary, $\|Ty\| = 0$ or $Ty = 0$ for all $y \in X$, and it follows that $T = 0 : T_1 = T_2$.

(ii) Such unique T is called the F-derivative of f at x, in symbols $f'(x) = T$, and $f'(x)h$ is called the F-differential of f at x, in symbols $df(x, h)$.

(iii) In (3.1), we may let $r(h) = \frac{o(\|h\|)}{\|h\|}$. Then $r(h) = o(1)$, and $o(\|h\|) = \|h\|r(h)$.

Definition 3.3 Let $U(x)$ be a neighborhood of x in X, and consider $f : U(x) \to Y$. f is G-differentiable at x if there is $T \in L(X, Y)$ such that

$$f(x + tk) - f(x) = tTk + o(t) \text{ as } t \to 0$$

for all k in X with $\|k\| = 1$.

Remark 3.4

(i) Such T is unique, and is called the G-derivative of f at x, in symbols $f'(x) = T$, and $f'(x)k$ is called the G-differential of f at x, in symbols $df(x, k)$;

(ii) The G-derivative of f at x can be defined equivalently through

$$f'(x)k = \lim_{t \to 0} \frac{f(x + tk) - f(x)}{t}.$$

Definition 3.5 Let A be a set in X. If the F-derivatives (respectively G-derivatives) $f'(x)$ exist for all $x \in A$, then the map $f' : A \to L(X, Y)$ defined by $x \to f'(x)$ is called the F-derivative (respectively G-derivative) of f on A.

Definition 3.6 Higher derivatives are defined successively. Thus, $f''(x)$ is the derivative of f' at x, and also the second derivative of f at x.

As the heading implies, "F-derivative" and "G-derivative" are abbreviations for the "Fréchet derivative" and the "Gâteaux derivative".

Example 3.7 Let x be in $\mathbb{R}^N$, $U(x)$ an open neighborhood of x in $\mathbb{R}^N$, and $f : U(x) \to \mathbb{R}$.

(i) The F-differentiable of f at x and the total differentiable of f at x are the same;

(ii) f is G-differentiable at x if and only if some $a(x) = (a_1(x), \cdots,$ there is $a_N(x))$ in $\mathbb{R}^N$ such that it is true for all $k \in \mathbb{R}^N$ that

$$\lim_{t \to 0} \frac{f(x + tk) - f(x)}{t} = \sum_{i=1}^{N} a_i k_i = Tk;$$

(iii) *If f is F-differentiable (or G-differentiable) at x, and as in (ii), $k = e_i = (0, 0, \cdots, 1, \cdots, 0)$ for $i = 1, \cdots, N$, then there are all partial derivatives $D_i f(x)$, and $a_i(x) = D_i f(x)$ so that*

$$df(x, h) = f'(x)h = \sum_{i=1}^{N} D_i f(x) h_i,$$

i.e., the F-differential agrees with the total differential;

(iv) *If there are all the partial derivatives $D_i f$ in $U(x)$ and are continuous at x, then f is totally differentiable at x, and hence is F-differentiable at x;*

(v) *If $f \in L(X, Y)$, then $f'(x) = f$ for all $x \in X$.*

Proof. It suffices to prove (iv): $N = 2$, $x = (\xi, \eta)$, $h = (\alpha, \beta)$,

$$f(\xi + \alpha, \eta + \beta) - f(\xi, \eta) = f(\xi + \alpha, \eta + \beta) - f(\xi, \eta + \beta) + f(\xi, \eta + \beta) - f(\xi, \eta)$$
$$= f_\xi(\xi + t_1\alpha, \eta + \beta)\alpha + f_\eta(\xi, \eta + t_2\beta)\beta, 0 < t_1, t_2 < 1.$$

By the continuity of partial derivatives at x, then

$$f(x + h) - f(x) = f_\xi(x)\alpha + f_\eta(x)\beta + o(\|h\|)$$
$$= f'(x)h + o(\|h\|) \text{ as } h \to 0. \qquad \blacksquare$$

Theorem 3.8 (*Relationship of F- and G-derivatives*)

(i) *Every F-derivative at x is also a G-derivative at x;*

(ii) *A G-derivative at x, for which*

$$f'(x)k = \lim_{t \to 0} \frac{f(x + tk) - f(x)}{t}$$

uniformly for k, $\|k\| = 1$, is also an F-derivative at x;

(iii) *If there is f' as a G-derivative in some neighborhood $U(x)$ of x, and if f' is continuous at x, then $f'(x)$ is also an F derivative at x;*

(iv) *If there is $f'(x)$ as an F-derivative at x, then f is continuous at x.*

Proof.

(i) f has an F-derivative at x :

$$f(x + h) - f(x) = f'(x)h + o(\|h\|) \text{ as } h \to 0.$$

Letting $h = tk$, $\|k\| = 1$, then

$$f(x + tk) - f(x) = tf'(x)k + o(t) \text{ as } t \to 0.$$

(ii) $f'(x)k = \lim_{t \to 0} \frac{f(x+tk)-f(x)}{t}$ uniformly for k, $\|k\| = 1$: for $\epsilon > 0$ there is some $\delta > 0$ such that $|t| < \delta$ implies

$$\|f(x+tk) - f(x) - tf'(x)k\| < \epsilon|t|.$$

For $h \neq 0$, let $t = \|h\|$, $k = \frac{h}{\|h\|}$, if $\|h\| < \delta$, then

$$\|f(x+h) - f(x) - f'(x)h\| = \|f(x+tk) - f(x) - tf'(x)k\|$$
$$< \epsilon|t| = \epsilon\|h\|,$$

so

$$f(x+h) - f(x) = f'(x)h + o(\|h\|).$$

(iii) For $h \neq 0$, let $g(t) = f(x+tk)$, $k = \frac{h}{\|h\|}$. Applying the Mean Value Theorem in $\mathbb{R}$, and the Chain Rule to obtain, for some c, $0 < c < \|h\|$,

$$g(\|h\|) - g(0) = g'(c)\|h\| = f'(x+ck)\|h\|k = f'(x+ck)h,$$

then

$$\|f(x+h) - f(x) - f'(x)h\| = \|g(\|h\|) - g(0) - f'(x)h\|$$
$$= \|f'(x+ck)h - f'(x)h\|$$
$$\leq \|f'(x+ck) - f'(x)\|\|h\|.$$

Thus, $\overline{\lim}_{\|h\| \to 0} \frac{\|f(x+h)-f(x)-f'(x)h\|}{\|h\|} = 0$ because f' is continuous at x. We obtain

$$\lim_{\|h\| \to 0} \frac{\|f(x+h) - f(x) - f'(x)h\|}{\|h\|} = 0,$$

and consequently, f has an F-derivative at x.

(iv) $\lim\sup_{\|h\| \to 0} \|f(x+h) - f(x)\|$

$$\leq \lim\sup_{\|h\| \to 0} [\frac{\|f(x+h) - f(x) - f'(x)h\|}{\|h\|}\|h\| + \|f'(x)h\|] = 0$$

and it follows at once that $\lim_{\|h\| \to 0} \|f(x+h) - f(x)\| = 0$.   ∎

3.2   Sum Rule, Chain Rule, and Product Rule

Theorem 3.9 (*Sum Rule*) *Suppose that the maps f, $g : U(x) \subset X \to Y$ are F-differentiable (respectively, G-differentiable) at x, and that α, $\beta \in \mathbb{R}$, then $(\alpha f + \beta g)$ is F-differentiable (respectively, G-differentiable) at x satisfying*

$$(\alpha f + \beta g)'(x) = \alpha f'(x) + \beta g'(x).$$

Proof. We prove only the F-differentiation case:

$$f(x+h) - f(x) = f'(x)h + o(\|h\|) \text{ as } h \to 0,$$
$$g(x+h) - g(x) = g'(x)h + o(\|h\|) \text{ as } h \to 0.$$

Then

$$(\alpha f + \beta g)(x+h) - (\alpha f + \beta g)(x) = (\alpha f'(x) + \beta g'(x))h + o(\|h\|)$$

as $h \to 0$. ∎

Theorem 3.10 (*Chain Rule*) *Let x be fixed and set $y = f(x)$. Assume we are given $f : U(x) \subset X \to Y$ and $g : U(y) \subset Y \to Z$ with $f(U(x)) \subset U(y)$, where X, Y, and Z are B-spaces, and $U(x)$, $U(y)$ are open neighborhoods of x, y, respectively. Suppose that there are $f'(x)$ and $g'(f(x))$ as F-derivatives. Then $H = g \circ f$ is F-differentiable at x, and*

$$(g \circ f)'(x) = g'(f(x)) \circ f'(x).$$

Proof. By hypothesis, $g(y+k) = g(y) + g'(y)k + \|k\|r_1(k)$, where $r_1(k) \to 0$ as $k \to 0$. We choose $k = f(x+h) - f(x) = f'(x)h + \|h\|r_2(h)$, where $r_2(h) \to 0$ as $h \to 0$. Note that $k \to 0$ as $h \to 0$. This implies that

$$g(f(x+h)) = g(f(x)) + g'(f(x))f'(x)h + \|h\|r(h),$$

where $r(h) = g'(f(x))r_2(h) + \frac{\|k\|}{\|h\|}r_1(k)$. We must prove $r(h) = o(1)$. In fact, $\| g'(f(x))r_2(h)\| \le \|g'(f(x))\|\|r_2(h)\| \to 0$ as $h \to 0$, and

$$\frac{\|k\|}{\|h\|}\|r_1(k)\| \le \frac{\|f'(x)h\| + \|h\|\|r_2(h)\|}{\|h\|}\|r_1(k)\|$$
$$\le \|f'(x)\|\|r_1(k)\| + \|r_2(h)\|\|r_1(k)\| \to 0, \text{ as } h \to 0.$$

We obtain $r(h) = o(1)$, and consequently, $(g \circ f)'(x) = g'(f(x)) \circ f'(x)$. ∎

Theorem 3.11 (*Product Rule*) *Suppose X, X_1, X_2, and Y are B-spaces, $B : X_1 \times X_2 \to Y$ is bilinear and bounded, $f_i : U_i(x) \subset X \to X_i$, $i = 1, 2$, are F-differentiable at x, and $H(x) = B(f_1(x), f_2(x))$. Then, H is F-differentiable at x, and for $h \in X$*

$$H'(x)h = B(f_1'(x)h, f_2(x)) + B(f_1(x), f_2'(x)h).$$

Proof. Write $f_i(x+h) = f_i(x) + f_i'(x)h + \|h\|r_i(h)$, where $r_i(h) \to 0$ as $h \to 0$.

Then,

$$H(x+h) - H(x) = B(f_1(x+h), f_2(x+h)) - B(f_1(x), f_2(x))$$
$$= B(f_1(x) + f_1'(x)h + \|h\|r_1(h), f_2(x) + f_2'(x)h$$
$$+ \|h\|r_2(h)) - B(f_1(x), f_2(x))$$
$$= B(f_1(x), f_2'(x)h) + B(f_1'(x)h, f_2(x))$$
$$+ B(f_1(x), \|h\|r_2(h)) + B(f_1'(x)h, f_2'(x)h)$$
$$+ B(f_1'(x)h, \|h\|r_2(h)) + B(\|h\|r_1(h), f_2(x))$$
$$+ B(\|h\|r_1(h), f_2'(x)h) + B(\|h\|r_1(h), \|h\|r_2(h))$$
$$= B(f_1(x), f_2'(x)h) + B(f_1'(x)h, f_2(x)) + r(h)$$

where

$$r(h) = B(f_1(x), \|h\|r_2(h)) + B(f_1'(x)h, f_2'(x)h) + B(f_1'(x)h, \|h\|r_2(h))$$
$$+ B(\|h\|r_1(h), f_2(x)) + B(\|h\|r_1(h), f_2'(x)h)$$
$$+ B(\|h\|r_1(h), \|h\|r_2(h))$$
$$\leq \|f_1(x)\|\|h\|\|r_2(h)\| + \|f_1'(x)\|\|h\|\|f_2'(x)\|\|h\| + \|f_1'(x)\|\|h\|^2\|r_2(h)\|$$
$$+ \|h\|\|r_1(h)\|\|f_2(x)\| + \|h\|^2\|r_1(h)\|\|f_2'(x)\| + \|h\|^2\|r_1(h)\|\|r_2(h)\|$$
$$= o(\|h\|) \ as \ \|h\| \to 0.$$

Hence, $H'(x)h = B(f_1'(x)h, f_2(x)) + B(f_1(x), f_2'(x)h)$. ∎

3.3 Mean Value Theorems

Theorem 3.12 *Let $[a,b]$ be a bounded interval in $\mathbb{R}$, and E a Banach space. Let $f : [a,b] \to E$ and $g : [a,b] \to \mathbb{R}$ be continuous on $[a,b]$, and differentiable on (a,b). Suppose that $\|f'(t)\| \leq g'(t)$ for $a < t < b$. Then,*

$$\|f(b) - f(a)\| \leq g(b) - g(a).$$

Proof. It suffices to prove $\|f(d) - f(c)\| \leq g(d) - g(c)$ for $a < c < d < b$. Suppose that there are $a_0, b_0 \in (a, b)$, $a_0 < b_0$, with

$$\|f(b_0) - f(a_0)\| - [g(b_0) - g(a_0)] = M > 0.$$

Set $m = \frac{a_0+b_0}{2}$. Then,

$$\|f(b_0) - f(m)\| - [g(b_0) - g(m)] \geq \frac{M}{2}$$

or

$$\|f(m) - f(a_0)\| - [g(m) - g(a_0)] \geq \frac{M}{2}.$$

Denote by $[a_1, b_1]$ one of the intervals on which the inequality holds. Repeating the procedure we obtain a sequence of intervals $[a_n, b_n]$ such that

$$a_0 \leq \cdots \leq a_n \leq b_n \leq \cdots \leq b_0,$$
$$|b_n - a_n| = (b_0 - a_0)/\, 2^n,$$
$$\|f(b_n) - f(a_n)\| - [g(b_n) - g(a_n)] \geq \frac{M}{2^n}.$$

It follows that a_n and b_n converge to the same point $w \in (a, b)$ and that

$$
\begin{aligned}
\frac{M}{2^n} &\leq \|f(b_n) - f(w)\| + \|f(w) - f(a_n)\| \\
&\quad - [g(b_n) - g(w)] - [g(w) - g(a_n)] \\
&\leq \|f'(w)(b_n - w)\| + o(b_n - w) + \|f'(w)(w - a_n)\| + o(w - a_n) \\
&\quad - [g'(w)(b_n - w) + o(b_n - w)] - [g'(w)(w - a_n) + o(w - a_n)] \\
&\leq \|f'(w)\|(|b_n - w| + |w - a_n|) - g'(w)(|b_n - w| + |w - a_n|) \\
&\quad + o(b_n - w) + o(w - a_n) \\
&\leq [\|f'(w)\| - g'(w)](b_n - a_n) + o(b_n - a_n).
\end{aligned}
$$

Divide both sides by $b_n - a_n = (b_0 - a_0)/\, 2^n$, and let n tend to ∞. We obtain

$$\frac{M}{b_0 - a_0} \leq \|f'(w)\| - g'(w),$$

so $\|f'(w)\| > g'(w)$, a contradiction.   ∎

Corollary 3.13 *Let $[a, b]$ be a bounded interval in $\mathbb{R}$, E a Banach space, and $f : [a, b] \to E$ be continuous on $[a, b]$ and differentiable on (a, b). Suppose there is a constant k such that $\|f'(t)\| \leq k$ for all $t \in (a, b)$. Then,*

$$\|f(b) - f(a)\| \leq k(b - a).$$

Proof.   Take $g(t) = kt$ in Theorem 3.12.   ∎

Theorem 3.14 (*Mean Value Theorem*) *Let E, F be two Banach spaces, $U \subset E$ an open set, $[a, b] \subset U$ a segment, and let $f : U \to F$ be differentiable. Then,*

$$\|f(b) - f(a)\| \leq \sup_{0 \leq t \leq 1} \|f'((1 - t)a + tb)\|\|b - a\|.$$

Proof. Let $h(t) = f((1-t)a+tb)$, $t \in [0,1]$, $k = \sup_{0 \leq t \leq 1} \|f'((1-t)a+tb)\|\|b-a\|$. By the Chain Rule 3.10 we have $h'(t) = f'((1-t)a+tb)(b-a)$, and $\|h'(t)\| \leq k$ for all $t \in [0,1]$. By Corollary 3.13, we obtain

$$
\begin{aligned}
\|f(b) - f(a)\| &= \|h(1) - h(0)\| \\
&\leq k(1-0) \\
&= \sup_{0 \leq t \leq 1} \|f'((1-t)a+tb)\|\|b-a\|.
\end{aligned}
$$
∎

Theorem 3.15 *Let E, F be Banach spaces, $U \subset E$ a connected open set, and $f : U \to F$ a differentiable map. If $f'(x) = 0$ for all $x \in U$, then f is a constant function.*

Proof. Fix $a \in U$, and let $A := \{x \in U \mid f(x) = f(a)\}$. Because f is continuous, A is closed in U. On the other hand, if $x \in A$ then $f(x) = f(a)$. Take a ball $B_r(x) \subset U$. By the Mean Value Theorem 3.14, for $y \in B_r(x)$

$$
\|f(y) - f(x)\| \leq \sup_{0 \leq t \leq 1} \|f'((1-t)x+ty)\|\|y-x\| = 0,
$$

and consequently, $f(y) = f(x) = f(a) : B_r(x) \subset A$, or A is open. Thus, $A = U$, and it follows that f is a constant function. ∎

Theorem 3.16 (*Uniform Convergence Theorem*) *Let E, F be Banach spaces, $U \subset E$ a connected open set, and let $f_n : U \to F$ be a sequence of differentiable maps. Suppose*

(i) *there is a point $a \in U$ such that $f_n(a)$ converges;*

(ii) *the sequence $f'_n : U \to L(E, F)$ converges uniformly on each bounded subset of U to a map $g : U \to L(E, F)$.*

Then for each $x \in U$, the sequence $f_n(x)$ converges to a limit $f(x)$. This convergence is uniform on each bounded convex subset of U. Finally, f is differentiable with $Df = g$.

Proof. Let $D \subset U$ be a bounded open convex set with $d = diam\ D$. By the Mean Value Theorem 3.14, for $x \in D$

$$
\begin{aligned}
\|f_p(x) - f_q(x) - [f_p(a) - f_q(a)]\| &\leq \sup_{y \in D} \|f'_p(y) - f'_q(y)\|\|x-a\| \\
&\leq (\sup_{y \in D} \|f'_p(y) - f'_q(y)\|)d.
\end{aligned}
\tag{3.2}
$$

Because $f_n(a)$ converges and $f'_n(y)$ converges uniformly on D, we have for each $x \in D$, $f_n(x)$ is Cauchy in F. Therefore, there is $f : D \to F$, satisfying

$$
\lim_{n \to \infty} f_n(x) = f(x) \text{ uniformly in } D.
\tag{3.3}
$$

We conclude that the set $A = \{u \in U \mid f_n(u) \text{ converges}\}$ is nonempty and open. Moreover, by (3.3), A is closed in U, so $A = U$. In other words, $\lim_{n \to \infty} f_n(x) = f(x)$ for each $x \in U$. Letting $p \to \infty$ in (3.2),

$$\|f(x) - f(a) - [f_q(x) - f_q(a)]\| \leq \sup_{y \in D} \|g(y) - f_q'(y)\| \|x - a\|. \tag{3.4}$$

Given $\epsilon > 0$ there is $N > 0$ such that $q > N$ implies

$$\sup_{y \in D} \|g(y) - f_q'(y)\| < \epsilon, \tag{3.5}$$

for $q > N$, (3.4) can be rewritten as

$$\|f(x) - f(a) - [f_q(x) - f_q(a)]\| \leq \epsilon \|x - a\|. \tag{3.6}$$

Moreover, fix $q_0 > N$, and note that f_{q_0} is differentiable at a : there is s, $s \leq r$, such that $B_s(a) \subset D$, and $\| x - a \| < s$ implies

$$\|f_{q_0}(x) - f_{q_0}(a) - f_{q_0}'(a)(x - a)\| \leq \epsilon \|x - a\|. \tag{3.7}$$

By (3.5)–(3.7), for $\|x - a\| \leq s$

$$\begin{aligned}
\|f(x) - f(a) - g(a)(x - a)\| &\leq \|f(x) - f(a) - [f_{q_0}(x) - f_{q_0}(a)]\| \\
&\quad + \|f_{q_0}(x) - f_{q_0}(a) - f_{q_0}'(a)(x - a)\| \\
&\quad + \|f_{q_0}'(a)(x - a) - g(a)(x - a)\| \\
&\leq 3\epsilon \|x - a\|.
\end{aligned}$$

Thus, f is differentiable at a with $f'(a) = g(a)$.   ∎

3.4 Sard Theorem

Definition 3.17 *Let $U \subset \mathbb{R}^p$ be an open set, and $f : U \to \mathbb{R}^N$ a differentiable map. We say $a \in U$ is a critical point of f if rank $f'(a) < N$.*

Theorem 3.18 *(Sard Theorem) Let $U \subset \mathbb{R}^M$ be an open set, and $f : U \to \mathbb{R}^N$ a C^1 map. Then the image $f(S)$ of the set S of all critical points of f is of measure zero.*

Proof.   The proof is trivial for $M < N$, and see Milnor [35] for $M > N$. We prove the case $M = N$. Recall that every open set in $\mathbb{R}^N$ can be written as a countable union of nonoverlapping closed cubes, see Wheeden-Zygmund [51, p. 8]. Let $U = \cup_{i=1}^{\infty} Q_i$, Q_i be nonoverlapping closed cubes. Then $f(U) = \cup_{i=1}^{\infty} f(Q_i)$. It suffices to prove that $|f(S \cap Q_i)| = 0$. Let Q be any Q_i. Divide each of the n sides of Q into k equal parts. We obtain k^N cubes of side $\frac{1}{k}$, each of them of diameter $\frac{\sqrt{N}}{k}$. Let $\epsilon = \frac{\sqrt{N}}{k}$.

If one J of them intersect S, and $x \in S \cap J$. Because rank $f'(x) < N$, we have $\dim[f'(x)\mathbb{R}^N] < N$. There is a hyperplane Γ with

$$\{f(x) + f'(x)z \mid z \in \mathbb{R}^N\} \subset \Gamma.$$

Set $g(t) = f(t) - f'(x)t$, for $t \in J$. Then $g'(t) = f'(t) - f'(x)$, and for $y \in B$,

$$
\begin{aligned}
\|f(y) - f(x) - f'(x)(y-x)\| &= \|f(y) - f'(x)y - [f(x) - f'(x)x]\| \\
&= \|g(y) - g(x)\| \le \sup_{u \in J} \|g'(u)\|\|y-x\| \\
&= \sup_{u \in J} \|f'(u) - f'(x)\|\|y-x\| \\
&= b(\epsilon)\|y-x\| \le \epsilon b(\epsilon),
\end{aligned}
\tag{3.8}
$$

where $b(\epsilon) = \sup_{u \in J} \|f'(u) - f'(x)\| \to 0$ as $\epsilon \to 0$, because f' is continuous and is thus uniformly continuous on the compact set J. Because $f(x) + f'(x)(y-x) \in \Gamma$, we have $\mathrm{dist}(f(y), \Gamma) \le \epsilon b(\epsilon)$. Thus, $f(J)$ lies between two hyperplanes parallel to Γ, distant ϵ and $b(\epsilon)$, respectively, from Γ. Moreover, for $y \in J$, by the Mean Value Theorem 3.14,

$$\|f(y) - f(x)\| \le \sup_{u \in Q} \|f'(u)\|\|y-x\| = a\epsilon. \tag{3.9}$$

where $a = \sup_{u \in Q} \|f'(u)\|$. It follows that $f(J) \subset B_{a\epsilon}(f(x))$. To sum up, by (3.8) and (3.9), $f(J)$ lies in a right cylinder with base the intersection $\Gamma \cap B_{a\epsilon}(f(x))$, and height $2\epsilon b(\epsilon)$. Note that

$$|\Gamma \cap B_{a\epsilon}(f(x))| \le (2a\epsilon)^{N-1},$$

so

$$|f(J)| \le 2^{N-1}a^{N-1}\epsilon^{N-1}2\epsilon b(\epsilon) = 2^N a^{N-1}\epsilon^N b(\epsilon) = 2^N a^{N-1}\frac{N^{N/2}}{k^N}b\left(\frac{\sqrt{N}}{k}\right).$$

It follows that $S \cap Q$ is contained in the union of at most k^N cubes J, so

$$|f(S \cap Q)| \le 2^N a^{N-1} N^{N/2} b\left(\frac{\sqrt{N}}{k}\right) \to 0 \ \text{ as } \ k \to \infty.$$

Therefore, $f(S \cap Q)$ is of measure zero, and consequently $|f(S)| = 0$. ∎

3.5 Fundamental Theorem of Calculus

Definition 3.19 *Let E be a Banach space and $[a, b] \subset \mathbb{R}$ a bounded closed interval.*

(i) *A function $f : [a, b] \to E$ is called a step function if $f = \sum_{i=0}^{n-1} c_i \chi_{[a_i, \, a_{i+1}]}$ where $P = \{a = a_0 < a_1 < \cdots < a_n = b\}$ is a partition of $[a, b]$ and $c_i \in E$ for $i = 0, 1,$ $\cdots, n - 1$. The space $S[a, b]$ of all step functions forms a normed linear space under the supreme norm.*

(ii) *A function $f : [a, b] \to E$ is called a regulated function if it is the uniform limit of a sequence of step functions. A continuous function $h : [a, b] \to E$ is regulated. The space $R[a, b]$ of all regulated functions forms a Banach space under the supreme norm.*

(iii) *If $f = \sum_{i=0}^{n-1} c_i \chi_{[a_i, \, a_{i+1}]}$ is a step function on $[a, b]$, then we define the integral of f by*

$$T(f) = \int_a^b f(x)dx = \sum_{i=0}^{n-1}(a_{i+1} - a_i)c_i,$$

and we have

(a) *T is linear on $S[a, b]$ with, for $f \in S[a, b]$,*

$$\|T(f)\| \le (b - a)\|f\|_\infty;$$

(b) *For $c \in (a, b)$,*

$$\int_a^b f(x)dx = \int_a^c f(x) + \int_c^b f(x)dx;$$

(c) *If we write $\int_u^v = -\int_v^u$ when $v < u$, and $\int_u^u = 0$, then the Charles relation holds:*

$$\int_u^v f(x)dx = \int_u^w f(x)dx + \int_w^v f(x)dx \text{ for } u, v, w \in [a, b];$$

(iv) *If there are $f \in R[a, b]$, and $(f_n) \subset S[a, b]$ such that $f_n \to f$ uniformly, then there is $\lim_{n \to \infty} T(f_n)$ and is independent of the choice of (f_n). We define the integral of f by*

$$T(f) = \int_a^b f(x)dx = \lim_{n \to \infty} T(f_n),$$

and we have

(a) *T is linear on $R[a, b]$, with $f \in R[a, b]$*

$$\|T(f)\| \le (b - a)\|f\|_\infty;$$

(b) *The Charles relation holds for regulated functions.*

Definition 3.20 *Let* $(a, b) \subset \mathbb{R}$ *be an unnecessarily bounded open interval, and let* $g : (a, b) \to E$ *be regulated on every closed bounded interval in* (a, b). *Fix* $x_0 \in (a, b)$. *The function* G *defined by*

$$G(x) = \int_{x_0}^{x} g(t)dt, \text{ for } x \in (a, b)$$

is called a primitive of g. *If* G_1, G_2 *are two primitives of* g, *then*

$$G_2(x) - G_1(x) = \int_{x_2}^{x} g(t)dt - \int_{x_1}^{x} g(t)dt$$

$$= \int_{x_2}^{x_1} g(t)dt = c, a \text{ constant.}$$

We conclude that $G_2(x) = G_1(x) + c$ *for* $x \in (a, b)$.

Theorem 3.21 *Let* $g : (a, b) \to E$ *be continuous, and* G *a primitive of* g. *Then* $G \in C^1((a, b))$ *satisfies* $G' = g$.

Proof.

$$G(x + h) - G(x) - g(x)h = \int_{x_0}^{x+h} g(t)dt - \int_{x_0}^{x} g(t)dt - \int_{x}^{x+h} g(x)dt$$

$$= \int_{x}^{x+h} [g(t) - g(x)]dt.$$

Now

$$\frac{\| \int_{x}^{x+h} [g(t) - g(x)]dt \|}{\|h\|} \leq \sup_{t \in [x, x+h]} \|g(t) - g(x)\| \to 0 \text{ as } h \to 0,$$

because g is continuous. The theorem follows. ∎

Theorem 3.22 (*Fundamental Theorem of Calculus*) *Let* E, F *be Banach spaces,* $U \subset E$ *an open set, and let* $f : U \to F$ *be a* C^1 *map. If* $x + ty \in U$ *for* $t \in [0, 1]$, *then*

$$f(x + y) = f(x) + \int_{0}^{1} f'(x + ty)ydt.$$

Proof.   Set $g(t) = f(x + ty)$ for $0 \leq t \leq 1$. The Chain Rule 3.10 implies $g'(t) = f'(x + ty)y$ for $0 < t < 1$. Define $h(t) = f(x) + \int_{0}^{t} f'(x + sy)yds$, for $0 \leq t \leq 1$. By Theorem 3.21, $h'(t) = f'(x + ty)y$ for $0 < t < 1$, and so $g'(t) - h'(t) = 0$ for $0 < t < 1$. Hence, there is a constant c such that $g(t) = h(t) + c$ for $0 < t < 1$. Because g and h are continuous, $g(t) = h(t) + c$ for $0 \leq t \leq 1$. For $g(0) = h(0) = f(x)$, so $c = 0$. Thus, $g(1) = h(1)$, or $f(x + y) = f(x) + \int_{0}^{1} f'(x + ty)ydt$. ∎

3.6 Inverse Function Theorem

Let E, F be Banach spaces.

Definition 3.23 *Let $U \subset E$, $V \subset F$ be open sets. A map $f : U \to V$ is a diffeomorphism of U onto V if f is a bijection, and f, f^{-1} are C^1 maps.*

Remark 3.24 *If U is nonempty, and if $f : U \to V$ is a diffeomorphism onto, then E and F are isomorphic. In particular, their dimensions are equal if they are finite.*

Proof. Applying the Chain Rule 3.10 to $f^{-1} \circ f = I_U$ and $f \circ f^{-1} = I_V$ implies that $(f^{-1})'(f(u)) \circ f'(u) = I_E$ for $u \in U$, $f'(u) \circ (f^{-1})'(f(u)) = (f)'(f^{-1}(w)) \circ (f^{-1})'(w) = I_F$ for $w = f(u) \in V$. Thus E and F are isomorphic. ∎

Notation. Let $GL(E, F)$ denote all isomorphisms from E onto F.

Theorem 3.25 (*Inversion of an Isomorphism between Banach Spaces*) *The set $GL(E, F)$ forms an open subset of $L(E, F)$, and the inverse map $J : u \to u^{-1}$ of $GL(E, F)$ to $GL(F, E)$ is continuous.*

Proof.

(*i*) Assume $E = F$. Let $u \in GL(E, E)$. We claim that $u + h \in GL(E, E)$ whenever $h \in L(E, E)$ satisfies $\|h\| < \frac{1}{\|u^{-1}\|}$. Denote I_E by 1. Recall that for $x \in \mathbb{R}$ with $|x| < 1$, we have

$$(1 - x)^{-1} = 1 + x + x^2 + \cdots.$$

This formula suggests the consideration of the sequence, for $v \in L(E, E)$, $\|v\| < 1$,

$$a_0 = 1,$$
$$a_1 = 1 + v,$$
$$a_2 = 1 + v + v^2,$$
$$\vdots$$

Then, for $q > 1$,

$$\begin{aligned}
\|a_{p+q} - a_p\| &= \|v^{p+q} + \cdots + v^{p+1}\| \\
&\leq \|v\|^{p+q} + \cdots + \|v\|^{p+1} \\
&\leq \|v\|^{p+1}(1 + \|v\| + \|v\|^2 + \cdots) \\
&= \|v\|^{p+1}(1/(1 - \|v\|)) \to 0 \text{ as } p \to \infty.
\end{aligned}$$

Thus $\{a_n\}$ is a Cauchy sequence in $L(E, E)$. Therefore, there is $a \in L(E, E)$ such that $a_n \to a$ in $L(E, E)$. Letting $n \to \infty$ in $(1 - v) \circ a_n = 1 - v^{n+1}$, we obtain $(1 - v) \circ a = 1 \circ a = (1 - v)^{-1}$. Hence, $(1 - v)^{-1} = 1 + v + v^2 + \cdots$. If we let $v = -u^{-1} \circ h$, then $\| v \| = \| -u^{-1} \circ h \| \leq \| u^{-1} \| \| h \| < 1$, and $u + h = u \circ [1 + u^{-1} \circ h] = u \circ (1 - v) \in GL(E, E)$. Moreover,

$$
\begin{aligned}
J(u + h) - J(u) &= [(1 - v)^{-1} - 1] \circ u^{-1} \\
&= [\lim_{n \to \infty} (a_n - 1)] \circ u^{-1} \\
&= [\lim_{n \to \infty} (v + \cdots + v^n)] \circ u^{-1} \\
&= \lim_{n \to \infty} [(v + \cdots + v^n) \circ u^{-1}].
\end{aligned}
$$

Because

$$
\begin{aligned}
&\| (v + \cdots + v^n) \circ u^{-1} \| \\
&\leq (\| v \| + \cdots + \| v \|^n) \| u^{-1} \| \\
&\leq \| v \| \| u^{-1} \| / 1 - \| v \| \leq \| h \| \| u^{-1} \|^2 / (1 - \| u^{-1} \circ h \|) \\
&\to 0 \text{ as } h \to 0,
\end{aligned}
$$

we have $\lim_{h \to 0} \| J(u + h) - J(u) \| = 0$.

(ii) Assume $E \neq F$. If we fix $w \in GL(E, F)$, then the map $T_w : u \to w^{-1} \circ u$ of $GL(E, F)$ into $L(E, E)$ is continuous and

$$
T_w(GL(E, F)) = GL(E, E).
$$

Therefore, $GL(E, F)$ is open in $L(E, F)$, and $u \to u^{-1}$ can be decomposed as follows:

$$
u \to w^{-1} \circ u \to (w^{-1} \circ u)^{-1} = u^{-1} \circ w \to (u^{-1} \circ w \circ w^{-1}) = u^{-1},
$$

so $u \to u^{-1}$ is continuous from $GL(E, F)$ to $GL(F, E)$.   ■

Theorem 3.26 (*Inverse Function Theorem*) *Let $U \subset E$ be an open set, and $f : U \to F$ a C^1 map. If, for some $a \in U$, $f'(a)$ is an isomorphism of E onto $F : f'(a) \in GL(E, F)$, then there is an open neighborhood V of a and an open neighborhood W of $f(a)$ such that $f|_V : V \to W$ is a diffeomorphism onto. Moreover, the derivative of the inverse map f^{-1} at $y = f(x) \in W$ is given by*

$$
(f^{-1})'(y) = (f'(x))^{-1}.
$$

Proof.

(i) We may assume $a = 0$, $f(0) = 0$, and $E = F$, $f'(0) = I$, the identity map: it is enough to replace $f(x)$ by $h(x) = [f'(a)]^{-1}[f(a + x) - f(a)]$, and note that $h(0) = 0$, $h'(0) = [f'(a)]^{-1} f'(0) = I$, $h : U(0) \subset E \to E$.

(ii) $f^{-1} : B_r(0) \to B_{2r}(0)$: set $g(x) = x - f(x)$ for $x \in U$. We have $g(0) = 0$, $g'(0) = I - f'(0) = I - I = 0$. Because g' is continuous, there is $r > 0$ such that $\| g'(u) \| \leq \frac{1}{2}$ for $u \in B_{2r}(0)$. Apply the Mean Value Theorem 3.14 to obtain, for $x \in B_{2r}(0)$,

$$\|g(x)\| = \|g(x) - g(0)\| \leq \frac{\|x\|}{2} < r.$$

Hence, $g[B_{2r}(0)] \subset B_r(0)$. Let $y \in B_r(0)$; a unique $x \in B_{2r}(0)$ is required such that $f(x) = y$, but

$$f(x) = y \text{ if and only if } x - g(x) = y \text{ if and only if } x = y + g(x).$$

Setting $h(x) = y + g(x)$, then for $x \in \overline{B_{2r}(0)}$, $\| h(x) \| \leq \|y\| + \|g(x)\| < r + r = 2r$

$$\|h(u) - h(v)\| = \|g(u) - g(v)\|$$
$$\leq \frac{1}{2}\|u - v\| \quad for \ u, v \in B_{2r}(0).$$

Thus, $h : \overline{B_{2r}(0)} \to B_{2r}(0)$ is a 1/2-contraction. By the Banach Fixed Point Theorem 2.1, there is a unique $x \in B_{2r}(0)$ such that $h(x) = x$, and consequently, $f(x) = y$. It follows that there is $f^{-1} : B_r(0) \to B_{2r}(0)$.

(iii) $\| f^{-1}(x) - f^{-1}(y) \| \leq 2\|x - y\|$ for $x, y \in B_r(0)$: by (ii), for $x, y \in B_r(0)$, there are $u, v \in B_{2r}(0)$ such that $x = f(u), and \ y = f(v)$. Because $w = g(w) + f(w)$ for $w \in U$, we obtain

$$\|u - v\| = \|g(u) + f(u) - g(v) - f(v)\|$$
$$\leq \|g(u) - g(v)\| + \|f(u) - f(v)\|$$
$$\leq \frac{1}{2}\|u - v\| + \|f(u) - f(v)\|,$$

and consequently, $\|u - v\| \leq 2\|f(u) - f(v)\|$, and $\| f^{-1}(x) - f^{-1}(y) \| \leq 2\|x - y\|$.

(iv) f^{-1} is of class C^1 : because $x \to (f'(x))^{-1}$ is continuous at 0, choose r so small that there is $(f'(x))^{-1}$ on $B_{2r}(0)$, and $\| (f'(x))^{-1} - I \| < 1$ for $x \in B_{2r}(0)$. It follows that $\| (f'(x))^{-1} \| \leq 1 + \|I\| = c$ on $B_{2r}(0)$. Let $y, y + k \in B_r(0)$ with $x = f^{-1}(y)$, and $x + h = f^{-1}(y + k)$ for some $x, x + h \in B_{2r}(0)$. Now,

$$\|f^{-1}(y + k) - f^{-1}(y) - (f'(x))^{-1}k\|$$
$$= \|(x + h) - x - (f'(x))^{-1}[f(x + h) - f(x)]\|$$
$$= \|(f'(x))^{-1}[f(x + h) - f(x) - f'(x)h]\|$$
$$\leq c\|f(x + h) - f(x) - f'(x)h\|.$$

Note that by (iii), $\frac{\|h\|}{\|k\|} = \frac{\|(x+h)-x\|}{\|k\|} = \frac{\|f^{-1}(y+k)-f^{-1}(y)\|}{\|k\|} \leq 2$, and consequently,

$\frac{\|f^{-1}(y+k)-f^{-1}(y)-(f'(x))^{-1}k\|}{\|k\|} \leq c\frac{\|f(x+h)-f(x)-f'(x)h\|}{\|h\|}\frac{\|h\|}{\|k\|} \to 0$ as $h \to 0$. It follows

that f^{-1} is differentiable and that

$$(f^{-1})'(y) = (f'(x))^{-1}. \tag{3.10}$$

By (3.10), write $(f^{-1})' = J \circ Df \circ f^{-1}$, where $J(u) = u^{-1}$, so $(f^{-1})'$ is continuous in $B_r(0)$ for small $r > 0$. Finally, let $V = f^{-1}(B_r(0)) \cap B_{2r}(0)$, and $W = B_r(0)$ we have $f|_V : V \to W$ is a diffeomorphism onto. ∎

Theorem 3.27 *(Invariance of Domain Theorem) Let $U \subset E$ be open, $f : U \to F$ is of class C^1, and $f'(u) \in GL(E, F)$ for all $u \in U$. Then $f(U) = V$ is open in F. In particular, if $f : U \to V$ is a diffeomorphism onto, then $f(G)$ is open for every open set G in U.*

Proof. By the Inverse Function Theorem 3.26, and Remark 3.24. ∎

3.7   Implicit Function Theorem

Theorem 3.28 *(Implicit Function Theorem) Let E, F and G be Banach spaces, and let $U \subset E$, $V \subset F$ be open sets, and $f : U \times V \to G$ a C^1 map. Assume that at $(a, b) \in U \times V$, $f(a, b) = 0$ and the partial derivative $D_2 f(a, b) \in GL(F, G)$. Then, there are an open neighborhood A of a, an open neighborhood B of b, and a unique C^1 map $g : A \to B$ such that*

$$\begin{cases} g(a) = b, \\ f(x, g(x)) = 0 \text{ for } x \in A \end{cases}$$

and $g'(x) = -[D_2 f(x, g(x))]^{-1} \circ D_1 f(x, g(x))$ for all $x \in A$.

Proof. Consider the map $\Phi : U \times V \to E \times G$ defined by

$$\Phi(x, y) = (x, f(x, y)) = (\Phi_1(x, y), \Phi_2(x, y)).$$

Then $\Phi \in C^1$, and for $(h, k) \in E \times F$,

$$\Phi'(a, b)(h, k) = \begin{pmatrix} \dfrac{\partial \Phi_1}{\partial x} & \dfrac{\partial \Phi_1}{\partial y} \\[2mm] \dfrac{\partial \Phi_2}{\partial x} & \dfrac{\partial \Phi_2}{\partial y} \end{pmatrix} \binom{h}{k}$$

$$= \begin{pmatrix} I & 0 \\ D_1 f(a, b) & D_2 f(a, b) \end{pmatrix} \begin{pmatrix} h \\ k \end{pmatrix}$$

$$= (h, D_1 f(a, b)h + D_2 f(a, b)k)$$

If $\Phi'(a, b)(h, k) = 0$, then $(h, D_1 f(a, b)h + D_2 f(a, b)k) = 0$, and consequently $h = 0$, $D_2 f(a, b)k = 0$. Because $D_2 f(a, b)$ is an isomorphism, $k = 0$, $\Phi'(a, b)$ is injective. Moreover, for $(h, s) \in E \times G$ and noting that

$$D_1 f(a,b)h + D_2 f(a,b)k = s \iff D_2 f(a,b)k = s - D_1 f(a,b)h,$$

and that $D_2 f(a, b)$ is surjective, then $\Phi'(a, b)$ is surjective. Therefore, $\Phi'(a, b)$ is an isomorphism of $E \times F$ onto $E \times G$, with the inverse $\Phi'(a, b)^{-1}$:

$$\Phi'(a,b)(h, k) = (h, D_1 f(a,b)h + D_2 f(a,b)k) = (h, s),$$

that is,

$$\Phi'(a,b)^{-1}(h, s) = (h, k) = (h, D_2 f(a,b)^{-1}[s - D_1 f(a,b)h]).$$

By the Inverse Function Theorem 3.26, there are open neighborhoods $D \subset U$ of x, V of (a, b), $A \subset U$ of a, and $W \subset G$ of $f(a, b) = 0$ such that $\Phi : D \to A \times W$ is a diffeomorphism onto. For $x \in A$ and $w \in W$, there is a unique $(x', y) \in D$ such that $\Phi^{-1}(x, w) = (x', y)$ or $(x, w) = \Phi(x', y) = (x', f(x', y))$. This implies that $x = x'$ and $f(x, y) = w$. We conclude that for $(x, w) \in A \times W$, there is an open set B in V, a unique $y \in B$, and a function $h : A \times W \to B$ defined by $h(x, w) = y$ satisfying $w = f(x, h(x, w))$. If we choose $w = 0 = f(a, b)$, and define the function $g : A \to B$ by $g(x) = h(x, 0)$, we have $0 = f(x, g(x))$ for all $x \in A$, $g(a) = h(a, 0)$, and $0 = f(a, h(a, 0)) = f(a, b) : \phi(a, b) = (a, f(a, b)) = (a, f(a, h(a, 0))) = \phi(a, h(a, 0))$, so we have $g(a) = b$. Applying the Chain Rule 3.10 to $f(x, g(x)) = 0$, we have

$$D_1 f(x, y) + D_2 f(x, y) \circ g'(x) = 0,$$

but $D_2 f$ is continuous and $D_2 f(a, b)$ is invertible. We may choose A and B so that if $(x, y) \in A \times B$, then $D_2 f(x, g(x))$ is invertible. Hence,

$$g'(x) = -[D_2 f(x, g(x))]^{-1} \circ D_1 f(x, g(x)) \quad \text{for} \ \ x \in A. \qquad \blacksquare$$

Theorem 3.29 *Let E, F and G be Banach spaces, $(0, 0) \in A \subset E \times F$ an open set, and let $f : A \to G$ be of C^1. Assume that:*

(i) $f(0, 0) = 0$;

(ii) $D_1 f(0, 0)$: $E \to G$ *is surjective;*

(iii) $E_1 = Ker D_1 f(0, 0)$ *and* $E = E_1 \oplus E_2$ *for some closed subspace E_2 of E.*

Then there are $0 \in U \subseteq E_1$, $0 \in V \subseteq F$ open sets, and a unique C^1 map $u : U \times V \to E_2$ such that $u(0, 0) = 0$ and $f(x_1 + u(x_1, y), y) = 0$ for $x_1 \in U$, $y \in V$.

Proof. Set $F_1 = E_1 \times F$, $0 \in B \subset E_2 \times F_1$ as an open set. The map $\varphi : B \to G$ defined by $\varphi(x_2, (x_1, y)) = f(x_1 + x_2, y)$ is of C^1 and satisfies

$$\varphi(0,0) = 0,$$
$$D_1\varphi(0,0) = D_1 f(0,0) : E_2 \to G \text{ is onto,}$$
$$Ker D_1\varphi(0,0) = Ker D_1 f(0,0) \cap E_2 = E_1 \cap E_2 = \{0\}.$$

By the Implicit Function Theorem 3.28, there are $0 \in U \subseteq E_1$, $0 \in V \subset F$ open sets, and a unique C^1 map $u : U \times V \to E_2$ such that

$$u(0,0) = 0,$$
$$f(x_1 + u(x_1, y), y) = 0 \text{ for } x_1 \in U, y \in V. \qquad \blacksquare$$

Theorem 3.30 (*Lyapunov-Schmidt Procedure*) *Let E, Λ, and G be B-spaces, with* $\dim \Lambda = n_1$ (*think of Λ as the parameter space*), $(0,0) \in A \subset E \times \Lambda$ *an open set, and* $f : A \to G$ *a C^1 map satisfying:*

  (*i*) $f(0,\ 0) = 0;$

 (*ii*) *Ker* $D_1 f(0,\ 0) = E_1$, $\dim\ E_1 = n_2 < \infty;$

(*iii*) *range* $D_1 f(0,\ 0) = G_1$, *codim* $G_1 = m < \infty.$

Then $f(x,\ \lambda) = 0$ if and only if $(x,\ \lambda)$ satisfies m equations with $n_1 + n_2$ unknowns.

Proof. Decompose E, F by

$$E = E_1 \oplus E_2, \dim E_1 = n_2,$$
$$G = G_1 \oplus G_2, \dim G_2 = m.$$

Consider the projections $P : G \to G_1$, $Q = I - P : G \to G_2$. Note that $(x, \lambda) \in A$, so we have

$$f(x, \lambda) = 0 \iff \begin{cases} P \circ f(x, \lambda) = 0 \\ Q \circ f(x, \lambda) = 0. \end{cases} \qquad (3.11)$$

Because

$$\text{range} D_1 f(0,0) = G_1,$$

then

$$\text{range} D_1(P \circ f)(0,0) = \text{range } P \circ (D_1 f(0,0)) = G_1.$$

Therefore, $P \circ f : A \to G_1$ is of C^1 satisfying

$$(P \circ f)(0,0) = 0,$$
$$D_1(P \circ f)(0,0) : E \to G_1 \text{ is onto,}$$
$$Ker D_1(P \circ f)(0,0) = E_1.$$

By Theorem 3.29, $0 \in U \subset E_1$, $0 \in V \subset \Lambda$ are open sets, and there is a unique C^1 map $u : U \times V \to E_2$ such that

$$(P \circ f)(x_1 + u(x_1, \lambda), \lambda) = 0 \text{ for } (x_1, \lambda) \in U \times V.$$

By (3.11) together with these observations, we have, for $(x, \lambda) \in U \times V$,

$$f(x_1 + u(x_1, \lambda), \lambda) = 0 \text{ if and only if } (Q \circ f)(x_1 + u(x_1, \lambda), \lambda) = 0.$$

Note that we have $(Q \circ f)(x_1 + u(x_1, \lambda), \lambda) = 0$ is a system of m equations of $n_1 + n_2$ unknowns. $\blacksquare$

Let E be a Banach space, and let f, $g : E \to \mathbb{R}$ be differentiable functions. For $c \in \mathbb{R}$, let $M_c = \{x \in E \mid g(x) = c\}$. Then M_c is a level surface that may be empty or disconnected. For example, if $g(x) = \|x\|$, $R > 0$ then $M_R = S_R = \{x \in E \mid \|x\| = R\}$.

Theorem 3.31 (*Lagrange Multiplier Theorem*) *Let E be a Banach space, let f, $g : E \to \mathbb{R}$ be differentiable, and let $g \in C^1$, and $M_c = \{x \in E \mid g(x) = c\}$. Assume that when $a \in M_c$, $Dg(a) \neq 0$ is a relatively extreme point of f with respect to M_c, then there is a real number λ, called the Lagrange multiplier, such that*

$$Df(a) = \lambda Dg(a).$$

Proof. Write

$$f(a + h) - f(a) = Df(a)h + r(h),$$
$$g(a + h) - g(a) = Dg(a)h + s(h),$$

where $r(h) = o(\|h\|)$, $s(h) = o(\|h\|)$ as $h \to 0$. We may assume $\dim E > 1$, and then take v, $w \in E$, $v \neq 0$, $w \neq 0$ such that $Dg(a)v = 1$ and $Dg(a)w = 0$. Consider the function $\varphi : \mathbb{R} \times \mathbb{R} \to \mathbb{R}$ defined by

$$\varphi(t, \epsilon) = g(a + tv + \epsilon w) - c.$$

Then $\varphi \in C^1$ satisfies $\varphi(0, 0) = 0$ and $D_1\varphi(0, 0) = Dg(a)v = 1$. By the Implicit Function Theorem 3.28, there are open intervals I, J of 0, respectively, and a unique C^1 map $t : J \to I$ such that

$$t(0) = 0,$$
$$t(\epsilon) \to 0 \text{ as } \epsilon \to 0,$$
$$\varphi(t(\epsilon), \epsilon) = 0 \text{ for } \epsilon \in J.$$

Now,

$$
\begin{aligned}
0 &= \varphi(t(\epsilon), \epsilon) \\
&= g(a + t(\epsilon)v + \epsilon w) - g(a) \\
&= Dg(a)[t(\epsilon)v + \epsilon w] + s(t(\epsilon)v + \epsilon w) \\
&= t(\epsilon) + s(t(\epsilon)v + \epsilon w).
\end{aligned}
$$

For $0 < \delta < \|v\|$, there is $\epsilon_0 > 0$ such that $|\epsilon| < \epsilon_0$ implies

$$
\frac{\|s(t(\epsilon)v + \epsilon w)\|}{\|t(\epsilon)v + \epsilon w\|} < \frac{\delta}{2\|v\|} < \frac{1}{2}.
$$

Therefore,

$$
\frac{|t(\epsilon)|}{|\epsilon|} \leq \frac{\|s(t(\epsilon)v + \epsilon w)\|}{\|t(\epsilon)v + \epsilon w\|} \frac{\|t(\epsilon)v + \epsilon w\|}{|\epsilon|} < \frac{1}{2}\frac{|t(\epsilon)|}{|\epsilon|} + \frac{\delta}{2}\frac{\|w\|}{\|v\|}
$$

or $\frac{|t(\epsilon)|}{|\epsilon|} < c\delta$. Hence, we have $t(\epsilon) = o(|\epsilon|)$ as $\epsilon \to 0$. We claim next that $Df(a)w = 0$. In fact, because $g(a + t(\epsilon)v + \epsilon w) = c$ for $\epsilon \in J$, $\{a + t(\epsilon)v + \epsilon w \mid |\epsilon| < \delta\}$, for some δ, it represents a curve on M_c through a. Now

$$
\begin{aligned}
f(a + t(\epsilon)v + \epsilon w) - f(a) &= Df(a)(t(\epsilon)v) + Df(a)(\epsilon w) + o(t(\epsilon)v + \epsilon w) \\
&= t(\epsilon)Df(a)v + \epsilon(Df(a)w) + o(\epsilon) \\
&= \epsilon Df(a)w + o(\epsilon).
\end{aligned}
$$

Note that

$$
\frac{o(t(\epsilon)v + \epsilon w)}{\epsilon} = \frac{o(t(\epsilon)v + \epsilon w)}{t(\epsilon)v + \epsilon w} \frac{t(\epsilon)v + \epsilon w}{\epsilon} \to 0 \ \ as \ \epsilon \to 0,
$$

because $t(\epsilon) = o(\epsilon)$, and therefore $o(t(\epsilon)v + \epsilon w) = o(\epsilon)$. Because f attains its extreme at a with respect to M_c,

$$
Df(a)w = \lim_{\epsilon \to 0} \frac{f(a + t(\epsilon)v + \epsilon w) - f(a)}{\epsilon} = 0.
$$

We conclude that, for $w \in E$,

$$
Dg(a)w = 0 \Rightarrow Df(a)w = 0.
$$

Now $Dg(a)$ and $Df(a) \in L(E, \mathbb{R})$ satisfy $Ker Dg(a) \subset Ker\ Df(a)$. Recall that if $T : E \to \mathbb{R}$ is linear, then either $KerT = E$ or $E =< KerT, x_0 >$ for some $x_0 \in E \backslash KerT$. Because $Dg(a) \neq 0$, there is $x_0 \notin KerDg(a)$, with $E = KerDg(a) \oplus \mathbb{R}x_0$. For any $x \in E$, $x = y + \beta x_0$ where $y \in KerDg(a)$. Therefore, for $\lambda = \frac{Df(a)(x_0)}{Dg(a)(x_0)}$,

$$
\begin{aligned}
Df(a)(x) &= \beta Df(a)(x_0) = \beta \lambda Dg(a)(x_0) \\
&= \lambda Dg(a)(\beta x_0) = \lambda Dg(a)(x),
\end{aligned}
$$

or $Df(a) = \lambda Dg(a)$.   ∎

3.8   Taylor's Theorem

We study successive derivatives, and the Schwarz Theorem.

Definition 3.32

(i) *Let E and F be Banach spaces, $U \subset E$ be open, and let $f : U \to F$ be twice differentiable at $a \in U$ if Df is differentiable at a. The derivative of Df at a is denoted by $D^2 f(a) \in L(E; L(E, F))$, called the second derivative of f at a;*

(ii) *If $Df : U \to L(E, F)$ is differentiable at each point of U, we say that f is twice differentiable on U, and $D^2 f : U \to L(E; L(E, F))$ is called the second derivative of f. If $D^2 f$ is continuous on U, we say that f is of class C^2 on U.*

Note that, if denoted by

$$L^n(E; F) = L(E, \cdots, E; F)$$
$$= \{T : E \times \cdots \times E \to F \mid T \text{ is continuous multilinear}\},$$

$L^n(E; F)$ is a Banach space under the norm

$$\|T\| = \sup\{\|T(x_1, \cdots, x_n)\| \mid \|x_1\| = \cdots = \|x_n\| = 1\},$$

and so we have

$$L(\mathbb{R}; F) \cong F,$$
$$L(E; L^k(E; F)) \cong L^{k+1}(E; F),$$

where $\cong$ means canonically isomorphic. Because $L(E; L(E; F))$ is canonically isomorphic to $L^2(E; F)$, we may consider $D^2 f(a) \in L^2(E; F)$, and define

$$(h, k) \to D^2 f(a)(h, k) = (D^2 f(a)h)k$$

so that each $h \in E$ corresponds to the element $D^2 f(a)h$ of $L(E; F)$, and the image of $k \in E$ under this last map is $(D^2 f(a)h)k \in F$.

Theorem 3.33 (*Schwarz Theorem*) *If $f : U \subset E \to F$ is twice differentiable at $a \in U$, then*

$$\lim_{\substack{\|u\| \to 0 \\ \|v\| \to 0}} \frac{\|f(a + u + v) - f(a + u) - f(a + v) + f(a) - D^2 f(a)(u, v)\|}{(\|u\| + \|v\|)^2} = 0.$$

In particular, $D^2 f(a)$ is a bilinear symmetric form:

$$D^2 f(a)(u, v) = D^2 f(a)(v, u) \text{ for each } u, v \in E.$$

Proof.   Because Df is differentiable at a, given $\epsilon > 0$, there is $\delta > 0$ such that if $\|x\| < 2\delta$, then

$$\|Df(a+x) - Df(a) - D^2 f(a)x\| \leq \epsilon\|x\|.$$

Choose $\|u\| < \delta$, $\|v\| < \delta$, and set

$$g(u) = f(a+u+v) - f(a+u) - f(a+v) + f(a) - D^2 f(a)(u,v).$$

We have

$$\begin{aligned}
Dg(u) &= Df(a+u+v) - Df(a+u) - D^2 f(a)v \\
&= [Df(a+u+v) - Df(a) - D^2 f(a)(u+v)] \\
&\quad - [Df(a+u) - Df(a) - D^2 f(a)u].
\end{aligned}$$

Therefore, $\| Dg(u)\| \leq 2\epsilon(\|u\| + \|v\|)$. Apply the Mean Value Theorem 3.14 and $g(0) = 0$ to obtain

$$\|g(u)\| = \|g(u) - g(0)\| \leq 2\epsilon(\|u\| + \|v\|)\|u\| \leq 2\epsilon(\|u\| + \|v\|)^2,$$

and the first part of the theorem follows. Because $f(a+u+v) - f(a+u) - f(a+v) + f(a)$ is symmetric in u and v, so is $D^2 f(u,\, v)$. Denote the spaces

$$\begin{aligned}
C^n(U, F) &= \{f : U \subset E \to F \mid f \text{ is of } C^n\}, \\
C^\infty(U, F) &= \cap_{n>0} C^n(U, F).
\end{aligned}$$

$C^\infty(U,\, F)$ consists of infinitely differentiable functions.   ∎

Theorem 3.33 can be generalized to C^n maps.

Theorem 3.34 (*Schwarz Theorem*) *If $f : U \subset E \to F$ is n times differentiable at $a \in U$, then $D^n f(a) \in L^n(E;\, F)$ is a symmetric n-linear map. In other words, for all $h_1, \cdots, h_n \in E$, and for any permutation s of $\{1, 2, \cdots, n\}$ we have*

$$D^n f(a)(h_1, \cdots, h_n) = D^n f(a)(h_{s(1)}, \cdots, h_{s(n)}).$$

Theorem 3.35 (*Product Rule*) *Let E, F_1, F_2 and G be Banach spaces, let $U \subset E$ be open, let $f : U \to F_1$, $g : U \to F_2$ of C^k, and let $b : F_1 \times F_2 \to G$ be a continuous bilinear map. Define $p : U \to G$ by $p(x) = b(f(x),\, g(x))$. Then p is of C^k.*

Proof.   Suppose that f and g are C^1. Then p is differentiable at $a \in U$, and if $h \in E$ we have

$$Dp(a)h = b(Df(a)h, g(a)) + b(f(a), Dg(a)h). \tag{3.12}$$

Thus, Dp is continuous and p is C^1. By induction, suppose that this is true for $k - 1$. Suppose f and g are C^k. The map $w : E \to L(E,\, G)$ defined by

$$w(h) = b(A(h), \ell) \text{ for } A \in L(E, F_1), \ell \in F_2 \tag{3.13}$$

is continuous and bilinear. Moreover, because Df and g are C^{k-1}, it follows that $x \to b(Df(x), g(x))$ is C^{k-1}. Similarly, $x \to b(f(x), Dg(x))$ is C^{k-1}. By (3.12) and (3.13), Dp is C^{k-1} and p is of C^k. ∎

Let $f, g : U \subset E \to \mathbb{R}$ of C^n. By Theorem 3.35, $fg \in C^n$, and

$$(fg)^{(n)} = \sum_{i=0}^{n} A_{in} f^{(n-i)} g^{(i)},$$

choose $f(x) = e^{ax}$, $g(x) = e^{bx}$ where $a, b \in \mathbb{R}$. Then we have from the above equation $(a + b)^n = \sum A_{in} a^{n-i} b^i$. By the binomial formula, $A_{in} = C_i^n = \frac{n!}{(n-i)!i!}$. Hence, $(fg)^{(n)} = \sum_{i=0}^{n} C_i^n f^{(n-i)} g^i$.

Theorem 3.36 (*Chain Rule*) *Let* $U \subset E$ *and* $V \subset F$ *be open,* $f : U \to V$ *and* $g : V \to G$ *be of* C^n. *Then* $g \circ f$ *is* C^n.

Proof. We already know that if f, g are differentiable, then so is $g \circ f$, and $D(g \circ f)(x) = Dg(f(x)) \circ Df(x)$. Hence, if f, g are C^1, then $g \circ f$ is C^1. By induction, suppose that this is true for $n - 1$. Assume that f and g are C^n, and Dg and Df are C^{n-1}. By the Product Rule, $x \to Dg(f(x)) \circ Df(x)$ is C^{n-1} and $g \circ f$ is C^n. ∎

We intend to extend to "order n" Taylor's Formula of the Fundamental Theorem of Calculus

$$f(a + h) - f(a) = \int_0^1 Df(a + th)h\,dt,$$

which uses the first derivative of f.

Lemma 3.37 *If* $u : \mathbb{R} \to F$ *is of* C^{n+1}, *then*

$$D[u(t) + (1 - t)Du(t) + \cdots + \frac{1}{n!}(1 - t)^n D^n u(t)]$$

$$= \frac{1}{n!}(1 - t)^n D^{n+1} u(t).$$

Proof. We apply the Product Rule 3.35 with $E = F_1 = \mathbb{R}$, $F_2 = F = G$, $f(t) = (1 - t)^k$, $g(t) = D^k u(t)$, and $0 \le k \le n$. Let $b : F_1 \times F_2 = \mathbb{R} \times F \to F$ be a continuous bilinear map defined by $b(r, y) = ry$. We have

$$D[u(t) + (1 - t)Du(t) + \frac{(1 - t)^2}{2!}D^2 u(t) + \cdots + \frac{1}{n!}(1 - t)^n D^n u(t)]$$

$$= Du(t) - Du(t) + (1 - t)D^2 u(t) - (1 - t)D^2 u(t) + \frac{(1 - t)^2}{2!}D^3 u(t)$$

$$+ \cdots - \frac{(1 - t)^{n-1}}{(n - 1)!}D^n u(t) + \frac{1}{n!}(1 - t)^n D^{n+1} u(t)$$

$$= \frac{1}{n!}(1 - t)^n D^{n+1} u(t). \qquad ∎$$

Lemma 3.38 *If we let $U \subset \mathbb{R}$ be open, and let $[0, 1] \subset U$ and $u : U \to F$ be C^{n+1}, then*

$$u(1) - u(0) - Du(0) - \frac{1}{2}D^2u(0) - \cdots - \left(\frac{1}{n!}\right)D^nu(0)$$

$$= \int_0^1 \frac{(1-t)^n}{n!}D^{n+1}u(t)dt.$$

Proof. We apply the Fundamental Theorem of Calculus to the function

$$f(t) = [u(t) + (1-t)Du(t) + \cdots + \frac{1}{n!}(1-t)^nD^nu(t)].$$

Then,

$$f(1) - f(0) = \int_0^1 Df(t)dt,$$

$$u(1) - u(0) - Du(0) - \frac{1}{2}D^2u(0) - \cdots - \frac{1}{n!}D^nu(0)$$

$$= \int_0^1 \frac{1}{n!}(1-t)^nD^{n+1}u(t)dt. \qquad \blacksquare$$

Lemma 3.39 *Given $U \subset \mathbb{R}$ and $[0, 1] \subset U$, let $u : U \to F$ be of C^{n+1}, and assume there is a constant c with $\|D^{n+1}u(t)\| \leq c$ for $0 \leq t \leq 1$. Then*

$$\|u(1) - u(0) - Du(0) - \cdots - \frac{1}{n!}D^nu(0)\| \leq \frac{c}{(n+1)!}.$$

Proof. Let $[a, b] = [0, 1]$,

$$f(t) = u(t) + (1-t)Du(t) + \cdots + \frac{1}{n!}(1-t)^nD^nu(t),$$

$$g(t) = \frac{-c(1-t)^{(n+1)}}{(n+1)!}.$$

Then

$$\|Df(t)\| = \|\frac{1}{n!}(1-t)^nD^{n+1}u(t)\| \leq \frac{c}{n!}(1-t)^n = g'(t).$$

We apply the Mean Value Theorem 3.14 to obtain

$$\|f(1) - f(0)\| \leq g(1) - g(0) = \frac{c}{(n+1)!},$$

which is the inequality required. $\blacksquare$

Theorem 3.40 (*Taylor's Formula with Integral Remainder*) *Assume that $U \subset E$ is open, $f : U \to F$ is C^{n+1}, and $[a, a + h] \subset U$. Then*

$$f(a + h) - f(a) - Df(a)h - \cdots - \frac{1}{n!} D^n f(a)(h)^n$$

$$= \int_0^1 \frac{1}{n!} (1 - t)^n D^{(n+1)} f(a + th)(h)^{n+1} dt$$

where

$$D^k f(a)(h)^k = D^k f(a)(h, \cdots, h).$$

Proof. The function $u(t) = f(a + th)$, $0 \le t \le 1$ is of C^{n+1} with

$$D^k u(t) = D^k f(a + th)(h)^k.$$

It is now enough to use Lemma 3.38. ∎

Theorem 3.41 (*Taylor's Formula with Lagrange's Remainder*) *Let $U \subset E$ be open, let $f : U \to F$ be $(n + 1)$ times differentiable and assume there is a constant c with $\|D^{n+1} f(x)\| \le c$, for $x \in U$. Then*

$$\|f(a + h) - f(a) - Df(a)h - \cdots - \frac{1}{n!} D^n f(a)(h)^n\| \le \frac{c}{(n + 1)!} \|h\|^{n+1}.$$

Proof. The function $u(t) = f(a + th)$ is $(n + 1)$ times differentiable on $[0, 1]$ and

$$\|D^{n+1} u(t)\| = \|D^{n+1} f(a + th)(h)^{n+1}\|$$
$$\le \|D^{n+1} f(a + th)\| \|h\|^{n+1} \le c \|h\|^{n+1}.$$

The result then follows from Lemma 3.39. ∎

Notes Most material in this chapter was adapted from Dieudonne [18]. For related materials see Avez [4], Cartan [12], Minor [35], Nirenberg [36], Schwartz [44], and Wheeden-Zygmund [51].

Chapter 4

BIFURCATION THEORY

In this chapter we study bifurcation theory for operators on Banach spaces.

4.1 Morse Theory

Let E and F be two real Banach spaces, let $U \subset E$ be an open set, assume $x_0 \in U$, and let $f : U \to F$ satisfy:

(i) $f(x_0) = 0$;

(ii) f is a $C^k(U)$ map, $k \geq 1$;

(iii) $\dim Ker\ f'(x_0) = m < \infty$, and range $f'(x_0) = Y$ is closed in F such that

$$codim\ Y = \dim\ F/Y = n < \infty.$$

In this section, we will study the structure of solution sets for finite m and n.
Case 1. $m = n = 0$, then $f'(x_0) \in GL(E, F)$. In this case we will see from the following Inverse Function Theorem that there is an open set $V \subset U$ containing x_0 such that the set S of solutions in V consists of only x_0, where

$$S = \{x \in V \mid f(x) = 0\},$$

that is, x_0 is the isolated solution of the equation $f(x) = 0$.

Theorem 4.1 (*Inverse Function Theorem*) *Let $U \subset E$ be an open set, and let $f : U \to F$ be a C^1 map. If, for some $x_0 \in U$, $f'(x_0)$ is an isomorphism of E onto $F : f'(x_0) \in GL(E, F)$, then there is an open neighborhood V of x_0 and an open neighborhood W of $f(x_0)$ such that $f|_V : V \to W$ is a diffeomorphism onto. Moreover, the derivative of the inverse map f^{-1} at $y = f(x) \in W$ is given by*

$$(f^{-1})'(y) = (f'(x))^{-1}.$$

Proof. See Theorem 3.27. ∎

Case 2. $m > 0$, and $n = 0$. In this case, the set S of solutions in U contains a manifold of dimension m, where

$$S = \{x \in U \mid f(x) = 0\}.$$

This is a consequence of the following result.

Theorem 4.2 *Let E and F be two real Banach spaces, let $U \subset E$ be an open set, assume $x_0 \in U$, and let $f : U \to F$ satisfy:*

(i) $f(x_0) = 0$;

(ii) f *is of a $C^k(U)$ map, $k \geq 1$;*

(iii) $\dim Ker\ f'(x_0) = m < \infty$, *range* $f'(x_0) = F$.
  Let $S = \{x \in U \mid f(x) = 0\}$. Then S contains a manifold of dimension m.

Proof.   Assume $X = Ker\ f'(x_0)$; there is a closed subspace Y of E such that $E = X \oplus Y$. Write $x_0 = x_0^1 + x_0^2$, where $x_0^1 \in X$, $x_0^2 \in Y$. Choose a neighborhood $W \subset U$ of x_0, define $\varphi : W \longrightarrow F$ by $\varphi(x, y) = f(x + y)$. Then

$$D_2\varphi(x_0^1, x_0^2) = f'(x_0),$$
$$\text{range } D_2\varphi(x_0^1, x_0^2) = \text{range } f'(x_0) = F,$$
$$Ker\ D_2\varphi(x_0^1, x_0^2) = (Ker\ f'(x_0)) \cap Y = X \cap Y = \{0\}.$$

Thus, $D_2\varphi(x_0^1, x_0^2) \in GL(Y, F)$. By the Implicit Function Theorem 3.28, there is a neighborhood A of x_0^1 in X, a neighborhood B of x_0^2 in Y, and a unique $C^k(A, B)$ map u such that $u(x_0^1) = x_0^2$, $f(x + u(x)) = \varphi(x, u(x)) = 0$ for $x \in A$. Therefore, the set S of solutions in U contains A, which is an open set in X. Therefore, A is of dimension m.   ∎

Case 3. $m > 0$, and $n = 1$. Let E, F be two real Banach spaces, let $V \subset E$ be an open set, assume $x_0 \in V$, and let $g : V \to F$ satisfy:

(i) $g(x_0) = 0$;

(ii) g is a C^k map, $k \geq 2$;

(iii) $\dim\ Ker\ g'(x_0) = m > 0$, range $g'(x_0) = Y$ is closed in F, and codim $Y = \dim\ F/Y = 1$.

Consider the canonical map $\pi : F \longrightarrow F/Y$ defined by $\pi(x) = \bar{x} = x + Y$. Because $\dim\ F/Y = 1$, we have an isomorphism onto $\alpha : F/Y \to \mathbb{R}$. If we set $T = \alpha \circ \pi$, then T is a continuous linear functional with $Ker\ T = Y$. Suppose $x_0 = 0$; $f(0) = 0$. By the Lyapunov-Schmidt Procedure 3.30, the local study of the equation $g(x) = 0$ reduces to the single bifurcation equation

$$Tg(x + u(x)) = 0 \text{ for } x \in U, \tag{4.1}$$

where $X = Ker\ g'(x_0)$, $E = X \oplus Z$, U is an open neighborhood of 0 in X, and $u : U \to Z$, a C^k map, $k \geq 1$, $u(0) = 0$. Equation (4.1) is a single equation with m unknowns. Set $f(x) = Tg(x + u(x))$ for $x \in U$, where $U \subset \mathbb{R}^m = X$ is an open set containing 0, and $f : U \longrightarrow \mathbb{R}$ satisfies:

(i) $f(0) = 0$;

(ii) f is a C^k map, $k \geq 2$.

There are two important cases.

Case 3.1. $f(0) = 0$, f' is nonsingular: if we set $X = Ker f'(0)$, then $\dim X = m - 1$, range $f'(0) = \mathbb{R}$. By Theorem 4.2, the solution set $S = \{x \in U\,|\,f(x) = 0\}$ contains a C^k-hypersurface: a C^k-manifold of dimension $m - 1$.

Case 3.2. $f(0) = 0$, $f'(0) = 0$, $f''(0)$ is nonsingular. In this case we have the important result:

Lemma 4.3 (*Morse Lemma*) *Let $\Omega \subset \mathbb{R}^N$ be an open set. Assume $0 \in \Omega$, and $f : \Omega \to \mathbb{R}$ is a C^k map, $k \geq 2$ satisfying:*

(i) $f(0) = 0$;

(ii) $f'(0) = 0$;

(iii)

$$f''(0) = \begin{bmatrix} \dfrac{\partial^2 f}{\partial x_1 \partial x_1}(0) & \dfrac{\partial^2 f}{\partial x_1 \partial x_2}(0) & \dfrac{\partial^2 f}{\partial x_1 \partial x_N}(0) \\[2mm] \dfrac{\partial^2 f}{\partial x_2 \partial x_1}(0) & \dfrac{\partial^2 f}{\partial x_2 \partial x_2}(0) & \dfrac{\partial^2 f}{\partial x_2 \partial x_N}(0) \\[2mm] \vdots & \vdots & \vdots \\[2mm] \dfrac{\partial^2 f}{\partial x_N \partial x_1}(0) & \dfrac{\partial^2 f}{\partial x_N \partial x_2}(0) & \dfrac{\partial^2 f}{\partial x_N \partial x_N}(0) \end{bmatrix}$$

is nonsingular.

Then there are neighborhoods U of 0 in Ω and V of 0 in $\mathbb{R}^N$, and a C^{k-2} map $g : V \to U$ with inverse such that $f \circ g : V \to \mathbb{R}$ satisfies

$$f \circ g(y) = -(y_1^2 + \cdots + y_\lambda^2) + (y_{\lambda+1}^2 + \cdots + y_N^2)$$

for $y = (y_1, \cdots, y_N) \in V$, where λ is a fixed integer, $0 \leq \lambda \leq N$.

Proof. For $x = (x_1, \cdots, x_N) \in \Omega$, write

$$f(x) = \int_0^1 \frac{df(tx)}{dt} dt = \sum_{i=1}^N x_i \int_0^1 \frac{\partial f}{\partial x_i}(tx) dt$$
$$= \sum_{i=1}^N x_i g_i(x),$$

(4.2)

where for $i = 1, \cdots, N$, $g_i(x) = \int_0^1 \frac{\partial f}{\partial x_i}(tx) dt$, $g_i(x) \in C^{k-1}$, and $g_i(0) = \int_0^1 \frac{\partial f}{\partial x_i}(0) dt = \frac{\partial f}{\partial x_i}(0) = 0$. Again we have

$$g_i(x) = \int_0^1 \frac{dg_i(tx)}{dt} dt = \sum_{j=1}^N x_j \int_0^1 \frac{\partial g_i(tx)}{\partial x_j} dt$$
$$= \sum_{j=1}^N x_j h_{ij}(x)$$

(4.3)

where, for $i = 1, ..., N$, $h_{ij}(x) = \int_0^1 \frac{\partial g_i(tx)}{\partial x_j} dt$, and $h_{ij}(x) \in C^{k-2}$. Because $g_i(x) = \int_0^1 \frac{\partial f}{\partial x_i}(tx) dt$, we have $\frac{\partial g_i}{\partial x_j}(x) = \int_0^1 t \frac{\partial^2 f}{\partial x_i \partial x_j}(tx) dt$, that is, $\frac{\partial g_i}{\partial x_j}(0) = \frac{\partial^2 f}{\partial x_i \partial x_j}(0) \int_0^1 t \, dt = \frac{1}{2} \frac{\partial^2 f}{\partial x_i \partial x_j}(0)$. Therefore,

$$h_{ij}(0) = \int_0^1 \frac{\partial g_i(0)}{\partial x_j} dt = \frac{\partial g_i}{\partial x_j}(0) = \frac{1}{2} \frac{\partial^2 f}{\partial x_i \partial x_j}(0).$$

Thus, $(h_{ij}(0))_{i,j} = \frac{1}{2}\left(\frac{\partial^2 f}{\partial x_i \partial x_j}(0)\right)_{i,j}$ is nonsingular. If necessary, set $\bar{h}_{ij} = \frac{1}{2}(h_{ij} + h_{ji})$. We may assume $h_{ij}(x) = h_{ji}(x)$ for i, j. Now, by (4.2) and (4.3),

$$f(x) = \sum_{i,j=1}^N x_i x_j h_{ij}(x),$$

(4.4)

where $h_{ij}(x) \in C^{k-2}$, $h_{ij}(x) = h_{ji}(x)$ for i, j , and $(h_{ij}(0))_{i,j}$ is nonsingular. Diagonalize the quadratic form $f(x)$ by induction: suppose in a neighborhood W of 0, f can be written as

$$f(x) = \pm x_1^2 \pm \cdots \pm x_{r-1}^2 + \sum_{i,j \geq r} x_i x_j H_{ij}(x),$$

(4.5)

where $r \geq 1$, H_{ij} is real symmetric, and $(H_{ij}(0))_{r \leq i,j \leq N}$ is nonsingular. By (4.4), (4.5) is true for $r = 1$. Because $H = (H_{ij}(0))_{r \leq i,j \leq N}$ is real symmetric and nonsingular, there is an orthogonal matrix P such that

$$P^{-1}HP = \begin{pmatrix} \lambda_r & \cdots & 0 \\ \cdots & \cdots & \cdots \\ 0 & \cdots & \lambda_N \end{pmatrix} = A,$$

where $\lambda_i \neq 0$, for $i = r, \cdots, N$. Therefore, $H = PAP^{-1}$. Now, for $x = (x_r, \cdots, x_N)$, and $z = (z_r, \cdots, z_N) = P^{-1}x$,

$$\sum_{i,j \geq r} x_i x_j H_{ij}(0) = (Hx, x)$$

$$= (PAP^{-1}x, x)$$
$$= (AP^{-1}x, P^{-1}x)$$
$$= \sum_{i,j \geq r} z_i z_j A_{ij},$$

where $A = (A_{ij})$. If necessary, we replace $H_{rr}(0)$ by A_{rr}, so we may assume that $H_{rr}(0) \neq 0$. Denote $u(x) = |H_{rr}(x)|^{1/2}$. Choose $U \subset W$, a neighborhood of 0, such that $u \in C^{k-2}(U)$, $u(x) \neq 0$ for $x \in U$ because $u(0) \neq 0$. Define $y_i = x_i$, $i \neq r$,

$$y_r(x) = u(x) \left[x_r + \sum_{I > r} \frac{x_i H_{ir}(x)}{H_{rr}}(x) \right] \quad \text{for } x \in U.$$

Consider the Jacobian $\frac{\partial(y_1, \cdots, y_N)}{\partial(x_1, \cdots, x_N)}$ at 0:

$$\frac{\partial(y_1, \cdots, y_N)}{\partial(x_1, \cdots, x_N)}\Big|_{x=0} = \det \begin{bmatrix} 1 & 0 & \cdots & \cdots & \cdots & \cdots & \cdots & 0 \\ 0 & 1 & \cdots & \cdots & \cdots & \cdots & \cdots & 0 \\ \vdots & \vdots & & & & & & \vdots \\ 0 & 0 & \cdots & 1 & & & & 0 \\ \frac{\partial y_r}{\partial x_1} & \frac{\partial y_r}{\partial x_1} & & \frac{\partial y_r}{\partial x_1} & u(0) & & & \frac{\partial y_r}{\partial x_1} \\ \vdots & \vdots & & & & \ddots & & \\ 0 & 0 & & 0 & 0 & 0 & & 1 \end{bmatrix} = u(0).$$

The Jacobian is nonsingular, because $u(0) \neq 0$. By the Inverse Function Theorem 4.1, there are a neighborhood V of 0 and a C^{k-2} bijective map $g : V \to U$ such that

$g(y) = x$, where $x_i = y_i$ if $i \neq r$, $x_r = \frac{y_r}{u} - \sum\limits_{i>r} \frac{x_i H_{ir}}{H_{rr}}$. Therefore,

$$x_r x_r H_{rr} + 2 \sum_{j=r+1}^{N} x_j x_r H_{jr}$$

$$= \left[\frac{y_r}{u} - \sum_{i>r} \frac{x_i H_{ir}}{H_{rr}} \right]^2 H_{rr} + 2 \sum_{j=r+1}^{N} x_j \left[\frac{y_r}{u} - \sum_{i>r} \frac{x_i H_{ir}}{H_{rr}} \right] H_{jr}.$$

$$= \frac{y_r^2 H_{rr}}{u^2} - 2 \sum_{i>r} \frac{y_r x_i H_{ir}}{u \, H_{rr}} H_{rr} + \sum_{i,j>r} \frac{x_i x_j H_{ir} H_{jr}}{H_{rr}}$$

$$+ 2 \sum_{j>r} \frac{y_j y_r H_{jr}}{u} - 2 \sum_{i,j>r} \frac{x_j x_i H_{ir} H_{jr}}{H_{rr}}$$

$$= \pm y_r^2 - 2 \sum_{i>r} y_i y_r H'_{ir} - \sum_{i,j>r} y_i y_j H'_{ir} H'_{jr}$$

$$+ 2 \sum_{j>r} y_j y_r H'_{jr}$$

$$= \pm y_r^2 + \sum_{i,j>r} y_i y_j H''_{ij},$$

where $u = |H_{rr}|^{1/2}$, $H'_{ir} = \frac{H_{ir}}{\sqrt{|H_{rr}|}}$, and $H''_{ij} = -H'_{ir} H'_{jr}$. Note that H''_{ij} is real symmetric. Moreover, $(H''_{ij}(0))$ is nonsingular. In fact, $\sum_{i,j>r} a_i a_j H_{ir} H_{jr} > 0$ for $a = (a_{r+1}, ..., a_N) \neq 0$. Thus,

$$f(x) = \pm x_1^2 \pm \cdots \pm x_{r-1}^2 + \sum_{i,j \geq r} x_i x_j H_{ij}(x)$$

$$= \pm x_1^2 \pm \cdots \pm x_{r-1}^2 + x_r x_r H_{rr} + 2 \sum_{j=r+1}^{N} x_i x_r H_{jr} + \sum_{i,j \geq r} x_i x_j H_{ij}(x)$$

$$= \pm y_1^2 \pm \cdots \pm y_{r-1}^2 \pm y_r^2 + \sum_{i,j>r} y_i y_j H''_{ij}(y) + \sum_{i,j>r} y_i y_j H_{ij}(x),$$

$$= \pm y_1^2 \pm \cdots \pm y_r^2 + \sum_{i,j>r} y_i y_j G_{ij}(y),$$

where $G_{ij}(y) = H_{ij}(g(y)) + H''_{ij}(y)$. By induction the theorem follows. ∎

Remark 4.4 *In the Morse Lemma 4.3, $0 \leq \lambda \leq N$ and λ is called the index of the critical point 0 of f, where*

$$f \circ g(y) = -[y_1^2 + \cdots + y_\lambda^2] + [y_{\lambda+1}^2 + \cdots + y_N^2], \ 0 \leq \lambda \leq N.$$

f is positive-definite if $\lambda = 0$, negative-definite if $\lambda = N$, and indefinite if $0 < \lambda < N$.

Corollary 4.5 *Under the conditions of the Morse Lemma 4.3, if $N = 2$ and the quadratic form $(f''(0)y, \, y)$ is indefinite, then the set of solutions of $f(x)$ near 0 consists of two C^{k-2} curves intersecting only at 0. If $N > 2$ and $(f''(0)y, \, y)$ is indefinite, then the set of solutions of $f(x)$ resembles a deformed cone.*

Proof.

(i) $N = 2 : f(x) = -y_1^2 + y_2^2 = 0$, then $(y_1 - y_2)(y_1 + y_2) = 0$.

(ii) $N = 3 : f(x) = -(y_1^2 + y_2^2) + y_3^2 = 0 : y_3^2 = y_1^2 + y_2^2$ ∎

Remark 4.6 *If $N = 2$ or 3, the Morse Lemma can be calculated directly.*

(i) $N = 2 : A \neq 0$

$$f(x, y) = Ax^2 + 2Bxy + Cy^2$$
$$= A\left[x + \frac{B}{A}y\right]^2 + \frac{AC - B^2}{A}y^2$$
$$= \pm u^2 \pm v^2,$$

where $u = \sqrt{|A|}\left[x + \frac{B}{A}y\right]$, and $v = \sqrt{\left|\frac{AC - B^2}{A}\right|}\,y$.

(ii) $N = 3 :$

$$f(x_1, x_2, x_3) = x_1^2 + 4x_2^2 + x_3^2 + 2x_1 x_2 - 6x_1 x_3 + 8x_2 x_3$$
$$= (x_1 + x_2 - 3x_3)^2 + 3x_2^2 - 8x_3^2 + 14x_2 x_3$$
$$= (x_1 + x_2 - 3x_3)^2 - 2(2x_3 - \frac{7}{4}x_2)^2 + \frac{73}{8}x_2^2$$
$$= u^2 - v^2 + w^2,$$

where $u = x_1 + x_2 - 3x_3$, $v = \sqrt{2}(2x_3 - \frac{7}{4}x_2)$ and $w = \sqrt{\frac{73}{8}}x_2$.

Example 4.7 $N = 2 :$

(i) $u_3 = f(u_1, u_2) = u_1^2 + u_2^2$ *minimum (index 0);*

(ii) $u_3 = f(u_1, u_2) = -u_1^2 + u_2^2$ *saddle point (index 1);*

(iii) $u_3 = f(u_1, u_2) = -u_1^2 - u_2^2$ *maximum (index 2).*

Hormander [26] generalized the Morse Lemma as follows.

Lemma 4.8 (*Generalized Morse Lemma*) *Suppose* $\Omega \subset \mathbb{R}^d \times \mathbb{R}^k$ *is an open set,* $0 \in \Omega$, *and* $g : \Omega \to \mathbb{R}$ *is a* C^k *map,* $k \geq 2$ *satisfying:*

(*i*) $g(0, 0) = 0$;

(*ii*) $g_x(0, 0) = 0$;

(*iii*) $g_{xx}(0, 0) = Q$ *is nonsingular.*

Then there is a neighborhood U *of 0 in* $\mathbb{R}^k$, *there is a* C^k *map* $\varphi : U \to \mathbb{R}^d$ *satisfying* $\varphi(0) = 0$, $g_x(\varphi(y), y) = 0$ *for* $y \in U$, *and there is a function* $\xi : \Omega \to \mathbb{R}^d$ *of* C^{k-2} *satisfying*

$$\xi(x, y) = x - \varphi(y) + O(|x - \varphi(y)|^2),$$

$$g(x, y) = g(\varphi(y), y) + \frac{1}{2}(Q(y)\xi, \xi),$$

where $Q(y) = g_{xx}(\varphi(y), y)\,|_{x=\varphi(y)}$.

Proof.   By the Implicit Function Theorem 3.28, there is a neighborhood U of 0 in $\mathbb{R}^k$, a function $\varphi : U \to \mathbb{R}^d$ of C^k satisfying $\varphi(0) = 0$, and $g_x(\varphi(y), y) = 0$ for $y \in U$. Fix y near 0, replace x by $x - \varphi(y)$. Then

$$x = \varphi(y) \quad \text{if and only if} \quad x - \varphi(y) = 0.$$

Consider $\varphi(y) = 0$. We may assume $g_x(0, y) \equiv 0$ for $y \in U$. Apply integration by parts to obtain

$$
\begin{aligned}
g(x, y) - g(0, y) &= \int_0^1 \frac{d}{dt} g(tx, y) dt \\
&= \int_0^1 g_x(tx, y)x \; dt \\
&= t \; g_x(tx, y)x \mid_0^1 - \int_0^1 t \frac{d}{dt} g_x(tx, y)x \; dt \\
&= t \; g_x(tx, y)x \mid_0^1 - \int_0^1 t(g_{xx}(tx, y)x, x) dt \\
&= g_x(x, y)x - g_x(0, y)x - \int_0^1 t(g_{xx}(tx, y)x, x) dt \\
&= \int_0^1 \frac{d}{dt} g_x(tx, y)x \; dt - \int_0^1 t(g_{xx}(tx, y)x, x) dt \\
&= \int_0^1 (g_{xx}(tx, y)x, x) dt - \int_0^1 t(g_{xx}(tx, y)x, x) dt \\
&= \int_0^1 (1 - t)(g_{xx}(tx, y)x, x) dt
\end{aligned}
$$

$$= \frac{1}{2}(B(x,y)x, x),$$

where

$$B(x,y) = 2\int_0^1 (1-t)g_{xx}(tx, y)dt$$

$$= 2\int_0^1 (1-t) \begin{bmatrix} \dfrac{\partial^2 g}{\partial x_1 \partial x_1} & \cdots & \dfrac{\partial^2 g}{\partial x_1 \partial x_N} \\ \vdots & & \vdots \\ \dfrac{\partial^2 g}{\partial x_N \partial x_1} & \cdots & \dfrac{\partial^2 g}{\partial x_N \partial x_N} \end{bmatrix}\Big|_{(tx,y)} dt$$

$$= \text{a symmetric matrix in } M_{d\times d}.$$

Note that

$$B(0,y) = 2\int_0^1 (1-t)g_{xx}(0, y)dt$$

$$= 2\, g_{xx}(0, y)\int_0^1 (1-t)dt$$

$$= g_{xx}(0, y).$$

If we set $Q(y) = g_{xx}(0, y)$, then Q is a C^{k-2} map that satisfies $B(0, y) = Q(y)$. Consider the map $f : M_{d\times d} \longrightarrow M_{d\times d}$ defined by $f(R) = R^*Q(y)R$. Then $f(I) = Q(y)$, and $f(M_{d\times d}) \subset S_{d\times d}$, where $S_{d\times d}$ is the space of all symmetric matrices in $M_{d\times d}$. Now,

$$\frac{|f(I+R) - f(I) - [R^*Q(y) + Q(y)R]|}{|R|}$$

$$= \frac{|(I+R)^*Q(y)(I+R) - Q(y) - R^*Q(y) - Q(y)R|}{|R|}$$

$$= \frac{|R^*Q(y)R|}{|R|} \leq \frac{|R^*|\,|Q(y)|\,|R|}{|R|}$$

$$= |R^*|\,|Q(y)| = |R|\,|Q(y)| \to 0 \quad as \quad |R| \to 0.$$

Thus, f is Fréchet-differentiable at I, and $Df(I)R = R^*Q(y) + Q(y)R$. If we let $A = R^*Q(y) + Q(y)R$, then

$$A^* = (R^*Q(y) + Q(y)R)^* = Q(y)R + R^*Q(y) = A,$$

so A is symmetric. We claim that $Df(I) : M_{d\times d} \longrightarrow S_{d\times d}$ is onto. In fact, for the symmetric matrix A, let $R = \frac{1}{2}Q^{-1}(y)A$, then $A = 2Q(y)R$, and consequently,

$$A = \frac{A^* + A}{2} = \frac{2R^*Q(y) + 2Q(y)R}{2} = R^*Q(y) + Q(y)R = Df(I)R.$$

Because dim $S_{d\times d} <$ dim $M_{d\times d}$, $Df(I) : M_{d\times d} \longrightarrow M_{d\times d}$ is neither surjective nor injective. If we let $X = Ker\ Df(I)$, then $0 < \dim\ X < d^2$. There is a subspace Y of $E = M_{d\times d}$ such that $E = X \oplus Y$. Let $h : Y \times X \times S_{dxd} \longrightarrow S_{d\times d}$ be defined by $h(x_1, x_2, z) = f(x_1 + x_2) - z$, then $h(I, I, Q(y)) = 0$, and $D_{x_2}h(I, I, Q(y))I = Df(I + I)I = Q(y)$ is nonsingular. By the Implicit Function Theorem 3.28, there are $\delta > 0$, $r > 0$, $B_\delta(I)$ in X, $B_r(Q(y))$ in $S_{d\times d}$, and a unique C^1 map $u : B_\delta(I) \times B_r(Q(y)) \to Y$ such that $u(I, Q(y)) = I$, $f(A_1 + u(A_1, L)) = L$, for $A_1 \in B_\delta(I)$, $L \in B_r(Q(y))$. Note that $B(x, y) \to B(0, y) = Q(y)$ as $x \to 0$. There is a neighborhood V of 0 such that $x \in V$ implies $|B(x, y) - B(0, y)| < r$. Therefore, for $x \in V$, $A_1 = I$, we have

$$f(I + u(I, B(x, y))) = B(x, y).$$

Let $R(x, y) = I + u(I, B(x, y))$. Then

$$\begin{aligned}
R(0, y) &= I + u(I, B(0, y)) \\
&= I + u(I, Q(y)) \\
&= I + I \\
&= I.
\end{aligned}$$

$$R^*(x, y)Q(y)R(x, y) = f(R(x, y)) = f(I + u(I, B(x, y))) = B(x, y) \qquad (4.6)$$

Suppose

$$\xi(x, y) = R(x, y)x = [I + u(I, B(x, y))]x = x + u(I, B(x, y))x. \qquad (4.7)$$

Then

$$\begin{aligned}
|\xi - x| &= |u(I, B(x, y))x| \\
&= |u(I, B(x, y))x - u(I,\ Q(y))0| \\
&= |(D_2 u(I, B(\theta x,\ y)\theta x,\ x) + u(I, B(\theta x, y)\theta x, x)| \\
&\le c\, |x|^2.
\end{aligned}$$

By (4.6) and (4.7),

$$\begin{aligned}
g(x, y) - g(0, y) &= \frac{1}{2}(B(x, y)x, x) \\
&= \frac{1}{2}(R^*(x, y)Q(y)R(x, y)x, x) \\
&= \frac{1}{2}(Q(y)R(x, y)x, R(x, y)x) \\
&= \frac{1}{2}(Q(y)\xi, \xi). \qquad \blacksquare
\end{aligned}$$

Theorem 4.9 *Let E, F be two Banach spaces, $\Omega \subset E$ an open set, $0 \in \Omega$, and let $f : \Omega \to F$ satisfy:*

(i) $f(0) = 0$;

(ii) f is a C^k map, $k \geq 2$;

(iii) Ker $f'(0) = E_1$, dim $E_1 = n < \infty$, range $f'(0) = F_1$, $F_1 \subset F$ closed subspace of codimension 1: then $0 \neq y^* \in F^*$ with $F_1 = \{y \in F \mid y^*(y) = 0\}$;

(iv) $f \mid_{E_1}$ satisfies the condition that the $n \times n$ matrix $y^* f_{x_1 x_1}(0)$ is nonsingular and indefinite.

Then in a neighborhood of 0, the set of solutions of $f(x) = 0$ consists of a deformed cone of dimension $n - 1$ with its vertex at 0. In particular, if $n = 2$ then the set consists of two C^{k-2} curves crossing only at 0.

Proof.   By the Lyapunov-Schmidt Procedure, there is a closed subspace E_2 of E such that $E = E_1 \oplus E_2$, a neighborhood U of 0 in E_1, and a unique C^k map $u : U \to E_2$ with $u(0) = 0$ such that

$$f(x_1 + u(x_1)) = 0, \ x_1 \in U \tag{4.8}$$

if and only if

$$\begin{cases} Pf(x_1 + u(x_1)) = 0 \ for \ \ x_1 \in U, \\ g(x_1) = y^* f(x_1 + u(x_1)) = 0 \ for \ \ x_1 \in U, \end{cases}$$

where $P : F \to F_1$ is the projection. If we identify E_1 with $\mathbb{R}^n$, then the single equation (4.8) has n variables. We obtain $g : U \subset \mathbb{R}^n \to \mathbb{R}$, $g(0) = 0$, and g is a C^k map, $k \geq 2$.

(a) Differentiate $Pf(x_1 + u(x_1)) = 0$ to obtain

$$Pf'(x_1 + u(x_1)) \mid_{x_1=0} (x_1 + u'(x_1) \mid_{x_1=0} x_1) = 0,$$

so

$$Pf'(0)(x_1 + u'(0)x_1) = 0 : Pf'(0)x_1 + Pf'(0)(u'(0)x_1) = 0.$$

However, Ker $f'(0) = E_1$, so $f'(0)x_1 = 0 : Pf'(0)x_1 = 0$. We obtain $Pf'(0)(u'(0)x_1) = 0$, but $f'(0)(u'(0)x_1) \in$ range $f'(0) = F_1$, so

$$0 = Pf'(0)(u'(0)x_1) = f'(0)(u'(0)x_1),$$

because $u : U \to E_2$, and $u'(0) : E_1 \to E_2$. Thus, $u'(0)x_1 \in E_2$, but $f'(0) : E_2 \to F$ is an isomorphism into. Therefore, $f'(0)(u'(0)x_1) = 0$ implies $u'(0)x_1 = 0$ for $x_1 \in U$.

(b) Differentiate $g(x_1) = y * f(x_1 + u(x_1)) = 0$ to obtain

$$g'(x_1) \mid_{x_1=0} x_1 = y^* f'(x_1 + u(x_1)) \mid_{x_1=0} (x_1 + u'(x_1) \mid_{x_1=0} x_1) = 0,$$

so $g'(0)x_1 = y^* f'(0)(x_1 + u'(0)x_1) = 0$. Thus, $g'(0)x_1 = y^* f'(0)(x_1)$ for $x_1 \in U$. We have $g'(0) = y^* f'(0)$. Continuing this process, we obtain $g_{x_1 x_1}(0) = y^* f_{x_1 x_1}(0)$.

$$g(0) = 0,$$
$$g'(0) = y^* f'(0) = 0 \ \ on \ \ E_1,$$
$$g_{x_1 x_1}(0) \ \text{is nonsingular and indefinite,}$$

and the theorem then follows from Corollary 4.5. $\blacksquare$

Remark 4.10 *In Theorem 4.9, if $y^* f_{x_1 x_1}(0)$ is definite, then $x = 0$ is the only solution of $f(x) = 0$.*

Suppose E and F are two Banach spaces, $\Omega \subset E \times \mathbb{R}$ is an open set containing $(0, \lambda)$ for $|\lambda| \leq \eta$, and $f : \Omega \to F$ is of C^k, $k \geq 1$ satisfying $f(0, \lambda) = 0$ for $|\lambda| \leq \eta$. The point $(0, \lambda_0)$ is called a bifurcation point of f, if every neighborhood of $(0, \lambda_0)$ in $E \times \mathbb{R}$ contains a solution (x, λ) of $f(x, \lambda) = 0$, and $x \neq 0$.

Theorem 4.11 *Suppose E and F are two Banach spaces, $\Omega \subset E \times \mathbb{R}$ is an open set, $(0, \lambda_0) \in \Omega$, and $f : \Omega \to F$ of C^k, $k \geq 1$ satisfies $f(0, \lambda) = 0$ for $|\lambda - \lambda_0| \leq \eta$. If $f_x(0, \lambda_0) \in GL(E, F)$, then $(0, \lambda_0)$ is not a bifurcation point.*

Proof. Because $f(0, \lambda) = 0$ for $|\lambda - \lambda_0| \leq \eta$, we have $f_\lambda(0, \lambda_0) \equiv 0$. Set $\hat{z} = (x, \lambda)$ and $\hat{x} = (x, 0)$, $\hat{\lambda} = (0, \lambda)$, $f_{\hat{z}}(0, \lambda_0)(x, \lambda) = f_x(0, \lambda_0)x \oplus f_\lambda(0, \lambda_0)\lambda = f_x(0, \lambda_0)x$. Because $f_x(0, \lambda_0) \in GL(E, F)$, dim $Ker \ f_{\hat{z}}(0, \lambda_0) = 1$, and $dim \ coker \ f_{\hat{z}}(0, \lambda_0) = codim \ range \ f_{\hat{z}}(0, \lambda_0) = 0$. Therefore, the index of f at $(0, \lambda_0)$ is 1, and there is a neighborhood U of $(0, \lambda_0)$ such that the set $\{(x, \lambda) \in U \mid f(x, \lambda) = 0\}$ is a 1-dimensional manifold, which should be $\{0\} \times \Lambda$ for Λ open in $\mathbb{R}$. $\blacksquare$

Theorem 4.12 *Let E and F be two Banach spaces, let $\Omega \subset E \times \mathbb{R}$ be open, assume $(0, \lambda_0) \in \Omega$, and let a C^k map $f : \Omega \to F$, $k \geq 2$, satisfy:*

(i) $f(0, \lambda_0) = 0$;

(ii) $f_\lambda(0, \lambda_0) = 0$;

(iii) $Ker \ f_x(0, \lambda_0) =< x_0 >$ *is of dimension 1, and range $f_x(0, \lambda_0) = F_1$ is of codimension 1;*

(iv) $f_{\lambda\lambda}(0, \lambda_0) \in F_1$, *and* $f_{\lambda x}(0, \lambda_0)x_0 \notin F_1$.

Then $(0, \lambda_0)$ is a bifurcation point of f. Furthermore, the set of solutions of $f(x, \lambda)$ near $(0, \lambda_0)$ consists of two C^{k-2} curves Γ_1, Γ_2 intersecting only at $(0, \lambda_0)$. If $k > 2$, then Γ_1 is tangent to the λ axis at $(0, \lambda_0)$,

$$\Gamma_1 = \{(x(\lambda), \lambda) \mid |\lambda - \lambda_0| \leq \epsilon\},$$
$$\Gamma_2 = \{(sx_0 + x_2(s), \lambda(s)) \mid |s| \leq \epsilon\},$$

with $x_2(0)(0) = 0$, $\lambda(0) = \lambda_0$.

Proof.   Suppose $\lambda_0 = 0$. Let $Z = E \times \mathbb{R}$. Denote $\hat{z} = (x, \lambda)$, $\hat{x} = (x, 0)$, $\hat{\lambda} = (0, \lambda)$, and imbed E and $\mathbb{R}$ as subspaces of Z, then $f_{\hat{z}}(0, 0)(x, \lambda) = f_x(0, 0)x \oplus f_\lambda(0, 0)\lambda$. Recall that $f_\lambda(0, 0) = 0$, and Ker $f_x(0, 0) = <x_0>$, so we have

$$(x, \lambda) \in Ker\ f_{\hat{z}}(0, 0),$$

if and only if

$$f_{\hat{z}}(0, 0)(x, \lambda) = 0,$$

if and only if

$$f_x(0, 0)x \oplus f_\lambda(0, 0)\lambda = 0,$$

if and only if

$$(x, \lambda) = \alpha(x_0, 0) + \lambda(0, 1),$$

if and only if

$$(x, \lambda) \in <(x_0, 0), (0, 1)> \text{ a subspace of dimension 2.}$$

Let $X = <(x_0, 0), (0, 1)> = Ker\ f_{\hat{z}}(0, 0)$. Note that

$$\text{range} f_{\hat{z}}(0, 0) = \text{range } f_x(0, 0) \oplus \text{range } f_\lambda(0, 0)$$
$$= \text{range } f_x(0, 0) = F_1$$
$$= \{y \in F \mid y^*(y) = 0\},$$

for some $y^* \neq 0$ in F^*. Consider the restriction of the map $f : Z \to F$ to the 2-dimensional subspace $X = Ker\ f_{\hat{z}}(0, 0)$. Now, for $\hat{x}_0 = (x_0, 0)$, $\hat{1} = (0, 1)$, $z_1 = (x_0, 1)$,

$$y^* f_{z_1 z_1}(0, 0) = \left[\begin{array}{cc} y^* f_{\hat{x}_0 \hat{x}_0}(0, 0) & y^* f_{\hat{1} \hat{x}_0}(0, 0) \\ y^* f_{\hat{x}_0 \hat{1}}(0, 0) & y^* f_{\hat{1}\hat{1}}(0, 0) \end{array} \right].$$

By the hypothesis, we obtain

$$f_{\lambda\lambda}(0, 0) \in F_1 \text{ implies } y^* f_{\hat{1}\hat{1}}(0, 0) = 0,$$
$$f_{x\lambda}(0, 0) \notin F_1 \text{ implies } y^* f_{\hat{x}_0 \hat{1}}(0, 0) \neq 0,$$
$$\det\ y * f_{z_1 z_1}(0, 0) = -[y^* f_{\hat{x}_0 \hat{1}}(0, 0)]^2 < 0.$$

Recall that

$$\begin{pmatrix} a_{11} a_{12} \\ a_{21} a_{22} \end{pmatrix} \text{ positive-definite} \quad \text{if and only if} \quad a_{11} > 0, \begin{vmatrix} a_{11} a_{12} \\ a_{21} a_{22} \end{vmatrix} > 0$$

$$\begin{pmatrix} a_{11} a_{12} \\ a_{21} a_{22} \end{pmatrix} \text{ negative-definite} \quad \text{if and only if} \quad a_{11} < 0, \begin{vmatrix} a_{11} a_{12} \\ a_{21} a_{22} \end{vmatrix} > 0.$$

$$\begin{pmatrix} a_{11} a_{12} \\ a_{21} a_{22} \end{pmatrix} \text{ indefinite} \quad \text{if and only if} \quad \begin{vmatrix} a_{11} a_{12} \\ a_{21} a_{22} \end{vmatrix} < 0.$$

Now det $y^* f_{z_1 z_1}(0, 0) < 0$, so $y^* f_{z_1 z_1}(0, 0)$ is nonsingular and indefinite. By Theorem 4.9, $(0, 0)$ is a bifurcation point of f. Furthermore, because

$$\text{index} = \dim \text{Ker } f_{\hat{z}}(0,0) - \dim \text{coker } f_{\hat{z}}(0,0) = 2 - 1 = 1,$$

the set of solutions of $f(x, \lambda)$ near $(0, 0)$ consists of two C^{k-2} curves Γ_1, Γ_2 intersecting only at $(0, 0)$.   ∎

Remark 4.13 *In Theorem 4.12, if $f(0, \lambda) = 0$ for each λ, the curve Γ_1 is the λ-axis*

4.2   Krasnoselskii Theorem and Rabinowitz Theorem

In this section, let E be a Banach space and $\Omega \subset E \times \mathbb{R}$ an open domain, assume $(0, \lambda_0) \in \Omega$, and let $f : \Omega \to E$ be defined by

$$f(x, \lambda) = x - \lambda T x + g(x, \lambda). \tag{4.9}$$

Definition 4.14 *f admits the property (H) if f satisfies:*

(i) *$0 \neq \lambda_0$, a characteristic value of T, $(0, \lambda_0) \in \Omega$;*

(ii) *$T : E \to E$ is a compact linear operator;*

(iii) *$g : \Omega \to E$ is a compact nonlinear operator satisfying $g(0, \lambda) = 0$ for all λ with $(0, \lambda) \in \Omega$, and $g(x, \lambda) = o(\|x\|)$ uniformly for λ in any bounded interval.*

Proposition 4.15 *If g is smooth, and λ_0 is a characteristic value of T with multiplicity 1, then $(0, \lambda_0)$ is a bifurcation point of $f(x, \lambda)$.*

Proof.   Method 1. By the Krasnoselskii Theorem 4.18 below.
Method 2. Let I_{λ_0} be a bounded neighborhood of λ_0 such that $(x, \lambda) \in \Omega$ for x near 0, and $\lambda \in I_{\lambda_0}$. For $\lambda \in I_{\lambda_0}$, write $g(x, \lambda) = g(0, \lambda) + g_x(\xi, \lambda)x = g_x(\xi, \lambda)x$, where ξ is in the line segment between 0 and x. Because

$$\frac{\|g_x(\xi, \lambda)x\|}{\|x\|} = \frac{\|g(x, \lambda)\|}{\|x\|} \to 0 \quad as \;\; x \to 0,$$

we have

$$\frac{\|g_x(0, \lambda)x\|}{\|x\|} \leq \frac{\|g_x(\xi, \lambda)x - g_x(0, \lambda)x\|}{\|x\|} + \frac{\|g_x(\xi, \lambda)x\|}{\|x\|} \to 0 \quad as \;\; x \to 0.$$

Suppose $g_x(0, \lambda) : E \longrightarrow E$ is nonzero for $\lambda \in I_{\lambda_0}$, then take $x_0 \neq 0$ such that $\|g_x(0, \lambda)x_0\| = 1$ to obtain

$$\lim_{t \longrightarrow 0} \frac{\|g_x(0, \lambda)t x_0\|}{\|t x_0\|} = \frac{\|g_x(0, \lambda)x_0\|}{\|x_0\|} = \frac{1}{\|x_0\|} \neq 0,$$

a contradiction. Thus, $g_x(0, \lambda) \equiv 0$, for $\lambda \in I_{\lambda_0}$. Similarly, $g_\lambda(0, \lambda) \equiv 0$, for $\lambda \in I_{\lambda_0}$. We have:

(i) $f(0, \lambda_0) = 0$;

(ii) $f_\lambda(0, \lambda_0) = -Tx \mid_{x=0} + g_\lambda(0, \lambda_0) = 0$;

(iii) $f_x(0, \lambda_0)x = x - \lambda Tx \mid_{\lambda=\lambda_0} + g_x(0, \lambda_0)x = x - \lambda_0 Tx$, hence, by the hypothesis, $Ker\, f_x(0, \lambda_0) = Ker(I - \lambda_0 T) = < \bar{x} >$ is of dimension 1;

(iv) T is compact linear, and by (iii),

$$\dim\, Ker(I - \lambda_0 T) = \dim\, coker(I - \lambda_0 T) = 1.$$

Thus, range $f_x(0, \lambda_0) = F_1$ is of codimension 1.

(v) $f_{\lambda\lambda}(0, \lambda_0) \in F_1$;

(vi) $f_{\lambda x}(0, \lambda_0)\bar{x} = -T\bar{x} = c\bar{x} \notin F_1$, where $c = -1/\lambda_0$. In fact, if $c\bar{x} \in F_1$, then take y such that $f_x(0, \lambda_0)y = c\bar{x}$, so we have $y - \lambda_0 Ty = c\bar{x}$, or $(I - \lambda_0 T)^2 y = 0$, so $y = \alpha \bar{x}$ because λ_0 is of multiplicity 1. Therefore, $c\bar{x} = f_x(0, \lambda_0)(\alpha \bar{x}) = 0$, a contradiction. ∎

By Proposition 4.15, $(0, \lambda_0)$ is a bifurcation point of $f(x, \lambda)$. Furthermore, the set of solutions of $f(x, \lambda) = 0$ near $(0, \lambda_0)$ consists of two C^{k-2} curves Γ_1, Γ_2 intersecting only at $(0, \lambda_0)$.

Remark 4.16 *In Proposition 4.15, if the characteristic value λ_0 of T has multiplicity greater than 1, then Γ_1, Γ_2 may not intersect as above. For example, let $E = \mathbb{R}^3$, $T = I$, $\lambda_0 = 1$, $g(x) = h(\frac{x}{|x|})e^{-1/|x|^2}$, where $h : S^2 \to \mathbb{R}^3$ such that $h(\frac{y}{|y|})$ is perpendicular to y for each y, and $h(y) = 0$ implies that y is the north pole. If $x \neq 0$ satisfies*

$$f(x, \lambda) = x - \lambda x + g(x) = (1 - \lambda)x + g(x) = 0,$$

then $(1 - \lambda)(x, x) + (g(x), x) = 0$. Because $h(\frac{x}{|x|}) \perp x$, we have $(g(x), x) = 0$, and $\lambda = 1$. We conclude that $g(x) = 0$, and $h(\frac{x}{|x|}) = 0 : \frac{x}{|x|}$ is the north pole. Set $x = (0, 0, x_3)$, $x_3 > 0$. Then the set of nontrivial solutions is $\{(x, \lambda) \mid x = (0, 0, x_3), x_3 > 0, \lambda = 1\}$.

Proposition 4.17 *If $(0, \lambda_0)$ is a bifurcation point of $f(x, \lambda)$, then λ_0 is a characteristic value of T.*

Proof.   Because T is compact linear, the set of all characteristic values of T is discrete. Suppose λ_0 is not a characteristic value of T, then there is a neighborhood J of λ_0 such that no characteristic value of T exists in J. Consequently, $I - \lambda T$ is invertible on J, $\lambda \to (I - \lambda T)^{-1}$ is continuous on J, and $\|(I - \lambda T)^{-1}\| \leq c$, for $\lambda \in J$. Suppose

$$x \neq 0, \lambda \in J \text{ with } f(x, \lambda) = 0 : x - \lambda Tx + g(x, \lambda) = 0,$$

that is, $x = (I - \lambda T)^{-1}(-g(x, \lambda))$, and

$$1 = \frac{\|x\|}{\|x\|} = \frac{\|(I - \lambda T)^{-1}(-g(x, \lambda))\|}{\|x\|} \le \frac{c\,\|g(x, \lambda)\|}{\|x\|} \longrightarrow 0,$$

a contradiction. Thus, λ_0 is a characteristic value of T.   ■

Theorem 4.18 (*Krasnoselskii Theorem*) *Suppose λ_0 is a characteristic value of T with odd multiplicity, then $(0, \lambda_0)$ is a bifurcation point of $f(x, \lambda)$.*

Proof.   Suppose $(0, \lambda_0)$ is not a bifurcation point of $f(x, \lambda)$, then $\epsilon > 0$, there is $\eta > 0$ such that, in $\overline{B_\epsilon(0)} \times \overline{B_\eta(\lambda_0)}$, the only solutions of $f(x, \lambda) = 0$ are $\{(0, \lambda) \mid |\lambda - \lambda_0| < \eta\}$, and the only characteristic value of T in the interval $(\lambda_0 - \eta,\ \lambda_0 - \eta)$ is λ_0. By the Leray-Schauder Degree Theory, $\deg(f(\cdot, \lambda), B_r(0), 0)$ is defined and independent of λ. By the Leray-Schauder Index Theorem, take λ_1 and λ_2 satisfying $\lambda_0 - \eta < \lambda_2 < \lambda_0 < \lambda_1 < \lambda_0 + \eta$.

(*i*) If $\lambda_0 < \lambda_1 < \lambda_0 + \eta$. Then

$$\deg(f(\cdot, \lambda_1), B_r(0), 0) = \deg((I - \lambda_1 T), B_r(0), 0) = (-1)^{\beta(\lambda_1)},$$

where $\beta(\lambda_1)$ is the sum of the multiplicities of the eigenvalues of T that are greater than $\frac{1}{\lambda_1}$.

(*ii*) If $\lambda_0 - \eta < \lambda_2 < \lambda_0$, then similarly,

$$\deg(f(\cdot, \lambda_2), B_r(0), 0) = (-1)^{\beta(\lambda_2)},$$

where $\beta(\lambda_2)$ is the sum of the multiplicities of the eigenvalues of T that are greater than $\frac{1}{\lambda_2}$. Let $2k + 1$ be the multiplicity of the eigenvalue $\frac{1}{\lambda_0}$, $\beta(\lambda_2) - \beta(\lambda_1) = 2k + 1$. Therefore,

$$\beta(\lambda_2) = \beta(\lambda_1) + 2k + 1,$$

and consequently,

$$(-1)^{\beta(\lambda_2)} = (-1)^{\beta(\lambda_1)+2k+1} = -(-1)^{\beta(\lambda_1)}.$$

By (*i*) and (*ii*), we have

$$\deg(f(\cdot, \lambda_2), B_r(0), 0) = -\deg(f(\cdot, \lambda_1), B_r(0), 0).$$

This contradicts the assumption that $\deg(f(\cdot, \lambda), B_r(0), 0)$ is independent of λ, and that $\deg(f(\cdot, \lambda_1), B_r(0), 0) \ne 0$.   ■

Example 4.19 *If $\frac{1}{\lambda_0}$ is an eigenvalue of T with even multiplicity, then the conclusion of the Krasnoselskii Theorem 4.18 may fail as follows: let $E = \mathbb{R}^2$, $x = \binom{x_1}{x_2}$, $T = I$, $g(x, \lambda) = \binom{-x_2^3}{x_1^3}$, and $\lambda_0 = 1$. Then*

$$f(x, \lambda) = \left(\begin{array}{c} x_1 \\ x_2 \end{array} \right) - \lambda \left(\begin{array}{c} x_1 \\ x_2 \end{array} \right) + \left(\begin{array}{c} -x_2^3 \\ x_1^3 \end{array} \right) = 0$$

$$\Longleftrightarrow \left(\begin{array}{c} (1-\lambda)x_1 - x_2^3 \\ (1-\lambda)x_2 + x_1^3 \end{array} \right) = \left(\begin{array}{c} 0 \\ 0 \end{array} \right)$$

$$\Longleftrightarrow \left\{ \begin{array}{l} (1-\lambda)x_1 = x_2^3 \\ (1-\lambda)x_2 = x_1^3 \end{array} \right.$$

$$\Longleftrightarrow \left\{ \begin{array}{l} (1-\lambda)x_1x_2 = x_2^4 \\ (1-\lambda)x_1x_2 = x_1^4 \end{array} \right. \quad \text{if } x_1x_2 \neq 0$$

$$\Rightarrow x_1^4 - x_2^4 = 0$$

$$\Rightarrow x_1 = x_2 = 0.$$

Therefore the only solution of $f(x, \lambda) = 0$ near $(0, 1)$ is $(0, \lambda)$, so $(0, 1)$ is not a bifurcation point of $f(x, \lambda)$. In this case,

$$Ker(I - T) = Ker(I - I) = Ker\ 0 = \mathbb{R}^2,$$

so the multiplicity of $\lambda_0 = 1$ is 2.

Lemma 4.20 *(Ize Lemma) Suppose the property (H) in Definition 4.14 holds. Choose $\eta > 0$ such that $\frac{1}{\lambda}$ is not an eigenvalue of T for $0 < |\lambda - \lambda_0| \leq \eta$. Let*

$$i_- = \text{index of } (I - \lambda T) \text{ at } 0$$
$$= \deg(I - \lambda T), B_r(0), 0) \text{ for } \lambda_0 - \eta \leq \lambda < \lambda_0,$$

$$i_+ = \text{index of } (I - \lambda T) \text{ at } 0$$
$$= \deg(I - \lambda T), B_r(0), 0) \text{ for } \lambda_0 < \lambda \leq \lambda_0 + \eta,$$

where $r = r(\lambda)$ is small. For a neighborhood U of $(0,0)$ in $E \times \mathbb{R}$, consider $H_r : U \to E \times \mathbb{R}$ defined by

$$H_r(x, \lambda) = (y, \tau) = ((x - \lambda T)x - g(x, \lambda), \|x\|^2 - r^2).$$

Then for small η and $r > 0$, we have

$$\deg(H_r(x, \lambda_0), \|x\|^2 + \lambda^2 < r^2 + \eta^2, (0, 0)) = i_- - i_+.$$

Proof. In the interval $[\lambda_0 - \eta, \lambda_0 + \eta]$, the only characteristic value of T is λ_0, so there is $(I - (\lambda_0 \pm \eta)T)^{-1}$ and is bounded. As before, for small $\|x\|$, we prove as in the proof of Proposition 4.17 that the only solution x of

$$(I - (\lambda_0 \pm \eta)T)x - g(x, \lambda_0 \pm \eta) = 0 \tag{4.10}$$

is 0. If (x, λ) satisfy

$$\|x\|^2 + \lambda^2 = r^2 + \eta^2, \tag{4.11}$$

and

$$H_r(x, \lambda) = ((x - \lambda T)x - g(x, \lambda), \|x\|^2 - r^2) = (0, 0),$$

then we have $\|x\|^2 - r^2 = 0 : \|x\| = r$. By (4.11) $\lambda^2 = \eta^2 : \lambda = \pm\eta$. By (4.10) $x = 0$. Thus, the only solution (x, λ) of $H_r(x, \lambda) = (0, 0)$, satisfying $\|x\|^2 + \lambda^2 = r^2 + \eta^2$, is $x = 0$. Consider the deformation, $0 \le t \le 1$,

$$H_r^t(x, \lambda) = (y^t, \tau^t) = ((I - \lambda T)x - tg(x, \lambda), t(\|x\|^2 - \|r\|^2) + (1 - t)(\eta^2 - \lambda^2)).$$

As above, the only solution (x, λ) of $H_r^t(x, \lambda) = (0, 0)$ satisfying $\|x\|^2 + \lambda^2 = r^2 + \eta^2$ is $x = 0$, so the degree

$$\deg(H_r^t(x, \lambda), \|x\|^2 + |\lambda|^2 < r^2 + \eta^2, (0, 0))$$

is well defined and independent of t. For $t = 0$, if

$$H_r^0(x, \lambda) = ((I - \lambda T)x, \eta^2 - \lambda^2) = (0, 0),$$

then $\lambda = \pm\eta$ and $x = 0$, which implies that the only solutions (x, λ) of $H_r^0(x, \lambda)$ are $(0, \eta)$, and $(0, -\eta)$. The Fréchet derivative of $H_r^0(x, \lambda)$ at $(0, \lambda)$ is as follows:

$$H_r^0(x, \lambda) = ((I - \lambda T)x, \eta^2 - \lambda^2),$$

we have

$$DH_r^0(0, \lambda)(x, \lambda') = ((I - \lambda T)x, -2\lambda\lambda')$$
$$= (I - \lambda T, -2\lambda)(x, \lambda'),$$

and consequently $DH_r^0(0, \lambda) = ((I - \lambda T), -2\lambda I)$, a product map. By the Leray-Schauder Degree Theory,

$$\deg(H_r^0(x, \lambda_0), \|x\|^2 + \lambda^2 < r^2 + \eta^2, (0, 0))$$
$$= \deg(I - (\lambda_0 + \eta)T, \|x\|^2 + \lambda^2 < r^2 + \eta^2, (0, 0))$$
$$\cdot \deg(-2\lambda_0 I, \|x\|^2 + \lambda^2 < r^2 + \eta^2, (0, 0))$$
$$+ \deg(I - (\lambda_0 - \eta)T, \|x\|^2 + \lambda^2 < r^2 + \eta^2, (0, 0))$$
$$\cdot \deg(2\eta I, \|x\|^2 + \lambda^2 < r^2 + \eta^2, (0, 0))$$
$$= -i_+ + i_- = i_- - i_+$$
$$= \deg(H_r^1(x, \lambda_0), \|x\|^2 + \lambda^2 < r^2 + \eta^2, (0, 0))$$
$$= \deg(H_r(x, \lambda_0), \|x\|^2 + \lambda^2 < r^2 + \eta^2, (0, 0)) \qquad \blacksquare$$

Let N be the closure of the set of all nontrivial solutions of Equation (4.9) $f(x, \lambda) = 0$ in Ω, and $\gamma(T)$ the set of all characteristic values of T.

Lemma 4.21 *Let $\lambda_0 \in \gamma(T)$ be of odd multiplicity, and let C be the connected component of $N \cup \{(0, \lambda_0)\}$, containing $(0, \lambda_0)$. If C is bounded, and does not meet any point $(0, \lambda)$, $\lambda_0 \neq \lambda \in \gamma(T)$, then there is an open bounded set $G \subset \Omega \subset \mathbb{R} \times E$ satisfying:*

(i) $C \subset G$;

(ii) $\partial G \cap N = \emptyset$;

(iii) $G \cap (\{0\} \times \mathbb{R}) = \{0\} \times (\lambda_0 - \epsilon, \lambda_0 + \epsilon)$, *where* $0 < \epsilon < \frac{\epsilon_0}{2}$, *and* $\epsilon_0 = dist(C, \{0\} \times \{\gamma(T) \backslash \{\lambda_0\}\})$;

(iv) *There is $\alpha > 0$ such that if $(x, \lambda) \in G$, and $|\lambda - \lambda_0| \geq \epsilon$, then $\|x\| > \alpha$.*

Proof. Because C is closed and bounded in $N_1 = N \cup \{(0, \lambda_0)\}$, as in the proof of the Leray-Schauder Theorem 10.31, C is compact. For $0 < \delta < \epsilon_0$, let $U = N_\delta(C)$, then $K = \bar{U} \cap N_1$ is a compact metric space. Let $K_1 = C$, and $K_2 = \partial U \cap N_1$ be two disjoint compact sets in K. By Lemma 9.24, there are two compact sets $\bar{K}_1, \bar{K}_2$ such that $K_i \subset \bar{K}_i$, $i = 1, 2$, $\bar{K}_1 \cap \bar{K}_2 = \Phi$, $K = \bar{U} \cap N_1 = \bar{K}_1 \cup \bar{K}_2$. Let $\epsilon = \frac{1}{2}$ min (dist $(\bar{K}_1, \bar{K}_2)$, dist $(\bar{K}_1, \partial U)$), $A = N_\epsilon(\bar{K}_1)$. We obtain $0 < \epsilon < \frac{\epsilon_0}{2}$, $C \subset A$, and $\partial A \cap N_1 = \Phi$, $A \supset \{0\} \times (\lambda_0 - \epsilon, \lambda_0 + \epsilon)$. Let

$$\alpha = \frac{1}{2} \text{dist} \left(K, \{0\} \times \{\mathbb{R} \backslash (\lambda_0 - \epsilon, \lambda_0 + \epsilon)\}\right).$$

Because the only possible bifurcation point in A is $(0, \lambda_0)$, $\alpha > 0$. Let

$$Z = \overline{B_\alpha(0)} \times \{\mathbb{R} \backslash (\lambda_0 - \epsilon, \lambda_0 + \epsilon)\}, G = A - Z.$$

The open bounded set $G \subset \mathbb{R} \times E$ is our candidate. ∎

Theorem 4.22 *(Rabinowitz Theorem) Let E be a Banach space, let $\Omega \subset E \times \mathbb{R}$ be open, assume $(0, \lambda_0) \in \Omega$, and let $f : \Omega \to E$ be defined by*

$$f(x, \lambda) = (I - \lambda T)x - g(x, \lambda)$$

satisfying the property (H) in Definition 4.14:

(i) *$0 \neq \lambda_0$ is a characteristic value of T, $(0, \lambda_0) \in \Omega$;*

(ii) *$T : E \to E$ is a compact linear operator;*

(iii) *$g : \Omega \to E$ is a compact nonlinear operator satisfying $g(0, \lambda) = 0$ for all λ with $(0, \lambda) \in \Omega$, and $g(x, \lambda) = o(\|x\|)$ uniformly for $|\lambda - \lambda_0| < \epsilon$ for small ϵ.*

Let λ_0 be a characteristic value of T of odd multiplicity, and C is the connected component of $N \cup \{(0, \lambda_0)\}$ containing $(0, \lambda_0)$. Then, either
(a) C is not compact in Ω (if $\Omega = E \times \mathbb{R}$, C is unbounded); or
(b) C contains a finite number of points $(0, \lambda_j)$, with λ_j a characteristic value of T. Furthermore, the number of such points $(0, \lambda_j)$ having odd multiplicities, including $(0, \lambda_0)$, is even.

Proof.   Method 1. Suppose C is compact in Ω, because the only possible accumulation point of eigenvalues of a compact linear T is 0. Then by Proposition 4.17, C contains at most a finite number of values $(0, \lambda_j)$, $j = 0, 1, \cdots, k$. By Lemma 4.21, let $G \subset \Omega \subset E \times \mathbb{R}$ be open, $C \subset G$, such that there is no nontrivial solution (x, λ) of $f(x, \lambda) = 0$ on ∂G, and $G \cap (\{0\} \times \gamma(T)) = \{(0, \lambda_j) \mid j = 0, 1, ..., k\}$. For $r > 0$, consider $f_r(\cdot, \lambda) : \bar{G} \to E \times \mathbb{R}$ defined by

$$f_r(x, \lambda) = (f(x, \lambda), \|x\|^2 - r^2).$$

Because on ∂G, there are no nontrivial solutions of $f(x, \lambda) = 0$, the degree $\deg(f_r(x, \lambda), G, (0, 0))$ is well-defined and independent of r.
Case 1. Let r be large such that $\|x\|^2 - r^2 < 0$ for $x \in G$, then $(0, 0) \notin f_r(G)$, and consequently, $\deg(f_r(x, \lambda), G, (0, 0)) = 0$.
Case 2. Let r be small. If $f_r(x, \lambda) = (f(x, \lambda), \|x\|^2 - r^2) = (0, 0)$, then $\|x\| = r$, and $f(x, \lambda) = x - \lambda T x - g(x, \lambda) = 0$. Then λ is close to one of the λ_j, for some j, $j = 0$, $1, \cdots, k$. (However, if this is not the case then $(I - \lambda T)^{-1}$ is bounded and $x = 0$). By the Excision Propertyin Corollary 9.18, and Cases 1, 2, and the Ize Lemma 4.20,

$$0 = \deg(f_r(x, \lambda), G, (0, 0)) \text{ for large } r,$$
$$= \deg(f_r(x, \lambda), G, (0, 0)) \text{ for small } r,$$
$$= \sum_{j=0}^{K} \deg(f_r(x, \lambda), \lambda_j, (0, 0)), \tag{4.12}$$
$$= \sum_{j=0}^{K} (i_-(j) - i_+(j)),$$

but $i_+(j) = (-1)^{m_j} i_-(j)$ where m_j is the multiplicity of μ_j. Therefore, $i_-(j) - i_+(j) = 0$ if m_j is even. By (4.12), $0 = \sum_{m_j \text{ odd}} (i_-(j) - i_+(j))$. Therefore, the terms in $\sum$ must be even numbers, and consequently, the number of such points $(0, \lambda_j)$ having odd multiplicity including $(0, \lambda_0)$ is even.
Method 2. Let $\Omega = E \times \mathbb{R}$. By contradiction, suppose the component C of $N \cup \{(0, \lambda_0)\}$, containing $(0, \lambda_0)$ is bounded and does not meet $(0, \lambda)$, $\lambda_0 \neq \lambda \in \gamma(T)$. Assume G, ϵ and α satisfy Lemma 4.21 $(i) - (iv)$. Let $s_\lambda = \frac{1}{2} \text{ dist } (G, \lambda)$. Then $s_\lambda > 0$, for $0 < |\lambda - \lambda_0| < \epsilon$, and $s_\lambda > \frac{1}{2}\alpha$ for $|\lambda - \lambda_0| \geq \epsilon$. Now, $\overline{B_{s_\lambda}(0)} \cap \bar{G}_\lambda = \Phi$

for $|\lambda - \lambda_0| \geq \epsilon$, and on the other hand, Equation (4.9) does not have solutions in $\partial(G_\lambda - \overline{B_{s_\lambda}(0)}) \times \{\lambda\}$ for $0 < |\lambda - \lambda_0| < \epsilon$. Thus, the degree $\deg(f(\cdot, \lambda), G_\lambda - \overline{B_{s_\lambda}(0)},$ $0)$ is defined for $\lambda \neq \lambda_0$. We claim that $\deg(f(\cdot, \lambda), G_\lambda - \overline{B_{s_\lambda}(0)}, 0) = 0$ for $\lambda \neq \lambda_0$. In fact, for $\lambda > \lambda_0$, choose $\lambda^* > \lambda$ large with $G_{\lambda^*} = \Phi$, and set $s = \inf\{s_\theta \mid \lambda \leq \theta \leq \lambda^*\}$. Note that there is no bifurcation point on $(\lambda_0, \lambda_0 + \epsilon)$, $s > 0$. Let $\Omega = G \cap ([E - \overline{B_s(0)}] \times [\lambda, \lambda^*]) \subset E \times [\lambda, \lambda^*]$, be bounded open such that f does not annihilate on $\partial\Omega$. By Homotopy Invariance in Theorem 9.16, $\deg(f(\cdot, \theta), G_\lambda - \overline{B_s(0)},$ $0)$ is constant for $\theta \in [\lambda, \lambda*]$, so $\deg(f(\cdot, \lambda), G_\lambda - \overline{B_s(0)}, 0) = 0$. There are no solutions of (4.9) in $\overline{B_{s_\lambda}(0)} - B_s(0)$, by the Excision Property

$$\deg(f(\cdot, \lambda), G_\lambda - \overline{B_{s_\lambda}(0)}, 0) = 0.$$

Similarly for $\lambda < \lambda_0$. On the other hand, thanks to the generalized property of Homotopy Invariance, the degree $\deg(f(\cdot, \lambda), G_\lambda, 0)$ is constant for $\lambda \in (\lambda_0 - \epsilon, \lambda_0 + \epsilon)$. Choose $\underline{\lambda}, \bar{\lambda} \in \mathbb{R}$ such that $\lambda_0 - \epsilon < \underline{\lambda} < \lambda_0 < \bar{\lambda} < \lambda_0 + \epsilon$. Applying the Excision property and Additivity Property in Theorem 9.16 of degree, and noting that 0 is the only zero of $f(\cdot, \lambda)$ contained in $\overline{B_{s_\lambda}(0)}$ for $0 < |\lambda - \lambda_0| < \epsilon$, we have

$$\deg(f(\cdot, \underline{\lambda}), G_{\underline{\lambda}}, 0) = i(f(\cdot, \underline{\lambda}), 0, 0) + \deg(f(\cdot, \underline{\lambda}), G_{\underline{\lambda}} - \overline{B_{s_\lambda}(0)}, 0),$$
$$\deg(f(\cdot, \bar{\lambda}), G_{\bar{\lambda}}, 0) = i(f(\cdot, \bar{\lambda}), 0, 0) + \deg(f(\cdot, \bar{\lambda}), G_{\bar{\lambda}} - \overline{B_{s_{\bar{\lambda}}}(0)}, 0).$$

Thus, $i(f(\cdot, \underline{\lambda}), 0, 0) = i(f(\cdot, \bar{\lambda}), 0, 0)$. However, the index $i(f(\cdot, \lambda), 0, 0) = \pm 1$ for $\lambda \notin \gamma(T)$, and changes of sign occur when λ passes through characteristic values of T with odd multiplicity, a contradiction. ∎

Remark 4.23 *Theorem 4.22 still holds if we replace λT by $T(\lambda)$, if a family of compact linear operators of E, $\lambda \to T(\lambda)$ is continuous, and if the index $i(I - T(\lambda), 0, 0)$ is defined for $|\lambda - \lambda_0| > 0$ is sufficiently small. Note that*

$$i(I - T(\lambda) - g(\cdot, \lambda), 0, 0) = i(I - T(\lambda), 0, 0)$$

and $i(I - T(\lambda), 0, 0)$ changes sign when λ passes through λ_0.

Remark 4.24 *There are two possible cases in Theorem 4.22:*

(i) *If $g = 0$, then, for every $\lambda \in \gamma(T)$, $(0, \lambda)$ is a bifurcation point and $C = F_\lambda x\{\lambda\}$, where F_λ is a nontrivial subspace associated to λ.*

(ii) *$E = \mathbb{R}^2$, and consider the equation*

$$x = \lambda T x + T H(x),$$

where

$$x = \begin{pmatrix} x_1 \\ x_2 \end{pmatrix}, H(x) = \begin{pmatrix} -x_2^3 \\ x_1^3 \end{pmatrix}, T = \begin{pmatrix} 1 & 0 \\ 0 & \frac{1}{2} \end{pmatrix}.$$

We claim that $\gamma(T) = \{1, 2\}$, and every characteristic value is simple. In fact, if $\lambda T x = x$, then

$$\lambda \begin{pmatrix} 1 & 0 \\ 0 & \frac{1}{2} \end{pmatrix} \begin{pmatrix} x_1 \\ x_2 \end{pmatrix} = \begin{pmatrix} x_1 \\ x_2 \end{pmatrix},$$

$$\begin{cases} \lambda x_1 = x_1, \\ \dfrac{1}{2} \lambda x_2 = x_2, \end{cases}$$

*Take $x = \begin{pmatrix} 1 \\ 0 \end{pmatrix}$, and $x = \begin{pmatrix} 0 \\ 1 \end{pmatrix}$ so that $\lambda = 1$, and 2. Moreover, let $I - T = \begin{bmatrix} 0 & 0 \\ 0 & \frac{1}{2} \end{bmatrix}$,
$(I - T)^k = \begin{pmatrix} 0 & 0 \\ 0 & c_k \end{pmatrix}$, and $c_k \neq 0$ for $k = 1, 2, \cdots$. Suppose $(I - T)^k \begin{pmatrix} x_1 \\ x_2 \end{pmatrix} = \begin{pmatrix} 0 \\ 0 \end{pmatrix}$,
then $x_2 = 0$, and consequently, $\cup_{k=1}^{\infty} Ker(I - T)^k = \mathbb{R}$, and $\dim \cup_{k=1}^{\infty} Ker(I - T)^k = 1$.
Then 1 is simple. Similarly, 2 is simple.*

Note that

$$\mathcal{C} = \{((x_1, x_2), \lambda) \mid 1 \leq \lambda \leq 2, x_1 = \pm(2 - \lambda)^{3/8}(\lambda - 1)^{1/8},$$
$$x_2 = \pm(2 - \lambda)^{1/8}(\lambda - 1)^{3/8}\}.$$

In fact, let $A = T^{-1} = \begin{pmatrix} 1 & 0 \\ 0 & 2 \end{pmatrix}$, then

$$x = \lambda T x + T H x$$

if and only if

$$A x = \lambda x + H x$$

if and only if

$$\begin{pmatrix} 1 & 0 \\ 0 & 2 \end{pmatrix} \begin{pmatrix} x_1 \\ x_2 \end{pmatrix} = \lambda \begin{pmatrix} x_1 \\ x_2 \end{pmatrix} + \begin{pmatrix} -x_2^3 \\ x_1^3 \end{pmatrix}$$

if and only if

$$\begin{cases} (1 - \lambda)x_1 = -x_2^3 \\ (2 - \lambda)x_2 = x_1^3 \end{cases}$$

if and only if

$$\begin{cases} (\lambda - 1)(2 - \lambda)^3 = x_1^8 \\ -(\lambda - 1)^3(2 - \lambda) = x_2^8 \end{cases}$$

if and only if

$$\begin{cases} x_2 = \pm(2 - \lambda)^{1/8}(\lambda - 1)^{3/8} \\ x_1 = \pm(2 - \lambda)^{3/8}(\lambda - 1)^{1/8} \end{cases}$$

Notes Most of the material in this chapter is adapted from Nirenberg [36] and Rabinowitz [38]. For related material, see Crandall-Rabinowitz [15], Dancer [16], Krasnoselskii [27], Nirenberg [36], Rabinowitz [38], [39], and Sattinger [42].

Chapter 5

SOBOLEV SPACES $W^{1,p}(I)$

In this chapter we study various properties of Sobolev spaces on an interval I in a real line, which are useful in both Analysis and Partial Differential Equations.

5.1　Motivation

Before providing motivations for studying the Sobolev spaces, we recall definitions of classical derivatives and weak derivatives of a function. Let f be a function defined on a (bounded or unbounded) interval (a, b).

Definition 5.1

(i) If for each $c \in (a, b)$, the limit

$$\lim_{x \to c} \frac{f(x) - f(c)}{x - c} = f'(c)$$

exists, f' is called the classical derivative of f;

(ii) If there is a function g on (a, b) such that

$$\int_a^b f \, \varphi' = - \int_a^b g \, \varphi$$

for each $\varphi \in C_c^1(a, b)$, then g is called the weak derivative or the derivative in the distribution sense of f, denoted also by f';

(iii) If for some positive integer k,

$$\int_a^b f \, \varphi^{(k)} = (-1)^k \int_a^b g \, \varphi$$

for each $\varphi \in C_c^k(a, b)$, then g is called the kth weak derivative of f, in symbols $g = f^{(k)}$.

Remark 5.2 *In the definitions of the weak derivatives (ii) and (iii), we may replace the test function φ in $C_c^1(a,b)$ or $C_c^k(a,b)$ by functions in $C_c^\infty(a,b)$. In fact, let (ρ_n) be a regularity sequence: $\rho_n \in C_c^\infty(\mathbb{R})$, $\operatorname{supp} \rho_n \subset B(0, \frac{1}{n})$, $\int \rho_n = 1$, $\rho_n \geq 0$ on $\mathbb{R}$. By Theorem 1.35, $\rho_n * \varphi \in C_c^\infty(\mathbb{R})$ satisfying $\rho_n * \varphi \to \varphi$, $\rho_n * \varphi' \to \varphi'$, and $\rho_n * \varphi^{(k)} \to \varphi^{(k)}$ uniformly, where φ is considered as 0 outside (a,b).*

Definition 5.3 *Consider the following problem. Given $f \in L^2(a,b)$, find a function $u(x)$ satisfying*

$$\begin{cases} -u'' + u = f \text{ in } (a,b), \\ u(a) = u(b) = 0. \end{cases} \tag{5.1}$$

(i) *If $f \in C(a,b)$, then a function $u \in C^2[a,b]$ satisfying (5.1) is called a classical solution of (5.1).*

(ii) *If u, u' and $u'' \in L^2[a,b]$, satisfy*

$$\begin{cases} -u'' + u = f \text{ a.e. in } (a,b), \\ u(a) = u(b) = 0, \end{cases}$$

then u is called a strong solution of (5.1).
Multiply (5.1) by $\varphi \in C^1[a,b]$, $\varphi(a) = \varphi(b) = 0$, and then use integration by parts to obtain

$$\int_a^b u'\,\varphi' + \int_a^b u\,\varphi = \int_a^b f\,\varphi. \tag{5.2}$$

(iii) *A function $u \in C^1[a,b]$, or a function u, $u' \in L^2[a,b]$ satisfying (5.2) is called a weak solution of (5.1).*

Variational methods of Partial Differential Equations can be described as follows:

(i) Make precise the notations of weak solution of (5.1: introduce Sobolev spaces;

(ii) Use variational methods such as the Lax-Milgram Theorem, Stampacchia Theorem, Fredholm Alternative Theorem, and the Mountain Pass Lemma, to prove the existence and the uniqueness of weak solutions;

(iii) Regularity: prove the weak solutions are in C^2;

(iv) Return to the classical solutions: prove that weak solutions in C^2 are classical solutions.

5.2 The Sobolev Space $W^{1,p}(I)$

Let $I = (a, b)$ be a bounded or an unbounded interval, $1 \leq p \leq \infty$.

Definition 5.4 *The Sobolev space $W^{1,p}(I)$ is defined by*

$$W^{1,p}(I) = \{u \in L^p(I) \,|\, u' \in L^p(I)\}.$$

It is denoted by $H^1(I) = W^{1,2}(I)$.

Remark 5.5 *If $u \in C^1(I)$ such that u and its derivative u' are in $L^p(I)$, then $u \in W^{1,p}(I)$. Moreover, the derivative u' coincides with the weak derivative of u. In particular, if I is bounded, then $C^1(\bar{I}) \subset W^{1,p}(I)$ for all $1 \leq p \leq \infty$.*

Example 5.6 *Let $I = (-1, 1)$.*

(i) *Define $h(x) = \frac{1}{2}(|x| + x)$ for $x \in I$. Then $h \in W^{1,p}(I)$ for $1 \leq p \leq \infty$ with $u' = H \in L^p(I)$, where*

$$H(x) = \begin{cases} 0 & \text{if } -1 \leq x < 0 \\ 1 & \text{if } 0 < x \leq 1 \end{cases}$$

is the Heaviside function on I;

(ii) *The Heaviside function $H(x) \notin W^{1,p}(I)$, for any $1 \leq p \leq \infty$.*

Proof.

(i) For $\varphi \in C_c^\infty(I)$,

$$\int_{-1}^1 h(x)\varphi'(x) = \int_{-1}^1 \frac{1}{2}(|x| + x)\varphi' = \int_{-1}^0 0 + \int_0^1 x\,\varphi'$$

$$= \int_0^1 x\,\varphi' = x\,\varphi(x)|_0^1 - \int_0^1 \varphi\,dx$$

$$= -\int_{-1}^1 H(x)\varphi(x)\,dx.$$

Thus, $h' = H$.

(ii) Let $\varphi \in C_c^\infty(I)$ with $\varphi(0) = -1$. Then

$$\int_{-1}^1 H\,\varphi' = \int_0^1 \varphi' = -\varphi(0) = 1.$$

Suppose $H \in W^{1,p}(I)$, and $g = H' \in L^p(I)$. Let $\varphi \in C_c^\infty(I)$, $\varphi(0) = -1$, $|\varphi| \leq 1$ on I, and $|\operatorname{supp}\varphi|^{1/p'} \|g\|_{L^p} < 1$. Now,

$$1 = \left| \int_{-1}^1 H\,\varphi' \right| = \left| -\int_{-1}^1 g\,\varphi \right| \leq \int_{\operatorname{supp}\varphi} |g| < |\operatorname{supp}\varphi|^{1/p'} \|g\|_{L^p} < 1,$$

a contradiction. Therefore, $H \notin W^{1,p}(I)$ for $1 \leq p \leq \infty$. ∎

For $u \in W^{1,p}(I)$, define the norm

$$\|u\|_{W^{1,p}} = \|u\|_{L^p} + \|u'\|_{L^p} \approx \left(\|u\|_{L^p}^p + \|u'\|_{L^p}^p\right)^{1/p}.$$

In addition, for $u, v \in H^1(I)$, define the scalar product

$$(u, v)_{H^1} = (u, v)_{L^2} + (u', v')_{L^2}.$$

Theorem 5.7 *The space* $(W^{1,p}(I), \|\cdot\|_{W^{1,p}(I)})$ *is a Banach space for* $1 \leq p \leq \infty$. *Moreover,* $W^{1,p}$ *is reflexive for* $1 < p < \infty$ *and separable for* $1 \leq p < \infty$. *The space* $H^1(I)$ *is a separable Hilbert space.*

Proof.

(i) Let (u_n) be a Cauchy sequence in $W^{1,p}(I)$. Then (u_n), (u'_n) are Cauchy in $L^p(I)$. Thus, $u_n \to u$, and $u'_n \to g$ strongly in L^p for some $u, g \in L^p(I)$. By the definition,

$$\int_I u_n \varphi' = \int_I u'_n \varphi \quad \text{for } \varphi \in C_c^1(I).$$

Letting $n \to \infty$, we obtain

$$\int_I u \varphi' = -\int_I g \varphi \quad \text{for } \varphi \in C_c^1(I).$$

Therefore, $u \in W^{1,p}(I)$, $u' = g$ and $\|u_n - u\|_{W^{1,p}(I)} \to 0$.

(ii) $W^{1,p}(I)$ is reflexive for $1 < p < \infty$: it is known that the product space $E = L^p(I) \times L^p(I)$ is reflexive. The operator $T : W^{1,p}(I) \to E$ defined by $Tu = (u, u')$ is an isometry of $W^{1,p}(I)$ into E, and consequently, $T(W^{1,p}(I))$ is a closed subspace of E. Therefore $T(W^{1,p}(I))$ is reflexive: $W^{1,p}(I)$ is reflexive.

(iii) $W^{1,p}(I)$ is separable for $1 \leq p < \infty$: as in (ii), $T(W^{1,p}(I))$ is a subspace of E, and E is separable, so $T(W^{1,p}(I))$ is separable: $W^{1,p}(I)$ is separable.

■

Lemma 5.8 *Let* $f \in L^1_{\text{loc}}(I)$ *such that*

$$\int_I f\varphi = 0 \quad \text{for } \varphi \in C_c(I).$$

Then $f = 0$ *a.e. on* I.

Proof.

(i) Suppose $f \in L^1(I)$, $|I| < \infty$. For $\epsilon > 0$ take $f_1 \in C_c(I)$ such that $\|f - f_1\|_{L^1} < \epsilon$. Let

$$K_1 = \{x \in I \,|\, f_1(x) \geq \epsilon\},$$

$$K_2 = \{x \in I \,|\, f_1(x) \leq -\epsilon\}.$$

Then K_1, K_2 are two disjoint compact subsets. By the Tietze-Urysohn Extension Theorem 1.24, there is $\varphi_0 \in C_c(I)$ such that

$$\varphi_0(x) = \begin{cases} 1 & \text{for } x \in K_1, \\ -1 & \text{for } x \in K_2, \end{cases}$$

$$|\varphi_0(x)| \leq 1 \quad \text{for } x \in I.$$

Set $K = K_1 \cup K_2$, then

$$\int_I f_1\, \varphi_0 = \int_{I \backslash K} f_1\, \varphi_0 + \int_K f_1\, \varphi_0.$$

However, by the assumption $\int_I f\, \varphi_0 = 0$, we have

$$\left| \int_I f_1\, \varphi_0 \right| \leq \left| \int_I (f_1 - f)\, \varphi_0 \right| + \left| \int_I f\, \varphi_0 \right| \leq \int_I |f_1 - f|\, |\varphi_0| \leq \|f_1 - f\|_{L^1} < \epsilon,$$

$$\int_K |f_1| = \int_K f_1\, \varphi_0 = \int_I f_1\, \varphi_0 - \int_{I \backslash K} f_1\, \varphi_0$$

$$\leq \left| \int_I f_1\, \varphi_0 \right| + \left| \int_{I \backslash K} f_1\, \varphi_0 \right| \leq \epsilon + \int_{I \backslash K} |f_1\, \varphi_0| \leq \epsilon + \int_{I \backslash K} |f_1|,$$

or

$$\int_I |f_1| = \int_K |f_1| + \int_{I \backslash K} |f_1| \leq \epsilon + \int_{I \backslash K} |f_1| + \int_{I \backslash K} |f_1|$$

$$\leq \epsilon + 2 \int_{I \backslash K} |f_1| \leq \epsilon + 2\epsilon\, |I| \quad \text{because } |f_1| \leq \epsilon \text{ on } I \backslash K.$$

Therefore,

$$\|f\|_{L^1(I)} \leq \|f - f_1\|_{L^1(I)} + \|f_1\|_{L^1(I)} \leq 2\epsilon + 2\epsilon\, |I| \quad \text{for } \epsilon > 0.$$

Hence,

$$f = 0 \quad \text{a.e. on } I.$$

(ii) Let I be an unbounded interval, say $I = \bigcup_{n=1}^{\infty} \bar{I}_n$, where the I_n are open intervals, $I_n \subset\subset I$. Apply (i) to obtain $f = 0$ a.e. on I_n, so $f = 0$ a.e. on I.

■

Lemma 5.9 *Let* $f \in L^1_{\text{loc}}(I)$ *such that*

$$\int_I f\,\varphi' = 0 \quad \text{for } \varphi \in C^1_c(I).$$

Then f *is constant a.e. on* $I = (a, b)$.

Proof. Fix $\psi \in C_c(I)$ and set $\int_I \psi = 1$. For $w \in C_c(I)$ we can find $\varphi \in C^1_c(I)$ such that

$$\varphi' = w - \left(\int_I w \right) \psi.$$

In fact, let $h = w - \left(\int_I w \right) \psi$. Then $h \in C_c(I)$, $\int_I h = 0$. Set

$$\varphi(x) = \int_a^x h(t)\,dt \quad \text{for } x \in I.$$

Then $\varphi \in C^1_c(I)$ with

$$\varphi' = w - \left(\int_I w \right) \psi.$$

Now,

$$\int_I f\,\varphi' = \int_I f \left[w - \left(\int_I w \right) \psi \right] = 0,$$

and we have

$$\int_I \left[f - \int_I f\psi \right] w = 0 \quad \text{for } w \in C_c(I).$$

Lemma 5.8 implies that $f = c$ a.e. on I, where $c = \int_I f\psi$.   ■

Lemma 5.10 *Let* $f \in L^1_{\text{loc}}(I)$, *fix* $x_0 \in I = (a, b)$, *and let*

$$u(x) = \int_{x_0}^x f(t)\,dt \quad \text{for } x \in I.$$

Then $u \in C(I)$ *and*

$$\int_I u\,\varphi' = - \int_I f\,\varphi \quad \text{for } \varphi \in C^1_c(I).$$

Proof. Let $I = (a, b)$. If $a \in \mathbb{R}$, then take a sequence $\{x_n\}$ such that $x_n \searrow a$. Then

$$|u(x_n) - u(x_m)| = \left| \int_{x_m}^{x_n} f(t)\, dt \right| \to 0, \text{ as } n, m \to 0,$$

because $f \in L^1_{\text{loc}}$, so

$$u(x_n) \to \alpha = u(a).$$

Similarly, if $b \in \mathbb{R}$, then $u(y_n) \to \beta = u(b)$. It is easy to see that $u \in C(I)$. Therefore $u \in C(I)$. Moreover,

$$
\begin{aligned}
\int_I u\, \varphi' &= \int_I \varphi'(x)\, dx \int_{x_0}^x f(t)\, dt = \int_a^b dx \int_{x_0}^x [f(t)\varphi'(x)]\, dt \\
&= -\int_a^{x_0} dx \int_x^{x_0} [f(t)\varphi'(x)]\, dt + \int_{x_0}^b dx \int_{x_0}^x [f(t)\varphi'(x)]\, dt \\
&= -\int_a^{x_0} dt \int_a^t [f(t)\varphi'(x)]\, dx + \int_{x_0}^b dt \int_t^b [f(t)\varphi'(x)]\, dx \\
&= -\int_a^{x_0} f(t)\, dt \int_a^t \varphi'(x)\, dx + \int_{x_0}^b f(t)\, dt \int_t^b \varphi'(x)\, dx \\
&= -\int_a^{x_0} f(t)\varphi(t)\, dt - \int_{x_0}^b f(t)\varphi(t)\, dt = -\int_I f(t)\varphi(t)\, dt \\
&= -\int_I f\varphi \quad \text{for } \varphi \in C_c^1(I).
\end{aligned}
$$

We conclude that $\int_I u\, \varphi' = -\int_I f\, \varphi$. ∎

Roughly speaking, a function in $W^{1,p}(I)$ is the primitive of some function in $L^p(I)$, which is a consequence of the following theorem.

Theorem 5.11 (*Continuous Representation Theorem*) *For $u \in W^{1,p}(I)$, $1 \leq p \leq \infty$. Then there is a function $\tilde{u} \in C(I)$ such that $u = \tilde{u}$ a.e. on I, and*

$$\tilde{u}(x) - \tilde{u}(y) = \int_y^x u'(t)\, dt \quad \text{for } x, y \in I.$$

In particular, if $u \in W^{1,p}(I)$, and $u' \in C(\bar{I})$, then $u \in C^1(\bar{I})$.

Proof. Fix $x_0 \in I$, and set

$$\bar{u}(x) = \int_{x_0}^x u'(t)\, dt.$$

By Lemma 5.10, $\bar{u}$ is continuous on $\bar{I}$ with $\int_I \bar{u}\, \varphi' = -\int_I u'\, \varphi$ for $\varphi \in C_c^1(I)$. Moreover, by the definition

$$-\int_I u'\, \varphi = \int_I u\, \varphi' \quad \text{for } \varphi \in C_c^1(I).$$

Therefore, $\int_I (u - \bar{u}) \, \varphi' = 0$ for $\varphi \in C^1_c(I)$. By Lemma 5.9, $u - \bar{u} = c$ a.e. on I. Set $\tilde{u} = \bar{u} + c$. Then $\tilde{u} \in C(\bar{I})$ and $u = \tilde{u}$ a.e. on $\bar{I}$,

$$\tilde{u}(x) - \tilde{u}(y) = \bar{u}(x) - \bar{u}(y) = \int_y^x u'(t) \, dt \quad \text{for } x, y \in \bar{I}. \qquad \blacksquare$$

Remark 5.12 *By Lemma 5.10, if $g \in L^p(I)$, with its primitive $u \in L^p(I)$, then $u' = g$, or $u \in W^{1,p}(I)$.*

Theorem 5.13 *Let $u \in L^p(I)$, $1 < p \leq \infty$. The following are equivalent:*

(i) $u \in W^{1,p}$;

(ii) There is $c > 0$ such that

$$\left| \int_I u \, \varphi' \right| \leq c \, \|\varphi\|_{L^{p'}(I)} \quad \text{for } \varphi \in C^\infty_c(I);$$

(iii) There is $c > 0$ such that for every open set $\omega \subset\subset I$ and for every $h \in \mathbb{R}$ with $|h| < \operatorname{dist}(\omega, I^c)$, we have $\|\tau_h u - u\|_{L^p(\omega)} \leq c|h|$, where $\tau_h u(x) = u(x + h)$. Moreover, we can choose $c = \|u'\|_{L^p}$ in (ii) and (iii), where $\tau_h u(x) = u(x + h)$.

Proof. $(i) \Rightarrow (ii)$: Let $u \in W^{1,p}(I)$. Then $u' \in L^p(I)$, or u' is a bounded linear operator on $L^{p'}(I)$. Thus,

$$\left| \int_I u \varphi' \right| = \left| - \int_I u' \varphi \right| \leq \|u'\|_{L^p} \|\varphi\|_{L^p} \quad \text{for } \varphi \in C^\infty_c(I).$$

$(ii) \Rightarrow (i)$: the linear functional

$$F : \varphi \in C^\infty_c(I) \to \int u \, \varphi'$$

is bounded on $L^{p'}(I)$. Because $C^\infty_c(I)$ is dense in $L^{p'}(I)$, $1 \leq p' < \infty$, F can be extended to be a bounded linear functional on $L^{p'}(I)$. By the Riesz Representation Theorem, there is a function $g \in L^p(I)$ such that

$$F(\varphi) = \int g \varphi \quad \text{for } \varphi \in L^{p'}(I).$$

In particular,

$$\int u \varphi' = \int g \varphi \quad \text{for } \varphi \in C^\infty_c(I).$$

Therefore $u' = -g \in L^p(I)$, and consequently $u \in W^{1,p}$.

$(i) \Rightarrow (iii)$: by Theorem 5.11, for $x \in \omega$,

$$u(x+h) - u(x) = \int_0^1 \frac{du}{ds}(x+sh)ds = h\int_0^1 u'(x+sh)ds.$$

Thus,

$$|u(x+h) - u(x)| \leq |h| \int_0^1 |u'(x+sh)| \, ds.$$

If $p = \infty$, then $\|\tau_h u - u\|_{L^\infty} \leq c|h|$. If $1 < p < \infty$, by the Jensen inequality

$$|u(x+h) - u(x)|^p \leq |h|^p \int_0^1 |u'(x+sh)|^p \, ds$$

or

$$\int_\omega |u(x+h) - u(x)|^p \leq |h|^p \int_\omega dx \int_0^1 |u'(x+sh)|^p \, ds$$

$$= |h|^p \int_0^1 ds \int_\omega |u'(x+sh)|^p \, dx.$$

For $0 < s < 1$, we have

$$\int_\omega |u'(x+sh)|^p \, dx = \int_{\omega+sh} |u'(y)|^p \, dy \leq \int_I |u'(y)|^p \, dy.$$

Thus, $\|\tau_h u - u\|_{L^p(\omega)} \leq c|h|$, where $c = \int_I |u'(y)|^p \, dy$. Note that we may also apply the Minkowski Inequality of Integrals to obtain the above inequalities.

$(iii) \Rightarrow (ii)$: let $\varphi \in C_c^1(I)$. Choose $\omega \subset\subset I$ such that $\operatorname{supp}\varphi \subset \omega$. Suppose $h \in \mathbb{R}$, $|h| < \operatorname{dist}(\omega, I^c)$, then

$$\left| \int_I u(x) \left[\varphi(x-h) - \varphi(x) \right] dx \right| = \left| \int_I [u(x+h) - u(x)] \, \varphi(x) dx \right|$$

$$\leq \|u(x+h) - u(x)\|_{L^p} \|\varphi\|_{L^{p'}}$$

$$\leq c|h| \|\varphi\|_{L^{p'}}.$$

Thus,

$$\left| \int_I u(x) \frac{\varphi(x-h) - \varphi(x)}{h} dx \right| \leq c\|\varphi\|_{L^{p'}}.$$

Letting $h \to 0$, by the Lebesgue Dominated Convergence Theorem,

$$\left| \int_I u(x)\varphi'(x)dx \right| \leq c\|\varphi\|_{L^{p'}}.$$

∎

Remark 5.14 *If $p = 1$, we have $(i) \Rightarrow (ii) \Longleftrightarrow (iii)$.*

Proof. Let I be bounded. Then

(a) u satisfies $(i) \Longleftrightarrow u \in W^{1,1}(I) \Longleftrightarrow u$ is absolutely continuous on $I \Longleftrightarrow$ for every $\epsilon > 0$, there is $\delta > 0$ such that for any finite disjoint subintervals (a_k, b_k) in I with $\sum(b_k - a_k) < \delta$, we have $\sum |u(b_k) - u(a_k)| < \epsilon$.

(b) u satisfying (ii) or $(iii) \Longleftrightarrow u$ has bounded variation on I
$\Longleftrightarrow u = u_1 - u_2$, where u_1, u_2 are increasing functions on I
$\Longleftrightarrow$ there is $c > 0$ such that for any $t_0 < t_1 < \cdots < t_k$ in I,

$$\sum_{i=0}^{k-1} |u(t_{i+1}) - u(t_i)| \leq c$$

$\Longleftrightarrow u \in L^1(I)$ together with its weak derivative u' is of bounded measure. ∎

Corollary 5.15 *For $u \in L^\infty(I)$. Then $u \in W^{1,\infty}(I)$ if and only if there is a constant $c > 0$ such that*

$$|u(x) - u(y)| \leq c\,|x - y| \quad \text{a.e. } x, y \in I.$$

In particular, each function u in $W^{1,\infty}(I)$ is Lipschitz continuous.

Proof. By Theorem 5.13 (iii). ∎

Take $\eta \in C^1(\mathbb{R})$, $0 \leq \eta \leq 1$ such that

$$\eta(x) = \begin{cases} 1 & \text{for } x < \dfrac{1}{4}, \\[2mm] 0 & \text{for } x > \dfrac{3}{4}. \end{cases}$$

For a function $f : (0, 1) \to \mathbb{R}$, set

$$\widetilde{f}(x) = \begin{cases} f(x) & \text{for } 0 < x < 1, \\ 0 & \text{for } x \geq 1. \end{cases}$$

Lemma 5.16 *If $u \in W^{1,p}(I)$, then $\eta\,\widetilde{u} \in W^{1,p}(0, \infty)$ and $(\eta\widetilde{u})' = \eta'\,\widetilde{u} + \eta\,\widetilde{u}'$.*

Proof. For $\varphi \in C_c^1(0, \infty)$, we have

$$\int_0^\infty \eta\,\widetilde{u}\,\varphi' = \int_0^1 \eta\,u\,\varphi' = \int_0^1 u\,[(\eta\varphi)' - \eta'\varphi]$$

$$= -\int_0^1 u'\,\eta\,\varphi - \int_0^1 u\,\eta'\,\varphi \quad \text{because } \eta\varphi \in C_c^1(0, \infty)$$

$$= -\int_0^\infty (\widetilde{u'}\,\eta + \widetilde{u}\,\eta')\varphi.$$

Thus, $(\eta\,\widetilde{u})' = \eta'\,\widetilde{u} + \eta\,\widetilde{u}'$. Note that $\eta\,\widetilde{u}$, $\eta'\,\widetilde{u}$, $\eta\,\widetilde{u}' \in L^p(0, \infty)$, so $\eta\,\widetilde{u} \in W^{1,p}(0, \infty)$. ∎

Theorem 5.17 (*Extension Theorem*) *For $1 \le p \le \infty$, there is an extension $P :$ $W^{1,p}(I) \to W^{1,p}(\mathbb{R})$, linear and continuous, such that:*

(*i*) $Pu = u$ *for $u \in W^{1,p}(I)$;*

(*ii*) $\|Pu\|_{L^p(\mathbb{R})} \le c \|u\|_{L^p(I)}$ *for $u \in W^{1,p}(I)$;*

(*iii*) $\|Pu\|_{W^{1,p}(\mathbb{R})} \le c \|u\|_{W^{1,p}(I)}$ *for $u \in W^{1,p}(I)$, where c depends only on $|I|$.*

Proof.

(*a*) Let I be an unbounded interval in $\mathbb{R}$, say $I = (0, \infty)$. First note that by Theorem 5.11 we may assume $u \in C(\bar{I})$. Set by reflection,

$$(Pu)(x) = \bar{u}(x) = \begin{cases} u(x) & \text{for } x \ge 0, \\ u(x) & \text{for } x < 0. \end{cases}$$

Then

$$\|\bar{u}\|_{L^p(\mathbb{R})} \le 2 \|u\|_{L^p(I)}.$$

Set

$$v(x) = \begin{cases} u'(x) & \text{for } x > 0, \\ -u'(-x) & \text{for } x < 0. \end{cases}$$

Then

$$v \in L^p(\mathbb{R}), \quad \|v\|_{L^p(\mathbb{R})} \le 2 \|u'\|_{L^p(I)}.$$

For $x > 0$,

$$\bar{u}(x) - \bar{u}(0) = u(x) - u(0) = \int_0^x u'(t)\, dt = \int_0^x v(t)\, dt.$$

For $x < 0$,

$$\bar{u}(x) - \bar{u}(0) = u(-x) - u(0) = \int_0^{-x} u'(t)\, dt$$

$$= \int_0^x -u'(-t)\, dt = \int_0^x v(t)\, dt.$$

Thus,

$$\bar{u}(x) - \bar{u}(0) = \int_0^x v(t)\, dt \quad \text{for } x \in \mathbb{R}.$$

By Lemma 5.10, $\bar{u}' = v$. Therefore $\bar{u} \in W^{1,p}(\mathbb{R})$, and

$$\|\bar{u}\|_{W^{1,p}(\mathbb{R})} \le 2 \|u\|_{W^{1,p}(I)}.$$

(b) If I is a bounded interval, it suffices to consider the case $I = (0,1)$. For $u \in W^{1,p}(I)$, write

$$u = \eta\, u + (1 - \eta)u.$$

Extending ηu first by $\eta\widetilde{u}$ and then by reflection, we obtain a function $v_1 \in W^{1,p}(\mathbb{R})$, which is an extension of ηu such that

$$\|v_1\|_{L^p(\mathbb{R})} \le 2\, \|u\|_{L^p(I)},$$

$$\|v_1\|_{W^{1,p}(\mathbb{R})} \le c\, \|u\|_{W^{1,p}(I)},$$

where c depends on $\|\eta'\|_{L^\infty}$. Extending $(1 - \eta)u$ first by $(1 - \eta)\widetilde{u}$, where

$$\widetilde{u}(x) = \begin{cases} u(x) & 0 < x < 1, \\ 0 & x \le 0, \end{cases}$$

and then by reflection with respect to $x = 1$, we obtain a function $v_2 \in W^{1,p}(\mathbb{R})$, which is an extension of $(1 - \eta)u$ such that

$$\|v_2\|_{L^p(\mathbb{R})} \le 2\, \|u\|_{L^p(I)},$$

$$\|v_2\|_{W^{1,p}(\mathbb{R})} \le c\, \|u\|_{W^{1,p}(I)}.$$

Set $Pu = v_1 + v_2$, which satisfies properties (i)–(iii). ∎

Lemma 5.18 For $\varphi \in L^1(\mathbb{R})$, $v \in W^{1,p}(\mathbb{R})$, $1 \le p \le \infty$. Then $\varphi * v \in W^{1,p}(\mathbb{R})$, and $(\varphi * v)' = \varphi * v'$.

Proof.

(i) Suppose $\varphi \in C_c^1(I)$. Then $\varphi * v \in L^p(I)$, $\varphi * v' \in L^p(I)$, and

$$\int (\varphi * v)\psi' = \int v(\check{\varphi} * \psi') = \int v(\check{\varphi} * \psi)'$$

$$= -\int v'(\check{\varphi} * \psi) = -\int (\varphi * v')\psi \quad \text{for} \quad \psi \in C_c^1(\mathbb{R})$$

where $\check{\varphi}(x) = \varphi(-x)$. Hence, $\varphi * v \in W^{1,p}$, and $(\varphi * v)' = \varphi * v'$.

(ii) For $\varphi \in L^1(\mathbb{R})$, take a sequence (φ_n) in $C_c^1(\mathbb{R})$ such that $\varphi_n \to \varphi$ in L^1. Then

$$\varphi_n * v \in W^{1,p}, \ (\varphi_n * v)' = \varphi_n * v'.$$

Note that $\varphi_n * v \to \varphi * v$, and $\varphi_n * v' \to \varphi * v'$ strongly in L^p. Therefore $\varphi * v \in W^{1,p}$, and $(\varphi * v)' = \varphi * v'$. ∎

The Product Rule and the Chain Rule for the space $C^1(\mathbb{R})$ still hold for the Sobolev spaces $W^{1,p}(I)$: this is because of the following density property.

Theorem 5.19 (*First Density Theorem*) *For $u \in W^{1,p}(I)$, $1 \leq p < \infty$, there is a sequence (u_n) in $C_c^{\infty}(\mathbb{R})$ such that*

$$u_n|_I \to u \text{ in } W^{1,p}(I).$$

Proof. We use the methods of convolution and truncation.

(a) $I = \mathbb{R}$. Truncation: fix $\zeta \in C_c^{\infty}(\mathbb{R})$ such that $0 \leq \zeta \leq 1$ and

$$\zeta(x) = \begin{cases} 1 & \text{for } |x| \leq 1, \\ 0 & \text{for } |x| \geq 2. \end{cases}$$

Define a sequence of functions $\zeta_n(x) = \zeta\left(\dfrac{x}{n}\right)$ for $n = 1, 2, \cdots$. We call (ζ_n) a sequence of cut-off (truncation) functions. By the Lebesgue Dominated Convergence Theorem, for $1 \leq p < \infty$, $\zeta_n f \to f$ in L^p. Choose a regularity sequence (ρ_n) in $C_c^{\infty}(\mathbb{R})$, and a sequence (ζ_n) of cut-off functions. Define $u_n = \zeta_n(\rho_n * u)$, then $u_n \in C_c^{\infty}(\mathbb{R})$. Moreover, $u_n \to u$ in $W^{1,p}$. In fact, $u_n - u = \zeta_n\left[(\rho_n * u) - u\right] + \left[\zeta_n u - u\right]$, so

$$\|u_n - u\|_{L^p} \leq \|\rho_n \star u - u\|_{L^p} + \|\zeta_n u - u\|_{L^p} \to 0 \text{ as } n \to \infty.$$

By Lemma 5.16,

$$u_n' = \zeta_n'(\rho_n * u) + \zeta_n(\rho_n * u').$$

Then

$$\|u_n' - u'\|_{L^p} \leq \|\zeta_n'(\rho_n * u)\|_{L^p} + \|\zeta_n(\rho_n * u') - \zeta_n u'\| + \|\zeta_n u' - u'\|_{L^p}$$
$$\leq \frac{c}{n}\|u\|_{L^p} + \|\rho_n * u' - u'\|_{L^p} + \|\zeta_n u' - u'\|_{L^p}$$
$$\to 0 \text{ as } n \to \infty,$$

where $c = \|\zeta_n'\|_{L^{\infty}}$.

(b) $I \subset \mathbb{R}$, $u \in W^{1,p}(I)$. By the Extension Theorem 5.17, we extend u to $Pu \in W^{1,p}(\mathbb{R})$ and then by part (a), there is a sequence (u_n) in $C_c^{\infty}(\mathbb{R})$ such that, as $n \to \infty$,

$$\|u_n - u\|_{L^p(I)} \leq \|u_n - Pu\|_{L^p(\mathbb{R})} \to 0,$$
$$\|u_n' - u'\|_{L^p(I)} \leq \|u_n' - (Pu')\|_{L^p(\mathbb{R})}$$
$$= \|u_n' - (Pu)'\|_{L^p(\mathbb{R})} \text{ because } p \text{ is continuous and linear}$$
$$\to 0.$$

Thus, $u_n|_I \to u$ in $W^{1,p}(I)$. $\blacksquare$

Remark 5.20 *In general, for $u \in W^{1,p}(I)$ we cannot choose a sequence (u_n) in $C_c^\infty(I)$ that converges to u. In other words, in general, $C_c^\infty(I)$ is not dense in $W^{1,p}(I)$ unless $I = \mathbb{R}$.*

Theorem 5.21 *(Embedding Theorem) The injection $W^{1,p}(I) \hookrightarrow L^\infty(I)$ is continuous for $1 \le p \le \infty$. That is, there is a constant $c > 0$, independent of function u in $W^{1,p}(I)$ such that*

$$\|u\|_{L^\infty(I)} \le c\|u\|(I) \quad \text{for } u \in W^{1,p}(I).$$

Moreover, if I is bounded, then:

(i) *the injection $W^{1,p}(I) \hookrightarrow C(\bar{I})$ is compact for $1 < p \le \infty$;*

(ii) *the injection $W^{1,p}(I) \hookrightarrow L^q(I)$ is compact for $1 \le q < \infty$.*

Proof. Clearly, $\|u\|_{L^\infty(I)} \le \|u\|_{W^{1,\infty}(I)}$. Assume $I = \mathbb{R}$, $v \in C_c^1(\mathbb{R})$. Set $G(s) = |s|^{p-1} s$. Then $G \in C^1(\mathbb{R})$, $G'(s) = p|s|^{p-1}$ for $s \in \mathbb{R}$. By the Chain Rule 5.25 below, we have

$$w = G(v) \in C_c^1(\mathbb{R}),$$
$$w' = G'(v)v' = p|v|^{p-1} v'.$$

For $x \in \mathbb{R}$,

$$G(v(x)) = \int_{-\infty}^{x} p|v(t)|^{p-1} v'(t)\, dt.$$

By the Hölder inequality, for $p' = \frac{p}{p-1}$

$$|v(x)|^p = |G(v(x))| = \left| p \int_{-\infty}^{x} |v(x)|^{p-1} v'(t)dt \right|$$
$$\le p \left(\int_{-\infty}^{\infty} |v(t)|^{p'(p-1)} \right)^{1/p'} \left(\int_{-\infty}^{\infty} |v'(t)|^p \right)^{1/p}$$
$$\le p\|v\|_{L^p}^{p-1} \|[4pt]v'\|_{L^p}.$$

Therefore,

$$|v(x)| \le p^{1/p}\|v\|_{L^p}^{(p-1)/p} \|v'\|_{L^p}^{1/p} \quad = p^{1/p}\|v\|_{L^p}^{1/p'} \|v'\|_{L^p}^{1/p}$$
$$\le c\|v\|_{W^{1,p}}^{1/p'} \|v'\|_{W^{1,p}}^{1/p} \quad = c\|v\|_{W^{1,p}} \quad \text{for } x \in \mathbb{R}.$$

We have $\|v\|_{L^\infty(\mathbb{R})} \le c\|v\|_{W^{1,p}(R)}$ for $v \in W^{1,p}(\mathbb{R})$. For $u \in W^{1,p}(\mathbb{R})$, there is $(u_n) \in C_c^1(\mathbb{R})$ such that $u_n \to u$ in $W^{1,p}(\mathbb{R})$, and a.e. in $\mathbb{R}$. Because

$$|u_n(x) - u_m(x)| \le c\|u_n - u_m\|_{W^{1,p}} \quad \text{for } x \in \mathbb{R},$$

(u_n) is a Cauchy sequence in L^∞. Thus,

$$u_n \to u \text{ in } L^\infty$$

or

$$|u(x)| \le c \|u\|_{W^{1,p}} \quad \text{for } x \in \mathbb{R}.$$

For general I, $u \in W^{1,p}(I)$. If we extend u to Pu in $W^{1,p}(\mathbb{R})$, then

$$\|u\|_{L^\infty(I)} \le \|Pu\|_{L^\infty(\mathbb{R})} \le c\|Pu\|_{W^{1,p}(\mathbb{R})} \le c\|u\|_{W^{1,p}(I)}.$$

(i) Let F be the unit ball in $W^{1,p}(I)$, which is contained in $C(\bar{I})$, $1 < p \le \infty$. For $u \in F$

$$|u(x) - u(y)| = \left| \int_y^x u'(t)\,dt \right| \le \|u'\|_{L^p} |x - y|^{1/p'} \le |x - y|^{1/p'} \quad \text{for } x, y \in \bar{I}.$$

Therefore F satisfies the uniform equicontinuity. Apply the Ascoli Theorem 1.31 to show that F is relatively compact in $C(\bar{I})$.

(ii) By (i), it suffices to consider the case $W^{1,1}(I)$. Let F be the unit ball of $W^{1,1}(I)$. For $\omega \subset\subset I$, $u \in F$, $|h| < \text{dist}(\omega, I^c)$. By Remark 5.14,

$$\|\tau_h u - u\|_{L^1(\omega)} \le \|u'\|_{L^1(I)} |h| \le |h|.$$

Hence,

$$\int_w |u(x + h) - u(x)|^q \, dx = \int_w |u(x + h) - u(x)|^{q-1} |u(x + h) - u| \, dx$$

$$\le \left(2 \|u\|_{L^\infty(I)} \right)^{q-1} \int_w |u(x + h) - u(x)| \, dx$$

$$\le c|h|,$$

because

$$\|u\|_{L^\infty(I)} \le c \|u\|_{W^{1,1}(I)} \le c.$$

Now,

$$\left(\int_\omega |u(x + h) - u(x)|^q \, dx \right)^{1/q} \le c^{1/q} |h|^{1/q} < \epsilon,$$

if we take δ with $c^{1/q} |\delta|^{1/q} < \epsilon$ and $|h| < \delta$. For $u \in F$

$$\|u\|_{L^q(I\backslash\omega)} \le \|u\|_{L^\infty(I)} |I\backslash\omega|^{1/q} \le c|I\backslash\omega|^{1/q} < \epsilon,$$

if $|I\backslash\omega|$ is small enough. By the Fréchet-Kolmogorov Theorem 1.37, F is relatively compact in $L^q(I)$, $1 \le q < \infty$. ∎

Remark 5.22

(i) If I is unbounded, the injection $W^{1,1}(I) \hookrightarrow C(\bar{I})$ is continuous but not compact. However, if (u_n) is bounded in $W^{1,1}(I)$, there is a subsequence (u_{n_k}) of (u_n) such that $u_{n_k}(x)$ converges for all $x \in I$ (Helly Theorem);

(ii) If I is unbounded and $1 < p \le \infty$, the injection $W^{1,p}(I) \hookrightarrow L^\infty(I)$ is continuous but not compact. For any bounded sequence (u_n) in $W^{1,p}(I)$, $1 < p \le \infty$ there is a subsequence (u_{n_k}) of (u_n) and $u \in W^{1,p}(I)$ such that $u_{n_k} \to u$ in $L^\infty(J)$ for each bounded $J \subset I$;

(iii) If I is bounded and $1 \le q \le \infty$,

$$\|u\|_{L^q} + \|u'\|_{L^p} \approx \|u\|_{L^p} + \|u'\|_{L^p};$$

(iv) If I is unbounded, $u \in W^{1,p}(I)$, then $u \in L^q(I)$ for $q \in [p, \infty]$ because

$$\int_I |u|^q \le \|u\|_{L^\infty(I)}^{q-p} \|u\|_{L^p(I)}^p,$$

but in general $u \notin L^q(I)$ for $q \in [1, p)$.

Theorem 5.23 *Suppose I is unbounded, $u \in W^{1,p}(I)$, $1 \le p < \infty$. Then*

$$\lim_{x \in I, |x| \to \infty} u(x) = 0.$$

Proof.   By the First Density Theorem 5.19, for $u \in W^{1,p}(I)$, there is a sequence (u_n) in $C_c^1(\mathbb{R})$ such that $u_n |_I \to u$ in $W^{1,p}(I)$. By Theorem 5.21 (i)

$$\|u_n - u\|_{L^\infty(I)} \le c\,\|u_n - u\|_{W^{1,p}(I)} \to 0 \text{ as } n \to \infty.$$

For $\epsilon > 0$, there is $N > 0$ such that

$$\|u_N - u\|_{L^\infty(I)} < \epsilon.$$

There is $M > 0$ such that $u_N(x) = 0$ for $|x| \ge M$, so $|u(x)| < \epsilon$ for $|x| \ge M$.   ∎

Theorem 5.24 *(Product Rule) For $u, v \in W^{1,p}(I)$, $1 \le p \le \infty$. Then $uv \in W^{1,p}(I)$ and $(uv)' = u'v + uv'$. Moreover, integration by parts holds:*

$$\int_y^x u'v = u(x)v(x) - u(y)v(y) - \int_y^x uv' \quad \text{for } x, y \in \bar{I}.$$

Proof. By Theorem 5.21 (i), $u \in L^\infty(I)$, so $u\,v \in L^p(I)$. In case $1 \le p < \infty$, take (u_n), (v_n) in $C_c^1(\mathbb{R})$ such that

$$u_n|_I \to u \text{ in } W^{1,p}(I),$$
$$v_n|_I \to v \text{ in } W^{1,p}(I).$$

By Theorem 5.21 (i) $u_n \to u$, $v_n \to v$ in $L^\infty(I)$, and in $L^p(I)$. Hence,

$$u_n\,v_n \to u\,v \text{ in } L^\infty(I) \text{ and in } L^p(I).$$

Because

$$\|u_n\,v_n - u\,v\|_{L^p} \le \|u_n\|_{L^\infty}\|v_n - v\|_{L^p} + \|u_n - u\|_{L^p}\|v\|_{L^\infty} \to 0, \quad 1 \le p < \infty.$$

Similarly, we have

$$(u_n\,v_n)' = u_n'\,v_n + u_n\,v_n' \to u'\,v + u\,v' \text{ in } L^p(I).$$

For $\varphi \in C_c^\infty(I)$,

$$\int (u_n\,v_n)'\,\varphi = -\int (u_n\,v_n)\varphi' \to -\int (u\,v)\varphi',$$
$$\int (u_n\,v_n)'\,\varphi \to \int (u'\,v + u\,v')\varphi.$$

Thus, $(u\,v)' = u'\,v + u\,v' \in L^p$, and consequently, $u\,v \in W^{1,p}(I)$. Therefore $u\,v \in W^{1,p}(I)$, and $(u\,v)' = u'\,v + u\,v'$. Now $u'\,v = (u\,v)' - u\,v'$, or

$$\int_y^x u'\,v = u(x)v(x) - u(y)v(y) - \int_y^x u\,v' \quad \text{for } x,\,y \in I.$$

Finally, if we suppose $u, v \in W^{1,\infty}(I)$, then $u\,v \in L^\infty(I)$, $u'\,v + u\,v' \in L^\infty(I)$. We require

$$\int_I u\,v\,\varphi' = -\int_I (u'\,v + u\,v')\varphi.$$

For $\varphi \in C_c^1(I)$, fix a bounded interval $J \subset I$, $\operatorname{supp}\varphi \subset J$. Then $u, v \in W^{1,p}(J)$ for $p < \infty$, and by the previous part

$$\int_J u\,v\,\varphi' = -\int_J (u'\,v + u\,v')\varphi$$

or

$$\int_I u\,v\,\varphi' = -\int_I (u'\,v + u\,v')\varphi. \qquad \blacksquare$$

Theorem 5.25 (*Chain Rule*) *For* $1 \le p \le \infty$, $u \in W^{1,p}(I)$. *If* $G \in C^1(\mathbb{R})$, $G(0) = 0$, *then* $G \circ u \in W^{1,p}(I)$, *and* $(G \circ u)' = (G' \circ u)u'$.

Proof. Set $M = \|u\|_{L^\infty}$. Let c_1, c_2 be such that $|G'(y)| \le c_1$ for $|y| \le M$. Assume $\|u\|_{L^\infty(I)} \le c \, \|u\|_{W^{1,p}(I)} = c_2$. Then

$$|G(s)| = |G(s) - G(0)| = |G'(\xi)s| \le c_1 \, |s| \quad \text{for } s \in [-M, \, M].$$

That is,

$$|G \circ u| \le c_1 \, |u| \, .$$

Thus $G \circ u \in L^p(I)$. Now G' is continuous on $[-M, \, M]$, and $|G' \circ u| \le c_1$. We have

$$|(G' \circ u)u'| \le c_1 \, |u'| \, . \tag{5.3}$$

Therefore $(G' \circ u)u' \in L^p(I)$. We require

$$\int_I (G \circ u)\varphi' = - \int_I (G' \circ u)u' \, \varphi \quad \text{for } \varphi \in C_c^1(I). \tag{5.4}$$

If $1 \le p < \infty$, take a sequence (u_n) in $C_c^\infty(\mathbb{R})$ such that $u_n|_I \to u$ in $W^{1,p}(I)$, and in $L^\infty(I)$. Moreover, by Lemma 1.38 we may assume that $u_n \to u$ a.e. in I, and $u_n' \to u'$ a.e. in I, $|u_n| \le g$, $|u_n'| \le h$ a.e. in I for some g, h in $L^p(I)$. By the Lebesgue Dominated Convergence Theorem, (5.3), (5.4) and $\|u_n\|_{L^\infty} \le c$, we have

$$G \circ u_n \to G \circ u, \ \text{and} \ (G' \circ u_n)u_n' \to (G' \circ u)u' \ \text{in } L^p(I).$$

Now,

$$\int (G \circ u_n)\varphi' = - \int (G' \circ u_n)u_n' \, \varphi \quad \text{for } \varphi \in C_c^1(I).$$

Letting $n \to \infty$ we obtain

$$\int (G \circ u)\varphi' = - \int (G' \circ u)u' \, \varphi \quad \text{for } \varphi \in C_c^1(I).$$

For $p = \infty$, the proof is similar to that of Theorem 5.23.   ∎

5.3   Sobolev Spaces $W^{m,p}(I)$

For any positive integer $m \ge 2$, real p, $1 \le p \le \infty$, define

$$W^{m,p}(I) = \left\{ u \in L^p(I) \, \Big| \, u', u'', \cdots, u^{(m)} \in L^p(I) \right\}$$

$$= \left\{ u \in W^{m-1,p}(I) \, \Big| \, u' \in W^{m-1,p}(I) \right\}.$$

Set $H^m(I) = W^{m,2}(I)$. For $u \in W^{m,p}(I)$, define

$$\|u\|_{W^{m,p}} = \sum_{k=0}^{m} \left\| u_{L^p}^{(k)} \right\|.$$

For $u, v \in H^m(I)$, define

$$(u,v)_{H^m} = \sum_{k=0}^{m} \left(u^{(k)}, v^{(k)} \right)_{L^2}$$

From the proof of Theorem 5.7, we can prove the following result.

Theorem 5.26 *For $m \geq 1$ and $1 \leq p \leq \infty$, $(W^{m,p}(I), \|\cdot\|_{W^{m,p}(I)})$ forms a Banach space, and $(H^m(I), \|\cdot\|_{H^m(I)})$ forms a Hilbert space.*

The property

$$\|u\|_{W^{m,p}} = \sum_{k=0}^{m} \left\| u^{(k)} \right\|_{L^p} \approx \|u\|_{L^p} + \left\| u^{(m)} \right\|_{L^p}.$$

follows from:

Theorem 5.27 *For a positive integer j, $1 \leq j \leq m-1$, and real p, $1 \leq p < \infty$, there is a constant $\epsilon_0 = \epsilon_0(I, p, m)$, and for every $\epsilon_0 \geq \epsilon > 0$ there is a constant $c = c(I, p, m, \epsilon)$ such that*

$$\left\| u^{(j)} \right\|_{L^p} \leq \epsilon \left\| u^{(m)} \right\|_{L^p} + c\|u\|_{L^p} \quad \text{for } u \in W^{m,p}(I).$$

Proof. It suffices to prove the inequality

$$\int_I \left| u^{(j)} \right|^p \leq \epsilon \int_I \left| u^{(m)} \right|^p + \frac{c}{\epsilon^{j/(m-j)}} \int_I |u|^p \quad j < m, \tag{5.5}$$

where $0 < \epsilon \leq \epsilon_0$.

(i) $j = 1$, $m = 2$, $|I| < \infty$, $u \in C^m(\bar{I})$.

Method 1. Divide I into subintervals I_k with $\frac{\epsilon^{1/p}}{2} \leq |I_k| \leq \epsilon^{1/p}$. For each such subinterval $I_k = (a, b)$, set $\alpha = \frac{b-a}{4}$, and let $x_1 \in (a, a+\alpha)$, $x_2 \in (a+3\alpha, b)$. By the Mean Value Theorem,

$$\frac{u(x_2) - u(x_1)}{x_2 - x_1} = u'(x_{12}) \quad x_1 < x_{12} < x_2.$$

Write, for any $x \in (a, b)$

$$u'(x) = u'(x_{12}) + \int_{x_{12}}^{x} u''(t)\, dt.$$

We then obtain

$$|u'(x)| \leq |u'(x_{12})| + \int_{a}^{b} |u''(t)|\, dt$$

$$\leq \frac{|u(x_1)| + |u(x_2)|}{2\alpha} + \int_{a}^{b} |u''(t)|\, dt,$$

or

$$\int_{a}^{a+\alpha} |u'(x)|\, dx_1 \leq \frac{\displaystyle\int_{a}^{a+\alpha} |u(x_1)|\, dx_1 + \int_{a}^{a+\alpha} |u(x_2)|\, dx_1}{2\alpha} + \int_{a}^{a+\alpha} dx_1 \int_{a}^{b} |u''(t)|\, dt$$

$$\alpha\, |u'(x)| \leq \frac{\displaystyle\int_{a}^{a+\alpha} |u(x)|\, dx + \alpha\, |u(x_2)|}{2\alpha} + \alpha \int_{a}^{b} |u''(t)|\, dt.$$

Similarly, integrating with respect to x_2 on $(a + 3\alpha, b)$, we obtain

$$\alpha^2\, |u'(x)| \leq \frac{1}{2} \int_{a}^{b} |u(x)|\, dx + \alpha^2 \int_{a}^{b} |u''(t)|\, dt.$$

By the Hölder inequality, and $b - a = 4\alpha$,

$$\alpha^2\, |u'(x)| \leq \frac{1}{2}(b-a)^{1/p'} \left(\int_{a}^{b} |u|^p \right)^{1/p} + \alpha^2 (b-a)^{1/p'} \left(\int_{a}^{b} |u''|^p \right)^{1/p}$$

$$\leq c\alpha^{1/p'} \left(\int_{a}^{b} |u|^p \right)^{1/p} + c\alpha^2 \alpha^{1/p'} \left(\int_{a}^{b} |u''|^p \right)^{1/p},$$

$$|u'(x)| \leq c\alpha^{-2} \alpha^{1/p'} \left(\int_{a}^{b} |u|^p \right)^{1/p} + c\alpha^{1/p'} \left(\int_{a}^{b} |u''|^p \right)^{1/p},$$

or

$$|u'(x)|^p \leq c\alpha^{[(1/p')-2]p} \left(\int_{a}^{b} |u|^p \right) + c\alpha^{p/p'} \left(\int_{a}^{b} |u''|^p \right)$$

$$= c\alpha^{-(p+1)} \left(\int_{a}^{b} |u|^p \right) + c\alpha^{p-1} \left(\int_{a}^{b} |u''|^p \right).$$

Therefore,

$$\int_a^b |u'(x)|^p \, dx \leq \frac{c}{\alpha^p} \int_a^b |u|^p + c\alpha^p \int_a^b |u''|^p$$

$$\leq \frac{c}{\epsilon} \int_a^b |u|^p + c\epsilon \int_a^b |u''|^p$$

$$\leq \epsilon \int_a^b |u''|^p + \frac{c}{\epsilon} \int_a^b |u|^p \, ,$$

where $\frac{\epsilon^{1/p}}{2} \leq 4\alpha \leq \epsilon^{1/p}$. Summing over all the subintervals of I, we obtain

$$\int_I |u'|^p \leq \epsilon \int_I |u''|^p + \frac{c}{\epsilon} \int_I |u|^p \, .$$

Method 2. Let $I = [0,1]$. By Taylor's Theorem,

$$u(x \pm \frac{1}{2}) = u(x) \pm \frac{1}{2} \, u'(x) + \int_x^{x \pm 1/2} u''(t)(x \pm \frac{1}{2} - t) \, dt.$$

If $x \in (0, \frac{1}{2})$, then

$$|u'(x)|^p \leq c(|u(x)|^p + \left| u(x + \frac{1}{2}) \right|^p + \int_0^1 |u''(t)|^p \, dt),$$

if $x \in (\frac{1}{2}, 1)$, then

$$|u'(x)|^p \leq c(|u(x)|^p + \left| u(x + \frac{1}{2}) \right|^p + \int_0^1 |u''(t)|^p \, dt),$$

and we obtain

$$\int_0^1 |u'(x)|^p \, dx \leq c \left(\int_0^1 |u(x)|^p \, dx + \int_0^1 |u''(t)|^p \, dt \right).$$

If we let $I = [a,b]$, and $v(x) = u((b-a)x + a)$, then $v \in C^2[0,1]$, and we have

$$\int_0^1 |v(x)|^p \, dx \leq c \left(\int_0^1 |v(x)|^p \, dx + \int_0^1 |v''(x)|^p \, dt \right),$$

so

$$\int_0^1 |u'|^p \, |b-a|^{p-1} \, dx \leq c \left(\int_0^1 |u|^p \, |b-a|^{-1} \, dx + \int_0^1 |u''(t)|^p \, |b-a|^{2p-1} \, dt \right),$$

or

$$\int_0^1 |u'(x)|^p \, dx \le c \left((b-a) \int_0^1 |u(x)|^p \, dx + (b-a)^p \int_0^1 |u''(t)|^p \, dt \right).$$

Divide $[a, b]$ into n equal subintervals and then take the summation

$$\int_0^1 |u'(x)|^p \, dx \le c \left(\left(\frac{b-a}{n} \right)^{-p} \int_0^1 |u(x)|^p \, dx + \left(\frac{b-a}{n} \right)^p \int_0^1 |u''(t)|^p \, dt \right).$$

If we let $\epsilon/2 \le (b-a)/n \le \epsilon$, then

$$\int_0^1 |u'(x)|^p \, dx \le c \left(\epsilon^{-p} \int_0^1 |u(x)|^p \, dx + \epsilon^p \int_0^1 |u''(t)|^p \, dt \right).$$

(ii) For general m, $|I| < \infty$, $u \in C^m(\bar{I})$. Suppose (5.5) holds for all $j \le k < m$, $i < j$,

$$\int_I \left| u^{(i)} \right|^p \le \epsilon \int_I \left| u^{(j)} \right|^p + \frac{c}{\epsilon^{i/(j-i)}} \int_I |u|^p.$$

By (i),

$$\int_I | u^{(k)} |^p \le \frac{\epsilon}{2} \int_I | u^{(k+1)} |^p + \frac{c}{\epsilon} \int_I | u^{(k-1)} |^p,$$

and by induction,

$$\int_I | u^{(k-1)} |^p \le \delta \int_I | u^{(k)} |^p + \frac{c}{\delta^{k-1}} \int_I | u |^p.$$

Taking $\delta = \frac{\epsilon}{2\delta}$, we obtain

$$\int_I | u^{(k)} |^p \le \epsilon \int_I | u^{(k+1)} |^p + \frac{c}{\epsilon^k} \int_I | u |^p.$$

We obtain the result with $i = k$, $m = k+1$. If $i < k$, by the inductive assumption,

$$\int_I | u^{(i)} |^p \le \delta \int_I | u^{(k)} |^p + \frac{c}{\delta^{i/(k-1)}} \int_I | u |^p.$$

By the previous part,

$$\int_I | u^{(k)} |^p \le a \int_I | u^{(k+1)} |^p + \frac{c}{a^k} \int_I | u |^p.$$

Take $\delta = \epsilon^{(k-i)/(k+1-i)}$ and $a = \epsilon^{1/(k+1-i)}$, and we obtain (5.5) for $j = k+1$.

(iii) For general I, $u \in W^{m,p}(I)$. From the First Density Theorem 5.19, for $\epsilon > 0$, there is $f \in C_c^\infty(\mathbb{R})$ such that

$$\|u^{(i)} - f^{(i)}\|_{L^p(I)} < \epsilon \|u^{(i)}\|_{L^p(I)} \quad \text{for } i = 1, \cdots, m.$$

By (i), (ii),

$$\|f^{(j)}\|_{L^p} \leq \epsilon \|f^{(m)}\|_{L^p} + c\|f\|_{L^p} \quad \text{for } j = 0, 1, \cdots, m-1.$$

Now,

$$\begin{aligned}
\|u^{(j)}\|_{L^p} &\leq \|u^{(j)} - f^{(j)}\|_{L^p} + \|f^{(j)}\|_{L^p} \\
&< \epsilon \|u^{(j)}\|_{L^p} + \epsilon \|f^{(m)}\|_{L^p} + c\|f\|_{L^p} \\
&< \epsilon \|u^{(j)}\|_{L^p} + \epsilon[\|f^{(m)} - u^{(m)}\|_{L^p} + \|u^{(m)}\|_{L^p}] + c[\|f - u\|_{L^p} + \|u\|_{L^p}].
\end{aligned}$$

Hence, for suitable ϵ we have

$$\|u^{(j)}\|_{L^p} \leq \epsilon \|u^{(m)}\|_{L^p} + c\|u\|_{L^p}. \qquad \blacksquare$$

Similarly to the case $m = 1$ of Theorem 5.19, we can prove

Theorem 5.28 *For $m \geq 1$, $1 < p \leq \infty$, we have*

$$W^{m,p}(I) \hookrightarrow C^{m-1}(\bar{I}) \quad \text{for } 1 < p \leq \infty.$$

5.4 The Space $W_0^{1,p}(I)$

For $1 \leq p < \infty$, define

$$W_0^{1,p}(I) = \overline{C_c^1(I)}^{W^{1,p}(I)}.$$

Denote

$$H_0^1(I) = W_0^{1,2}(I).$$

Theorem 5.29 *The space $(W_0^{1,p}(I), \|\cdot\|_{W^{1,p}})$ is a separable Banach space, and is reflexive for $1 < p < \infty$. The space $(H_0^1(I), (\,,\,)_{H^1})$ is a separable Hilbert space.*

Remark 5.30 *It is known that $C_c^1(\mathbb{R})$ is dense in $W^{1,p}(\mathbb{R})$. Therefore*

$$W_0^{1,p}(\mathbb{R}) = W^{1,p}(\mathbb{R}).$$

Remark 5.31 *If we apply a regularity sequence $\{\rho_n\}$ to function u in $C_c^1(I)$, we find that:*

(i) $C_c^\infty(I)$ is dense in $W_0^{1,p}(I)$;

(ii) If $u \in W^{1,p}(I) \cap C_c(I)$, then $u \in W_0^{1,p}(I)$.

Proof.

(ii) Extend u to Pu in $W^{1,p}(\mathbb{R})$, and take a sequence (u_n) in $C_c^\infty(\mathbb{R})$ such that

$$u_n \to Pu \text{ in } W^{1,p}(\mathbb{R}).$$

Because $u \in C_c(I)$, take a function $\zeta \in C_c^\infty(I)$ such that

$$\zeta = 1 \text{ on } \operatorname{supp} u.$$

Then,

$$\zeta u_n \to \zeta(Pu) = u.$$

Note that $\zeta u_n \in C_c^\infty(I)$ implies that $u \in W_0^{1,p}(I)$. ∎

Theorem 5.32 *Assume I is bounded and $u \in W^{1,p}(I)$. Then $u \in W_0^{1,p}(I)$ if and only if $u = 0$ on ∂I.*

Proof.   For $u \in W_0^{1,p}(I)$, take a sequence (u_n) in $C_c^1(I)$ such that $u_n \to u$ in $W^{1,p}(I)$. Hence, $u_n \to u$ uniformly in $\bar{I}$, so $u = 0$ on ∂I.
Conversely, let $u \in W^{1,p}(I)$ such that $u = 0$ on ∂I. Fix $G \in C^1(\mathbb{R})$, $|G(t)| \le |t|$ for $t \in \mathbb{R}$, and

$$G(t) = \begin{cases} 0 & \text{for } |t| \le 1, \\ t & \text{for } |t| \ge 2. \end{cases}$$

Set $u_n = \frac{1}{n}G(nu)$, by Theorem 5.25, $u_n \in W^{1,p}(I)$. Moreover,

$$\operatorname{supp} u_n \subset \left\{ x \in I \,\middle|\, |u(x)| \ge \frac{1}{n} \right\}.$$

Because $u = 0$ on ∂I and $u(x) \to 0$ as $x \in I$, $|x| \to \infty$, we have $\operatorname{supp} u_n \subset\subset I$ or $u_n \in W_0^{1,p}$. We claim that

$$u_n \to u \text{ in } W^{1,p}.$$

In fact,

$$|u_n| = \left| \frac{1}{n}G(nu) \right| \le \left| \frac{1}{n}nu \right| \le |u| \in L^p.$$

Moreover,

$$u_n = \frac{1}{n}G(nu) = \begin{cases} 0 & |u(x)| \le \frac{1}{n}, \\ u(x) & |u(x)| \ge \frac{2}{n}. \end{cases}$$

Therefore for any $u(x) \ne 0$ and large n, we have $u_n(x) = u(x)$. Thus,

$$u_n \to u \quad \text{a.e. in } L^P,$$

Furthermore,

$$u_n' = G'(nu)u'.$$

Similarly,

$$\mid u_n' \mid \le c \mid u' \mid \in L^p,$$
$$u_n' = G'(nu)u' \to u' \quad \text{a.e. in } L^p.$$

By the Lebesgue Dominated Convergence Theorem,

$$u_n \to u \text{ in } W^{1,p}. \qquad \blacksquare$$

Theorem 5.33

(i) *For $1 < p < \infty$, $u \in L^p(I)$. Then $u \in W_0^{1,p}(I)$ if and only if there is a constant $c > 0$ such that*

$$\mid \int_I u\,\varphi' \mid \le c\,\|\varphi\|_{L^{p'}(I)} \quad \text{for } \varphi \in C_c^1(\mathbb{R});$$

(ii) *For $1 \le p < \infty$, $u \in L^p(I)$, define*

$$\bar{u}(x) = \begin{cases} u(x) & \text{for } x \in I, \\ 0 & \text{for } x \in \mathbb{R}\backslash I. \end{cases}$$

Hence, $u \in W_0^{1,p}(I)$ if and only if $\bar{u} \in W^{1,p}(\mathbb{R})$.

Proof. See Theorem 6.35. $\blacksquare$

Theorem 5.34 (*Poincaré Inequality*) *Suppose I is bounded and $1 \le p < \infty$. Then there is a constant c (depending on $|I|$) such that*

$$\|u\|_{L^p} \le c\|u'\|_{L^p} \quad \text{for } u \in W_0^{1,p}(I).$$

In other words, in $W_0^{1,p}(I)$

$$\|u\|_{W^{1,p}} \approx \|u'\|_{L^p}.$$

Proof. Method 1. For $u \in W_0^{1,p}(I)$, $I = [a, b]$, then $u \in C(\bar{I})$, and

$$\mid u(x) \mid = \mid u(x) - u(a) \mid = \mid \int_a^x u'(t)dt \mid \le \int_a^b \mid u'(t) \mid dt$$

$$\le (b-a)^{1/p'} \left(\int_a^b \mid u'(t) \mid^p \right)^{1/p} c\|u'\|_{L^p} \quad \text{for } x \in I.$$

Thus,

$$\mid u \mid_{L^p} \le c\|u\|_{L^\infty} \le c\|u'\|_{L^p}$$

or

$$\|u'\|_{L^p} \le \|u\|_{W^{1,p}} \le c\|u'\|_{L^p}.$$

Method 2. For $u \in C_0^1[a,b]$, if we assume $u = 0$ is outside $[a,b]$, then

$$\|u\|_{L^p}^p = \int_a^b | u(x) |^p \, dx = | \int_{-\infty}^\infty | u(x) |^p \, d(x - x_0) |$$

$$= | \, p \int_{-\infty}^\infty | u(x) |^{p-2} \, u(x)u'(x)(x - x_0) \, dx \, |$$

$$\le p\|u\|_{L^p}^{p-1} \, | \, b - a \, | \, \|u'\|_{L^p}.$$

Therefore,

$$\|u\|_{L^p} \le p \, | \, b - a \, | \, \|u'\|_{L^p}. \qquad \blacksquare$$

Theorem 5.35 (*Poincaré-Wirtinger Inequality*) *Suppose I is bounded and $1 \le p < \infty$, then there is a constant c (depending on $|I|$) such that*

$$\|u - u_I\|_{L^p(I)} \le c\|u'\|_{L^p(I)} \quad \text{for } u \in W^{1,p}(I),$$

where $u_I = \frac{1}{b-a} \int_a^b u(x) \, dx$. In particular, if $u_I = 0$, then $\|u\|_{L^p(I)} \le c\|u'\|_{L^p(I)}$.

Proof. For $x, y \in [a,b]$, $u \in C^1(\mathbb{R})$

$$u(x) - u(y) = \int_y^x u'(t) \, dt.$$

Integrate $\frac{1}{b-a} \int_a^b dy$ on both sides to obtain

$$u(x) - u_I = \frac{1}{b-a} \int_a^b dy \int_y^x u'(t) \, dt$$

$$= \frac{1}{b-a} [\int_a^x dy \int_y^x u'(t) \, dt - \int_x^b dy \int_x^y u'(t) \, dt]$$

$$= \frac{1}{b-a} [\int_a^x dt \int_a^t u'(t) \, dy - \int_x^b dt \int_t^b u'(t) \, dy]$$

$$= \frac{1}{b-a} [\int_a^x (a - t)u'(t) \, dt + \int_x^b (b - t)u'(t) \, dt],$$

$$| u(x) - u_I | \le 2 \int_a^b | u'(t) | \, dt \le 2(b-a)^{1/p'} \left(\int_a^b | u'(t) |^p \, dt \right)^{1/p} \le c\|u'\|_{L^p}.$$

Now

$$\|u(x) - u_I\|_{L^p(I)} \le c\|u(x) - u_I\|_{L^\infty} \le c\|u'\|_{L^p}.$$

For general $u \in W^{1,p}(I)$, use the First Density Theorem 5.19 to obtain the result. $\blacksquare$

Remark 5.36 *For a positive integer $m \geq 2$, real p, $1 \leq p < \infty$, define the space $W_0^{m,p}(I)$ by*

$$W_0^{m,p}(I) = \overline{C_c^m(I)}^{W^{m,p}(I)}.$$

We have

$$W_0^{m,p}(I) = \{u \in W^{m,p}(I) \mid u = u' = \cdots = u^{(m-1)} = 0 \text{ on } \partial I\}.$$

It is better to distinguish the following:

$$W_0^{2,p}(I) = \{u \in W^{2,p}(I) \mid u = u' = 0 \text{ on } \partial I\},$$

$$W^{2,p}(I) \cap W_0^{1,p}(I) = \{u \in W^{2,p}(I) \mid u = 0 \text{ on } \partial I\}.$$

5.5   The Dual Space of $W_0^{1,p}(I)$

Notation: denote the dual space of $W_0^{1,p}(I)$ by the space $W^{-1,p'}(I)$, $1 \leq p < \infty$, $\frac{1}{p} + \frac{1}{p'} = 1$, and define $H^{-1}(I)$ as the dual of $H_0^1(I)$. We ask in the following Identify or not identify? Let H be a Hilbert space. By the Riesz Representation Theorem, H can be identified with its dual H'. However, sometimes we do not want to identify H with H'. Let V be a dense vector subspace of H such that $(V, \|\cdot\|_V)$ is a reflexive Banach space. Suppose the injection $(V, \|\cdot\|_V) \hookrightarrow (H, \|\cdot\|_H)$ is continuous, that is $\|u\|_H \leq c \|u\|_V$ for $u \in V$. We identify H with H'. We can extend H to V' as follows. For $f \in H \cong H'$,

$$v \in V \to (f, v)$$

is a continuous linear functional on V, and can be considered as a continuous linear functional on H. This is denoted by $Tf \in V'$, where

$$(Tf, v)_{(V', V)} = (f, v) \quad \text{for } f \in H \text{ for } v \in H.$$

$T : H \to V'$ possesses the following properties:

(i) $\|Tf\|_{V'} \leq c\|f\|_H$ for $f \in H$;

(ii) T is injective;

(iii) $T(H)$ is dense in V',

and we have

$$V \hookrightarrow_{\text{continuous,dense}} H = H' \hookrightarrow_{\text{continuous,dense}} V'.$$

After we have identified $(\varphi, v)_{(V', V)}$ with (φ, v) of $\varphi \in H$, $v \in V$, we consider H a pivot space. If V is itself a Hilbert space, we must choose one of the two identifications: the previous identification or $V = V'$. For example:

(i)

$$H = \ell^2 = \{u = (u_n) \mid \sum u_n^2 < \infty\}$$

with $(u, v) = \sum u_n \, v_n$.

$$V = \{u = (u_n) \mid \sum n^2 \, u_n^2 < \infty\}$$

with $(u, v) = \sum n^2 u_n v_n$. Then $V \hookrightarrow_{\text{continuous,dense}} H = H' \hookrightarrow_{\text{continuous,dense}} V$.

(ii) If we identity L^2 with its dual, but not H_0^1 with its dual, we have

$$H_0^1 \hookrightarrow L^2 \hookrightarrow H^{-1}$$

with continuous injection and dense. If I is bounded, we have

$$W_0^{1,p} \hookrightarrow L^2 \hookrightarrow W^{-1,p'} \quad \text{for } 1 \leq p < \infty$$

with continuous injection and dense. If I is unbounded, we have only

$$W_0^{1,p} \hookrightarrow L^2 \hookrightarrow W^{-1,p'} \quad \text{for } 1 \leq p \leq 2$$

with continuous injection and dense.

Theorem 5.37 *For $F \in W^{-1,p'}$, there are $f_0, f_1 \in L^{p'}$ such that*

$$(F, v) = \int f_0 \, v + \int f_1 \, v' \quad \text{for } v \in W_0^{1,p},$$
$$\|F\| = \text{Max}\{\|f_0\|_{L^{p'}}, \|f_1\|_{L^{p'}}\}.$$

If I is bounded we may take $f_0 = 0$.

Proof.   Consider the space $E = L^p \times L^p$ under the sum norm

$$\|h\| = \|h_0\|_{L^p} + \|h_1\|_{L^p} \quad \text{where } h = (h_0, h_1).$$

The operator $T : u \in W_0^{1,p} \to (u, u') \in E$ is an isometry of $W_0^{1,p}$ into E. Let $G = T(W_0^{1,p})$ with the norm induced from E, and $S = T^{-1} : G \to W_0^{1,p}$. The map

$$h \in G \to (F, Sh)$$

is a continuous linear functional on G. By the Hahn-Banach Theorem 1.1, it can be extended to be a continuous linear functional on E, denoted by Φ with $\|\Phi\|_{E'} = \|F\|$. By the Riesz Representation Theorem, there are $f_0, f_1 \in L^{p'}$ such that

$$(\Phi, h) = \int f_0 \, h_0 + \int f_1 \, h_1 \quad \text{for } h \in E$$

with

$$\|\Phi\|_{E'} = \mathrm{Max}\{\|f_0\|_{L^{p'}}, \|f_1\|_{L^{p'}}\}.$$

In particular,

$$(F, v) = \int f_0\, v + \int f_1\, v' \quad \text{for } v \in W_0^{1,p}.$$

If I is bounded in the space, we may replace the norm in $W_0^{1,p}$ by the norm $\|u'\|_{L^p}$. In the previous situation, we identify $E = L^p$ and $T : u \in W_0^{1,p} \to u' \in L^p$ to obtain $f_0 = 0$. ∎

Notes Most results in this Chapter are taken from Brezis [7, Chapter 8]. For related materials see Adams [1], Friedman [23], and Mazja [34].

Chapter 6

SOBOLEV SPACES $W^{1,p}(\Omega)$

In this chapter let $\Omega \subset \mathbb{R}^N$ be an open set, and p a real number, $1 \leq p \leq \infty$. We intend to study the Sobolev spaces $W^{1,p}(\Omega)$, which are useful in Analysis and Partial Differential Equations.

6.1 Elementary Properties of Sobolev Spaces $W^{1,p}(\Omega)$

Let $\Omega \subset \mathbb{R}^N$ be open, $N > 1$ an integer, p a real number, and $1 \leq p < \infty$.

Definition 6.1 *The Sobolev space $W^{1,p}(\Omega)$ is defined by*

$$W^{1,p}(\Omega) = \{u \in L^p(\Omega) \,|\, u^\alpha \in L^p(\Omega), \ |\alpha| \leq 1\}.$$

Recall that if there is $g_i \in L^p(\Omega)$ such that

$$\int_\Omega u \frac{\partial \varphi}{\partial x_i} = -\int_\Omega g_i \varphi \text{ for } \varphi \in C_c^\infty(\Omega)$$

then we say that g_i is the $\alpha_i = (0,0,\cdots,1,0,\cdots,0)$ weak derivative of u, in symbols $u^{\alpha_i} = \frac{\partial u}{\partial x_i} = g_i$.

Denote $H^1(\Omega) = W^{1,2}(\Omega)$. For $u \in W^{1,p}(\Omega)$, denote the gradient of u by

$$Du = \nabla u = (\frac{\partial u}{\partial x_1}, \cdots, \frac{\partial u}{\partial x_N}).$$

Thanks to the following lemma, g_i is unique.

Lemma 6.2 *If we let $f \in L^1_{\ell oc}(\Omega)$ with $\int_\Omega f\varphi = 0$ for $\varphi \in C_c(\Omega)$, then $f = 0$ a.e. on Ω.*

Proof.   We prove this in two steps.

(i) Suppose $f \in L^1(\Omega)$, $|\Omega| < \infty$. Given $\epsilon > 0$, there is $f_1 \in C_c(\Omega)$ such that $\|f - f_1\|_{L^1(\Omega)} < \epsilon$, then for $\varphi \in C_c(\Omega)$,

$$| \int_\Omega f_1 \varphi | \leq | \int_\Omega (f_1 - f)\varphi | + | \int_\Omega f\varphi | \leq \epsilon \|\varphi\|_{L^\infty}$$

Let $K_1 = \{x \in \Omega : f_1(x) \geq \epsilon\}$, $K_2 = \{x \in \Omega : f_1(x) \leq -\epsilon\}$. Then K_1 and K_2 are disjoint compact sets, and by the Tietze-Urysohn Extension Theorem 1.24, there is $\varphi_0 \in C_c(\Omega)$ such that $|\varphi_0(x)| \leq 1$ for $x \in \Omega$, and

$$\varphi_0(x) = \begin{cases} 1 \text{ if } x \in K_1, \\ -1 \text{ if } x \in K_2. \end{cases}$$

If we let $K = K_1 \cup K_2$, then

$$\int_\Omega f_1 \varphi_0 = \int_{\Omega \setminus K} f_1 \varphi_0 + \int_K f_1 \varphi_0,$$

$$\int_K |f_1| = \int_K f_1 \varphi_0 \leq \int_{\Omega \setminus K} |f_1 \varphi_0| + \left| \int_\Omega f_1 \varphi_0 \right| \leq \int_{\Omega \setminus K} |f_1| + \epsilon.$$

Thus,

$$\int_\Omega |f_1| = \int_K |f_1| + \int_{\Omega \setminus K} |f_1|$$

$$\leq \epsilon + 2 \int_{\Omega \setminus K} |f_1|$$

$$\leq \epsilon + 2\epsilon |\Omega| \text{ because } |f_1| \leq \epsilon \text{ on } \Omega \setminus K.$$

Hence,

$$\|f\|_{L^1(\Omega)} \leq \|f - f_1\|_{L^1(\Omega)} + \|f_1\|_{L^1(\Omega)} \leq 2\epsilon + 2\epsilon |\Omega|,$$

or $f = 0$ a.e. on Ω.

(ii) General Ω. Let $\Omega = \bigcup_{n=1}^{\infty} \Omega_n$ with Ω_n open, $\bar{\Omega}_n$ compact, $\bar{\Omega}_n \subset \Omega$. For example,

$$\Omega_n = \{x \in \Omega : \text{dist}\,(x, \Omega^c) > \frac{1}{n}, |x| < n\}$$

$f \in L^1_{\ell oc}(\Omega)$, so $f \mid_{\Omega_n} \in L^1(\Omega_n)$ for each n. By (i), $f = 0$ a.e. on Ω_n for each n. Thus $f = 0$ a.e. on Ω.   ∎

In the space $W^{1,p}(\Omega)$, the norm

$$\|u\|_{W^{1,p}} = \|u\|_{L^p} + \sum_{i=1}^{N} \left\| \frac{\partial u}{\partial x_i} \right\|_{L^p}$$

is equivalent to the norm

$$\left(\|u\|_{L^p}^p + \sum_{i=1}^{N} \left\| \frac{\partial u}{\partial x_i} \right\|_{L^p}^p \right)^{1/p}.$$

In the space $H^1(\Omega)$, the scalar product

$$(u,v)_{H^1} = (u,v)_{L^2} + \sum_{i=1}^{N} \left(\frac{\partial u}{\partial x_i}, \frac{\partial v}{\partial x_i} \right)_{L^2}$$

associates the norm

$$\|u\|_{H^1} = \left(\|u\|_{L^2}^2 + \sum_{i=1}^{N} \left\| \frac{\partial u}{\partial x_i} \right\|_{L^2}^2 \right)^{1/2}.$$

Proposition 6.3 *The space $W^{1,p}(\Omega)$ is a Banach space for $1 \leq p \leq \infty$; $W^{1,p}(\Omega)$, is reflexive for $1 < p < \infty$, and separable for $1 \leq p < \infty$. The space $H^1(\Omega)$ is a separable Hilbert space.*

Proof. The proof is similar to that for Theorem 5.7. ∎

Remark 6.4

(i) *In the definition of $W^{1,p}(\Omega)$, instead of $C_c^\infty(\Omega)$ functions, we can use $C_c^1(\Omega)$ functions as test functions through the regularity sequence.*

(ii) *If $u \in C^1(\Omega) \cap L^p(\Omega)$, and $\frac{\partial u}{\partial x_i} \in L^p(\Omega)$, $i = 1, \cdots, N$ where $\frac{\partial u}{\partial x_i}$ denotes the usual derivatives, then $u \in W^{1,p}(\Omega)$. Moreover, the usual derivatives coincide with weak derivatives. In particular, if Ω is bounded, then $C^1(\bar{\Omega}) \subset W^{1,p}(\Omega)$ for $1 \leq p \leq \infty$.*

Conversely,

Theorem 6.5 *Suppose $u \in W^{1,p}(\Omega) \cap C(\Omega)$, $1 \leq p \leq \infty$, and the weak derivatives are $\frac{\partial u}{\partial x_i} \in C(\Omega)$ for $i = 1, 2, \cdots, N$. Then $u \in C^1(\Omega)$.*

Proof. Assume $f \in C^1(\Omega)$. Fix $x \in \Omega$. Assume $0 \leq r \leq r_0$ such that $y = x + re_i \in \Omega$. By the Fundamental Theorem of Calculus,

$$f(y) - f(x) = \int_0^1 \frac{df}{dt}(x + t(y - x))dt = \int_0^1 \nabla f(x + t(y-x)) \cdot (y - x)dt$$

$$= \int_0^1 r \frac{\partial f}{\partial x_i}(x + t(y-x))dt,$$

and we have

$$\frac{f(y) - f(x)}{r} - \frac{\partial f}{\partial x_i}(x) = \int_0^1 \left[\frac{\partial f}{\partial x_i}(x + t(y - x)) - \frac{\partial f}{\partial x_i}(x)\right] dt. \tag{6.1}$$

For $u \in W^{1,p}(\Omega)$, by the Meyers-Serrin Theorem 6.10 below, there is a sequence (u_n) in $C^1(\Omega)$ such that

$$u_n \to u \text{ in } L^p(\Omega),$$

$$\frac{\partial u_n}{\partial x_i} \to \frac{\partial u}{\partial x_i} \text{in } L^p(\Omega), \text{ for } i = 1, \cdots, N.$$

We may then choose a subsequence of (u_n), still denoted by (u_n), and $h \in L^p(\Omega)$ such that

$$u_n \to u \text{ a.e. on } \Omega$$

$$\frac{\partial u_n}{\partial x_i} \to \frac{\partial u}{\partial x_i} \text{ a.e. on } \Omega \text{ for } i = 1, \cdots, N,$$

$$\left|\frac{\partial u_n}{\partial x_i}(x)\right| \leq h(x). \text{ a.e. on } \Omega,$$

By (6.1), we have

$$\frac{u_n(y) - u_n(x)}{r} - \frac{\partial u_n}{\partial x_i}(x) = \int_0^1 \left[\frac{\partial u_n}{\partial x_i}(x + t(y - x)) - \frac{\partial u_n}{\partial x_i}(x)\right] dt.$$

The Lebesgue Dominated Convergence Theorem implies

$$\frac{u(y) - u(x)}{r} - \frac{\partial u}{\partial x_i}(x) = \int_0^1 \left[\frac{\partial u}{\partial x_i}(x + t(y - x)) - \frac{\partial u}{\partial x_i}(x)\right] dt,$$

because $\frac{\partial u}{\partial x_i} \in C(\Omega)$. Letting $r \to 0$, we obtain

$$\lim_{r \to 0} \frac{u(y) - u(x)}{r} - \frac{\partial u}{\partial x_i}(x) = 0,$$

or

$$\lim_{r \to 0} \frac{u(y) - u(x)}{r} = \frac{\partial u}{\partial x_i}(x).$$

This proves that the usual derivatives and the weak derivatives are the same. Therefore $u \in C^1(\Omega)$. ∎

Remark 6.6

(i) *Assume* $(u_n) \subset W^{1,p}(\Omega)$, $1 \leq p \leq \infty$, $u \in L^p(\Omega)$ *such that* $u_n \to u$ *in* L^p, *and* ∇u_n *converges in* $(L^p)^N$. *Then* $u \in W^{1,p}(\Omega)$, *and* $\|u_n - u\|_{W^{1,p}(\Omega)} \to 0$;

(ii) *Let $(u_n) \subset W^{1,p}(\Omega)$, $1 < p \leq \infty$, $u \in L^p(\Omega)$ such that*

$$u_n \longrightarrow u \text{ in } L^p, (\nabla u_n) \text{ is bounded in } (L^p)^N.$$

Then $u \in W^{1,p}(\Omega)$, and there is a subsequence of (u_n), still denoted by (u_n) such that $\|u_n - u\|_{W^{1,p}(\Omega)} \to 0$;

(iii) *Given a function $f : \Omega \to \mathbb{R}$, define*

$$\widetilde{f}(x) = \begin{cases} f(x) \ x \in \Omega, \\ 0 \ x \in \mathbb{R}^N \backslash \Omega. \end{cases}$$

If $u \in W^{1,p}(\Omega)$, $\alpha \in C_c^1(\Omega)$, then

$$(\alpha u)^\sim \in W^{1,p}(\mathbb{R}^N),$$

$$\frac{\partial}{\partial x_i}(\alpha u)^\sim = \left(\alpha \frac{\partial u}{\partial x_i} + \frac{\partial \alpha}{\partial x_i} u \right)^\sim.$$

Proof.

(i) Set $g = (g_1, \cdots, g_N) \in (L^p(\Omega))^N$ such that $\nabla u_n \to g$. Note that

$$\int_\Omega \frac{\partial u_n}{\partial x_i} \varphi = - \int_\Omega u_n \frac{\partial \varphi}{\partial x_i} \text{ for } \varphi \in C_c^\infty(\Omega),$$

$$\lim_{n \to \infty} \int_\Omega \frac{\partial u_n}{\partial x_i} \varphi = \int_\Omega g_i \varphi,$$

$$\lim_{n \to \infty} \int_\Omega u_n \frac{\partial \varphi}{\partial x_i} = \int_\Omega u \frac{\partial \varphi}{\partial x_i}.$$

Thus,

$$\int g_i \varphi = - \int u \frac{\partial \phi}{\partial x_i},$$

or $\frac{\partial u}{\partial x_i} = g_i$ for $i = 1, \cdots, N$.

(ii) By the Banach-Alaoglu Theorem 1.5.

(iii) For $\varphi \in C_c^1(\mathbb{R}^N)$,

$$\int_{\mathbb{R}^N} (\alpha u)^\sim \frac{\partial \varphi}{\partial x_i} = \int_\Omega \alpha u \frac{\partial \varphi}{\partial x_i}$$

$$= \int_\Omega u \left[\frac{\partial}{\partial x_i}(\alpha \varphi) - \frac{\partial \alpha}{\partial x_i} \varphi \right]$$

$$= - \int_{\mathbb{R}^N} \left(\frac{\partial u}{\partial x_i}(\alpha \varphi) + u \frac{\partial \alpha}{\partial x_i} \varphi \right)$$

$$= - \int_{\mathbb{R}^N} \left(\alpha \frac{\partial u}{\partial x_i} + \frac{\partial \alpha}{\partial x_i} u \right)^\sim \varphi.$$

Therefore,

$$\frac{\partial}{\partial x_i}(\alpha u)^{\sim} = \left(\alpha \frac{\partial u}{\partial x_i} + \frac{\partial \alpha}{\partial x_i} u\right)^{\sim}.$$

Note that if supp $\alpha \subset \mathbb{R}^N \backslash \Gamma$, we may replace $\alpha \in C_c^1(\Omega)$ by $\alpha \in C^1(\mathbb{R}^N) \cap L^\infty(\mathbb{R}^N)$, $\nabla \alpha \in (L^\infty(\mathbb{R}))^N$.   ∎

Lemma 6.7 *If $\rho \in L^1(\mathbb{R}^N)$, $u \in W^{1,p}(\mathbb{R}^N)$, $1 \leq p \leq \infty$, then $\rho * u \in W^{1,p}(\mathbb{R}^N)$, and*

$$\frac{\partial}{\partial x_i}(\rho * u) = \rho * \frac{\partial u}{\partial x_i} \text{ for } i = 1, 2, \cdots, N.$$

Proof.   First, recall that
(a) If $f \in L^1(\mathbb{R}^N)$, $g \in L^p(\mathbb{R}^N)$, $h \in L^{p'}(\mathbb{R}^N)$, $\frac{1}{p} + \frac{1}{p'} = 1$, $1 \leq p, p' \leq \infty$, then

$$\int (f * g)h = \int g(\check{f} * h),$$

where $\check{f}(x) = f(-x)$;
(b) If $f \in C_c^k(\mathbb{R}^N)$, $g \in L_{loc}^1(\mathbb{R}^N)$, k is a positive integer, then $f * g \in C^k(\mathbb{R}^N)$, and $D^\alpha(f * g) = (D^\alpha f) * g$ for $|\alpha| \leq k$.

(i) Suppose ρ has compact support: for $\varphi \in C_c^1(\mathbb{R}^N)$, we have, for $i = 1, 2, \cdots, N$

$$\int (\rho * u)\frac{\partial \varphi}{\partial x_i} = \int u\left(\check{\rho} * \frac{\partial \varphi}{\partial x_i}\right) = \int u\frac{\partial}{\partial x_i}(\check{\rho} * \varphi)$$

$$= -\int \frac{\partial u}{\partial x_i}(\check{\rho} * \varphi) = -\int \left(\rho * \frac{\partial u}{\partial x_i}\right)\varphi.$$

Because $\rho * u$, and $\rho * \frac{\partial u}{\partial x_i} \in L^p(\Omega)$, we obtain $\rho * u \in W^{1,p}(\Omega)$, $\frac{\partial}{\partial x_i}(\rho * u) = \rho * \frac{\partial u}{\partial x_i}$.

(ii) For general $\rho \in L^1(\mathbb{R}^N)$, take a sequence (ρ_n) in $C_c^\infty(\mathbb{R}^N)$ such that $\rho_n \to \rho$ in L^1; we then have

$$[c]l\rho_n * u \to \rho * u \text{ in } L^p(\Omega)$$

$$\rho_n * \frac{\partial u}{\partial x_i} \to \rho * \frac{\partial u}{\partial x_i} \text{ in } L^p(\Omega) \text{ for } i = 1, 2, \cdots, N.$$

By (i) $\rho_n * u \in W^{1,p}(\Omega)$, and consequently,

$$\rho * u \in W^{1,p}(\Omega),$$

$$\frac{\partial}{\partial x_i}(\rho * u) = \rho * \left(\frac{\partial u}{\partial x_i}\right) \text{ for } i = 1, 2, \cdots, N.$$ ∎

We come to the following result. Note that $w \subset\subset \Omega$ means that w is open, $\bar{w}$ is compact, $\bar{w} \subset \Omega$.

Theorem 6.8 (*Second Density Theorem, Friedrichs Theorem*) *Assume* $u \in W^{1,p}(\Omega)$, $1 \leq p < \infty$. *There is a sequence* (u_n) *in* $C_c^\infty(\mathbb{R}^N)$ *such that*

$$u_n \mid_\Omega \to u \text{ in } L^p(\Omega),$$

$$\nabla u_n \mid_w \to \nabla u \mid_w \text{ in } L^p(w)^N,$$

for every $w \subset\subset \Omega$.

Proof. Denote

$$\widetilde{u}(x) = \begin{cases} u(x) \text{ if } x \in \Omega, \\ 0 \text{ if } x \in \mathbb{R}^N \backslash \Omega. \end{cases}$$

Let (ρ_n) be a regularity sequence: $\rho_n \in C_c^\infty(\mathbb{R}^N)$, supp $\rho_n \subset B\left(0, \frac{1}{n}\right)$, $\int_{\mathbb{R}^N} \rho_n = 1$, and $\rho_n \geq 0$ on $\mathbb{R}^N$, see Definition 1.33. Let $v_n = \rho_n * \widetilde{u}$. Then

$$\begin{cases} v_n \in C^\infty(\mathbb{R}^N) \\ v_n \to \widetilde{u} \text{ in } L^p(\mathbb{R}^N). \end{cases}$$

We require: $\nabla v_n \mid_w \longrightarrow \nabla u \mid_w$ in $L^p(w)$ for $w \subset\subset \Omega$. For fixed $w \subset\subset \Omega$, choose $\alpha \in C_c^1(\Omega)$, $0 \leq \alpha \leq 1$, $\alpha = 1$ near w. If n is large enough, then

$$\rho_n * \widetilde{\alpha} u = \rho_n * \widetilde{u} \text{ on } w. \tag{6.2}$$

In fact,

$$\text{supp}(\rho_n * \widetilde{u} - \rho_n * (\alpha u)^\sim)$$
$$= \text{supp}[\rho_n * (1 - \widetilde{\alpha})\widetilde{u}] \subset \text{supp}\rho_n + \text{sup}p(1 - \widetilde{\alpha})\widetilde{u}$$
$$\subset B\left(0, \frac{1}{n}\right) + \text{supp}(1 - \widetilde{\alpha}) \subset w^c \text{ for large } n, \text{because } \alpha = 1 \text{ near } \omega.$$

However,

$$\frac{\partial}{\partial x_i}(\rho_n * (\alpha u)^\sim) = \rho_n * \left(\alpha \frac{\partial u}{\partial x_i} + \frac{\partial \alpha}{\partial x_i} u\right)^\sim$$

$$\to \left(\alpha \frac{\partial u}{\partial x_i} + \frac{\partial \alpha}{\partial x_i} u\right)^\sim \text{ in } L^p(\mathbb{R}^N).$$

In particular,

$$\frac{\partial}{\partial x_i}(\rho_n * (\alpha u)^\sim) \to \frac{\partial u}{\partial x_i} \text{ in } L^p(w), \text{ because } \alpha = 1 \text{ on } w.$$

Thanks to (6.2), $\frac{\partial}{\partial x_i}(\rho_n * \widetilde{u}) \to \frac{\partial u}{\partial x_i}$ in $L^p(w)$. Assume $\zeta \in C_c^\infty(\mathbb{R}^N)$, $0 \leq \zeta \leq 1$ such that

$$\zeta(x) = \begin{cases} 1 \text{ if } |x| \leq 1, \\ 0 \text{ if } |x| \geq 2. \end{cases}$$

Assume $\zeta_n(x) = \zeta\left(\frac{x}{n}\right)$, $n = 1, 2, \cdots$. By the Lebesgue Dominated Convergence Theorem, for $f \in L^p(\mathbb{R}^N)$,

$$\zeta_n f \to f \text{ in } L^p(\mathbb{R}^N), 1 \leq p < \infty.$$

Let $u_n = \zeta_n v = \zeta_n(\rho_n * \widetilde{u})$. Then $u_n \in C_c^\infty(\mathbb{R}^N)$ such that

$$u_n \to u \ L^p(\Omega), \ L^p(\Omega),$$
$$\nabla u_n \to \nabla u \ L^p(w)^N.$$

In fact, because $u_n - \widetilde{u} = \zeta_n[(\rho_n * \widetilde{u}) - \widetilde{u}] + [\zeta_n\widetilde{u} - \widetilde{u}]$, we have

$$\|u_n - u\|_{L^p(\Omega)} \leq \|u_n - \widetilde{u}\|_{L^p(\mathbb{R}^N)} \leq \|\rho_n * \widetilde{u} - \widetilde{u}\|_{L^p(\mathbb{R}^N)} + \|\zeta_n\widetilde{u} - \widetilde{u}\|_{L^p(\mathbb{R}^N)} \to 0.$$

Moreover, because $\frac{\partial u_n}{\partial x_i} = \left(\frac{\partial \zeta_n}{\partial x_i}\right)(\rho_n * \widetilde{u}) + \zeta_n\left(\rho_n * \frac{\partial \widetilde{u}}{\partial x_i}\right)$, we have

$$\|\frac{\partial u_n}{\partial x_i} - \frac{\partial u}{\partial x_i}\|_{L^p(w)}$$

$$\leq \|\frac{\partial \zeta_n}{\partial x_i}(\rho_n * \widetilde{u})\|_{L^p(\mathbb{R}^N)} + \|\zeta_n\left(\rho_n\frac{\partial \widetilde{u}}{\partial x_i}\right) - \frac{\partial \widetilde{u}}{\partial x_i}\|_{L^p(w)}$$

$$\leq \frac{c}{n}\|\widetilde{u}\|_{L^p(\mathbb{R}^N)} + \|\zeta_n\left(\rho_n * \frac{\partial \widetilde{u}}{\partial x_i}\right) - \zeta_n\frac{\partial \widetilde{u}}{\partial x_i}\|_{L^p(w)}$$

$$+ \|\zeta_n\frac{\partial \widetilde{u}}{\partial x_i} - \frac{\partial \widetilde{u}}{\partial x_i}\|_{L^p(w)}$$

$$\leq \frac{c}{n}\|\widetilde{u}\|_{L^p(\mathbb{R}^N)} + \|\rho_n * \frac{\partial \widetilde{u}}{\partial x_i} - \frac{\partial \widetilde{u}}{\partial x_i}\|_{L^p(w)} + \|\zeta_n\frac{\partial \widetilde{u}}{\partial x_i} - \frac{\partial \widetilde{u}}{\partial x_i}\|_{L^p(w)}$$

$$\to 0 \text{ as } n \to \infty.$$

Therefore we have $\nabla v_n \mid_w \to \nabla u \mid_w$ in $L^p(w)$.   ■

Lemma 6.9 (*Partition of Unity Lemma*) *Let $E \subset \mathbb{R}^N$ be a set and O an open covering of E. Then there is a partition of unity subordinate to O: a collection of functions F in $C_c^\infty(\mathbb{R}^N)$ such that*

(*i*) $0 \leq \psi \leq 1$ *for each $\psi \in F$;*

(*ii*) *For any compact set $K \subset\subset E$, all but possibly finitely many $\psi \in F$ vanish identically on K;*

(*iii*) *For any $\psi \in F$, there is $U \in O$ such that supp $\psi \subset U$;*

(*iv*) *For every $x \in E$, $\sum\limits_{x \in \mathcal{F}} \psi(x) = 1$.*

We come to the following result.

Theorem 6.10 (*Third Density Theorem, Meyers-Serrin Theorem*) *For* $u \in W^{1,p}(\Omega)$, $1 \leq p < \infty$, *there is a sequence* (u_n) *in* $C^\infty(\Omega) \cap W^{1,p}(\Omega)$ *such that*

$$u_n \to u \text{ in } W^{1,p}(\Omega).$$

Proof. For $u \in W^{1,p}(\Omega)$, $\epsilon > 0$. Let

$$\Omega_k = \{x \in \Omega \mid |x| < k, \ dist(x, \partial\Omega) > \frac{1}{k}\}, k = 1, 2, \cdots.$$

$$\Omega_0 = \Omega_{-1} = \phi.$$

Let

$$\mathcal{O} = \{U_k \mid U_k = \Omega_{k+1} \cap (\bar{\Omega}_{k-1})^c, k = 1, 2, \cdots\}.$$

Then $\mathcal{O}$ is an open covering of Ω. Let F be a partition of unity for Ω subordinate to $\mathcal{O}$. Let ψ_k be the sum of the finitely many functions $\psi \in F$ with supp $\psi \subset U_k$. Then $\psi_k \in C_c^\infty(U_k)$ satisfies

$$\sum_{k=1}^\infty \psi_k(x) = 1 \text{ on } \Omega.$$

Let (ρ_{ϵ_k}) be a regularity sequence, with supp $\rho_{\epsilon_k} \subset B(0, \epsilon_k)$ where $0 < \epsilon_k < \frac{1}{(k+1)(k+2)}$, then

$$\text{supp} \rho_{\epsilon_k} * (\psi_k u) \subset \Omega_{k+2} \cap (\Omega_{k-2})^c = V_k \subset\subset \Omega,$$

and $\psi_k u \in W^{1,p}(\Omega)$. We may choose ϵ_k to be small, such that

$$\|\rho_{\epsilon_k} * (\psi_k u) - \psi_k u\|_{W^{1,p}(\Omega)} = \|\rho_{\epsilon_k} * (\psi_k u) - \psi_k u\|_{W^{1,p}(V_k)} < \frac{\epsilon}{2^k}.$$

Let $\Phi = \sum_{k=1}^\infty \rho_{\epsilon_k} * (\psi_k u)$. For any compact set $K \subset\subset \Omega$, only finitely many terms in the sum can fail to vanish. Thus $\Phi \in C^\infty(\Omega)$. For $x \in \Omega_k$, we have

$$u(x) = \sum_{j=1}^{k+2} \psi_j(x) u(x), \text{ and } \Phi(x) = \sum_{j=1}^{k+2} \rho_{\epsilon_j} * (\psi_j u)(x),$$

and we have

$$\|u - \Phi\|_{W^{1,p}(\Omega_k)} \leq \Phi(x) = \sum_{j=1}^{k+2} \|\rho_{\epsilon_j} * (\psi_j u) - \psi_j u\|_{W^{1,p}(\Omega)}$$

$$\leq \left(\frac{1}{2} + \frac{1}{2^2} + \cdots + \frac{1}{2^{k+2}}\right) \epsilon$$

$$< \epsilon.$$

However,

$$0 \leq |\, u - \Phi \,|\, \chi_{\Omega_k} \nearrow |\, u - \Phi \,|,$$
$$0 \leq |\, D^\alpha u - D^\alpha \Phi \,|\, \chi_{\Omega_k} \nearrow |\, D^\alpha u - D^\alpha \Phi \,| \quad \text{for} \quad |\, \alpha \,| \leq 1.$$

By the Monotone Convergence Theorem,

$$\|u - \Phi\|_{W^{1,p}(\Omega)} = \lim_{k \to \infty} \|u - \Phi\|_{W^{1,p}(\Omega_k)} \leq \epsilon. \qquad \blacksquare$$

Remark 6.11 *The Meyers-Serrin Theorem 6.10 fails for $p = \infty$. Let $\Omega = (-1, 1)$, $u(x) = |x|$. Then $u \in W^{1,\infty}(\Omega)$ and*

$$u' = \left\{ \begin{array}{l} 1 \ \text{on} \ (0, 1) \\ -1 \ \text{on} \ (-1, 0) \end{array} \right.$$

In fact, for $\varphi \in C_c^\infty(-1, 1)$,

$$\int_{-1}^1 u \varphi' = \int_0^1 x \varphi' + \int_{-1}^0 (-x) \varphi'$$
$$= x\varphi(x) \,\big|_0^1 - \int_0^1 \varphi(x) dx - x\varphi \,\big|_{-1}^0 + \int_{-1}^0 \varphi(x) dx$$
$$= -\int_0^1 \varphi(x) dx + \int_{-1}^0 \varphi(x) dx = -\int_{-1}^1 g(x)\varphi(x) dx,$$

where

$$g = \left\{ \begin{array}{l} 1 \ \text{on} \ (0, \ 1) \\ -1 \ \text{on} \ (-1, \ 0), \end{array} \right.$$

and therefore $u' = g$. It is easy to see that for any $\Phi \in C^1(\Omega)$,

$$\|\Phi' - u'\|_\infty > \frac{1}{2}.$$

6.2   Characterizations

Theorem 6.12 (*Characterization Theorem*) *Let $1 < p \leq \infty$ and $u \in L^p(\Omega)$. Then the following are equivalent:*

(*i*) $u \in W^{1,p}(\Omega)$;

(*ii*) *There is $c > 0$ such that*

$$\left| \int_\Omega u \frac{\partial \varphi}{\partial x_i} \right| \leq c \, \|\varphi\|_{L^{p'}} \ \text{for} \ \varphi \in C_c^\infty(\Omega), i = 1, 2, \cdots, N;$$

(iii) *Given* $c > 0$, *for* $w \subset\subset \Omega$, $h \in \mathbb{R}^N$, $|h| < dist(w, \Omega^c)$ *we have*

$$\|\tau_h u - u\|_{L^p(w)} \leq c|h|.$$

Moreover, $c = \|\nabla u\|_{L^p}$ *in* (ii) *and* (iii).

Proof.   (i) $\Longrightarrow$ (ii) : note that

$$\left| \int_\Omega u \frac{\partial \phi}{\partial x_i} \right| = \left| \int_\Omega \phi \frac{\partial u}{\partial x_i} \right|$$

$$\leq \left\| \frac{\partial \phi}{\partial x_i} \right\|_{L^p} \|\varphi\|_{L^{p'}}$$

$$\leq \|\nabla u\|_{L^p} \|\varphi\|_{L^{p'}}, \text{ for } \varphi \in C_c^\infty(\Omega).$$

(ii) $\Rightarrow$ (i): by the assumption of (ii), the linear maps

$$\varphi \in C_c^\infty(\Omega) \to \int u \frac{\partial \varphi}{\partial x_i}$$

are continuous on $L^{p'}(\Omega)$. Because $C_c^\infty(\Omega)$ is dense in $L^{p'}(\Omega)$, there is a linear continuous extension on $L^{p'}(\Omega)$. By the Riesz Representation Theorem, there is $g \in L^p$ such that

$$< F, \varphi >= \int g\varphi \text{ for } \varphi \in L^{p'}(\Omega).$$

In particular,

$$\int u \frac{\partial \varphi}{\partial x_i} = \int g\varphi \text{ for } \varphi \in C_c^\infty.$$

Thus $u \in W^{1,p}(\Omega)$.

(i) $\Rightarrow$ (iii) : suppose $u \in C_c^\infty(\mathbb{R}^N)$. For $h \in \mathbb{R}^N$, let

$$v(t) = u(x + th) \text{ for } t \in \mathbb{R}.$$

Then

$$u(x + h) - u(x) = v(1) - v(0)$$

$$= \int_0^1 v'(t)dt$$

$$= \int_0^1 h \cdot \nabla u(x + th)dt.$$

Hence,

$$\left(\int_w |\tau_h u - u|^p \, dx \right)^{1/p} \leq |h| \int_0^1 \|\nabla u(x + th)\|_{L^p(w)} \, dt$$

$$\leq \|\nabla u\|_{L^p(w')} |h| \leq \|\nabla u\|_{L^p(\Omega)} |h|,$$

where $|h| < dist(w, \Omega^c)$, some w', $w \subset w'$. For general $u \in W^{1,p}(\Omega)$, we have
$(a) 1 < p < \infty$. Take a sequence (u_n) in $C_c^\infty(\mathbb{R}^N)$ such that

$$u_n \to u \text{ in } L^p(\Omega)$$

$$\nabla u_n \to \nabla u \text{ in } L^p(w)^N \text{ for } w \subset\subset \Omega.$$

Now $\|\tau_h u_n - u_n\|_{L^p(w)} \leq \|\nabla u_n\|_{L^p(w')} |h|$. Letting $n \to \infty$,

$$\|\tau_h u - u\|_{L^p(w)} \leq \|\nabla u\|_{L^p(w')} |h| \leq \|\nabla u\|_{L^p(\Omega)} |h|.$$

(b) $p = \infty$: because $w \subset\subset w' \subset\subset \Omega$ and

$$\|\tau_h u - u\|_{L^p(w)} \leq \|\nabla u\|_{L^p(w')} |h|,$$

for every $1 < p < \infty$. Letting $p \to \infty$, we obtain

$$\|\tau_h u - u\|_{L^\infty(w)} \leq \|\nabla u\|_{L^\infty(w')} |h|$$
$$\leq \|\nabla u\|_{L^\infty(\Omega)} |h|.$$

$(iii) \Rightarrow (ii)$: for $\varphi \in C_c^\infty(\Omega)$, choose w such that

$$\text{supp}\varphi \subset w \subset\subset \Omega.$$

Assume $h \in \mathbb{R}^N$, $|h| < dist(w, \Omega^c)$. Then

$$\left| \int_\Omega (\tau_h u - u)\varphi \right| \leq \|\tau_h u - u\|_{L^p(w)} \|\varphi\|_{L^{p'}} \leq c |h| \|\varphi\|_{L^{p'}}.$$

On the other hand,

$$\int_\Omega (u(x+h) - u(x))\varphi(x)dx = \int_\Omega u(y)(\varphi(y-h) - \varphi(y))dy.$$

Thus,

$$\left| \int_\Omega u(y) \frac{(\varphi(y-h) - \varphi(y))}{|h|} dy \right| \leq c \|\varphi\|_{L^{p'}}.$$

Choose $h = te_i$, $t \in \mathbb{R}$. Let $t \to 0$, and by the Lebesgue Dominated Convergence Theorem,

$$\left| \int_\Omega u(y) \frac{\partial \varphi}{\partial y_i} dy \right| \leq c \|\varphi\|_{L^{p'}} \text{ for } i = 1, 2, \cdots, N. \qquad \blacksquare$$

Remark 6.13

(i) If $p = 1$, then $(i) \Rightarrow (ii) \Rightarrow (iii)$. The functions u in (ii) or (iii) are bounded variations: $u \in L^1(\Omega)$ with $\frac{\partial u}{\partial x_i} \in M(\Omega)$, $i = 1, 2, \cdots, N$, where $M(\Omega)$ is the space of bounded regular Borel measures. In the minimal surfaces theory, the bounded variation spaces are more important than the Sobolev space $W^{1,1}(\Omega)$, see Giusti [25];

(ii) Let F be the unit ball in $W^{1,p}(\Omega)$, $1 \leq p < \infty$, and Ω is open. By Theorem 1.36, and part (i), F is relatively compact in $L^p(w)$ for $w \subset\subset \Omega$. We will prove the Rellich-Kondrachov Theorem 6.40 below: if Ω is bounded and regular, then F is relatively compact in $L^p(\Omega)$. Note that the conclusion fails if Ω is unbounded or not regular.

(iii) By Theorem 6.12, (i) $\Rightarrow$ (iii). If Ω is a domain: an open connected set, and $u \in W^{1,\infty}(\Omega)$, then

$$|u(x) - u(y)| \leq \|\nabla u\|_{L^\infty} \ dist_\Omega(x, y) \text{ a.e. for } x, \ y \in \Omega,$$

where $dist_\Omega(x, y)$ is the geodesic distance between x and $y \in \Omega$;

(iv) By part (iii) of the Remark, if $u \in W^{1,p}(\Omega)$, $1 \leq p \leq \infty$, $\nabla u = 0$ a.e. on Ω, then u is constant on each connected component of Ω.

(v) By part (iii), if $u \in W^{1,\infty}(\Omega)$ and Ω is open and convex, then

$$|(u(x) - u(y)| \leq \|\nabla u\|_{L^\infty} |x - y| \text{ for } x, y \in \Omega.$$

Proof. (iii) Consider the polygon $(x_0 = x, \ x_1, \cdots, \ x_n = y)$, then

$$|(u(x_i) - u(x_{i+1})| \leq \|\nabla u\|_{L^\infty} |x_i - x_{i+1}|, i = 0, 1, \cdots, n - 1.$$

We have

$$|u(x) - u(y)| \leq \sum_{i=1}^{n-1} |u(x_i) - u(x_{i+1})| \leq \|\nabla u\|_{L^\infty} \sum_{i=1}^{n-1} |x_i - x_{i+1}|$$

Taking the infimum on both sides, we obtain

$$|(u(x) - u(y)| \leq \|\nabla u\|_{L^\infty} \ dist_\Omega(x, y).$$

(iv) Let C be a connected component of Ω, by part (iii)

$$|(u(x_i) - u(x_{i+1})| \leq \|\nabla u\|_{L^\infty} \ dist_\Omega(x, y) = 0 \text{ a.e. for } x, y \in C. \qquad \blacksquare$$

6.3   Derivatives

In the Sobolev space $W^{1,p}(\Omega)$, we have the same derivative formulas as those of $C^1(\Omega)$.

Theorem 6.14 (*Product Rule*) *Suppose* $u, \ v \in W^{1,p}(\Omega) \cap L^\infty(\Omega)$, $1 \leq p \leq \infty$, *then* $uv \in W^{1,p}(\Omega) \cap L^\infty(\Omega)$, *and* $\frac{\partial}{\partial x_i}(uv) = \frac{\partial u}{\partial x_i} v + u \frac{\partial v}{\partial x_i}$, $i = 1, 2, \cdots, N$.

Proof.

(i) $1 \leq p < \infty$. Take (u_n), (v_n) in $C_c^\infty(\mathbb{R}^N)$ such that, for a regularity sequence (ρ_n), $\|\rho_n\| = 1$, $u_n = \rho_n * u$, $v_n = \rho_n * v$, then

$$u_n \to u \text{ in } L^p, \text{ and } v_n \to u \text{ in } L^p, \text{ a.e. on } \Omega,$$

$$\nabla u_n \to \nabla u \text{ in } (L^p(\omega))^N, \text{ and } \nabla v_n \to \nabla v \text{ in } (L^p(\omega))^N \text{ for } \omega \subset\subset \Omega,$$

$$\|u_n\|_{L^\infty} \leq \|\rho_n\|_{L^1} \|u\|_{L^\infty} = \|u\|_{L^\infty},$$

$$\|v_n\|_{L^\infty} \leq \|\rho_n\|_{L^1} \|v\|_{L^\infty} = \|v\|_{L^\infty}.$$

Note that, for $\varphi \in C_c^1(\Omega)$, we may choose w with supp $\varphi \subset w \subset\subset \Omega$,

$$\int_\Omega u_n v_n \frac{\partial \varphi}{\partial x_i} = -\int_\Omega \left(\frac{\partial u_n}{\partial x_i} v_n + u_n \frac{\partial v_n}{\partial x_i}\right) \varphi,$$

$$\left|u_n v_n \frac{\partial \varphi}{\partial x_i}\right| \leq \|u\|_{L^\infty} \|v\|_{L^\infty} \left|\frac{\partial \varphi}{\partial x_i}\right| \leq c \left|\frac{\partial \varphi}{\partial x_i}\right| \in L^1(w),$$

$$\|u_n\|_{L^\infty} \leq \|u\|_{L^\infty}, \frac{\partial u_n}{\partial x_i} \to \frac{\partial u}{\partial x_i} \text{ in } L^p(w),$$

$$\|v_n\|_{L^\infty} \leq \|v\|_{L^\infty}, \frac{\partial v_n}{\partial x_i} \to \frac{\partial v}{\partial x_i} \text{ in } L^p(w).$$

By the Lebesgue Dominated Convergence Theorem, letting $n \to \infty$ in the above inequalities, we obtain

$$\int_\Omega uv \frac{\partial \varphi}{\partial x_i} = -\int_\Omega \left(\frac{\partial u}{\partial x_i} v + u \frac{\partial v}{\partial x_i}\right) \varphi.$$

Thus,

$$\frac{\partial(uv)}{\partial x_i} = \frac{\partial u}{\partial x_i} v + u \frac{\partial v}{\partial x_i}, \quad i = 1, 2, \cdots, N.$$

(ii) $p = \infty$. For $\varphi \in C_c^\infty(\Omega)$, choose w with supp $\varphi \subset w \subset\subset \Omega$. Because u, $v \in W^{1,\infty}(\Omega)$, u, $v \in W^{1,p}(w)$ for $p < \infty$. By part (i),

$$\int_w uv \frac{\partial \varphi}{\partial x_i} = -\int_w \left(\frac{\partial u}{\partial x_i} v + u \frac{\partial v}{\partial x_i}\right) \varphi.$$

supp $\varphi \subset w \subset\subset \Omega$, so

$$\int_\Omega uv \frac{\partial \varphi}{\partial x_i} = -\int_w \left(\frac{\partial u}{\partial x_i} v + u \frac{\partial v}{\partial x_i}\right) \varphi,$$

and we have

$$\frac{\partial(uv)}{\partial x_i} = \frac{\partial u}{\partial x_i} v + u \frac{\partial v}{\partial x_i} \quad i = 1, 2, \cdots, N. \qquad \blacksquare$$

Theorem 6.15 (*Chain Rule*) *Let* $G \in C^1(\mathbb{R})$, $G(0) = 0$, $|G'(s)| \leq M$ *for each* $s \in \mathbb{R}$. *For every* $u \in W^{1,p}(\Omega)$, *we have* $G \circ u \in W^{1,p}(\Omega)$, *and* $\frac{\partial}{\partial x_i}(G \circ u) = (G' \circ u)\frac{\partial u}{\partial x_i}$.

Proof. Note that

$$|G(s)| = |G(s) - G(0)| = |G'(t)s| \leq M\,|s|\,,$$

where $s \in \mathbb{R}$, t lies between 0 and s. Thus,

$$|G \circ u| \leq M\,|u|\,,$$

and we obtain $G \circ u \in L^p(\Omega)$. Because $G' \circ u$ is bounded, $(G' \circ u)\frac{\partial u}{\partial x_i} \in L^p(\Omega)$. For $\varphi \in C_c^1(\Omega)$, we require

$$\int_\Omega (G \circ u)\frac{\partial \varphi}{\partial x_i} = -\int_\Omega (G' \circ u)\frac{\partial u}{\partial x_i}\varphi.$$

(i) $1 \leq p < \infty$. Take a sequence (u_n) in $C_c^\infty(\mathbb{R}^N)$ such that

$$u_n \to u \text{ in } L^p(\Omega), \text{ a.e. on } \Omega,$$
$$\nabla u_n \to \nabla u \text{ in } L^p(w)^N \text{ for } w \subset\subset \Omega.$$

By integration by parts and the usual Chain Rule,

$$\int_\Omega (G \circ u_n)\frac{\partial \varphi}{\partial x_i} = -\int (G' \circ u_n)\frac{\partial u_n}{\partial x_i}\varphi.$$

By Theorem 1.35, we may choose (u_n) such that

$$u_n \to u \text{ in } L^p(\Omega), \text{ a.e. on } \Omega,$$
$$|u_n| \leq h \in L^p(\Omega) \text{ for } n = 1, 2, \cdots,$$
$$\frac{\partial u_n}{\partial x_i} \to \frac{\partial u}{\partial x_i} \text{ in } L^p(\Omega), \text{ a.e. on } \Omega,$$
$$\left|\frac{\partial u_n}{\partial x_i}\right| \leq k \in L^p(\Omega) \text{ for } n = 1, 2, \cdots.$$

Because $G \circ u_n \to G \circ u$ a.e. on Ω, we have

$$(G' \circ u_n)\frac{\partial u_n}{\partial x_i} \to (G'u)\frac{\partial u}{\partial x_i} \text{ a.e. on } \Omega.$$
$$|G \circ u_n| \leq M\,|u_n| \leq Mh \in L^p(\Omega),$$
$$\left|(G' \circ u_n)\frac{\partial u_n}{\partial x_i}\right| \leq C\left|\frac{\partial u_n}{\partial x_i}\right| \leq ck \in L^p(\Omega).$$

By the Lebesgue Dominated Convergence Theorem,

$$G \circ u_n \to G \circ u \text{ in } L^p(\Omega),$$

$$(G' \circ u_n)\frac{\partial u_n}{\partial x_i} \to (G' \circ u)\frac{\partial u}{\partial x_i} \text{ in } L^p(\Omega).$$

Letting $n \to \infty$, in the following identity:

$$\int_\Omega (G \circ u_n)\frac{\partial \varphi}{\partial x_i} = -\int (G' \circ u_n)\frac{\partial u_n}{\partial x_i}\varphi$$

we obtain

$$\int_\Omega (G \circ u)\frac{\partial \varphi}{\partial x_i} = -\int_\Omega (G' \circ u)\frac{\partial u}{\partial x_i}\varphi.$$

(ii) $p = \infty$. For $\varphi \in C_c^\infty(\Omega)$, choose w such that supp $\varphi \subset w \subset\subset \Omega$. Then $u \in W^{1,p}(w)$ for $p < \infty$. By part (i),

$$\int_w (G \circ u)\frac{\partial \varphi}{\partial x_i} = -\int_w (G' \circ u)\frac{\partial u}{\partial x_i}\varphi,$$

or

$$\int_\Omega (G \circ u)\frac{\partial \varphi}{\partial x_i} = -\int_\Omega (G' \circ u)\frac{\partial u}{\partial x_i}\varphi. \qquad \blacksquare$$

Remark 6.16 *In Theorem 6.15, if $p = \infty$ or Ω is bounded, we may forget the assumption $G(0) = 0$.*

Theorem 6.17 (*Formula for Changing Variables*) *Let Ω and Ω' be two open sets in $\mathbb{R}^N$, let $H : \Omega' \to \Omega$ be a bijection, and let $x = H(y)$ such that*

$$H \in C^1(\Omega'), \text{ and Jac } H \in (L^\infty(\Omega'))^{N \times N},$$

$$H^{-1} \in C^1(\Omega), \text{ and Jac } H^{-1} \in (L^\infty(\Omega'))^{N \times N}.$$

Suppose $u \in W^{1,p}(\Omega)$, then $u \circ H \in W^{1,p}(\Omega')$, and

$$\frac{\partial}{\partial y_j}(u \circ H)(y) = \sum_i \frac{\partial u}{\partial x_i}(H(y))\frac{\partial H_i}{\partial y_j}(y) \text{ for } j = 1, 2, \cdots, N,$$

where Jac $H = \left(\frac{\partial H_i}{\partial y_j}\right)$ is the Jacobian matrix.

Proof.   It suffices to prove the case $1 \leq p < \infty$. For $p = \infty$, the proof is similar to the proof of Remark 6.16. Take a sequence (u_n) in $C_c^\infty(\mathbb{R}^N)$ such that

$$u_n \to u \text{ in } L^p(\Omega), \text{ a.e. on } \Omega \text{ and } |u_n| \leq h \in L^p(\Omega),$$

$\nabla u_n \to \nabla u \ L^p(w)^N$ for $w \subset\subset \Omega$, a.e. on w, and $|\nabla u_n| \le k \in L^p(w)$ a.e. Then,

$$\int_{\Omega'} |h(H(y)|^p = \int_{\Omega} |h(x)|^p |\text{Jac } H| < \infty,$$
$$|u_n \circ H(y)| \le h(H(y)) \in L^p(\Omega'),$$
$$u_n \circ H \to u \circ H, \text{ a.e. on } \Omega'.$$

By the Lebesgue Dominated Convergence Theorem,

$$u_n \circ H \to u \circ H \text{ in } L^p(\Omega').$$

Moreover,

$$\left| \left(\frac{\partial u_n}{\partial x_i} \circ H \right) \frac{\partial H_i}{\partial y_j} \right| \le C \left| \frac{\partial u_n}{\partial x_i}(H(x)) \right| \text{ because Jac } H \in (L^\infty(\Omega'))^{N \times N}$$
$$\le ck(H(x)) \in L^p(w'),$$

where $w' = H^{-1}(w) \subset\subset \Omega'$, so we have

$$\left(\frac{\partial u_n}{\partial x_i} \circ H \right) \frac{\partial H_i}{\partial y_j} \to \left(\frac{\partial u}{\partial x_i} \circ H \right) \frac{\partial H_i}{\partial y_j} \text{ a.e. on } \omega.$$

By the Lebesgue Dominated Convergence Theorem,

$$\left(\frac{\partial u_n}{\partial x_i} \circ H \right) \frac{\partial H_i}{\partial y_j} \to \left(\frac{\partial u}{\partial x_i} \circ H \right) \frac{\partial H_i}{\partial y_j} \text{ in } L^p(w').$$

For $\psi \in C_c^1(\Omega')$, take w' such that supp $\psi \subset w' \subset\subset \Omega'$. We have, by the Chain Rule and integration by parts

$$\int_{\Omega'} (u_n \circ H) \frac{\partial \psi}{\partial y_j} dy = -\int_{\Omega'} \sum_{i=1}^N \left(\frac{\partial u_n}{\partial x_i} \circ H \right) \frac{\partial H_i}{\partial y_j} \psi dy.$$

Let $n \to \infty$ to obtain

$$\int_{\Omega'} (u \circ H) \frac{\partial \psi}{\partial y_j} dy = -\int_{\Omega'} \sum_{i=1}^N \left(\frac{\partial u_i}{\partial x_i} \circ H \right) \frac{\partial H_i}{\partial y_j} \psi dy.$$

Thus,

$$\frac{\partial}{\partial y_j}(u \circ H)(y) = \sum_i \frac{\partial u}{\partial x_i}(H(y)) \frac{\partial H_i}{\partial y_j}. \qquad \blacksquare$$

Theorem 6.18 *(Poincaré-Wirtinger Theorem) Let Ω be the cube $\{x \in \mathbb{R}^N \mid |x_i| < a, i = 1, \cdots, N\}$. Let $u \in H^1(\Omega)$, then*

$$\int_{\Omega} |u|^2 \le 2a^2 N \int_{\Omega} \sum_{i=1}^N \left| \frac{\partial u}{\partial x_i} \right|^2 + \frac{1}{(2a)^N} \left| \int_{\Omega} u(x)dx \right|^2.$$

In particular, if $\int_{\Omega} u(x)dx = 0$, then $\|u\|_{L^2} \le c \|\nabla u\|_{L^2}$.

Proof. By the Friedrichs Theorem 6.8, it suffices to consider the case $u \in C_c^\infty(\mathbb{R}^N)$. For x, $y \in \Omega$, $x = (x_1, \cdots, x_N)$, $y = (y_1, \cdots, y_N)$, then

$$
\begin{aligned}
u(x) - u(y) &= u(x_1, x_2, \cdots, x_N) - u(x_1, \cdots, x_{N-1}, y_N) \\
&\quad + u(x_1, \cdots, x_{N-1}, y_N) - u(x_1, \cdots, x_{N-2}, y_{N-1}, y_N) \\
&\quad + \cdots + u(x_1, y_2, \cdots, y_N) - u(y_1, y_2, \cdots, y_N) \\
&= \int_{y_N}^{x_N} \frac{\partial u}{\partial x_N}(x_1, \cdots, x_{N-1}, t_N) dt_N \\
&\quad + \int_{y_{N-1}}^{x_{N-1}} \frac{\partial u}{\partial x_{N-1}}(x_1, \cdots, x_{N-2} t_{N-1}, y_N) dt_{N-1} \\
&\quad + \cdots + \int_{y_1}^{x_1} \frac{\partial u}{\partial x_1}(t_1, y_2, \cdots, y_N) dt_1.
\end{aligned}
$$

Let $A_i = \int_{y_i}^{x_i} \frac{\partial u}{\partial t_i} dt_i$. By the Hölder inequality,

$$
|A_i| = \left| \int_{y_i}^{x_i} \frac{\partial u}{\partial x_i} dt_i \right| \le (2a)^{1/2} \left(\int_{-a}^{a} \left| \frac{\partial u}{\partial x_i} \right|^2 \right)^{1/2}.
$$

Now,

$$
\begin{aligned}
|u(x)|^2 &+ |u(y)|^2 - 2u(x)u(y) \\
&= |u(x) - u(y)|^2 \\
&= |A_1 + A_2 + \cdots + A_N|^2 \\
&= N(A_1^2 + \cdots + A_N^2) \\
&\le N \left[2a \int_{-a}^{a} \left| \frac{\partial u}{\partial x_1} \right|^2 + 2a \int_{-a}^{a} \left| \frac{\partial u}{\partial x_2} \right|^2 + \cdots + 2a \int_{-a}^{a} \left| \frac{\partial u}{\partial x_N} \right|^2 \right].
\end{aligned}
$$

Integrating $\int_{\Omega \times \Omega} dx dy$ on both sides, we obtain

$$
\begin{aligned}
\int_\Omega |u(x)|^2 \, dx &\int_\Omega dy + \int_\Omega dx \int_\Omega |u(y)|^2 \, dy - 2 \int_\Omega u(x) dx \int_\Omega u(y) dy \\
&\le 2aN \sum_{i=1}^{N} \int_{\Omega \times \Omega} dx dy \int_{-a}^{a} \left| \frac{\partial u}{\partial x_i} \right|^2.
\end{aligned}
$$

However,

$$\int_{\Omega \times \Omega} dx dy \int_{-a}^{a} \left| \frac{\partial u}{\partial x_1}(t_1, y_2, \cdots, y_N) \right|^2 dt_1$$

$$= \int_{\Omega} dx \int_{-a}^{a} dy_1 \int_{\Omega} \left| \frac{\partial u}{\partial x_1}(t_1, y_2, \cdots,) \right|^2 dt_1 dy_2, \cdots dy_N$$

$$= (2a)^{N+1} \int_{\Omega} \left| \frac{\partial u}{\partial x_1} \right|^2.$$

Similarly for other indices, we obtain

$$2(2a)^N \int_{\Omega} |u(x)|^2 \, dx - 2 \left(\int_{\Omega} u(x) dx \right)^2 \leq (2aN)(2a)^{N+1} \int_{\Omega} \sum_{i=1}^{N} \left| \frac{\partial u}{\partial x_i} \right|^2 dx,$$

or

$$\int_{\Omega} |u(x)|^2 \, dx \leq 2a^2 N \int_{\Omega} \sum_{i=1}^{N} \left| \frac{\partial u}{\partial x_i} \right|^2 + \frac{1}{(2a)^N} \left| \int_{\Omega} u(x) \right|^2. \qquad \blacksquare$$

6.4 Extension Operators

For $x = (x_1, \cdots, x_N) \in \mathbb{R}^N$, write $x = (x', x_N)$ with $x' = (x_1, \cdots, x_{N-1}) \in \mathbb{R}^{N-1}$, $|x'| = \left(\sum_{I=1}^{N-1} x_i^2 \right)^{1/2}$. Denote by

$$\mathbb{R}_+^N = \{x = (x', x_N) \mid x_N > 0\},$$
$$\mathbb{Q} = \{x = (x', x_N) \mid |x'| < 1, |x_N| < 1\},$$
$$\mathbb{Q}_+ = \mathbb{Q} \cap \mathbb{R}_+^N,$$
$$\mathbb{Q}_0 = \{x = (x', x_N) \mid |x'| < 1, x_N = 0\}.$$

Definition 6.19 *Let Ω be an open set in $\mathbb{R}^N$. Ω is of class C^1 if for each $x \in \Gamma = \partial\Omega$, there is a neighborhood U of x in $\mathbb{R}^N$, and a bijection $H : \mathbb{Q} \to U$ such that*

$$H \in C^1(\bar{\mathbb{Q}}) \text{ and } H^{-1} \in C^1(\bar{U}),$$
$$H(\mathbb{Q}_+) = U \cap \Omega \text{ and } H(\mathbb{Q}_0) = U \cap \Gamma.$$

Lemma 6.20 *(Reflexive Extension Lemma) Given $u \in W^{1,p}(\mathbb{Q}_+)$, define the reflexive extension $\tilde{u}$ by*

$$\tilde{u}(x', x_N) = \begin{cases} u(x', x_N) & \text{if } x_N > 0 \\ u(x', -x_N) & \text{if } x_N < 0. \end{cases}$$

Then

$$\tilde{u} \in W^{1,p}(Q),$$
$$\|\tilde{u}\|_{L^p(Q)} \le 2 \|u\|_{L^p(Q_+)},$$
$$\|\tilde{u}\|_{w^{1,p}(Q)} \le 2 \|u\|_{w^{1,p}(Q_+)},$$

Proof. It is clear that $\tilde{u} \in L^p(\mathbb{Q})$, $\|\tilde{u}\|_{L^p(\mathbb{Q})} \le 2 \|u\|_{L^p(\mathbb{Q}_+)}$. To prove $\tilde{u} \in W^{1,p}(\mathbb{Q})$, with $\|\tilde{u}\|_{w^{1,p}(\mathbb{Q})} \le 2 \|u\|_{W^{1,p}(\mathbb{Q}_+)}$. It suffices to prove

(i) $\dfrac{\partial \tilde{u}}{\partial x_i} = \left(\dfrac{\partial u}{\partial x_i} \right) \quad 1 \le i \le N - 1,$

(ii) $\dfrac{\partial \tilde{u}}{\partial x_N} = \left(\dfrac{\partial u}{\partial x_N} \right)^{\#}.$

Note that if $f : \mathbb{Q}_+ \to \mathbb{R}$, then we define

$$f^{\#}(x', x_N) = \begin{cases} f(x', x_N) & \text{if } x_N > 0, \\ -f(x', -x_N) & \text{if } x_N < 0. \end{cases}$$

Choose $\eta \in C^{\infty}(\mathbb{R})$, $0 \le \eta \le 1$ on $\mathbb{R}$, such that

$$\eta(t) = \begin{cases} 0 \text{ if } t < \dfrac{1}{2}, \\ 1 \text{ if } t > 1. \end{cases}$$

Let $\eta_k(t) = \eta(kt), t \in \mathbb{R}, k = 1, 2, \cdots$. We shall prove (i) and (ii).

(i) For $\varphi \in C_c^1(\mathbb{Q})$, $1 \le i \le N - 1$, we have

$$\int_{\mathbb{Q}} \tilde{u} \frac{\partial \varphi}{\partial x_i} = \int_{\mathbb{Q}_+} \tilde{u} \frac{\partial \varphi}{\partial x_i} + \int_{\mathbb{Q}_-} \tilde{u} \frac{\partial \varphi}{\partial x_i}$$

$$= \int_{\mathbb{Q}_+} u \frac{\partial \varphi}{\partial x_i} + \int_{\mathbb{Q}_-} u(x', -x_N) \frac{\partial \varphi}{\partial x_i}(x', x_N) dx$$

$$= \int_{\mathbb{Q}_+} u \frac{\partial \varphi}{\partial x_i} + \int_{|x'|<1} dx' \int_{-1}^{0} u(x', -x_N) \frac{\partial \varphi}{\partial x_i}(x', x_N) dx_N$$

$$= \int_{\mathbb{Q}_+} u \frac{\partial \varphi}{\partial x_i} + \int_{\mathbb{Q}_+} u(x', x_N) \frac{\partial \varphi}{\partial x_i}(x', -x_N) dx.$$

Let $\psi(x', x_N) = \varphi(x', x_N) + \varphi(x', -x_N)$, then

$$\frac{\partial \psi}{\partial x_i}(x', x_N) = \frac{\partial \varphi}{\partial x_i}(x', x_N) + \frac{\partial \varphi}{\partial x_i}(x', -x_N) \text{ for } x_N > 0.$$

We have

$$\int_{\mathbb{Q}} \tilde{u}\frac{\partial \varphi}{\partial x_i} = \int_{\mathbb{Q}_+} u\frac{\partial \psi}{\partial x_i}.$$

Note that in general $\psi \notin C_c^1(\mathbb{Q}_+)$, but

$$\eta_k(x_N)\psi(x', x_N) \in C_c^1(\mathbb{Q}_+) \text{ for large } k.$$

Thus,

$$\int_{\mathbb{Q}_+} u\frac{\partial}{\partial x_i}(\eta_k\psi) = -\int_{\mathbb{Q}_+} \frac{\partial u}{\partial x_i}\eta_k\psi.$$

Because η_k is independent of $x_i's$,

$$\frac{\partial}{\partial x_i}(\eta_k\psi) = \eta_k\frac{\partial \psi}{\partial x_i}, \quad i = 1, \cdots, N-1.$$

Thus,

$$\int_{\mathbb{Q}_+} u\eta_k\frac{\partial \psi}{\partial x_i} = -\int_{\mathbb{Q}_+} \frac{\partial u}{\partial x_i}\eta_k\psi.$$

Note that

$$u\eta_k\frac{\partial \psi}{\partial x_i} \to u\frac{\partial \psi}{\partial x_i} \text{ a.e. on } .\mathbb{Q}_+$$

$$\left| u\eta_k\frac{\partial \psi}{\partial x_i} \right| \leq c\,|u| \in L^1(\mathbb{Q}_+),$$

$$\frac{\partial u}{\partial x_i}\eta_k\psi \to \frac{\partial u}{\partial x_i}\psi \text{ a.e. on } \mathbb{Q}_+.$$

$$\left| \frac{\partial u}{\partial x_i}\eta_k\psi \right| \leq c\left| \frac{\partial \varphi}{\partial x_i} \right| \in L^1(\mathbb{Q}_+).$$

By the Lebesgue Dominated Convergence Theorem, letting $k \to \infty$ we obtain

$$\int_{\mathbb{Q}_+} u\frac{\partial \psi}{\partial x_i} = -\int_{\mathbb{Q}_+} \frac{\partial u}{\partial x_i}\psi.$$

Now,

$$\int_{\mathbb{Q}} \tilde{u}\frac{\partial \varphi}{\partial x_i} = \int_{\mathbb{Q}_+} u\frac{\partial \psi}{\partial x_i} = -\int_{\mathbb{Q}_+} \frac{\partial u}{\partial x_i}\psi$$

$$= -\left[\int_{\mathbb{Q}_+} \frac{\partial u}{\partial x_i}(x', x_N)\varphi(x', x_N) + \int_{\mathbb{Q}_+} \frac{\partial u}{\partial x_i}(x', x_N)\varphi(x', x_N) \right]$$

$$= -\left[\int_{\mathbb{Q}_+} \frac{\partial u}{\partial x_i}(x', x_N)\varphi(x', x_N) + \int_{\mathbb{Q}_-} \frac{\partial u}{\partial x_i}(x', x_N)\varphi(x', x_N) \right]$$

$$= -\int_{\mathbb{Q}} \left(\frac{\partial u}{\partial x_i} \right)\, \varphi(x', x_N),$$

and we conclude that

$$\frac{\partial \widetilde{u}}{\partial x_i} = \left(\frac{\partial u}{\partial x_i} \right) \widetilde{} , i = 1, 2, \cdots, N-1.$$

(ii) For $\varphi \in C_c^1(Q)$, as in part (i), we obtain

$$\int_Q \widetilde{u} \frac{\partial \varphi}{\partial x_N} = \int_{\mathbb{Q}_+} u \frac{\partial \zeta}{\partial x_N},$$

where $\zeta(x', x_N) = \varphi(x', x_N) - \varphi(x', -x_N)$. Now $\zeta(x', 0) = 0$, so

$$|\zeta(x', x_N)| = |\zeta(x', x_N) - \zeta(x', 0)|$$
$$= \left| \frac{\partial \zeta}{\partial x_N}(x', t) \right| |x_N| \text{ for some } t \text{ between } 0 \text{ and } x_N$$
$$\leq M |x_N|, \text{ where } M = \left\| \frac{\partial \zeta}{\partial x_N} \right\|_{L^\infty}.$$

Because $\eta_k \zeta \in C_c^1(\mathbb{Q}_+)$,

$$\int_{\mathbb{Q}_+} u \frac{\partial}{\partial x_N}(\eta_k \zeta) = -\int_{\mathbb{Q}_+} \frac{\partial u}{\partial x_N}(\eta_k \zeta)$$

or

$$\int_{\mathbb{Q}_+} u \eta_k \frac{\partial \zeta}{\partial x_N} + \int_{\mathbb{Q}_+} u k \eta'(k x_N) \zeta = -\int_{\mathbb{Q}_+} \frac{\partial u}{\partial x_N}(\eta_k \zeta).$$

Now $u \eta_k \frac{\partial \zeta}{\partial x_N} \to u \frac{\partial \zeta}{\partial x_N}$ a.e. on $\mathbb{Q}_+$, and

$$\left| u \eta_k \frac{\partial \zeta}{\partial x_N} \right| \leq c |u| \in L^1$$

$$\left| \int_{\mathbb{Q}_+} u k \eta'(k x_N) \zeta(x', x_N) \right| \leq kM \sup_{t \in [0,1]} |\eta'(t)| \int_{0 < x_N < \frac{1}{k}} |u| \, x_N dx$$

$$\leq M \sup_{t \in [0,1]} |\eta'(t)| \int_{0 < x_N < \frac{1}{k}} |u|$$

$$\to 0 \text{ as } k \to \infty,$$

$$\frac{\partial u}{\partial x_N}(\eta_k \zeta) \to \frac{\partial u}{\partial x_N} \zeta \text{ a.e. on } \mathbb{Q}_+.$$

$$\left| \frac{\partial u}{\partial x_N}(\eta_k \zeta) \right| \leq C \left| \frac{\partial u}{\partial x_N} \right| \in L^1(\mathbb{Q}_+).$$

Apply the Lebesgue Dominated Convergence Theorem to obtain

$$\int_{\mathbb{Q}_+} u\frac{\partial \zeta}{\partial x_N} = -\int_{\mathbb{Q}_+} \frac{\partial u}{\partial x_N}\zeta.$$

Thus,

$$\int_{\mathbb{Q}} \widetilde{u}\frac{\partial \varphi}{\partial x_N} = \int_{\mathbb{Q}_+} u\frac{\partial \zeta}{\partial x_N} = -\int_{\mathbb{Q}_+} \frac{\partial u}{\partial x_N}\zeta$$

$$= -\int_{\mathbb{Q}_+} \frac{\partial u}{\partial x_N}(x', x_N)\varphi(x', x_N) + \int_{\mathbb{Q}_+} \frac{\partial u}{\partial x_N}(x', x_N)\varphi(x', -x_N)$$

$$= -\int_{\mathbb{Q}_+} \frac{\partial u}{\partial x_N}(x', x_N)\varphi(x', x_N) + \int_{\mathbb{Q}_-} \frac{\partial u}{\partial x_N}(x', -x_N)\varphi(x', x_N)$$

$$= -\int_{\mathbb{Q}} \left(\frac{\partial u}{\partial x_N}\right)^{\#}\varphi.$$

Thus, $\dfrac{\partial \widetilde{u}}{\partial x_N} = \left(\dfrac{\partial u}{\partial x_N}\right)^{\#}$.   ∎

Remark 6.21

(i) *The proof of Lemma 6.20 will not change if we replace $\mathbb{Q}_+$ by $\mathbb{R}^N_+$;*

(ii) *Lemma 6.20 provides a method for easily constructing an extension operator for an open set that is not of C^1. For example, $\Omega = \{x \in \mathbb{R}^2 \mid 0 < x_1 < 1, 0 < x_2 < 1\}$.*

For $u \in w^{1,p}(\Omega)$, by the Reflexive Extension Lemma 6.20, with successive reflections, we obtain an extension $\widetilde{u} \in W^{1,p}(\widetilde{\Omega})$ of u in $\widetilde{\Omega}$,

$$\widetilde{\Omega} = \{x \in \mathbb{R}^N \mid -1 < x_1 < 3, -1 < x_2 < 3\}.$$

Fix $\varphi \in C_c^1(\widetilde{\Omega})$, $\varphi = 1$ on Ω. Define, for $u \in W^{1,p}(\Omega)$,

$$Pu = \begin{cases} \psi\widetilde{u} \text{ on } \widetilde{\Omega}, \\ 0 \text{ on } \mathbb{R}^2\backslash\widetilde{\Omega}, \end{cases}$$

then $Pu \in W^{1,p}(\mathbb{R}^N)$ such that

(i) $Pu\mid_{\Omega} = u$,

(ii) $\|Pu\|_{L^p(\mathbb{R}^N)} \leq C\|u\|_{L^p(\Omega)}$,

(iii) $\|Pu\|_{W^{1,p}(\mathbb{R}^N)} \leq C\|u\|_{W^{1,p}(\Omega)}$.

Lemma 6.22 (*Partition of Unity*) *If we let Γ be a compact set in $\mathbb{R}^N$, $U_1, \cdots, U_k$ an open covering of Γ, then there are the functions $\theta_0, \theta_1, \cdots, \theta_k \in C^\infty(\mathbb{R}^N)$ such that:*

(i) $0 \leq \theta_i \leq 1$ *for* $i = 0, 1, \cdots, k$

$$\sum_{i=0}^{k} \theta_i = 1 \text{ on } \mathbb{R}^N;$$

(ii) *Supp θ_i are compact, supp $\theta_i \subset U_i$ for $i = 1, 2, \cdots, k$, supp $\theta_0 \subset \mathbb{R}^N \backslash \Gamma$. If Ω is bounded open, $\Gamma = \partial\Omega$, then $\theta_0 \mid_\Omega \in C_c^\infty(\Omega)$.*

Theorem 6.23 (*Extension Operator Theorem*) *Suppose Ω is an open set in $\mathbb{R}^N$ of class C^1 with boundary Γ bounded, or $\Omega = \mathbb{R}_+^N$. There is a linear extension $P : W^{1,p}(\Omega) \to W^{1,p}(\mathbb{R}^N)$ such that for every $u \in W^{1,p}(\Omega)$,*

$$Pu|_\Omega = u;$$
$$\|Pu\|_{L^p(\mathbb{R}^N)} \leq c\|u\|_{L^p(\Omega)};$$
$$\|Pu\|_{W^{1,p}(\mathbb{R}^N)} \leq c\|u\|_{W^{1,p}(\Omega)}.$$

where c is dependent only on Ω.

Proof. We only prove the case where Ω is an open set in $\mathbb{R}^N$ of class C^1 with Γ bounded. For the case $\Omega = \mathbb{R}_+^N$, see Remark 6.21. Let $\Gamma = \partial\Omega$ be the boundary, which is a compact set. Assume $U_1, \cdots, U_k$ is an open covering of Γ, and $H_i : \mathbb{Q} \to U_i$ are bijections, and $i = 1, \cdots, k$ exist such that

$$H_i \in C^1(\bar{\mathbb{Q}}) \text{ and } H_i^{-1} \in C^1(\bar{U}_i),$$
$$H_i(\mathbb{Q}_+) = U_i \cap \Omega \text{ and } H_i(\mathbb{Q}_0) = U_i \cap \Gamma.$$

Let $\theta_0, \theta_1, \cdots, \theta_k$ be the partition of unity such that supp $\theta_0 \subset \mathbb{R}^N \backslash \Gamma$, supp $\theta_i \subset U_i$, $i = 1, \cdots, k$. For $u \in W^{1,p}(\Omega)$,

$$u = \sum_{i=0}^{k} \theta_i u = \sum_{i=0}^{k} u_i,$$

where $u_i = \theta_i u$.

(i) Extension of u_0 : Define

$$\tilde{u}_0(x) = \begin{cases} u_0(x) & \text{if } x \in \Omega, \\ 0 & \text{if } x \in \mathbb{R}^N \backslash \Omega. \end{cases}$$

Note that $\theta_0 \in C^1(\mathbb{R}^N) \cap L^\infty(\mathbb{R}^N)$, and $\sum_{i=0}^{k} \theta_i = 1$, or

$$\nabla \theta_0 = -\sum_{i=1}^{k} \nabla \theta_i \in L^\infty(\mathbb{R}^N),$$

$$\text{supp } \nabla \theta_0 \subset \bigcup_{i=1}^{k} \text{supp } \nabla \theta_i$$

$$\text{supp } \theta_0 \subset \mathbb{R}^N \backslash \Gamma.$$

By Remark 6.6 (iii), $\widetilde{u}_0 \in W^{1,p}(\mathbb{R}^N)$ such that $\frac{\partial \widetilde{u}_0}{\partial x_i} = \theta_0 \frac{\partial \widetilde{u}}{\partial x_i} + \frac{\partial \theta_0}{\partial x_i} \widetilde{u}$. Thus,

$$\|\widetilde{u}_0\|_{W^{1,p}(\mathbb{R}^N)} \leq c \|u\|_{W^{1,p}(\Omega)},$$

because $\theta_0, \frac{\partial \theta_0}{\partial x_i} \in L^\infty$.

(ii) Extension of u_i, $1 \leq i \leq k$: Let

$$v_i(y) = u(H_i(y)) \text{ for } y \in \mathbb{Q}_+.$$

Then $v_i \in W^{1,p}(\mathbb{Q}_+)$. By the Reflexive Extension Lemma 6.20, $\widetilde{v}_i \in W^{1,p}(\mathbb{Q})$, let

$$w_i(x) = \widetilde{v}_i(H_i^{-1}(x)) \text{ for } x \in U_i.$$

We have

$$w_i \in W^{1,p}(U_i), \ w_i \mid_{U_i \cap \Omega} = u,$$

$$\|w_i\|_{W^{1,p}(U_i)} \leq c \|u\|_{u_{W^{1,p}(U_i \cap \Omega)}}.$$

Define

$$\widetilde{u}_i(x) = \begin{cases} \theta_i(x) w_i(x) & \text{if } x \in U_i, \\ 0 & \text{if } x \in \mathbb{R}^N \backslash U_i. \end{cases}$$

Then $\widetilde{u}_i \in W^{1,p}(\mathbb{R}^N)$, $\widetilde{u}_i = u_i = \theta_i u$ on Ω, $\|\widetilde{u}_i\|_{W^{1,p}(\mathbb{R}^N)} \leq c \|u\|_{W^{1,p}(U_i \cap \Omega)}$. We conclude that the operator

$$Pu = \widetilde{u}_0 + \sum_{i=1}^{k} \widetilde{u}_i$$

possesses the desired properties.   ∎

Theorem 6.24 (*Forth Density Theorem*) *Suppose Ω is of class C^1. If $u \in W^{1,p}(\Omega)$, $1 \leq p < \infty$, then there is a sequence $(u_n) \subset C_c^\infty(\mathbb{R}^N)$ such that*

$$u_n \mid_\Omega \rightarrow u \text{ in } W^{1,p}(\Omega).$$

Proof.

(i) $\Gamma = \partial\Omega$ is bounded: by Theorem 6.23, there is an extension operator $P :$ $W^{1,p}(\Omega) \to W^{1,p}(\mathbb{R}^N)$. For $u \in W^{1,p}(\Omega)$, we have $Pu \in W^{1,p}(\mathbb{R}^N)$. As in the proof of the Second Density Theorem 6.8, we obtain

$$u_n = \zeta_n(\rho_n \star Pu) \to Pu \text{ in } W^{1,p}(\mathbb{R}^N),$$

where $(u_n) \subset C_c^\infty(\mathbb{R}^N)$. We have

$$u_n \to Pu \text{ in } L^p(\mathbb{R}^N),$$
$$\nabla u_n \to \nabla(Pu) \text{ in } (L^p(\mathbb{R}^N)^N \text{ in } (L^p(\mathbb{R}^N))^N,$$

or

$$u_n \mid_\Omega \to Pu \mid_\Omega = u \text{ in } L^p(\Omega),$$
$$\nabla u_n \mid_\Omega \to \nabla(Pu) \mid_\Omega = \nabla u \text{ in } (L^p(\Omega))^N.$$

(ii) $\Gamma = \partial\Omega$ is unbounded: for $u \in W^{1,p}(\Omega)$, let

$$\zeta\, |x| = \begin{cases} 1 \text{ if } |x| \leq 1 \\ 0 \text{ if } |x| \geq 2 \end{cases}, 0 \leq \zeta \leq 1, \zeta_n(x) = \zeta\left(\frac{x}{n}\right), n = 1, 2, \cdots,$$

$$\zeta_n u \to u \text{ in } L^p(\Omega).,$$

$$\left\| \frac{\partial(\zeta_n u)}{\partial x_i} - \frac{\partial u}{\partial x_i} \right\|_{L^p(\Omega)} = \left\| \zeta_n \frac{\partial u}{\partial x_i} + \frac{\partial \zeta_n}{\partial x_i} u - \frac{\partial u}{\partial x_i} \right\|_{L^p(\Omega)}$$

$$\leq \left\| \frac{\partial \zeta_n}{\partial x_i} u \right\|_{L^p(\Omega)} + \left\| \zeta_n \frac{\partial u}{\partial x_i} - \frac{\partial u}{\partial x_i} \right\|_{L^p(\Omega)}$$

$$\leq \frac{c}{n} \|u\|_{L^p(\Omega)} + \left\| \zeta_n \frac{\partial u}{\partial x_i} - \frac{\partial u}{\partial x_i} \right\|_{L^p(\Omega)}$$

$$\to 0 \text{ as } n \to \infty.$$

For $\epsilon > 0$, take a positive integer n_0 such that

$$\|\zeta_{n_0} u - u\|_{W^{1,p}(\Omega)} < \frac{\epsilon}{2}.$$

Then $\zeta_{n_0} u \in W^{1,p}(\Omega)$, supp $(\zeta_{n_0} u) \subset A \subset \bar{\Omega}$, ∂A is C^1 and bounded. By part (i), there is $v \in C_c^\infty(\mathbb{R}^N)$ such that

$$\|v - P(\zeta_{n_0} u)\|_{W^{1,p}(\mathbb{R}^N)} < \frac{\epsilon}{2}.$$

Now

$$|v - u|_{W^{1,p}(\Omega)} \le \|v - P(\zeta_{n_0} u)\|_{W^{1,p}(\mathbb{R}^N)} + \|P(\zeta_{n_0} u) - \zeta_{n_0} u\|_{W^{1,p}(\Omega)}$$
$$+ \|\zeta_{n_0} u - u\|_{W^{1,p}(\Omega)}$$
$$= \|v - P(\zeta_{n_0} u)\|_{W^{1,p}(\mathbb{R}^N)} + \|\zeta_{n_0} u - u\|_{W^{1,p}(\Omega)} < \epsilon. \qquad \blacksquare$$

Remark 6.25 *Denote by*

$$A^{1,p}(\Omega) = \overline{\{u|_\Omega|\, u \in C_c^\infty(\mathbb{R}^N)\}}^{W^{1,p}(\Omega)}$$

that $A^{1,p}(\Omega)$ is a Banach space under the $W^{1,p}(\Omega)$-norm. If Ω is of class C^1, then by the Forth Density Theorem 6.24,

$$A^{1,p}(\Omega) = W^{1,p}(\Omega).$$

However, if Ω is irregular, then in general,

$$A^{1,p}(L) \subset_{\ne} W^{1,p}(\Omega).$$

Let the domain $\Omega = \{(x,y) \in \mathbb{R}^2 \mid 0 \le x \le 1,\ y = x^r\}$ and take a function $u(x,y) = x^\alpha$ on $(x,y) \in \Omega$.

$$\|u\|_{L^2(\Omega)}^2 = \int_\Omega |u|^2 = \int_\Omega |x|^{2\alpha} = \int_0^1 |x|^{2\alpha}\, dx \int_0^{x^r} dy$$
$$= \int_0^1 |x|^{2\alpha+r}\, dx < \infty \text{ if } 2\alpha + r > -1,$$

$$\left\|\frac{\partial u}{\partial x}\right\|_{L^2(\Omega)}^2 = \int_\Omega \left|\frac{\partial u}{\partial x}\right|^2 = \alpha^2 \int_\Omega |x|^{2\alpha-2} = \alpha^2 \int_0^1 |x|^{2\alpha+2}\, dx \int_0^{x^r} dy$$
$$= \alpha^2 \int_0^1 |x|^{2\alpha-2+r}\, dx < \infty \text{ if } 2\alpha - 2 + \gamma > -1.$$

Thus, if $2\alpha + r > 1$, then $u \in W^{1,2}(\Omega)$. In particular, if r is large then α may be negative. We will prove later the following Sobolev inequality: if $\Omega \subset \mathbb{R}^2$ is bounded, $v \in A^{1,2}(\Omega)$, then $v \in L^p(\Omega)$ for each p, $1 \le p < \infty$. Hence, if $u \in A^{1,2}(\Omega)$, then $u \in L^p(\Omega)$ for each p, $1 \le p < \infty$. Now,

$$\|u\|_{L^p(\Omega)}^p = \int_\Omega |x|^{p\alpha} = \int_0^1 |x|^{p\alpha}\, dx \int_0^{x^r} dy = \int_0^1 |x|^{p\alpha+r}\, dx < \infty,$$

if $p\alpha + r > -1$ for each p, $1 \le p < \infty$. This implies that $\alpha \ge 0$. Therefore, if we take $\alpha < 0$, $r \in \mathbb{R}$ with $2\alpha + r > 1$, then $u \in W^{1,2}(\Omega)$ but $u \notin A^{1,2}(\Omega)$.

6.5   The space $\mathbf{W^{m,p}}(\Omega)$

For an integer m, $m \geq 2$ and a real p, $1 \leq p \leq \infty$, define

$$W^{m,p}(\Omega) = \{u \in W^{m-1,p}(\Omega) \mid \frac{\partial u}{\partial x_i} \in W^{m-1,p}(\Omega) \text{ for } i = 1, 2, \cdots, N\}$$
$$= \{u \in L^p(\Omega) \mid D^\alpha u \in L^p(\Omega) \text{ for } |\alpha| \leq m\}.$$

Under the norm

$$\|u\|_{W^{m,p}} = \sum_{0 \leq |\alpha| \leq m} \|D^\alpha u\|_{L^p},$$

$W^{m,p}(\Omega)$ is a Banach space. If we set

$$H^m(\Omega) = W^{m,2}(\Omega),$$

under the scalar product

$$(u, v)_{H^m} = \sum_{0 \leq |\alpha| \leq m} (D^\alpha u, D^\alpha v)_{L^2},$$

then $H^m(\Omega)$ is a Hilbert space.

Remark 6.26 *If Ω is regular with $\Gamma = \partial\Omega$ bounded, then the following two norms are equivalent:*

$$\|u\|_{W^{m,p}(\Omega)} \approx \|u\|_{L^p} + \sum_{|\alpha|=m} \|D^\alpha u\|_{L^p}.$$

More precisely, for α, $0 < |\alpha| < m$, for $\epsilon > 0$ there is a constant $c = c(\Omega, \epsilon, \alpha)$ such that

$$\|D^\alpha u\|_{L^p(\Omega)} \leq \epsilon \sum_{|\beta|=m} \|D^\beta u\|_{L^p} + C \|u\|_{L^p},$$

for $u \in W^{m,p}(\Omega)$.

6.6   Sobolev Inequalities

In Chapter 5, we proved that if $\Omega \subset \mathbb{R}^1$, then

$$W^{1,p}(\Omega) \hookrightarrow L^\infty(\Omega),$$

with continuous injection. For $N \geq 2$, this inclusion still holds, but only for $p > N$. When $p \leq N$ we can construct a function in $W^{1,p}$, but not in L^∞. However, we still have the Sobolev inequality: for $1 \leq p < N$, then $W^{1,p}(\Omega) \subset L^{p^*}(\Omega)$ for some p^* in (p, ∞).

Lemma 6.27 *Assume* $N \geq 2$ *and* $f_1, \cdots, f_N \in L^{N-1}(\mathbb{R}^{N-1})$. *For any* $x \in \mathbb{R}^N$, *and* $1 \leq i \leq N$, *denote* $\widehat{x}_i = (x_1, \cdots, x_{i-1}, x_{i+1}, \cdots, x_N) \in \mathbb{R}^{N-1}$. *Set the function* $f(x) = f_1(\widehat{x}_1) f_2(\widehat{x}_2) \cdots f_N(\widehat{x}_N)$. *Then* $f(x) \in L^1(\mathbb{R}^N)$, *and* $\|f\|_{L^1(\mathbb{R}^N)} \leq \Pi_{i=1}^N \|f_i\|_{L^{N-1}(\mathbb{R}^{N-1})}$.

Proof.

(i) $N = 2$, f_1, $f_2 \in L^1(\mathbb{R}^1)$, $f(x,y) = f_1(y) f_2(x)$. Then

$$\|f\|_{L^1(\mathbb{R}^2)} = \int |f(x,y)| \, dx dy = \int |f_1(y)| \, |f_2(x)| \, dx dy = \|f_1\|_{L^1(\mathbb{R})} \|f_2\|_{L^1(\mathbb{R})}.$$

(ii) $N = 3$, f_1, f_2, $f_3 \in L^2(\mathbb{R}^2)$, $f(x_1, x_2, x_3) = f_1(x_2, x_3) f_2(x_1, x_3) f_3(x_1, x_2)$. Then

$$\int_{\mathbb{R}} |f(x)| \, dx_3 = |f_3(x_1, x_2)| \int_{\mathbb{R}} |f_1(x_2, x_3)| \, |f_2(x_1, x_3)| \, dx_3$$
$$\leq |f_3(x_1, x_2)| \left(\int_{\mathbb{R}} |f_1(x_2, x_3)|^2 \, dx_3 \right)^{1/2} \left(\int_{\mathbb{R}} |f_2(x_1, x_3)|^2 \, dx_3 \right)^{1/2}.$$

If we let $g(x_1, x_2) = \left(\int_{\mathbb{R}} |f_1(x_2, x_3)|^2 \, dx_3 \right)^{1/2} \left(\int_{\mathbb{R}} |f_2(x_1, x_3)|^2 \, dx_3 \right)^{1/2}$, then

$$\int_{\mathbb{R}^3} |f(x)| \, dx_3 = \int_{\mathbb{R}^2} dx_1 dx_2 \int_{\mathbb{R}} |f(x)| \, dx_3$$
$$\leq \int_{\mathbb{R}^2} |f_3(x_1, x_2))| \, g(x_1, x_2) dx_1 dx_2$$
$$\leq \left(\int_{\mathbb{R}^2} |f_3(x_1, x_2)|^2 \, dx_1 dx_2 \right)^{1/2} \cdot \left(\int_{\mathbb{R}^2} g(x_1, x_2)^2 dx_1 dx_2 \right)^{1/2}$$
$$= \|f_3\|_{L^2(\mathbb{R}^2)} \|g\|_{L^2(\mathbb{R}^2)}$$
$$\leq \|f_3\|_{L^2(\mathbb{R}^2)} \|f_1\|_{L^2(\mathbb{R}^2)} \|f_2\|_{L^2(\mathbb{R}^2)} \text{ as in part } (i).$$

(iii) Suppose the theorem holds for $N = N$, and consider the case $N + 1$. Let f_1, $\cdots$, $f_{N+1} \in L^N(\mathbb{R}^N)$, and

$$f(x) = f_1(\widehat{x}_1) \cdots f_{N+1}(\widehat{x_{N+1}}).$$

Let $N' = \frac{N}{N-1}$ or $\frac{1}{N} + \frac{1}{N'} = 1$, and note that

$$\||f|^p\|_{L^q} = \|f\|_{L^{pq}}^p,$$

so that we have, by the recurrence hypothesis,

$$\int_{\mathbb{R}^N} |f(x)|\, dx_1, \cdots, dx_N$$

$$= \int_{\mathbb{R}^N} |f_1| \cdots |f_{N+1}|\, dx_1 \cdots dx_N$$

$$\leq \left(\int_{\mathbb{R}^N} |f_{N+1}|^N\, dx_1, \cdots, dx_N \right)^{1/N} \left(\int_{\mathbb{R}^N} |f_1|^{N'} \cdots |f_N|^{N'}\, dx_1, \cdots, dx_N \right)^{1/N}$$

$$\leq \|f_{N+1}\|_{L^N(\mathbb{R}^N)} \left[\prod_{i=1}^N \left\| |f_i|^{N'} \right\|_{L^{N-1}(\mathbb{R}^{N-1})} \right]^{1/N'}$$

$$= \|f_{N+1}\|_{L^N(\mathbb{R}^N)} \left[\prod_{i=1}^N \left\| |f_i|^{N'} \right\|_{L^{N'(N-1)}(\mathbb{R}^{N-1})}^{N'} \right]^{1/N'}$$

$$= \|f_{N+1}\|_{L^N(\mathbb{R}^N)} \cdot \prod_{i=1}^N \|f_i\|_{L^N(\mathbb{R}^{N-1})}.$$

Now by the generalized Hölder inequality

$$\int_{\mathbb{R}^{N+1}} |f(x)|\, dx_1, \cdots, dx_{N+1} \leq \|f_{N+1}\|_{L^N(\mathbb{R}^N)} \int_{\mathbb{R}} \prod_{i=1}^N \|f_i\|_{L^N(\mathbb{R}^{N-1})}\, dx_{N+1}$$

$$\leq \|f_{N+1}\|_{L^N(\mathbb{R}^N)} \prod_{i=1}^N \left[\left(\int_{\mathbb{R}} \|f_i\|_{L^N(\mathbb{R}^{N-1})}^N\, dx_{N+1} \right)^{1/N} \right]$$

$$= \|f_{N+1}\|_{L^N(\mathbb{R}^N)} \prod_{i=1}^N \|f_i\|_{L^N(\mathbb{R}^N)},$$

or

$$\|f\|_{L^1(\mathbb{R}^{N+1})} \leq \prod_{i=1}^{N+1} \|f_i\|_{L^N(\mathbb{R}^N)}. \qquad \blacksquare$$

Theorem 6.28 (*Sobolev, Gagliardo-Nirenberg*) *For $1 \leq p < N$, and $\frac{1}{p*} = \frac{1}{p} - \frac{1}{N}$, we have $W^{1,p}(\mathbb{R}^N) \subset L^{p*}(\mathbb{R}^N)$, with continuous injection. Moreover, there is $c = c(p, N)$ such that*

$$\|u\|_{L^{p*}} \leq c \|\nabla u\|_{L^p} \text{ for } u \in W^{1,p}(\mathbb{R}^N).$$

Proof.

(i) $p = 1$, $p* = \dfrac{N}{N-1}$ or $\dfrac{1}{p*} = 1 - \dfrac{1}{N}$. If we let $u \in C_c^1(\mathbb{R}^N)$, then

$$|u(x_1, \cdots, x_N)| = \left| \int_{-\infty}^{x_1} \frac{\partial u}{\partial x_1}(t, x_2, \cdots, x_N) dt \right|$$

$$\leq \int_{-\infty}^{\infty} \left| \frac{\partial u}{\partial x_1}(t, x_2, \cdots, x_N) \right| dt.$$

Similarly,

$$|u(x_1, \cdots, x_N)| = \int_{-\infty}^{\infty} \left| \frac{\partial u}{\partial x_i}(x_1, \cdots, x_{i-1}, t, x_{i+1}, \cdots, x_N) \right| dt.$$

Let

$$f_i(x_1, \cdots, x_{i-1}, x_{i+1}, \cdots, x_N) = \int_{-\infty}^{\infty} \left| \frac{\partial u}{\partial x_i}(x_1, \cdots, x_{i-1}, t, x_i, \cdots, x_N) \right| dt,$$

and we obtain

$$|u|^N \leq \prod_{i=1}^{N} f_i(\widehat{x_i}),$$

or

$$|u|^{N/(N-1)} \leq \prod_{i=1}^{N} (f_i(\widehat{x_i}))^{1/(N-1)}.$$

By Lemma 6.27,

$$\int |u|^{p*} = \int |u|^{N/(N-1)} \leq \prod_{i=1}^{N} \left\| f_i^{1/(N-1)} \right\|_{L^{N-1}(\mathbb{R}^{N-1})}$$

$$= \prod_{i=1}^{N} \|f_i\|_{L^1(\mathbb{R}^{N-1})}^{1/(N-1)} = \prod_{i=1}^{N} \left\| \frac{\partial u}{\partial x_i} \right\|_{L^1(\mathbb{R}^N)}^{1/(N-1)}.$$

Thus,

$$\|u\|_{L^{p*}} = \left(\int |u|^{N/(N-1)} \right)^{(N-1)/N} \leq \left(\prod_{i=1}^{N} \left\| \frac{\partial u}{\partial x_i} \right\|_{L^1(\mathbb{R}^N)}^{1/(N-1)} \right)^{(N-1)/N}$$

$$= \prod_{i=1}^{N} \left\| \frac{\partial u}{\partial x_i} \right\|_{L^1(\mathbb{R}^N)}^{1/N} \leq \prod_{i=1}^{N} \|\nabla u\|_{L^1(\mathbb{R}^N)}^{1/N} = \|\nabla u\|_{L^1(\mathbb{R}^N)}. \tag{6.3}$$

(ii) $1 < p < N$. For $t \geq 1$, replace u by $|u|^{t-1} u$ in (6.3), and we obtain

$$\|u\|^t_{L^{N/(N-1)t}(\mathbb{R}^N)} = \left\| |u|^{t-1} u \right\|_{L^{N/(N-1)}(\mathbb{R}^N)}$$

$$\leq \prod_{i=1}^{N} \left\| \frac{\partial}{\partial x_i} \left(|u|^{t-1} u \right) \right\|^{1/N}_{L^1(\mathbb{R}^N)}$$

$$= t \prod_{i=1}^{N} \left\| |u|^{t-1} \frac{\partial u}{\partial x_i} \right\|^{1/N}_{L^1(\mathbb{R}^N)},$$

where we use the formula

$$\frac{\partial |u|^{t-1} u}{\partial x_i} = \begin{cases} t |u|^{t-1} \dfrac{\partial |u|}{\partial x_i} & \text{for } u \geq 0, \\[2mm] -t |u|^{t-1} \dfrac{\partial |u|}{\partial x_i} & \text{for } u < 0. \end{cases}$$

Thus,

$$\|u\|^t_{L^{N/(N-1)t}(\mathbb{R}^N)} \leq t \prod_{i=1}^{N} \left\| |u|^{t-1} \right\|^{1/N}_{L^{p'}(\mathbb{R}^N)} \left\| \frac{\partial u}{\partial x_i} \right\|^{1/N}_{L^p(\mathbb{R}^N)}$$

$$= t \|u\|^{t-1}_{L^{p'(t-1)}(\mathbb{R}^N)} \prod_{i=1}^{N} \|\nabla u\|^{1/N}_{L^p(\mathbb{R}^N)}$$

$$= t \|u\|^{t-1}_{L^{p'(t-1)}(\mathbb{R}^N)} \|\nabla u\|_{L^p(\mathbb{R}^N)}.$$

Hence,

$$\|u\|^t_{L^{N/(N-1)t}} \leq t \|u\|^{t-1}_{L^{p'(t-1)}} \|\nabla u\|_{L^p}. \qquad (6.4)$$

Choose t with

$$\frac{tN}{N-1} = p'(t-1) \text{ if and only if } \frac{tN}{N-1} = \frac{p}{p-1}(t-1)$$

if and only if $t \left(\dfrac{N}{N-1} - \dfrac{p}{p-1} \right) = -\dfrac{p}{p-1}$

if and only if $t = \dfrac{-p/(p-1)}{N/(N-1) - p/(p-1)} = \dfrac{p(N-1)}{N-p} = \dfrac{N-1}{N} \dfrac{Np}{N-p}$

if and only if $t = \dfrac{N-1}{N} p* $ if and only if $\dfrac{N}{N-1} t = p* = p'(t-1)$.

Now,

$$\|u\|^{((N-1)/N)p*}_{L^{p*}} \leq t \|u\|^{((N-1)/N)p*-1}_{L^{p*}} \|\nabla u\|_{L^p}.$$

Thus,

$$\|u\|_{L^{p*}} \leq t \|\nabla u\|_{L^p} = \frac{(N-1)p}{N-p} \|\nabla u\|_{L^p}.$$

(iii) For $u \in W^{1,p}(\mathbb{R}^N)$, take a sequence (u_n) in $C_c^1(\mathbb{R}^N)$ such that

$$u_n \to u \text{ in } W^{1,p}(\mathbb{R}^N), \text{ a.e. on } \mathbb{R}^N.$$

Because, by (i) and (ii),

$$\|u_n\|_{L^{p*}} \leq C \|\nabla u_n\|_{L^p},$$

we have

$$\|u\|_{L^{p*}} = \left(\int |u|^{p*} \right)^{1/p*} = \left(\int \lim_{n \to \infty} |u_n|^{p*} \right)^{1/p*}$$

$$\leq \underline{\lim}_{n \to \infty} \left(\int |u_n|^{p*} \right)^{1/p*} \quad \text{by the Fatou Lemma}$$

$$\leq c \underline{\lim}_{n \to \infty} \|\nabla u_n\|_{L^p} = C \|\nabla u\|_{L^p}.$$

Hence, $u \in L^{p*}(\mathbb{R}^N)$, and $\|u\|_{L^{p*}} \leq c \|\nabla u\|_{L^p}$. ∎

Corollary 6.29 *Let $1 \leq p < N$. Then $W^{1,p}(\mathbb{R}^N) \subset L^q(\mathbb{R}^N)$ for $q \in [p, p*]$, with the continuous injection: $\|u\|_{L^q} \leq C \|u\|_{W^{1,p}}$.*

Proof. For any q, $p \leq q \leq p*$, write

$$\frac{1}{q} = \frac{\alpha}{p} + \frac{1-\alpha}{p*} \quad \text{for some } \alpha, \ 0 \leq \alpha \leq 1.$$

Then,

$$\|u\|_{L^q} \leq \|u\|_{L^p}^{\alpha} \|u\|_{L^{p*}}^{1-\alpha}$$

$$\leq \|u\|_{L^p} + \|u\|_{L^{p*}}$$

$$\leq C \|u\|_{W^{1,p}} \text{ for } u \in W^{1,p}(\mathbb{R}^N). \quad ∎$$

Remark 6.30 *By Theorem 6.28 and Corollary 6.29, we obtain $\|u\|_{L^{p*}} \leq c \|\nabla u\|_{L^p}$. Moreover, $\|u\|_{L^q} \leq c \|u\|_{W^{1,p}}$ for $q \in [p, p*]$. Actually, this is the only case for $p*$, because $u \in W^{1,p}(\mathbb{R}^N) : \|u\|_{L^q} \leq C \|\nabla u\|_{L^p}$ is false for $q \in [p, p*)$.*

Proof. Suppose for some q, $1 \leq q \leq \infty$, there is $C > 0$ such that

$$\|u\|_{L^q} \leq C \|\nabla u\|_{L^p} \text{ for } u \in W^{1,p}(\mathbb{R}^N).$$

By the homogeneity principle, take $u_\lambda(x) = u(\lambda x)$ for $\lambda > 0$,

$$\|u_\lambda\|_{L^q}^q = \int |u(\lambda x)|^q \, dx = \lambda^{-N} \int |u(x)|^q \, dx = \lambda^{-N} \|u\|_{L^q}^q,$$

$$\|\nabla u_\lambda\|_{L^p}^p = \int |\nabla u_\lambda(x)|^p \, dx = \lambda^p \int |\nabla u(\lambda x)|^p \, dx$$

$$= \lambda^p \lambda^{-N} \int |\nabla u(x)|^p \, dx = \lambda^{p-N} \|\nabla u\|_{L^p}^p.$$

Recall that $\|u_\lambda\|_{L^q} \le c\|\nabla u_\lambda\|_{L^p}$, so $\lambda^{-N/q}\|u\|_{L^q} \le C\lambda^{(p-N)/p}\|\nabla u\|_{L^p}$, or

$$\|u\|_{L^q} \le C\lambda^{1-(N/p)+(N/q)}\|\nabla u\|_{L^p}.$$

Because $\lambda > 0$ is arbitrary,

$$1 - \frac{N}{p} + \frac{N}{q} = 0, \text{ and } \frac{1}{q} = \frac{1}{p} - \frac{1}{N} \iff q = p*. \qquad \blacksquare$$

Theorem 6.31 *(The limit case $p = N$) We have*

$$W^{1,N}(\mathbb{R}^N) \subset L^q(\mathbb{R}^N) \text{ for } q \in [N, \infty),$$

with continuous injection.

Proof. By (6.4) for $u \in C_c^1(\mathbb{R}^N)$,

$$\|u\|_{L^{tN/(N-1)}}^t \le t\|u\|_{L^{p'(t-1)}}^{t-1}\|\nabla u\|_{L^p} \text{ for } t \ge 1.$$

Let $p = N$, then $p' = \frac{N}{N-1}$, and

$$\|u\|_{L^{tN/N-1}}^t \le t\|u\|_{L^{[N/(N-1)](t-1)}}^{t-1}\|\nabla u\|_{L^N} \text{ for } t \ge 1.$$

Alternatively,

$$\|u\|_{L^{tN/(N-1)}} \le t^{1/t}\|u\|_{L^{[N/(N-1)](t-1)}}^{(t-1)/t}\|\nabla u\|_{L^N}^{1/t} \text{ for } t \ge 1.$$

Because $\dfrac{1}{t/(t-1)} + \dfrac{1}{t/1} = 1$, $\lim\limits_{t\to\infty} t^{1/t} = 1$, and $\lim\limits_{t\to 1} t^{1/t} = 1$. Therefore,

$$\|u\|_{L^{tN/(N-1)}} \le c(\|u\|_{L^{(t-1)[N/(N-1)]}} + \|\nabla u\|_{L^N}).$$

Let $t = N$. Then

$$\|u\|_{L^{N^2/N-1}} \le C(\|u\|_{L^N} + \|\nabla u\|_{L^N}) \le C\|u\|_{W^{1,N}}.$$

By the Interpolation Property

$$\|u\|_{L^q} \le c\|u\|_{W^{1,N}} \text{ for } q \in \left[N, \frac{N^2}{N-1}\right].$$

If we take $t = N + 1$

$$\|u\|_{L^{[N(N+1)]/N-1}} \le c\left(\|u\|_{L^{N^2/(N-1)}} + \|\nabla u\|_{L^N}\right) \le C\|u\|_{W^{1,N}}.$$

By the Interpolation Property

$$\|u\|_{L^q} \le C\|u\|_{W^{1,N}} \text{ for } q \in \left[\frac{N^2}{N-1}, \frac{N(N+1)}{N-1}\right].$$

Continuing this process, we have

$$\|u\|_{L^q} \le C\|u\|_{W^{1,N}} \text{ for } q \in [N, \infty]$$

where $C = C(q, N) \to \infty$ as $q \to \infty$. $\qquad \blacksquare$

Theorem 6.32 (*Morrey Theorem*) *Let $p > N$. Then $W^{1,p}(\mathbb{R}^N) \subset L^\infty(\mathbb{R}^N)$ with continuous injection. Moreover, for $u \in W^{1,p}(\mathbb{R}^N)$*

$$|u(x) - u(y)| \leq c\,|x - y|^\alpha\,\|\nabla u\|_{L^p} \text{ a.e. for } x, y \in \mathbb{R}^N,$$

where $\alpha = 1 - \frac{N}{p}$, $c = C(p, N)$.

Proof.

(i) Let $u \in C_c^1(\mathbb{R}^N)$. For any x, $y \in \mathbb{R}^N$, let Q be an open cube centered at $\frac{x+y}{2}$ with side length $r = 2(x - y)$. For any $\eta \in Q$,

$$u(\eta) - u(x) = \int_0^1 \frac{d}{dt} u(x + t(\eta - x))dt$$
$$= \int_0^1 \nabla u(x + t(\eta - x)) \cdot (\eta - x)dt.$$

Now,

$$\frac{1}{|Q|} \int_Q u(\eta)d\eta - u(x) = \frac{1}{|Q|} \int_Q [u(\eta) - u(x)]d\eta$$
$$= \frac{1}{|Q|} \int_Q d\eta \int_0^1 \nabla u(x + t(\eta - x)) \cdot (\eta - x)dt$$
$$= \frac{1}{|Q|} \int_0^1 dt \int_Q \nabla u(x + t(\eta - x)) \cdot (\eta - x)d\eta$$
$$\leq \frac{1}{|Q|} \int_0^1 dt \int_Q |\nabla u(x + t(\eta - x))|\,|\eta - x|\,d\eta$$
$$\leq \frac{r}{|Q|} \int_0^1 dt \int_Q |\nabla u(x + t(\eta - x))|\,d\eta$$
$$\leq \frac{r}{|Q|} \int_0^1 dt \left(\int_Q |\nabla u(x + t(\eta - x)|^p\,d\eta \right)^{1/p} |Q|^{1-(1/p)}$$
$$= r\,|Q|^{-1/p} \int_0^1 dt \left(\int_{x+t(Q-x)} t^{-N} |\nabla u(\xi)|^p\,d\xi \right)^{1/p}$$
$$\leq r\,|Q|^{-1/p} \|\nabla u\|_{L^p} \int_0^1 t^{-N/p}\,dt = \frac{r^{1-(N/p)}}{1 - N/p} \|\nabla u\|_{L^p}$$
$$\leq C\,|x - y|^\alpha\,\|\nabla u\|_{L^p} \text{ where } \alpha = 1 - \frac{N}{p}.$$

Similarly,

$$\frac{1}{|Q|} \int_Q u(\eta)d\eta - u(y) \leq C\,|x - y|^\alpha\,\|\nabla u\|_{L^p}.$$

By the triangle inequality

$$|u(x) - u(y)| \le C\,|x - y|^{\alpha}\,\|\nabla u\|_{L^p}\,.$$

(ii) For general $u \in W^{1,p}(\mathbb{R}^N)$. Take a sequence $(u_n) \subset C_c^1(\mathbb{R}^N)$ such that $u_n \to u$ in $W^{1,p}(\mathbb{R}^N)$, a.e. on $\mathbb{R}^N$. Because

$$|u_n(x) - u_n(y)| \le c\,|x - y|^{\alpha}\,\|\nabla u_n\|_{L^p(\mathbb{R}^N)}\,.$$

Letting $n \to \infty$, we obtain

$$|u(x) - u(y)| \le c\,|x - y|^{\alpha}\,\|\nabla u\|_{L^p(\mathbb{R}^N)}\ \text{ a.e. for } x, y \in \mathbb{R}^N.$$

(iii) Let $u \in C_c^1(\mathbb{R}^N)$. For $x \in \mathbb{R}^N$, let Q be a cube centered at x with side length $r = 1$. As in part (i),

$$\left| \frac{1}{|Q|} \int_Q u(\eta) d\eta - u(x) \right| \le c\,\|\nabla u\|_{L^p(\mathbb{R}^N)}\,,$$

or

$$
\begin{aligned}
|u(x)| &\le \left| \frac{1}{|Q|} \int_Q u(\eta) d\eta \right| + c\,\|\nabla u\|_{L^p(\mathbb{R}^N)} \\
&\le \frac{1}{|Q|} \left(\int_Q |u|^p \right)^{1/p} |Q|^{1/p'} + c\,\|\nabla u\|_{L^p(\mathbb{R}^N)} \\
&\le c\,\|u\|_{L^p(Q)} + c\,\|\nabla u\|_{L^p(\mathbb{R}^N)}\ \text{ because } |Q| = \text{constant} \\
&\le c\,\|u\|_{L^p(\mathbb{R}^N)} + c\,\|\nabla u\|_{L^p(\mathbb{R}^N)} \\
&= c\,\|u\|_{W^{1,p}(\mathbb{R}^N)}\,,
\end{aligned}
$$

where $c = c(p, N)$. Hence,

$$\|u\|_{L^\infty(\mathbb{R}^N)} \le c\,\|u\|_{W^{1,p}(\mathbb{R}^N)}\,.$$

(iv) For $u \in W^{1,p}(\mathbb{R}^N)$, take $(u_n) \subset C_c^1(\mathbb{R}^N)$ such that

$$u_n \to u \quad \text{in } W^{1,p}(\mathbb{R}^N), \quad \text{a.e. on } \mathbb{R}^N.$$

Because $|u_n(x)| \le c\,\|u_n\|_{W^{1,p}(\mathbb{R}^N)}$. Letting $n \to \infty$,

$$|u(x)| \le c\,\|u\|_{W^{1,p}(\mathbb{R}^N)} \qquad \text{a.e. for } x \in \mathbb{R}^N,$$

or $\|u\|_{L^\infty(\mathbb{R}^N)} \le c\,\|u\|_{W^{1,p}(\mathbb{R}^N)}$.   ∎

Remark 6.33 *The inequality*

$$|u(x) - u(y)| \leq c\,|x - y|^{\alpha}\,\|\nabla u\|_{L^p} \text{ a.e. for } x,\ y \in \mathbb{R}^N$$

implies that there is $\widetilde{u} \in C(\mathbb{R}^N)$ such that

$$u = \widetilde{u} \text{ a.e. on } \mathbb{R}^N.$$

Proof.   Let $A \subset \mathbb{R}^N$, $|A| = 0$ such that $|u(x) - u(y)| \leq c\,|x - y|^{\alpha}\,\|\nabla u\|_{L^p}$ for $x, y \in \mathbb{R}^N \backslash A$. Note that $\mathbb{R}^N \backslash A$ is dense in $\mathbb{R}^N$, and u is continuous on $\mathbb{R}^N \backslash A$. For $x \in A$, take a sequence (x_n) in $\mathbb{R}^N \backslash A$ such that $x = \lim\limits_{n \to \infty} x_n$. Because

$$|u(x_n) - u(x_m)| \leq c\,|x_n - x_m|^{\alpha}\,\|\nabla u\|_{L^p} \to 0 \text{ as } n, m \to 0,$$

therefore $(u(x_n))$ is Cauchy. Define

$$\widetilde{u}(x) = \lim\limits_{n \to \infty} u(x_n).$$

We see that $\widetilde{u}$ is continuous on $\mathbb{R}^N$ and $\widetilde{u} = u$ on $\mathbb{R}^N \backslash A$. Therefore we may write:

$$W^{1,p}(\mathbb{R}^N) \subset C(\mathbb{R}^N) \cap L^{\infty}(\mathbb{R}^N), \text{ for } p > N. \qquad \blacksquare$$

Remark 6.34 *Let $u \in W^{1,p}(\mathbb{R}^N)$, then $\lim\limits_{|x| \to \infty} u(x) = 0 : W^{1,p}(\mathbb{R}^N) \hookrightarrow C_0(\mathbb{R}^N)$, $p > N$.*

Proof.   Take a sequence $(u_n) \subset C_c^1(\mathbb{R}^N)$ such that $u_n \to u$ in $W^{1,p}(\mathbb{R}^N)$, a.e. on $\mathbb{R}^N$. By Theorem 6.32

$$\|u_n - u\|_{L^{\infty}} \leq c\,\|u_n - u\|_{W^{1,p}(\mathbb{R}^N)}.$$

For $\epsilon > 0$ there is a constant $n_0 > 0$ with $\|u_{n_0} - u\|_{L^{\infty}} < \epsilon$. Take $M > 0$ such that $|x| \geq M \Rightarrow |u_{n_0}(x)| = 0$, or

$$|u(x)| \leq |u_{n_0}(x) - u(x)| + |u_{n_0}(x)| < \epsilon \text{ by Remark 6.33.}$$

Therefore $\lim\limits_{|x| \to \infty} u(x) = 0$. $\qquad \blacksquare$

Theorem 6.35 *Let m be a positive integer, $m \geq 1$, and p a real number, $1 \leq p < \infty$. Define $p*$ as $\frac{1}{p*} = \frac{1}{p} - \frac{m}{N}$. Then:*

(i) *If $1 \leq p < \frac{m}{N}$ then $W^{m,p}(\mathbb{R}^N) \hookrightarrow L^q(\mathbb{R}^N)$, for $q \in [p, p*]$, with continuous injection;*

(ii) *If $p = \frac{N}{m}$, then $W^{m,p}(\mathbb{R}) \hookrightarrow L^q(\mathbb{R}^N)$, for $q \in [p, \infty)$ with continuous injection;*

(iii) If $p > \frac{N}{m}$, then $W^{m,p}(\mathbb{R}^N) \hookrightarrow L^\infty(\mathbb{R}^N)$ with continuous injection;

(iv) Moreover, if $p > \frac{N}{m}$, $m - \frac{N}{p}$ is not an integer. Let

$$k = \left[m - \frac{N}{p} \right], \ \theta = \left(m - \frac{N}{p} \right) - k, 0 < \theta < 1.$$

For $u \in W^{m,p}(\mathbb{R}^N)$, we have, for α, $|\alpha| = k$,

$\|D^\alpha u\|_{L^\infty} \le C \|u\|_{W^{m,p}}$ for $|\alpha| \le k$

$|D^\alpha u(x) - D^\alpha u(y)| \le C |u|_{W^{m,p}} |x - y|$ a.e. for $x, y \in \mathbb{R}^N$, for $|\alpha| = k$.

In particular,

$$W^{m,p}(\mathbb{R}^N) \hookrightarrow C^k(\mathbb{R}^N).$$

Proof.

(i) (a) $m = 2$, $u \in W^{2,p}$. Then, for $i = 1, \cdots, N$, u, $\frac{\partial u}{\partial x_i} \in W^{1,p}$ or u, $\frac{\partial u}{\partial x_i} \in L^a$, $\frac{1}{a} = \frac{1}{p} - \frac{1}{N}$, we have $u \in W^{1,a}$ or $u \in L^{p*}$, where

$$\frac{1}{p*} = \frac{1}{a} - \frac{1}{N} = \frac{1}{p} - \frac{2}{N}.$$

(b) Suppose it is correct for $m = m$, if $u \in W^{m,p}$, then $u \in L^{p*}$, where $\frac{1}{p*} = \frac{1}{p} - \frac{m}{N}$.

(c) Let $u \in W^{m+1,p}$, then for $i = 1, \cdots, N$, u, $\frac{\partial u}{\partial x_i} \in W^{m,p}$, or u, $\frac{\partial u}{\partial x_i} \in L^b$, $\frac{1}{b} = \frac{1}{p} - \frac{m}{N}$. Thus, $u \in W^{1,b}$, or $u \in L^{p*}$, where

$$\frac{1}{p*} = \frac{1}{b} - \frac{1}{N} = \frac{1}{p} - \frac{m+1}{N}.$$

(ii) $m \ge 2$, $u \in W^{m,\frac{N}{m}}$. Then, for $i = 1, \cdots, N$, u, $\frac{\partial u}{\partial x_i} \in W^{m-1,\frac{N}{m}} \subset L^N$. Note that $p = \frac{N}{m} < \frac{N}{m-1}$, and

$$\frac{1}{c} = \frac{1}{p} - \frac{m-1}{N} = \frac{m}{N} - \frac{m-1}{N} = \frac{1}{N}.$$

Therefore, $u \in W^{1,N} \subset L^q$ for $q \in [N, \infty)$. However, $u \in W^{m,p}$, $p = \frac{N}{m}$, so $u \in L^p$. Consequently $u \in L^q$, $q \in [p, \infty)$.

(iii) The theorem is true for $m = 1$. Suppose it is also true for $m = m$. Let $p > \frac{N}{m+1}$, $u \in W^{m+1,p}$. Then u, $\frac{\partial u}{\partial x_i} \in W^{m,p}$, for $i = 1, \cdots, N$. If $p > \frac{N}{m}$ then $u \in W^{m,p} \subset L^\infty$. If $p \le \frac{N}{m}$, then u, $\frac{\partial u}{\partial x_i} \in W^{m,p} \subset L^{p*}$ where

$$\frac{1}{p*} = \frac{1}{p} - \frac{m}{N} < \frac{m}{N} < \frac{1}{N},$$

or $p* > N$. Now $u \in W^{m,p} \subset L^\infty$, so $u \in L^\infty(\mathbb{R}^N)$.

(iv) Let $k = \left[m - \frac{N}{p}\right]$, $\theta = \left(m - \frac{N}{p}\right) - k$, and we have

$$k + 1 = \left[m + 1 - \frac{N}{p}\right], \quad \theta = \left(m + 1 - \frac{N}{p}\right) - (k+1).$$

Suppose $u \in W^{m,p}$ implies

$$\|D^\alpha u\|_{L^\infty} \leq C \|u\|_{W^{m,p}} \text{ for } |\alpha| \leq k,$$

$$|D^\alpha u(x) - D^\alpha u(y)| \leq C \|u\|_{W^{m,p}} |x - y|^\theta \text{ a.e. for } x, y \in \mathbb{R}^N, |\alpha| = k,$$

$$W^{m,p} \hookrightarrow C^k.$$

Let $u \in W^{m+1,p}$. Then $u, \frac{\partial u}{\partial x_i} \in W^{m,p}$ for $i = 1, \cdots, N$. By the inductive hypothesis, replace u by $\frac{\partial u}{\partial x_i}$,

$$\|D^\alpha u\|_{L^\infty} \leq C \|u\|_{W^{m+1,p}}, \quad |\alpha| \leq k + 1,$$

$$|D^\alpha u(x) - D^\alpha u(y)| \leq C \|u\|_{W^{m+1,p}} |x - y|^\theta, \quad |\alpha| = k + 1.$$

Moreover, $u, \frac{\partial u}{\partial x_i} \subset C^k$, so $u \in C^{k+1}$. ∎

Remark 6.36 *By Theorem 6.35, $W^{N,1}(\mathbb{R}^N) \subset L^q(\mathbb{R}^N)$ for $q \in [1, \infty)$. Actually we have*

$$W^{N,1}(\mathbb{R}^N) \subset L^\infty(\mathbb{R}^N)$$

Proof. For $u \in C_c^\infty(\mathbb{R}^N)$,

$$u(x_1, \cdots, x_N) = \int_{-\infty}^{x_1} \int_{-\infty}^{x_2} \cdots \int_{-\infty}^{x_N} \frac{\partial^N u}{\partial x_1, \cdots, \partial x_N}(t_1, \cdots, t_N) dt_1, \cdots, dt_N$$

$$= \int_{-\infty}^{\infty} \cdots \int_{-\infty}^{\infty} \left| \frac{\partial^N u}{\partial x_1, \cdots, \partial x_N} \right| dt_1, \cdots, dt_N$$

$$\leq \|u\|_{W^{N,1}}.$$

Therefore $\|u\|_{L^\infty} \leq \|u\|_{W^{N,1}}$. By the Forth Density Theorem 6.24, $\|u\|_{L^\infty} \leq \|u\|_{W^{N,1}}$ for $u \in W^{N,1}(\mathbb{R}^N)$. ∎

Remark 6.37 *By Theorem 6.35*

$$W^{1,p}(I) \subset L^\infty(I),$$

where $I = (a, b)$ is an interval in $\mathbb{R}$, if $p > 1$. However, we proved in Chapter 5 that for $u \in W^{1,p}(I)$, $1 \leq p < \infty$, there is $\tilde{u} \in C(\bar{I})$ such that

$$u = \tilde{u} \text{ a.e. on } I,$$

$$\tilde{u}(x) - \tilde{u}(y) = \int_y^x u'(t) dt \text{ for } x, y \in \bar{I}.$$

From now on, we suppose $\Omega \subset \mathbb{R}^N$ is an open set of class C^1 with $\Gamma = \partial\Omega$ bounded, or $\Omega = \mathbb{R}^N_+$. If we consider the extension operator

$$P : W^{1,p}(\Omega) \to W^{1,p}(\mathbb{R}^N)$$

such that

$$Pu \mid_\Omega = u,$$
$$\|Pu\|_{L^p(\mathbb{R}^N)} \le c \|u\|_{L^p(\Omega)},$$
$$\|Pu\|_{W^{1,p}(\mathbb{R}^N)} \le c \|u\|_{W^{1,p}(\Omega)},$$

and apply Theorems 6.28, 6.31, 6.32 and 6.35, we obtain the following two theorems.

Theorem 6.38 *Let $1 \le p < \infty$. Then:*

(i) *If $1 \le p < N$, then $W^{1,p}(\Omega) \hookrightarrow L^q(\Omega)$ where $\frac{1}{p*} = \frac{1}{p} - \frac{1}{N}$, $q \in [p, p*]$.*

(ii) *If $p = N$, then $W^{1,p}(\Omega) \hookrightarrow L^q(\Omega)$, for $q \in [p, \infty)$.*

(iii) *If $p > N$, then $W^{1,p}(\Omega) \hookrightarrow L^\infty(\Omega)$.*
All injections in $(i) - (iii)$ are continuous.

(iv) *If $p > N$, $u \in W^{1,p}(\Omega)$. Then*

$$|u(x) - u(y)| \le c \|u\|_{W^{1,p}} |x - y|^\alpha \text{ a.e. for } x, \ y \in \Omega$$

with $\alpha = 1 - \frac{N}{p}$, $c = c(\Omega, p, N)$. In particular,

$$W^{1,p}(\Omega) \hookrightarrow C(\bar{\Omega}).$$

Theorem 6.39 *For an integer m, $m \ge 1$, and a real number p, $1 \le p < \infty$, then:*

(i) *If $1 \le p < \frac{N}{m}$ then $W^{m,p}(\Omega) \hookrightarrow L^q(\Omega)$, where*

$$\frac{1}{p*} = \frac{1}{p} - \frac{m}{N}, \quad q \in [p, p*];$$

(ii) *If $p = \frac{N}{m}$, then $W^{m,p}(\Omega) \hookrightarrow L^q(\Omega)$ for $q \in [p, \infty)$;*

(iii) *If $p > \frac{N}{m}$ then $W^{m,p}(\Omega) \hookrightarrow L^\infty(\Omega)$;*
All injections in (i)–(iii) are continuous.

(iv) *If $p > \frac{N}{m}$ and $m - \frac{N}{p}$ is not an integer, set*

$$k = \left[m - \frac{N}{p} \right], \text{ and } \theta = \left(m - \frac{N}{p} \right) - k, \, 0 < \theta < 1.$$

For $u \in W^{m,p}(\Omega)$, we have

$$\|D^\alpha u\|_{L^\infty} \leq c \|u\|_{W^{m,p}} \text{ for } \alpha, \ |\alpha| \leq k$$

$$|D^\alpha u(x) - D^\alpha u(y)| \leq c \|u\|_{W^{m,p}} |x-y|^\theta \text{ a.e. on } .x, \ y \in \Omega, \text{ for } \alpha, |\alpha| = k$$

$$W^{m,p}(\Omega) \subset C^k(\bar{\Omega}).$$

Theorem 6.40 (*Rellich-Kondrachov Theorem*) *Suppose $\Omega \subset \mathbb{R}^N$ is a bounded open set of class C^1. Then:*

(i) *If $1 \leq p < N$, then $W^{1,p}(\Omega) \subset L^q(\Omega)$ for $q \in [1, p*)$, $\frac{1}{p*} = \frac{1}{p} - \frac{1}{N}$, with compact injection;*

(ii) *If $p = N$, then $W^{1,p}(\Omega) \subset L^q(\Omega)$ for $q \in [1, \infty)$ with compact injection;*

(iii) *If $p > N$, then $W^{1,p}(\Omega) \subset C(\bar{\Omega})$ with compact injection.*

Proof.

(i) $1 \leq p < N$. Let $\mathcal{F}$ be the unit ball in $W^{1,p}(\Omega)$.

(a) For $\epsilon > 0$, $w \subset\subset \Omega$, $u \in F$, $|h| < dist(w, \Omega^c)$. For $q \in [1, p*)$, write

$$\frac{1}{q} = \frac{\alpha}{1} + \frac{1-\alpha}{p*} \text{ for some } \alpha, \ 0 < \alpha \leq 1.$$

By the Interpolation Property

$$\frac{1}{q} = \frac{1}{1/\alpha} + \frac{1}{p*/(1-\alpha)},$$

$$\|f\|_{L^q} \leq \left\| |f|^\alpha \right\|_{L^{1/\alpha}} \left\| |f|^{1-\alpha} \right\|_{L^{p*/(1-\alpha)}}$$

$$\leq \|f\|_{L^1}^\alpha \|f\|_{L^{p*}}^{1-\alpha}.$$

That is,

$$\|\tau_h u - u\|_{L^q(w)} \leq \|\tau_h u - u\|_{L^1(w)}^\alpha \|\tau_h u - u\|_{L^{p*}(w)}^{(1-\alpha)}$$

$$\leq \|\tau_h u - u\|_{L^1(w)}^\alpha \left(2 \|u\|_{L^{p*}(w)} \right)^{(1-\alpha)}$$

$$\leq C \|\tau_h u - u\|_{L^1(w)}^\alpha 2^{(1-\alpha)} \|\nabla u\|_{L^p(\Omega)}^{(1-\alpha)} \quad p < N.$$

Now Ω is bounded, so $u \in W^{1,p}(\Omega) \subset W^{1,1}(\Omega)$. By Remark 6.13 (i)

$$\|\tau_h u - u\|_{L^1(w)} \leq |h| \|\nabla u\|_{L^1(\Omega)}$$

$$\leq c|h| \|\nabla u\|_{L^p(\Omega)}.$$

Thus,

$$\|\tau_h u - u\|_{L^q(w)} \le C\,|h|^\alpha\,\|\nabla u\|_{L^p(\Omega)}^\alpha\,c\,2^{(1-\alpha)}\,\|\nabla u\|_{L^p(\Omega)}^{(1-\alpha)}$$
$$\le C\,|h|^\alpha\,\|\nabla u\|_{L^p(\Omega)}$$
$$\le c\,|h|^\alpha \quad \text{because } u \in \mathcal{F}.$$

If we take $\delta > 0$, $\delta^\alpha = \frac{\epsilon}{c}$, $\delta < dist(w, \Omega^c)$, then $|h| < \delta$ implies

$$\|\tau_h u - u\|_{L^q(w)} < \epsilon.$$

(b) For $\epsilon > 0$, $u \in \mathcal{F}$. $q \in [1, p*)$, we have

$$\|u\|_{L^q(\Omega \setminus w)} = \left(\int_{\Omega \setminus w} |u|^q \right)^{1/q}$$
$$\le \left(\int_{\Omega \setminus w} |u|^{qr} \right)^{1/qr} (|\Omega \setminus w|)^{1/r'}.$$

Take $r = \frac{p*}{q}$, then $r' = \frac{p*}{p*-q}$, so

$$\|u\|_{L^q(\Omega \setminus w)} \le \|u\|_{L^{p*}(\Omega \setminus w)} |\Omega \setminus w|^{1-q/p*}$$
$$\le \|\nabla u\|_{L^p(\Omega)} |\Omega \setminus w|^{1-q/p*}$$
$$\le |\Omega \setminus w|^{1-q/p*} \quad \text{because } u \in \mathcal{F}.$$

If we choose

$$w = \{x \in \Omega \mid \text{dist}\,(x, \partial\Omega) > \delta\}$$

with small δ, then

$$\|u\|_{L^q(\Omega \setminus w)} < \epsilon.$$

By Theorem 1.37, $\mathcal{F}$ is relatively compact in $L^p(\Omega)$.

(ii) $p = N$, $W^{1,p}(\Omega) \subset L^q(\Omega)$ for $q \in [1, \infty)$: for any $q \in [1, \infty)$, take $r > 1$ with $q \in [1, r]$ then it follows from part (i) that the injection

$$W^{1,p}(\Omega) \subset L^q(\Omega)$$

is compact.

(iii) $p > N$. Note

$$W^{1,p}(\Omega) \subset C(\bar{\Omega}),$$
$$|u(x) - u(y)| \le c\,\|u\|_{W^{1,p}} |x - y|^\alpha \quad \text{a.e. for } x,\, y \in \Omega,$$
$$\alpha = 1 - \frac{N}{p}, \quad c = c(\Omega, p, N).$$

Let $\mathcal{F}$ be a bounded set in $W^{1,p}(\Omega)$, then $\mathcal{F}$ is bounded in $C(\bar{\Omega})$, and $\mathcal{F}$ is uniformly equicontinuous. By the Ascoli Theorem 1.31, $\mathcal{F}$ is relatively compact in $C(\bar{\Omega})$. ∎

Remark 6.41 *In the proof of parts* (i) *and* (ii) *of Theorem 6.40, we may simplify as follows: because Ω is bounded in C^1, consider the extension operator*

$$P : W^{1,p}(\Omega) \to W^{1,p}(\mathbb{R}^N)$$

and $\mathcal{F}$ is a bounded set in $W^{1,p}(\Omega)$. If we take a bounded open set Ω' of C^1, $\Omega \subset\subset \Omega'$, and consider $P\mathcal{F} \mid_{\Omega'}$ and then prove the hypothesis in the Fréchet-Kolmogorov Theorem 1.36, we obtain the results.

Remark 6.42 *In general, if Ω is unbounded, then the injection $W^{1,p}(\Omega) \subset L^p(\Omega)$ is not compact, but if Ω satisfies the condition*

$$|\{x \in \Omega \mid |x| \geq r\}| \to 0 \text{ as } r \to \infty$$

then the injection is compact.

Proof. See Adams [1]. ∎

Theorem 6.43 *If Ω is bounded and regular, the injection*

$$W^{1,p}(\Omega) \subset L^{p*}(\Omega)$$

is never compact, where $\frac{1}{p*} = \frac{1}{p} - \frac{1}{N}$.

Proof. Choose a sequence of points (a_i) in Ω, and a sequence of positive numbers (r_i), $0 < r_i \leq 1$ such that for each i, $B_{r_i}(a_i) = \{x \in \mathbb{R}^N \mid |x - a_i| < r_i\} \subset \Omega$, and $B_{r_i}(a_i) \cap B_{r_j}(a_j) = \emptyset$ if $i \neq j$. Choose a function Φ in $C_c^\infty(B_1(0))$ such that

$$\|\Phi\|_{W^{1,p}(\mathbb{R}^N)} = A_{1,p} > 0,$$
$$\left|\frac{\partial\Phi}{\partial x_1}(a)\right| = B > 0,$$

for some $a \in B_1(0)$. Set

$$\Phi_i(x) = r_i^{1-(N/p)} \Phi\left(\frac{x - a_i}{r_i}\right).$$

Note that

$$|\Phi_i|_{L^p}^p = \int \left| r_i^{1-(N/p)} \Phi\left(\frac{x-a_i}{r_i}\right) \right|^p dx$$

$$= r_i^{p-N+N} \int |\Phi|^p = r_i^p \|\Phi\|_{L^p}^p \le \|\Phi\|_{L^p}^p ,$$

$$\|D^\alpha \Phi_i\|_{L^p}^p = \int \left| r_i^{-N/p} D^\alpha \Phi\left(\frac{x-a_i}{r_i}\right) \right|^p dx$$

$$= r_i^{-N+N} \int |\Phi|^p = \|\Phi\|_{L^p}^p .$$

Therefore, $\{\Phi_i\}$ is a bounded sequence in $W^{1,p}(\Omega)$, with

$$\|\Phi_i\|_{W^{1,p}} \le \|\Phi\|_{W^{1,p}} .$$

Now

$$\|\Phi_i\|_{L^{p*}}^{p*} = \int \left| r_i^{1-N/p} \Phi\left(\frac{x-a_i}{r_i}\right) \right|^{p^*} dx$$

$$= r_i^{-N} \int \left| \Phi\left(\frac{x-a_i}{r_i}\right) \right|^{p^*} dx, \quad \text{because} \quad p^*\left(1-\frac{N}{p}\right) = -N$$

$$= r_i^{-N+N} \int |\Phi|^{p*}$$

$$= \|\Phi\|_{L^{p*}}^{p*} ,$$

so

$$\|\Phi_i\|_{L^{p*}} = \|\Phi\|_{L^{p*}} \quad \text{for } i = 1, \ 2, \ \cdots .$$

Because the (Φ_i) have disjoint supports,

$$\|\Phi_i - \Phi_j\|_{L^{p*}} = 2^{1/p*} \|\Phi\|_{L^{p*}} .$$

We conclude that no subsequence of (Φ_i) converges in $L^{p*}(\Omega)$. This proves that the injection

$$W^{1,p}(\Omega) \subset L^{p*}(\Omega)$$

is not compact.   ∎

Remark 6.44 *If Ω is a bounded open set in $\mathbb{R}^N$ of class C^1, then*

$$\|u\|_{L^p} + \|\nabla u\|_{L^p} \approx \|u\|_{L^q} + \|\nabla u\|_{L^p}$$

where

(i) *if $1 \le p < N$, then $q \in [1, p*]$;*

(ii) if $p = N$, then $q \in [1, \infty)$;

(iii) if $p > N$, then $q \in [1, \infty]$.

Proof.   It suffices to prove

$$\|u\|_{L^q} + \|\nabla u\|_{L^p} \le c\|u\|_{L^{p*}} + \|\nabla u\|_{L^p} \le c\|\nabla u\|_{L^p} \le C(\|u\|_{L^p} + \|\nabla u\|_{L^p}).$$

On the other hand, if $p \le q$, then

$$\|u\|_{L^p} + \|\nabla u\|_{L^p} \le c\|u\|_{L^q} + \|\nabla u\|_{L^p} \le c(\|u\|_{L^q} + \|\nabla u\|_{L^p}).$$

If $1 \le q \le p \le p*$, then $\frac{1}{p} = \frac{1-a}{q} + \frac{a}{p*}$ for some a, $0 \le a \le 1$,

$$\|u\|_{L^p} \le \|u\|_{L^q}^{1-a} \|u\|_{L^{p*}}^{a} \le (1-a)\|u\|_{L^q} + a\|u\|_{L^{p*}}$$
$$\le \|u\|_{L^q} + C\|\nabla u\|_{L^p}$$

or

$$\|u\|_{L^p} + \|\nabla u\|_{L^p} \le c(\|u\|_{L^q} + \|\nabla u\|_{L^p}). \qquad \blacksquare$$

Remark 6.45 *If $\Omega \subset \mathbb{R}^N$ is a bounded open set of class C^1, then in general*

$$W^{1,N}(\Omega) \not\subseteq L^\infty(\Omega).$$

Proof.   Let $\Omega = \{x \in \mathbb{R}^N \mid |x| < \frac{1}{2}\}$. For $0 < \alpha < 1 - \frac{1}{N}$, define

$$u(x) = \left(\ln \frac{1}{|X|}\right)^\alpha \quad \text{for } x \in \Omega.$$

We claim that

$$u \in W^{1,N}(\Omega) \backslash L^\infty(\Omega).$$

In fact, note that

$$\lim_{r \to 0} \frac{\ln \frac{1}{r}}{\frac{1}{r}} = 0,$$

so $\ln\frac{1}{r} \le c\frac{1}{r}$ for $0 \le r \le \frac{1}{2}$, and we have

$$\int |u|^N = \int_\Omega \left|\ln \frac{1}{|x|}\right|^{\alpha N}$$
$$= w_N \int_0^{1/2} \left|\ln \frac{1}{r}\right|^{\alpha N} r^{N-1} dr$$
$$\le c \int_0^{1/2} r^{-\alpha N + N - 1} dr < \infty, \text{ where } -N\alpha + N - 1 > 0.$$

For $i = 1, 2, \cdots, N$,

$$\frac{\partial u}{\partial x_i} = -\alpha \left(\ln \frac{1}{|x|} \right)^{\alpha-1} \frac{x_i}{|x|^2},$$

$$\int_\Omega \left| \frac{\partial u}{\partial x_i} \right|^N = c \int_\Omega \left| \frac{x_i}{|x|^2} \right|^N \left| \ln \frac{1}{|x|} \right|^{N(\alpha-1)} dx$$

$$\leq c \int_0^{1/2} \frac{1}{r^N} \left(\ln \frac{1}{|r|} \right)^{N(\alpha-1)} r^{N-1} dr$$

$$\leq c \int_0^{1/2} r^{N-N\alpha-1} dr < \infty, \quad \text{because } N - N\alpha - 1 > 0.$$

Therefore $u \in W^{1,N}(\Omega)$. However, $u \notin L^\infty(\Omega)$ because

$$\lim_{|x| \to 0} u(x) = \infty. \qquad \blacksquare$$

From the above remark, we see that

$$W^{1,N}(\Omega) \nsubseteq L^\infty(\Omega).$$

However, we have:

Theorem 6.46 (*Trudinger Inequality*) *Let $\Omega \subset \mathbb{R}^N$ be a bounded open set of class C^1, then there are $c_1(N)$, $c_2(N) > 0$ such that*

$$\int_\Omega \exp \left(\frac{|u|}{c_1 \|\nabla u\|_{L^N}} \right)^{N/(N-1)} \leq c_2, \quad \text{for } u \in W^{1,N}(\Omega).$$

Proof.   See Theorem 6.66 for the proof. $\blacksquare$

6.7   The Space $\mathbf{W}_0^{1,p}(\Omega)$

Definition 6.47 *For $1 \leq p < \infty$, let*

$$W_0^{1,p}(\Omega) = \overline{C_c^1(\Omega)}^{W^{1,p}(\Omega)}, \quad \text{and } H_0^1(\Omega) = W_0^{1,2}(\Omega).$$

$W^{1,p}(\Omega)$ *is a separable Banach space under the $W^{1,p}$ norm, and is also reflexive if $1 < p < \infty$. Moreover, $H_0^1(\Omega)$ is a Hilbert space under the H^1 scalar product.*

Remark 6.48 *Because $C_c^1(\mathbb{R}^N)$ is dense in $W^{1,p}(\mathbb{R}^N)$,*

$$W_0^{1,p}(\mathbb{R}^N) = W^{1,p}(\mathbb{R}^N).$$

In general, $W_0^{1,p}(\Omega) \neq W^{1,p}(\Omega)$ for $\Omega \subsetneq \mathbb{R}^N$, but if $|\mathbb{R}^N \backslash \Omega| = 0$, satisfying the Polar condition, and $p < N$, then $W_0^{1,p}(\Omega) = W^{1,p}(\Omega)$. See Adams [1, p. 56].

Proposition 6.49 *Let* $\Omega = \mathbb{R}^N \backslash \{0\}$, $N \geq 2$. *Then* $H_0^1(\Omega) = H^1(\Omega)$.

Proof. It suffices to prove that $H^1(\Omega) \subset H_0^1(\Omega)$. For $w \in H^1(\Omega)$, $\epsilon > 0$, by the Third Density Theorem 6.10, there is $u \in H^1(\Omega) \cap C^1(\Omega)$ such that $\|w - u\|_{H^1(\Omega)} < \frac{\epsilon}{3}$. For $c > 0$, define

$$u_c = \begin{cases} u(x) & \text{if } |u(x)| \leq c \\ c & \text{if } u(x) > c \\ -c & \text{if } u(x) < -c, \end{cases}$$

then $u_c = (|u + c| - |u - c|)/2$, and

$$\nabla u_c = \begin{cases} \nabla u & \text{if } |u(x)| \leq c \\ 0 & \text{if } u(x) > c \\ 0 & \text{if } u(x) < -c. \end{cases}$$

Then

$$\|u - u_c\|_{H^1(\Omega)} = \int_\Omega (|u - u_c|^2 + |\nabla u - \nabla u_c|^2)$$

$$\leq 4 \int_{|u| > c} (u^2 + |\nabla u|^2) \to 0, \text{ as } c \to \infty,$$

because $u \in H^1(\Omega) \cap C^1(\Omega)$. Therefore $\|u - u_c\|_{H^1(\Omega)} < \frac{\epsilon}{3}$ for some large c. Let v be such that u_c with $\|v\|_{L^p(\Omega)} = M > 0$. By Adams [2], for $N \geq 2$, given r, $\epsilon > 0$, there is a function $\theta \in C^\infty(\mathbb{R}^N)$, $0 \leq \theta \leq 1$, with the properties:

(i) $\theta(x) = 0$ near 0;

(ii) $\theta(x) = 1$ for $|x| \geq r$;

(iii) $\|D^\alpha \theta\|_{L^2(\mathbb{R}^N)}^2 \leq \epsilon/6M$ for $|\alpha| = 1$.

Let $v_1 = v\theta$. Then $v_1 \in H_0^1(\Omega)$, and

$$\|v - v_1\|_{H^1(\Omega)}^2 = \int_\Omega (|v - v_1|^2 + |\nabla v - \nabla v_1|^2)$$

$$\leq \int_{|x| < r} |v|^2 + \int_\Omega |\nabla v - v\nabla\theta - \theta\nabla v|^2$$

$$\leq 2 \int_{|x| < r} (|v|^2 + |\nabla v|^2) + M \int_{\mathbb{R}^N} |\nabla\theta|^2$$

$$\leq 2 \int_{|x| < r} (|v|^2 + |\nabla v|^2) + \epsilon/6$$

$$< \epsilon/3 \text{ for large } r \text{ because } v \in H^1(\Omega).$$

Therefore, for each $\epsilon > 0$, there is $v_1 \in H_0^1(\Omega)$, such that

$$\|w - v_1\|_{H^1(\Omega)} < \epsilon,$$

and we have $w \in H_0^1(\Omega)$.   ∎

Remark 6.50 *By regularity, $C_c^\infty(\Omega)$ is dense in $W_0^{1,p}(\Omega)$.*

Roughly speaking, $W_0^{1,p}(\Omega)$ functions are those $W^{1,p}$ functions that vanish at the boundary $\Gamma = \partial\Omega$. More precisely, we have the following results:

Theorem 6.51 *Let $1 \leq p < \infty$, and $u \in W^{1,p}(\Omega)$. If the support of u is compact with supp $u \subset \Omega$, then $u \in W_0^{1,p}(\Omega)$.*

Proof.   Take w, supp $u \subset w \subset\subset \Omega$, and choose a function $\alpha \in C_c^1(w)$ such that $\alpha \mid_{\text{supp } u} = 1$. Then $\alpha u = u$. By Theorem 6.8, there is a sequence $(u_n) \subset C_c^\infty(\mathbb{R}^N)$ such that

$$u_n \to u \text{ in } L^p(\Omega), \text{ a.e. on } \Omega,$$

$$\nabla u \mid_w \to \nabla u \mid_w \text{ in } L^p(w)^N, \text{ a.e. on } w.$$

Because supp $(\alpha u_n) \subset w$, supp $(\alpha u) \subset w$, and

$$\frac{\partial(\alpha u_n)}{\partial x_i} = \begin{cases} \alpha \dfrac{\partial u_n}{\partial x_i} + \dfrac{\partial \alpha}{\partial x_i} u_n \to \alpha \dfrac{\partial u}{\partial x_i} + \dfrac{\partial \alpha}{\partial x_i} u = \dfrac{\partial(\alpha u)}{\partial x_i} \text{ on } \quad w \\ 0 \text{ on } \quad w^c, \end{cases}$$

we have

$$\alpha u_n \to \alpha u \text{ in } L^p(\Omega),$$

$$\nabla(\alpha u_n) \to \nabla(\alpha u) \text{ in } L^p(\Omega)^N.$$

Thus, $\|\alpha u_n - \alpha u\|_{W^{1,p}(\Omega)} \to 0$. Because $(\alpha u_n) \subset C_c^1(\Omega)$, $u = \alpha u \in W_0^{1,p}(\Omega)$.   ∎

Theorem 6.52 *Let Ω be of class C^1, $1 \leq p < \infty$, and $u \in W^{1,p}(\Omega) \cap C(\bar{\Omega})$. Then the following are equivalent:*

(i) $u = 0$ on $\Gamma = \partial\Omega$;

(ii) $u \in W_0^{1,p}(\Omega)$.

Proof. $(i) \Rightarrow (ii)$

(a) Suppose supp u is bounded. Choose $G \in C^1(\mathbb{R})$, $|G(t)| \leq |t|$ for $t \in \mathbb{R}$, and

$$G(t) = \begin{cases} 0 & \text{for } |t| \leq 1 \\ 1 & \text{for } |t| \geq 2 \end{cases}$$

We then have $|G'(t)| \leq M$ for $t \in \mathbb{R}$. For $n = 1, 2, \cdots$, let

$$u_n = \frac{1}{n} G(nu).$$

Note that if $u(x) \neq 0$, then $|n(u(x))| \geq 2$ for large n, or $u_n(x) = \frac{1}{n} G(n(u(x))) = u(x)$ for large n. If $u(x) = 0$, then $u_n(x) = G(nu(x)) = 0$, so

$$u_n \to u \quad a.e. \text{ in } \Omega.$$

Moreover,

$$|u_n| = \left| \frac{1}{n} G(nu) \right| \leq \left| \frac{1}{n} nu \right| \leq |u| \in L^p.$$

By the Lebesgue Dominated Convergence Theorem,

$$u_n \to u \text{ in } L^p(\Omega).$$

Moreover, because

$$\frac{\partial u}{\partial x_i} = 0 \qquad a.e. \text{ on } \{x \in \Omega \mid u(x) = 0\},$$

and if $u(x) \neq 0$, then for large n, $|nu(x)| > 2$ where

$$G'(nu(x)) = 1.$$

Hence,

$$\frac{\partial u_n}{\partial x_i} = G'(nu) \frac{\partial u}{\partial x_i} \to \frac{\partial u}{\partial x_i} \text{ a.e. on } \Omega,$$

$$\left| \frac{\partial u_n}{\partial x_i} \right| \leq M \left| \frac{\partial u}{\partial x_i} \right| \in L^p.$$

By the Lebesgue Dominated Convergence Theorem,

$$\frac{\partial u_n}{\partial x_i} \to \frac{\partial u}{\partial x_i} \text{ in } L^p(\Omega) \text{ for } i = 1, \cdots, N.$$

Therefore

$$u_n \to u \text{ in } W^{1,p}(\Omega).$$

Let $A_n = \{x \in \Omega \mid |u(x)| \geq \frac{1}{n}\}$. Then supp $u_n \subset A_n$, and A_n is closed in Ω. Thus supp $u_n \subset \Omega$ is compact. By Theorem 6.51, $u_n \in W_0^{1,p}(\Omega)$. Thus $u \in W_0^{1,p}(\Omega)$.

(b) Suppose supp u is unbounded. Choose $\zeta \in C_c^\infty(\mathbb{R}^N)$, $0 \le \zeta \le 1$,

$$\zeta(x) = \begin{cases} 1 & \text{for } |x| \le 1 \\ 0 & \text{for } |x| \ge 2. \end{cases}$$

For $n = 1, 2, \cdots$, let $\zeta_n(x) = \zeta\left(\frac{x}{n}\right)$. As before we have

$$\zeta_n u \to u \text{ in } W^{1,p}.$$

Because supp $(\zeta_n u)$ is bounded, by part (a), $\zeta_n u \in W_0^{1,p}(\Omega)$, or $u \in W_0^{1,p}(\Omega)$.
$(ii) \Rightarrow (i)$ By the local coordinates, it suffices to prove that if $u \in W_0^{1,p}(Q_+) \cap C(\bar{Q}_+)$, then $u = 0$ on Q_0. Let $u \in W_0^{1,p}(Q_+) \cap C(\bar{Q}_+)$. Take a sequence (u_n) in $C_c^1(Q_+)$ such that $u_n \to u$ in $W^{1,p}(Q_+)$. For $(x', x_N) \in Q_+$,

$$|u_n(x', x_N)| = |u_n(x', x_N) - u_n(x', 0)|$$
$$\le \int_0^{x_N} \left| \frac{\partial u_n}{\partial x_N}(x', t) \right| dt.$$

For $0 < \epsilon < 1$,

$$\frac{1}{\epsilon} \int_{|x'|<1} dx' \int_0^\epsilon |u_n(x', x_N)| \, dx_N \le \frac{1}{\epsilon} \int_{|x'|<1} dx' \int_0^\epsilon dx_N \int_0^{x_N} \left| \frac{\partial u_n}{\partial x_N}(x', t) \right| dt$$
$$= \frac{1}{\epsilon} \int_{|x'|<1} dx' \int_t^\epsilon dx_N \int_0^\epsilon \left| \frac{\partial u_n}{\partial x_N}(x', t) \right| dt$$
$$\le \int_{|x'|<1} dx' \int_0^\epsilon \left| \frac{\partial u_n}{\partial x_N}(x', t) \right| dt.$$

Letting $n \to \infty$, and by applying $\|\cdot\|_{L^1} \le c \|\cdot\|_{L^p}$, and the Lebesgue Dominated Convergence Theorem, we obtain

$$\frac{1}{\epsilon} \int_{|x'|<1} dx' \int_0^\epsilon |u_n(x', x_N)| \, dx_N \le \int_{|x'|<1} dx' \int_0^\epsilon \left| \frac{\partial u_n}{\partial x_N} \right| dt.$$

Because $\frac{\partial u}{\partial x_N} \in L^p(Q_+) \hookrightarrow L^1(Q_+)$,

$$\lim_{\epsilon \to 0} \int_{|x'|<1} \int_0^\epsilon \left| \frac{\partial u}{\partial x_n} \right| dx' dt = 0.$$

Moreover, $u \in C(\bar{Q}_+)$ implies

$$\lim_{\epsilon \to 0} \int_0^\epsilon |u(x', x_N)| \, dx_N = |u(x', 0)|.$$

Thus,

$$\int_{|x'|<1} |u(x', 0)| \, dx' = 0,$$

or $u(x', 0) = 0$ a.e. on Q_0. However, u is continuous on Q_0, so $u\,|_{Q_0} = 0$.   ∎

Remark 6.53 *The regularity assumption in Theorem 6.51 is only used for* $(ii) \Rightarrow (i)$. *Thus,* $(i) \Rightarrow (ii)$ *holds for any domain* Ω.

Theorem 6.54 *Let* $\Omega \subset \mathbb{R}^N$ *be an open set of class* C^1, $1 < p < \infty$, *and* $u \in L^p(\Omega)$. *Then the following are equivalent:*

(i) $u \in W_0^{1,p}(\Omega)$;

(ii) *There is a constant* $c > 0$ *with*

$$\left| \int_\Omega u \frac{\partial \varphi}{\partial x_i} \right| \le c \|\varphi\|_{L^{p'}} \text{ for } \varphi \in C_c^1(\mathbb{R}^N), \ i = 1, 2, \cdots, N;$$

(iii) *Define*

$$\widetilde{u}(x) = \begin{cases} u(x) & x \in \Omega \\ 0 & x \in \mathbb{R}^N \backslash \Omega. \end{cases}$$

Then $\widetilde{u} \in W^{1,p}(\mathbb{R}^N)$, $\frac{\partial \widetilde{u}}{\partial x_i} = \left(\frac{\partial u}{\partial x_i} \right)^\sim$, $i - 1, 2, \cdots, N$.

Proof. $(i) \Rightarrow (ii)$. For $u \in W_0^{1,p}(\Omega)$, take a sequence (u_n) in $C_c^1(\Omega)$ such that

$$u_n \to u \text{ in } W^{1,p}(\Omega).$$

For any $\varphi \in C_c^1(\mathbb{R}^N)$

$$\left| \int_\Omega u_n \frac{\partial \varphi}{\partial x_i} \right| = \left| \int_\Omega \frac{\partial u_n}{\partial x_i} \varphi \right| \le \|\nabla u_n\|_{L^p} \|\varphi\|_{L^{p'}}.$$

Let $n \to \infty$:

$$\left| \int_\Omega u \frac{\partial \varphi}{\partial x_i} \right| = c \|\varphi\|_{L^{p'}}.$$

$(ii) \Rightarrow (iii)$. For $\varphi \in C_c^1(\mathbb{R}^N)$,

$$\left| \int_{\mathbb{R}^N} \widetilde{u} \frac{\partial \varphi}{\partial x_i} \right| = \left| \int_\Omega u \frac{\partial \varphi}{\partial x_i} \right| \le c \|\varphi\|_{L^{p'}(\Omega)} \le c \|\varphi\|_{L^{p'}(\mathbb{R}^N)}.$$

By $\widetilde{u} \in W^{1,p}(\mathbb{R}^N)$, and for $i = 1, \cdots, N$,

$$\int_{\mathbb{R}^N} \frac{\partial \widetilde{u}}{\partial x_i} \varphi = - \int_{\mathbb{R}^N} \widetilde{u} \frac{\partial \varphi}{\partial x_i} = - \int_\Omega u \frac{\partial \varphi}{\partial x_i} = \int_\Omega \frac{\partial u}{\partial x_i} \varphi = \int_{\mathbb{R}^N} \left(\frac{\partial \varphi}{\partial x_i} \right)^\sim \varphi.$$

Thus,

$$\frac{\partial \widetilde{u}}{\partial x_i} = \left(\frac{\partial u}{\partial x_i} \right)^\sim.$$

$(iii) \Rightarrow (i)$. We may suppose that Ω is bounded, otherwise take the truncation $\zeta_n u$. By the local coordinates, it suffices to consider $u \in L^p(Q_+)$. In this case,

$$\widetilde{u}(x) = \begin{cases} u(x) & \text{for} \quad x \in Q, \ x_N > 0 \\ 0 & \text{for} \quad x \in Q, \ x_N \leq 0 \end{cases}$$

$\widetilde{u} \in W^{1,p}(Q)$. We require

$$\alpha u \in W_0^{1,p}(Q_+) \text{ for } \alpha \in C_c^1(Q).$$

Take $\rho \in C^\infty(\mathbb{R}^N)$ with

$$\text{supp } \rho \subset \{x \in \mathbb{R}^N \mid \frac{1}{2} < x_N < 1, \ |x^1| < 1\}$$

and for $n = 1, 2, \cdots$, let

$$\rho_n(x) = n^N \rho(nx).$$

Then (ρ_n) forms a regularity sequence with

$$\text{supp } \rho_n \subset \{x \in \mathbb{R}^N \mid \frac{1}{2n} < x_N < \frac{1}{n}, \ |x'| < 1\}.$$

Note that for large n

$$\text{supp } (\rho_n \star \alpha\widetilde{u}) \subset \text{supp } \rho_n + \text{supp } (\alpha\widetilde{u}) \subset Q_+.$$

Thus for large n, $\rho_n \star (\alpha\widetilde{u}) \in C_c^1(Q)$. Because $\rho_n \star (\alpha\widetilde{u}) \to \alpha\widetilde{u}$ in $W^{1,p}(\mathbb{R}^N)$, we have $\alpha u = \alpha\widetilde{u} \in W_0^{1,p}(Q_+)$, for every $\alpha \in C_c^1(Q)$. For small $\delta > 0$, let

$$Q_\delta = \{x \in Q \mid \text{dist}(x, \partial Q) \geq \delta\}.$$

Choose $\alpha \in C_c^1(Q)$ such that $0 \leq \alpha \leq 1$, $\alpha \mid_{Q_\delta} = 1$. Now

$$\|\alpha u - u\|_{L^p(Q)}^p \leq 2 \int_{Q \setminus Q_\delta} |u|^p \to 0 \text{ as } \delta \to 0,$$

$$\|\nabla(\alpha u) - \nabla u\|_{L^p(Q)}^p \leq 2 \int_{Q \setminus Q_\delta} |\nabla u|^p + \|\nabla \alpha\|_{L^\infty} \int_{Q \setminus Q_\delta} |u|^p \to 0 \text{ as } \delta \to 0,$$

and we obtain $u \in W_0^{1,p}(Q_+)$.   ∎

Remark 6.55 *In the proof of Theorem 6.54 $(i) \Rightarrow (iii)$ we did not use the regularity of Ω. Therefore there is an extension $P : W_0^{1,p}(\Omega) \to W^{1,p}(\mathbb{R}^N)$, where Ω is an open set in $\mathbb{R}^N$, and we can obtain the following results:*

Theorem 6.56 *(Sobolev Embedding Theorem) Let Ω be an open set in $\mathbb{R}^N$, $1 \leq p \leq \infty$. Then:*

(i) If $1 \le p < N$, then $W_0^{1,p}(\Omega) \subset L^q(\Omega)$ where $\frac{1}{p*} = \frac{1}{p} - \frac{1}{N}$, $q \in [p, p*]$;

(ii) If $p = N$, then $W_0^{1,p}(\Omega) \subset L^q(\Omega)$ for $q \in [p, \infty)$;

(iii) If $p > N$, then $W_0^{1,p}(\Omega) \subset L^\infty(\Omega)$ with continuous injections.

Theorem 6.57 (*Rellich-Kondrachov Theorem*) *Suppose Ω is a bounded open set in $\mathbb{R}^N$, so then we have:*

(i) If $1 \le p < N$, then $W_0^{1,p}(\Omega) \subset L^q(\Omega)$ for $q \in [1, p*)$, where $\frac{1}{p*} = \frac{1}{p} - \frac{1}{N}$;

(ii) If $p = N$, then $W_0^{1,p}(\Omega) \subset L^q(\Omega)$ for $q \in [1, \infty)$;

(iii) If $p > N$, then $W_0^{1,p}(\Omega) \subset C(\bar{\Omega})$,

with all compact injections.

Remark 6.58 *By the Sobolev Embedding Theorem 6.56, if Ω is an open set in $\mathbb{R}^N$, and $1 \le p < N$, then*

$$\|u\|_{L^{p*}} \le c(p, N) \|\nabla u\|_{L^p} \text{ for } u \in W_0^{1,p}(\Omega).$$

In particular, if Ω is bounded, and $1 \le p < N$

$$\|u\|_{L^p} \le c \|\nabla u\|_{L^p} \text{ for } u \in W_0^{1,p}(\Omega).$$

Actually, we have the following more general theorem, see Theorem 6.68.

Theorem 6.59 (*Poincaré Inequality*) *Let $\Omega \subset \mathbb{R}^N$ be a bounded open set, or an open set with $|\Omega| < \infty$, or an open set bounded in one direction. Let $1 \le p < \infty$. Then there is a constant $c = c(\Omega, p) > 0$ such that*

$$\|u\|_{L^p(\Omega)} \le c \|\nabla u\|_{L^p(\Omega)} \text{ for } u \in W_0^{1,p}(\Omega).$$

Thus, $\|\nabla u\|_{L^p} \approx \|u\|_{W^{1,p}}$ in $W_0^{1,p}(\Omega)$, and

$$\int_\Omega \nabla u \nabla v \approx \int_\Omega uv + \int_\Omega \nabla u \nabla v \text{ in } H_0^1(\Omega).$$

Remark 6.60 *If $\Omega \subset \mathbb{R}^N$ is bounded, then $W_0^{1,p}(\Omega) \subset L^2(\Omega)$, for $2 \in [1, p*]$ or $p \ge \frac{2N}{N+2}$. If $\Omega \subset \mathbb{R}^N$ is unbounded, then $W_0^{1,p}(\Omega) \subset L^2(\Omega)$ for $2 \in [p, p*]$ or $2 \ge p \ge \frac{2N}{N+2}$.*

Remark 6.61 *For a positive integer m, $m \ge 1$, and a real p, $1 \le p < \infty$, let*

$$W_0^{m,p}(\Omega) = \overline{C_c^m(\Omega)}.$$

We have $u \in W_0^{m,p}(\Omega)$ if $u \in W^{m,p}(\Omega)$ and the trace $D^\alpha u|_\Gamma = 0$ for $|\alpha| \le m - 1$.

Note that $W_0^{m,p}(\Omega)$ and $W^{m,p} \cap W_0^{1,p}(\Omega)$ are different for $m \geq 2$. Let $W^{-1,p'}(\Omega)$ be the dual of $W_0^{1,p}(\Omega)$, $1 \leq p < \infty$, and $H^{-1}(\Omega)$ the dual of $H_0^1(\Omega)$, $\frac{1}{p} + \frac{1}{p'} = 1$. If we identify $L^2(\Omega)$ with its dual, then we have

(i) $H_0^1(\Omega) \subset L^2(\Omega) \subset H^{-1}(\Omega)$;

(ii) $W_0^{1,p}(\Omega) \subset L^2(\Omega) \subset W^{-1,p'}(\Omega)$, where $\frac{2N}{N+2} \leq p < \infty$ if Ω is bounded, and $\frac{2N}{N+2} \leq p \leq 2$ if Ω is unbounded, with continuous and dense injection.

Theorem 6.62 *Let $F \in W^{-1,p'}(\Omega)$. Then there are $f_0, f_1, \cdots, f_N \in L^{p'}(\Omega)$ such that*

$$< F, v > = \int_\Omega f_0 v + \sum_{i=1}^N \int_\Omega f_i \frac{\partial v}{\partial x_i} \text{ for } v \in W_0^{1,p}(\Omega),$$

$$\|F\| = \max_{0 \leq i \leq N} \|f_i\|_{L^{p'}}.$$

If Ω is bounded, then we may take $f_0 = 0$.

Proof. Let $E = L^p \times \cdots \times L^p$ be the $N + 1$ copies of L^p space under the norm

$$\|h\| = \sum_{i=0}^N \|h_i\|_{L^p}$$

where $h = (h_0, h_1, \cdots, h_N)$. Consider the map $T : W_0^{1,p}(\Omega) \to E$ defined by

$$Tu = \left(u \frac{\partial u}{\partial x_1}, \cdots, \frac{\partial u}{\partial x_N} \right).$$

T is an isometry. Let $g = T(W_0^{1,p})$ under the induced norm, and $S = T^{-1} : G \to W_0^{1,p}(\Omega)$. For $F \in W^{-1,p'}(\Omega)$, consider the map $\varphi : G \to \mathbb{R}$ defined by $\varphi(h) = < F, Sh >$. Then φ is linear and continuous on G. By the Hahn-Banach Theorem 1.1, there is a continuous and linear extension Φ of φ on E such that

$$\|\Phi\|_{E'} = \|F\|.$$

By the Riesz Representation Theorem, there are $f_0, f_1, \cdots, f_N \in L^{p'}(\Omega)$ such that

$$< \Phi, h > = \int_\Omega f_0 h_0 + \sum_{i=1}^N \int_\Omega f_i h_i \text{ for } h = (h_0, h_1, \cdots, h_N) \in E,$$

or

$$< F, v > = < F, Sh > = < \Phi, \ h > = \int_\Omega f_0 v + \sum_{i=1}^N \int f_i \frac{\partial v}{\partial x_i}$$

and

$$\|\Phi\|_{E'} = Max\{\|f_0\|_{L^{p'}}, \cdots, \|f_N\|_{L^{p'}}\}.$$

If Ω is bounded,

$$\|u\|_{W_0^{1,p}(\Omega)} \approx \sum_{i=1}^{N} \left\| \frac{\partial u}{\partial x_i} \right\|_{L^p}.$$

We may therefore consider the map $T : W_0^{1,p}(\Omega) \to E$ by

$$T(u) = \left(\frac{\partial u}{\partial x_1}, \cdots, \frac{\partial u}{\partial x_N} \right)$$

and then similarly we reach the same conclusion with $f_0 = 0$.   ∎

6.8   Potential Estimates

Let $\Omega \subset \mathbb{R}^N$ be an open bounded set that is contained in a ball: $\Omega \subset B_R(0)$. Denote by ω_N and Ω_N the volumes of the unit sphere and the unit ball in $\mathbb{R}^N$, respectively. We see that $\omega_N = N\Omega_N$. For $\mu \in (0,1)$, define

$$(V_\mu f)(x) = \int_\Omega \frac{f(y)}{|x-y|^{N(1-\mu)}} dy.$$

Lemma 6.63 *For $1 \leq p \leq q \leq \infty$, $0 \leq \delta = \frac{1}{p} - \frac{1}{q} < \mu$. The operator V_μ is continuous on $L^p(\Omega)$ into $L^q(\Omega)$ such that*

$$\|V_\mu f\|_{L^q(\Omega)} \leq \left(\frac{1-\delta}{\mu-\delta} \right)^{1-\delta} \Omega_N^{1-\delta} R^{N(\mu-\delta)} \|f\|_{L^p(\Omega)}.$$

Proof.   Choose r, with $r \geq 1$, and

$$\left(1 - \frac{1}{r} \right) + \left(1 - \frac{1}{p} \right) = 1 - \frac{1}{q} : \frac{1}{r} = 1 + \frac{1}{q} - \frac{1}{p} = 1 - \delta.$$

Define

$$h(y) = \begin{cases} |y|^{N(\mu-1)} & \text{for} \quad y \in \Omega \\ 0 & \text{for} \quad y \notin \Omega. \end{cases}$$

Then

$$\|h\|_{L^r(\mathbb{R}^N)}^r = \int_{\mathbb{R}^N} |h|^r = \int_\Omega |y|^{Nr(\mu-1)} dy \leq \int_{B_R(0)} |y|^{Nr(\mu-1)} dy$$

$$= \omega_N \int_0^R s^{Nr(\mu-1)} s^{N-1} ds = \frac{\omega_N R^{Nr\mu-Nr+N}}{Nr\mu - Nr + N} = \frac{\omega_N R^{Nr(\mu-\delta)}}{Nr(\mu-\delta)}.$$

Thus,

$$\|h\|_{L^r(\mathbb{R}^N)} \leq \Omega_N^{1/r} \left(\frac{1}{r(\mu - \delta)} \right)^{1/r} R^{N(\mu-\delta)}$$

$$= \Omega_N^{1-\delta} \left(\frac{1-\delta}{\mu - \delta} \right)^{1-\delta} R^{N(\mu-\delta)}.$$

Define

$$\bar{f}(y) = \begin{cases} f(y) & \text{for} \quad y \in \Omega \\ 0 & \text{for} \quad y \notin \Omega, \end{cases}$$

$$\bar{h}(y) = \begin{cases} h(y) & \text{for } y \in \Omega \\ 0 & \text{for } y \notin \Omega, \end{cases}$$

$$(V_\mu f)^-(y) = \begin{cases} V_\mu f(y) & \text{for} \quad y \in \Omega \\ 0 & \text{for} \quad y \notin \Omega. \end{cases}$$

Then,

$$V_\mu f(x) = \int_\Omega \frac{f(y)}{|x-y|^{N(1-\mu)}} dy = \int_{\mathbb{R}^N} \bar{h}(x-y)\bar{f}(y)dy.$$

By Young's inequality

$$\|V_\mu f\|_{L^q(\Omega)} = \|(V_\mu f)^-\|_{L^q(\mathbb{R}^N)} \leq \|\bar{h}\|_{L^r(\Omega)}\|\bar{f}\|_{L^p(\mathbb{R}^N)}$$

$$= \|h\|_{L^r(\Omega)}\|f\|_{L^p(\Omega)} \leq \Omega_N^{1-\delta} \left(\frac{1-\delta}{\mu - \delta} \right)^{1-\delta} R^{N(\mu-\delta)}\|f\|_{L^p(\Omega)}. \qquad \blacksquare$$

Lemma 6.64 *Assume $f \in L^p(\Omega)$ for some p, $1 < p < \infty$, $\Omega \subset B_R(0)$, and $g = V_{1/p}f$. Then there are constants $c_1 = c_1(N,p) > 0$, $c_2 = c_2(N,p) > 0$ such that for $p' = \frac{p}{p-1}$,*

$$\int_\Omega \exp[\frac{|g|}{c_1\|f\|_{L^p}}]^{p'} dx \leq c_2.$$

Proof. Let q be such that $1 < p \leq q < \infty$. Then $0 \leq \delta = \frac{1}{p} - \frac{1}{q} < \mu = \frac{1}{p}$. By Lemma 6.63,

$$\|g\|_{L^q} \leq \Omega_N^{1-\delta} \left(\frac{1-\delta}{\mu-\delta} \right)^{1-\delta} R^{N(\mu-\delta)} \|f\|_{L^p}.$$

Because $\frac{1-\delta}{\mu-\delta} \leq \frac{1}{\frac{1}{q}} = q : \left(\frac{1-\delta}{\mu-\delta} \right)^{-\delta} \leq q^{1-(1/p)+(1/q)}$, we have

$$\|g\|_{L^q} \leq \Omega_N^{1-\delta} q^{1-(1/p)+(1/q)} R^{N(\mu-\delta)} \|f\|_{L^p},$$

or

$$\int_\Omega \cdot |g|^q \leq \Omega_N^{q(1-\delta)} q^{q[1-(1/p)+(1/q)]} R^{Nq(\mu-\delta)} \|f\|_{L^p}^q .$$

Now $q \geq p > p - 1$, so $p'q = \frac{p}{p-1}q \geq p$. Hence, we may replace q by $p'q$ in the above inequality and obtain

$$\int_\Omega |g|^{p'q} \leq \Omega_N^{p'q(1-\delta)} (p'q)^{p'q[1-(1/p)+(1/p'q)]} R^{Np'q(\mu-\delta)} \|f\|_{L^p}^{p'q} .$$

Now

$$p'q(1-\delta) = p'q \left(1 - \frac{1}{p} + \frac{1}{p'q} \right) = p'q \left(\frac{1}{p'} + \frac{1}{p'q} \right) = q + 1,$$

$$p'q \left(1 - \frac{1}{p} + \frac{1}{p'q} \right) = q + 1,$$

$$Np'q(\mu - \delta) = Np'q \left(\frac{1}{p} - \frac{1}{p} + \frac{1}{p'q} \right) = N.$$

We thus have

$$\int_\Omega |g|^{p'q} \leq \Omega_N \Omega_N^q (p')^q p'q^q q R^N \|f\|_{L^p}^{p'q} ,$$

or

$$\int_\Omega \left(\frac{|g|}{c_1 \|f\|_{L^p}} \right)^{p'q} \leq c \left(\frac{p'\Omega_N}{c_1^{p'}} \right)^q q^q q.$$

Choose $k = q \geq [p] + 1 = n_0$. Then

$$\int_\Omega \frac{1}{k!} \left(\frac{|g|}{c_1 \|f\|_{L^p}} \right)^{p'k} \leq c \left(\frac{p'\Omega_N}{c_1^{p'}} \right)^k \frac{k^k k}{k!} ,$$

and consequently,

$$\int_\Omega \sum_{k=n_0}^n \frac{1}{k!} \left(\frac{|g|}{c_1 \|f\|_{L^p}} \right)^{p'k} \leq c \sum_{k=n_0}^n \left(\frac{p'\Omega_N}{c_1^{p'}} \right)^k \frac{k^k}{(k-1)!} .$$

Choose c_1 so that $c_1^{p'} > e\Omega_N p'$ and $\alpha = \frac{c_1^{p'}}{p'\Omega_N} > e$. We have

$$\int_\Omega \sum_{k=n_0}^n \frac{1}{k!} \left(\frac{|g|}{c_1 \|f\|_{L^p}} \right)^{p'k} \leq c \sum_{k=n_0}^n \frac{\left(\frac{k}{\alpha} \right)^k}{(k-1)!} .$$

If we apply the ratio test,

$$\frac{\left(\frac{k+1}{\alpha} \right)^{k+1} (k-1)!}{k! \left(\frac{k}{\alpha} \right)^k} = \frac{1 + \frac{1}{k}}{\alpha} \left(1 + \frac{1}{k} \right)^k \to \frac{e}{\alpha} < 1 \text{ as } k \to \infty.$$

Therefore,

$$c \sum_{k=n_0}^{n} \frac{\left(\frac{k}{\alpha}\right)^k}{(k-1)!} \leq c_2' \text{ for all } n \geq n_0 .$$

We have

$$c \sum_{k=n_0}^{n} \frac{1}{k!} \left(\frac{|g|}{c_1 \|f\|_{L^p}} \right)^{p'k} \leq c_2' \text{ for all } n \geq n_0.$$

Letting $n \to \infty$, we apply the Monotone Convergence Theorem to obtain

$$\int_\Omega \exp \left(\frac{|g|}{c_1 \|f\|_{L^p}} \right)^{p'} dx \leq c_2. \hspace{2em} \blacksquare$$

Lemma 6.65 *Let $u \in W_0^{1,1}(\Omega)$. Then*

$$u(x) = \frac{1}{\omega_N} \int_\Omega \sum_{i=1}^{N} \frac{(x_i - y_i) D_i u(y)}{|x-y|^N} dy \text{ a.e. in } \Omega.$$

Proof.

(i) $u \in C_c^1(\Omega)$. Extend u to be 0 outside Ω. Let

$$S^{N-1} = \{ w \in \mathbb{R}^N \mid |w| = 1 \}.$$

For any $w \in S^{N-1}$, we have

$$u(x) = - \int_0^\infty D_r u(x + rw) dr.$$

Integrating both sides with respect to w,

$$\int_{S^{N-1}} u(x) dw = - \int_0^\infty dr \int_{S^{N-1}} D_r u(x + rw) dw$$

or

$$u(x) = -\frac{1}{\omega_N} \int_0^\infty \int_{S^{N-1}} D_r u(x + rw) dr dw$$

$$= -\frac{1}{\omega_N} \int_0^\infty \int_{S^{N-1}} \sum_i D_i u(x + rw) \cdot w_i dr dw$$

$$= -\frac{1}{\omega_N} \int_{\mathbb{R}^N} \sum_i \frac{D_i u(y) w_i}{r^{N-1}} dy,$$

where $y = x + rw$. We obtain

$$r = |x - y|, \text{ and } w_i = \frac{y_i - x_i}{r},$$

so
$$u(x) = \frac{1}{\omega_N} \int_{\mathbb{R}^N} \sum_i \frac{(x_i - y_i) D_i u(y)}{|x - y|^N} \, dy.$$

(ii) $u \in w_0^{1,1}(\Omega)$. Take a sequence (u_n) in $C_c^1(\Omega)$ such that

$$u_n \to u \text{ in } W^{1,1}(\Omega) \text{ a.e. on } \Omega.$$

Let
$$Tu_n(x) = \frac{1}{\omega_N} \int_\Omega \sum_i \frac{(x_i - y_i) D_i u_n(y)}{|x - y|^N} \, dy,$$

and we have
$$u_n(x) = Tu_n(x) \text{ for } n = 1, 2, \cdots .$$

Let
$$Tu(x) = \frac{1}{\omega_N} \int_{\Omega_i} \sum_i \frac{(x_i - y_i) D_i u(y)}{|x - y|^N} \, dy,$$

then
$$Tu(x) \leq \frac{1}{\omega_N} \int_\Omega \frac{[\sum (x_i - y_i)^2)^{1/2} [\sum (D_i u(y))^2]^{1/2}}{|x - y|^N} dy$$
$$= \frac{1}{\omega_N} \int_\Omega \frac{|\nabla u(y)|}{|x - y|^{N-1}} \, dy$$
$$= \frac{1}{\omega_N} V_{1/N}(|\nabla u|).$$

Because $u \in W_0^{1,1}(\Omega)$, $|\nabla u| \in L^1$. By Lemma 6.63, $V_{1/N}(|\nabla u|) \in L^1$, or $Tu \in L^1$. Moreover,

$$|Tu_n - Tu| = \left| \frac{1}{\omega_N} \int_\Omega \sum_i \frac{(x_i - y_i) D_i(u_n - u)(y)}{|x - y|^N} dy \right|$$
$$\leq |T(u_n - u)| \leq \frac{1}{\omega_N} V_{1/N}(|\nabla(u_n - u)|).$$

Thus,

$$\|Tu_n - Tu\|_{L^1} \leq c \left\| V_{1/N}(|\nabla(u_n - u)|) \right\|_{L^1} \leq c \|\nabla u_n - \nabla u\|_{L^1} \to 0$$

as $n \to \infty$. Hence, we have a subsequence of (u_n), still denoted by (u_n), such that

$$Tu_n \to Tu \text{ in } L^1, \text{ a.e. on } \Omega.$$

Letting $n \to \infty$ in the equality :

$$u_n(x) = \frac{1}{\omega_N} \int_\Omega \sum_i \frac{(x_i - y_i)D_i u_n(y)}{|x-y|^N} \, dy \text{ a.e. on } \Omega,$$

we have

$$u(x) = \frac{1}{\omega_N} \int_\Omega \sum_i \frac{(x_i - y_i)D_i u(y)}{|x-y|^N} dy \text{ a.e. on } \Omega. \qquad \blacksquare$$

Theorem 6.66 (*Trudinger Inequality*) *There are two constants* $c_1 = c_1(N) > 0$, $c_2 = c_2(N) > 0$ *such that*

$$\int_\Omega \exp \left(\frac{|u|}{c_1 \|\nabla u\|_{L^N}} \right)^{N/(N-1)} dx \leq c_2 \text{ for } u \in W_0^{1,N}(\Omega).$$

Proof. Because $W_0^{1,N}(\Omega) \subset W_0^{1,1}(\Omega)$, by Lemma 6.65

$$u(x) = \frac{1}{\omega_N} \int_\Omega \sum_i \frac{(x_i - y_i)D_i u(y)}{|x-y|^N} \, dy \text{ a.e. on } \Omega.$$

We have

$$|u| \leq \frac{1}{\omega_N} V_{1/N}(|\nabla u|).$$

By Lemma 6.64

$$\int_\Omega \exp \left[\frac{|g|}{c_1 \|f\|_{L^p}} \right]^{p'} \leq c_2,$$

and we set $p = N$, $g = V_{1/N}(|\nabla u|)$, $f = |\nabla u|$ and apply $|u| \leq \frac{1}{w_N} V_{1/N}(|\nabla u|)$. We obtain

$$\int_\Omega \exp \left[\frac{|u|}{\frac{1}{\omega_N} c_1 \|\nabla u\|_{L^N}} \right]^{N/(N-1)} \leq c_2. \qquad \blacksquare$$

Remark 6.67 *Why do we use the form*

$$\int_\Omega \exp \left[\frac{|u|}{c_1 \|\nabla u\|_{L^N}} \right]^{N/(N-1)}$$

in Theorem 6.66 to treat the space $W_0^{1,N}(\Omega)$? *The reason is as follows: for* $u \in W_0^{1,N}(\Omega)$, *let* $c = \left(\frac{1}{c_1 \|\nabla u\|_{L^N}} \right)^{1/(N-1)}$. *We assume* $u \geq 0$ (*otherwise, use* $|u|$ *instead* *of u), and $q \in [N, \infty)$. Because*

$$\lim_{u \to \infty} \frac{u^q}{e^{cu^{N/(N-1)}}} = \lim_{u \to \infty} \frac{qu^{q-1-1/(N-1)}}{c \cdot \frac{N}{N-1}e^{cu^{N/(N-1)}}}$$

$$= \lim_{u \to \infty} \frac{q(q-1)\cdots(q-m+1)u^{q-m-m/(N-1)}}{c^m(\frac{N}{N-1})^m e^{cu^{N/(N-1)}}}$$

after m times where $q - m - m/(N-1) < 0$.

$$= 0,$$

we have

$$u^q \leq c_1 e^{cu^{(N-1)/N}} \text{ for each } u,$$

$$\int_\Omega \exp\left[\frac{|u|}{c_1 \|\nabla u\|_{L^N}}\right]^{N/(N-1)} < \infty.$$

Therefore $u \in L^q(\Omega)$ for all $q \in [N, \infty)$. Hence, the space

$$\left\{ u \mid \int_\Omega \exp\left[\frac{|u|}{c_1 \|\nabla u\|_{L^N}}\right]^{N/(N-1)} < \infty \right\}$$

is the target space for the Sobolev Embedding Theorem on $W_0^{1,N}(\Omega)$.

Theorem 6.68 (*Poincaré Inequality*) *If $\Omega \subset \mathbb{R}^N$ is a bounded open set, or Ω is an open set of finite measure, or Ω is an open set bounded in one direction, $1 \leq p < \infty$, and $u \in W_0^{1,p}(\Omega)$, then*

$$\|u\|_{L^p} \leq \left(\frac{1}{\Omega_N}|\Omega|\right)^{1/N} \|\nabla u\|_{L^p}.$$

Proof. Method 1. $u \in W_0^{1,p}(\Omega)$, and $W_0^{1,p}(\Omega) \subset W_0^{1,1}(\Omega)$. By Lemma 6.65,

$$u(x) = \frac{1}{\omega_N}\int \sum_i \frac{(x_i - y_i)D_i u(y)}{|x-y|^N}dy$$

and consequently,

$$|u(x)| \leq \frac{1}{\omega_N}\int \frac{|x-y||\nabla u|}{|x-y|^N} = \frac{1}{\omega_N}V_{1/N}(|\nabla u|).$$

Because $0 = \delta = \frac{1}{p} - \frac{1}{p} < \frac{1}{N}$, by Lemma 6.63,

$$
\begin{aligned}
\|u\|_{L^p} &\leq \frac{1}{\omega_N} \left\| V_{1/N}(|\nabla u|) \right\|_{L^p} \\
&\leq \frac{1}{\omega_N} \left(\frac{1-\delta}{\mu-\delta} \right)^{1-\delta} (N\omega_N)^{1-\mu} |\Omega|^{\mu-\delta} \|\nabla u\|_{L^p} \\
&= \frac{1}{\omega_N} N\, N^{1-(1/N)} \omega_N^{1-(1/N)} |\Omega|^{1/N} \|\nabla u\|_{L^p} \quad \text{because } \mu = \frac{1}{N}, \delta = 0 \\
&\leq \left(\frac{|\Omega|}{\Omega_N} \right)^{1/N} \|\nabla u\|_{L^p}.
\end{aligned}
$$

Method 2. By the First Density Theorem we may consider $u \in C_c^\infty(\Omega)$. Extend u to be 0 outside Ω, and let $I = [a_1, b_1] \times \cdots \times [a_N, b_N]$ with $\Omega \subset I$. Now,

$$
\begin{aligned}
u(x) &= u(x) - u(a_1, x_2, \cdots, x_N) \\
&= \int_{a_1}^{x_1} \frac{\partial u}{\partial x_1}(t, x_2, \cdots, x_N)dt \\
&\leq \int_{a_1}^{b_1} \left| \frac{\partial u}{\partial x_1} \right| \leq c \left(\int_{a_1}^{b_1} \left| \frac{\partial u}{\partial x_1} \right|^p \right)^{1/p}.
\end{aligned}
$$

Further,

$$
\begin{aligned}
\int_\Omega |u|^p \, dx &\leq c \int_\Omega dx \int_{a_1}^{b_1} \left| \frac{\partial u}{\partial x_1} \right|^p dx_1 \\
&\leq c \int_{a_1}^{b_1} dx_1 \int_\Omega \left| \frac{\partial u}{\partial x_1} \right|^p dx \\
&\leq c \|\nabla u\|_{L^p}^p.
\end{aligned}
$$

Thus,

$$
\|u\|_{L^p} \leq c \|\nabla u\|_{L^p}. \qquad \blacksquare
$$

Lemma 6.69 *Let $\Omega \subset \mathbb{R}^N$ be a convex bounded open set, $u \in W^{1,1}(\Omega)$. Then*

$$
|u(x) - u_\Omega| \leq \frac{d^N}{N\,|\Omega|} \int_\Omega |x-y|^{1-N} |\nabla u| \, dy \quad \text{a.e. on } \Omega,
$$

where $d = \operatorname{diam} \Omega$, and $u_\Omega = \frac{1}{|\Omega|} \int_\Omega u(x)dx$.

Proof.   By the Meyers-Serrin Theorem 6.10, it suffices to consider the case $u \in C^1(\Omega)$. For $x, y \in \Omega$, $x \neq y$, and $w = \frac{y-x}{|y-x|}$, we obtain

$$
u(x) - u(y) = - \int_0^{|y-x|} D_r u(x + rw)dr.
$$

Integrating both sides with $\frac{1}{|\Omega|} \int_\Omega dy$, we have

$$
\begin{aligned}
|u(x) - u_\Omega| &= \left| -\frac{1}{|\Omega|} \int_\Omega dy \int_0^{|x-y|} D_r u(x + rw) dr \right| \\
&\leq \frac{1}{|\Omega|} \int_{B_d(x)} dy \int_0^{|x-y|} |D_r u(x + rw)| \, dr \\
&= \frac{1}{|\Omega|} \int_0^d \rho^{N-1} d\rho \int_\Sigma dw \int_0^\infty |A(x + rw)| \, dr,
\end{aligned}
$$

where $\sum$ is the unit sphere with center x in $\mathbb{R}^N$, and

$$
A(y) = \begin{cases} D_r u(y) & \text{if } y \in \Omega, \\ 0 & \text{if } y \notin \Omega. \end{cases}
$$

Thus,

$$
|u(x) - u_\Omega| \leq \frac{d^N}{N|\Omega|} \int_{\mathbb{R}^N} \frac{A(y)}{r^{N-1}} \, dy,
$$

where

$$
A(y) = D_r u(y) = D_r u(x + rw) = \nabla u(x + rw) \cdot w,
$$
$$
|A(y)| \leq |\nabla u(y)|.
$$

Thus,

$$
|u(x) - u_\Omega)| \leq \frac{d^N}{N|\Omega|} \int_\Omega \frac{|\nabla u(y)|}{|x - y|^{N-1}} dy. \qquad \blacksquare
$$

Theorem 6.70 (*Poincaré-Wirtinger Inequality*) *Let* $\Omega \subset \mathbb{R}^N$ *be a convex open bounded set,* $1 \leq p < \infty$, $u \in W^{1,p}(\Omega)$. *Then*

$$
\|u - u_\Omega\|_{L^p(\Omega)} \leq \left(\frac{\Omega_N}{|\Omega|} \right)^{1-(1/N)} d^N \|\nabla u\|_{L^p(\Omega)},
$$

where d is the diameter of Ω.

Proof. By Lemma 6.69,

$$
|u(x) - u_\Omega| \leq \frac{d^N}{N|\Omega|} \int_\Omega \frac{|\nabla u(y)|}{|x - y|^{N-1}} dy = \frac{d^N}{N|\Omega|} V_{1/N}(|\nabla u(x)|),
$$

where $N - 1 = N(1 - \mu)$, or $\mu = \frac{1}{N}$. Therefore

$$\|u(x) - u_\Omega\|_{L^p} \leq \frac{d^N}{N |\Omega|} \Omega \left\|V_{1/N}(|\nabla u|)\right\|_{L^p(\Omega)}$$

$$\leq \frac{d^N}{N |\Omega|} N \, \Omega_N^{1-(1/N)} |\Omega|^{1/N} \|\nabla u\|_{L^p(\Omega)}$$

$$= \left(\frac{\Omega_N}{|\Omega|}\right)^{1-(1/N)} d^N \|\nabla u\|_{L^p(\Omega)}. \qquad \blacksquare$$

Remark 6.71 *If $\Omega \subset \mathbb{R}^N$ is a convex open bounded set, $1 \leq p < \infty$, $u \in W^{1,p}(\Omega)$, $u_\Omega = \frac{1}{|\Omega|} \int_\Omega u = 0$, then by Theorem 6.70,*

$$\|u\|_{L^p(\Omega)} \leq c \|\nabla u\|_{L^p(\Omega)}.$$

6.9   Application of Chain Rules

In this section we give some useful applications of Chain Rules. The positive and negative parts of a function u are defined by $u^+ = \max(u, 0)$ and $u^- = \max(-u, 0)$. Then u^+, $u^- \geq 0$, $u = u^+ - u^-$, $|u| = u^+ + u^-$. Throughout this section we let $\Omega \subset \mathbb{R}^N$ be an open set, and $1 \leq p \leq \infty$.

Theorem 6.72 *If $u \in W^{1,p}(\Omega)$, then u^+, u^-, $|u| \in W^{1,p}(\Omega)$ such that, for $i = 1$, $\cdots, N$,*

$$\frac{\partial u^+}{\partial x_i} = \begin{cases} \dfrac{\partial u}{\partial x_i} & \text{if } u > 0, \\ 0 & \text{if } u \leq 0. \end{cases}$$

$$\frac{\partial u^+}{\partial x_i} = \begin{cases} 0 & \text{if } u \geq 0, \\ -\dfrac{\partial u}{\partial x_i} & \text{if } u < 0. \end{cases}$$

$$\frac{\partial |u|}{\partial x_i} = \begin{cases} \dfrac{\partial u}{\partial x_i} & \text{if } u > 0, \\ 0 & \text{if } u = 0, \\ -\dfrac{\partial u}{\partial x_i} & \text{if } u < 0. \end{cases}$$

In particular, $\left|\frac{\partial |u|}{\partial x_i}\right| = \left|\frac{\partial u}{\partial x_i}\right|$.

Proof.   For $\epsilon > 0$, define

$$f_\epsilon(u) = \begin{cases} (u^2 + \epsilon^2)^{1/2} - \epsilon & \text{if } u > 0, \\ 0 & \text{if } u < 0. \end{cases}$$

Then

$$f_\epsilon'(u) = \begin{cases} \dfrac{u}{\sqrt{(u^2 + \epsilon^2)}} & \text{if } u > 0, \\ 0 & \text{if } u < 0. \end{cases}$$

$$f_\epsilon'(0+) = \lim_{u \to 0+} \frac{f_\epsilon(u) - f_\epsilon(0)}{u} = \lim_{u \to 0+} \frac{(u^2 + \epsilon^2)^{1/2} - \epsilon}{u} = \lim_{u \to 0+} \frac{\frac{u}{\sqrt{u^2 + \epsilon^2}}}{1} = 0,$$

$$\lim_{u \to 0+} f_\epsilon'(u) = \lim_{u \to 0+} \frac{u}{(u^2 + \epsilon^2)^{1/2}} = 0.$$

We have $f_\epsilon \in C^1(\mathbb{R})$, $|f_\epsilon'| \leq 1$, $f_\epsilon(0) = 0$. By the Chain Rule 6.15,

$$f_\epsilon \circ u \in W^{1,p}(\Omega).$$

Now for $\varphi \in C_0^1(\Omega)$:

(i)

$$\int_\Omega (f_\epsilon \circ u) \frac{\partial \varphi}{\partial x_i} = -\int_\Omega (f_\epsilon' \circ u) \frac{\partial u}{\partial x_i} \varphi$$

$$= -\int_{u>0} \frac{u}{(u^2 + \epsilon^2)^{1/2}} \frac{\partial u}{\partial x_i} \varphi,$$

but

$$f_\epsilon \circ u \to u^+ \quad \text{a.e. as } \epsilon \to 0$$
$$|f_\epsilon \circ u| \leq |u| \in L^p,$$

so

$$\int_\Omega (f_\epsilon \circ u) \frac{\partial \varphi}{\partial x_i} \to \int_\Omega u^+ \frac{\partial u}{\partial x_i},$$

and

$$(f_\epsilon' \circ u) \frac{\partial u}{\partial x_i} \to \frac{\partial u}{\partial x_i} \quad \text{if } u > 0,$$

$$\left| (f_\epsilon' \circ u) \frac{\partial u}{\partial x_i} \right| \leq \left| \frac{\partial u}{\partial x_i} \right| \in L^p,$$

so

$$\int_{u>0} (f_\epsilon' \circ u) \frac{\partial u}{\partial x_i} \to \int_{u>0} \frac{\partial u}{\partial x_i} \varphi = \int_\Omega h\varphi,$$

where

$$h(x) = \begin{cases} \dfrac{\partial u}{\partial x_i} & \text{if } u > 0, \\ 0 & \text{if } u \leq 0. \end{cases}$$

Thus, letting $\epsilon \to 0$, we obtain

$$\int_\Omega u^+ \frac{\partial \varphi}{\partial x_i} = -\int_\Omega h\varphi,$$

or

$$\frac{\partial u^+}{\partial x_i} = \begin{cases} \dfrac{\partial u}{\partial x_i} & \text{if } u > 0, \\ 0 & \text{if } u \le 0. \end{cases}$$

(ii)

$$\int_\Omega u^- \frac{\partial \varphi}{\partial x_i} = \int_\Omega (-u)^+ \frac{\partial \varphi}{\partial x_i} = -\int_{-u>0} \frac{\partial(-u)}{\partial x_i} \varphi = -\int_{u<0} \left(-\frac{\partial u}{\partial x_i}\right)\varphi,$$

or

$$\frac{\partial u^-}{\partial x_i} = \begin{cases} 0 & \text{if } u \ge 0, \\ -\dfrac{\partial u}{\partial x_i} & \text{if } u < 0. \end{cases}$$

(iii) $|u| = u^+ + u^-$, so

$$\int_\Omega |u| \frac{\partial \varphi}{\partial x_i} = \int_\Omega u^+ \frac{\partial u}{\partial x_i} + \int_\Omega u^- \frac{\partial \varphi}{\partial x_i}$$

$$= -\int_{u>0} \frac{\partial \varphi}{\partial x_i}\varphi - \int_{u<0} \left(-\frac{\partial u}{\partial x_i}\right)\varphi$$

$$= -\int_\Omega g\varphi \text{ for } \varphi \in C_c^1(\Omega),$$

where

$$g = \begin{cases} \dfrac{\partial u}{\partial x_i} & \text{if } u > 0, \\ 0 & \text{if } u = 0, \\ -\dfrac{\partial u}{\partial x_i} & \text{if } u < 0. \end{cases}$$

Thus,

$$\frac{\partial |u|}{\partial x_i} = \begin{cases} \dfrac{\partial u}{\partial x_i} & \text{if } u > 0, \\ 0 & \text{if } u = 0, \\ \dfrac{\partial u}{\partial x_i} & \text{if } u < 0. \end{cases}$$

∎

Theorem 6.73 *If $u \in W_0^{1,p}(\Omega)$, then u^+, u^-, $|u| \in W_0^{1,p}(\Omega)$.*

Proof.

(i) Let f_ϵ be defined as in Theorem 6.72. Take a sequence (u_n) in $C_c^1(\Omega)$ such that

$$u_n \to u \text{ in } L^p, \text{ a.e. on } \Omega, \ |u_n| \leq h \in L^p,$$

$$\nabla u_n \to \nabla u \text{ in } (L^p)^N, \text{ a.e. on } \Omega, \ |\nabla u_n| \leq k \in L^p.$$

Because $f_\epsilon(u_n) \in W^{1,p}(\Omega) \cap C(\bar{\Omega})$, $f_\epsilon(u_n(x)) = 0$ for each $x \in \partial\Omega$, we have $f_\epsilon(u_n) \in W_0^{1,p}(\Omega)$. Now

$$f_\epsilon \circ u_n \to f_\epsilon \circ u \qquad \text{a.e.,}$$

$$|f_\epsilon \circ u_n| \leq c\,|u_n| \leq c\,h \in L^p,$$

$$f_\epsilon' \circ u_n \to f_\epsilon' \circ u \qquad \text{a.e.,}$$

$$\left| f_\epsilon' \circ u_n \frac{\partial u_n}{\partial x_i} \right| \leq c \left| \frac{\partial u_n}{\partial x_i} \right| \leq ck \in L^p.$$

Thus, $f_\epsilon \circ u_n \to f_\epsilon \circ u$ in $W^{1,p}(\Omega)$, or $f_\epsilon \circ u \in W_0^{1,p}(\Omega)$. However, as in the proof of Theorem 6.72,

$$f_\epsilon \circ u \to u^+ \quad \text{in } W^{1,p}, \text{ so } u^+ \in W_0^{1,p}(\Omega).$$

(ii) $u^- = (-u)^+ \in W_0^{1,p}(\Omega).$

(iii) $|u| = u^+ + u^- \in W_0^{1,p}(\Omega).$ ∎

Remark 6.74 *From the proof of Theorem 6.72, we have:*

(i) *For $u \geq 0$, $u \in W^{1,p}$, there is a sequence (u_n) in $W^{1,p}(\Omega)$ such that*

$$u_n = f_{1/n} \circ u \to u^+ \text{ in } W^{1,p}(\Omega),$$

$$u_n \geq 0;$$

(ii) $\frac{\partial |u|}{\partial x_i} = 0 \quad$ *a.e. on $\{u = 0\}$.*

Moreover, we have:

Theorem 6.75 *If $u \in W^{1,p}(\Omega)$, then $\frac{\partial u}{\partial x_i} = 0 \quad$ a.e. for $i = 1, \cdots, N$, on the level set $\{x \in \Omega \mid u(x) = c\}$, where c is any constant.*

Proof.

(i) Suppose $c = 0$. For $r \in [-1, 1]$, define for $\epsilon > 0$

$$\sigma_{\gamma\epsilon}(t) = \begin{cases} 1 & \epsilon(1 - \gamma) \le t < \infty, \\ \gamma + \dfrac{t}{\epsilon} & -\epsilon(1 + \gamma) < t < \epsilon(1 - \gamma), \\ -1 & -\infty < t < -\epsilon(1 + \gamma). \end{cases}$$

Now $\sigma_{\gamma\epsilon}(0) = \gamma$, and

$$\lim_{\epsilon \to 0} \sigma_{\gamma\epsilon}(t) = \sigma(t) = \begin{cases} 1 & t > 0, \\ \gamma & t = 0, \\ -1 & t < 0. \end{cases}$$

Define $Q_{\gamma\epsilon}(t) = \int_0^t \sigma_{\gamma\epsilon}(\tau)d\tau \quad for -\infty < t < \infty$. Then

$$Q_{\gamma\epsilon} \in C^1(\mathbb{R}), \ \left|Q'_{\gamma\epsilon}(t)\right| = |\sigma_{\gamma\epsilon}(t)| \le 1, \text{ and } Q_{\gamma\epsilon}(0) = 0.$$

By the Chain Rule 6.15,

$$Q_{\gamma\epsilon} \circ u \in W^{1,p}(\Omega),$$

$$\frac{\partial}{\partial x_i}(Q_{\gamma\epsilon} \circ u) = (\sigma_{\gamma\epsilon} \circ u)\frac{\partial u}{\partial x_i} \quad \text{for } i = 1, \cdots, N.$$

Furthermore, by the Lebesgue Dominated Convergence Theorem,

$$\lim_{\epsilon \to 0} Q_{\gamma\epsilon}(t) = \int_0^t \lim_{\epsilon \to 0} \sigma_{\gamma\epsilon}(\tau)d\tau = \begin{cases} t & \text{for } t > 0 \\ 0 & \text{for } t = 0 \\ -t & \text{for } t < 0 \end{cases} = |t|.$$

Now

$$Q_{\gamma\epsilon} \circ u \to |u| \quad \text{a.e.,}$$

$$|Q_{\gamma\epsilon} \circ u| \le \int_0^u |\sigma_{\gamma\epsilon}(\tau)| \, d\tau \le |u| \in L^p(\Omega),$$

$$\frac{\partial}{\partial x_i} Q_{\gamma\epsilon}(u) = \sigma_{\gamma\epsilon}(u)\frac{\partial u}{\partial x_i} \to \sigma\frac{\partial u}{\partial x_i} \quad \text{a.e.,}$$

$$\frac{\partial}{\partial x_i} Q_{\gamma\epsilon} \circ |u| = \left|\sigma_{\gamma\epsilon}(u)\frac{\partial u}{\partial x_i}\right| \le \left|\frac{\partial u}{\partial x_i}\right| \in L^P(\Omega).$$

Hence,

$$Q_{\gamma\epsilon} \circ u \to |u| \quad \text{in } L^p(\Omega),$$

$$\frac{\partial}{\partial x_i} Q_{\gamma\epsilon}(u) \to \sigma\frac{\partial u}{\partial x_i} \quad \text{in } L^p(\Omega).$$

Therefore, $|u| \in W^{1,p}(\Omega)$, $\frac{\partial}{\partial x_i}|u| = \sigma \frac{\partial u}{\partial x_i}$. In particular,

$$\frac{\partial}{\partial x_i}|u| = \gamma \frac{\partial u}{\partial x_i} \quad \text{a.e. on } \{x \in \Omega \mid u(x) = 0\}.$$

If we let $r = 1, -1$, we obtain

$$\frac{\partial}{\partial x_i}|u| = \frac{\partial u}{\partial x_i} = -\frac{\partial u}{\partial x_i} \quad \text{a.e. on } \{x \in \Omega \mid u(x) = 0\},$$

or

$$\frac{\partial u}{\partial x_i} = 0 \quad \text{a.e. on} \{x \in \Omega \mid u(x) = 0\}.$$

(ii) Let $v = u - c$. Then

$$\frac{\partial u}{\partial x_i} = \frac{\partial v}{\partial x_i} = 0 \quad \text{a.e. on} \{x \in \Omega \mid v(x) = 0\} = \{x \in \Omega \mid u(x) = c\}. \qquad \blacksquare$$

Theorem 6.76 *Let G be a piecewise smooth function on $\mathbb{R}$, $G' \in L^\infty(\mathbb{R})$, $G(0) = 0$. If $u \in W^{1,p}(\Omega)$, then $G \circ u \in W^{1,p}(\Omega)$. Furthermore, letting L denote the set of corner points of G, then*

$$\frac{\partial}{\partial x_i}(G \circ u) = \begin{cases} G' \circ u \dfrac{\partial u}{\partial x_i} & \text{if } u \notin L, \\ 0 & \text{if } u \in L. \end{cases}$$

Proof. By induction, it suffices to prove the case that L is a singleton $\{0\}$. Assume

$$G(u) = \begin{cases} G_1(u) & \text{for } u \geq 0, \\ G_2(u) & \text{for } u \leq 0, \end{cases}$$

where $G_1, G_2 \in C^1(\mathbb{R})$, $G_1', G_2' \in L^\infty(\mathbb{R})$, $G_1(0) = G_2(0) = 0$, and we have

$$G(u) = G_1(u^+) + G_2(u^-),$$

$$\frac{\partial G \circ u}{\partial x_i} = (G_1' \circ u^+)\frac{\partial u^+}{\partial x_i} + (G_2' \circ u^-)\frac{\partial u^-}{\partial x_i} = \begin{cases} G' \circ u \dfrac{\partial u}{\partial x_i} & \text{for } u > 0, \\ 0 & \text{for } u = 0, \\ G' \circ u \dfrac{\partial u}{\partial x_i} & \text{for } u < 0. \end{cases}$$

$\blacksquare$

Theorem 6.77 *Let $u \in W^{1,p}(\Omega)$, $u \geq 0$ a.e. on Ω. There is a sequence (u_n) in $W^{1,p}(\Omega) \cap C^\infty(\Omega)$ with $u_n \geq 0$ a.e. on Ω such that*

$$u_n \to u \text{ in } W^{1,p}.$$

Proof.   By the Meyer-Serrin Theorem 6.10, there is a sequence (v_n) in $C^\infty(\Omega) \cap$ $W^{1,p}(\Omega)$ such that

$$v_n \to u \text{ in } W^{1,p}(\Omega),$$

$$\frac{\partial v_n}{\partial x_i} \to \frac{\partial u}{\partial x_i} \quad \text{in } L^p, \text{ a.e. on } \left| \frac{\partial v_n}{\partial x_i} \right| \le h \in L^p.$$

Let $u_n = v_n^+$. Then $u_n \in C^\infty(\Omega) \cap W^{1,p}(\Omega)$, $u_n \ge 0$,

$$|u_n - u| = \left| v_n^+ - u^+ \right| \le |v_n - u|,$$

$$\|u_n - u\|_{L^p} \le \|v_n - u\|_{L^p} \to 0, \text{ as } n \to \infty.$$

Because

$$\frac{\partial u_n}{\partial x_i} = \begin{cases} \dfrac{\partial v_n}{\partial x_i} & v_n > 0, \\ 0 & v_n \le 0, \end{cases}$$

we have

$$\frac{\partial u_n}{\partial x_i} \to \frac{\partial u}{\partial x_i} \quad \text{a.e. on } \{u > 0\} = \Omega,$$

$$\left| \frac{\partial u_n}{\partial x_i} \right| \le \left| \frac{\partial v_n}{\partial x_i} \right| \le h \in L^p.$$

Therefore $u_n \to u$ in $W^{1,p}(\Omega)$. ∎

A similar proof implies:

Theorem 6.78 *Assume $u \in W_0^{1,p}(\Omega)$, $u \ge 0$ a.e. on Ω. There is a sequence (u_n) in $C^\infty(\Omega) \cap W_0^{1,p}(\Omega)$ such that*

$$u_n \ge 0 \quad \text{a.e. on } \Omega,$$

$$u_n \to u \quad \text{in } W^{1,p}(\Omega).$$

Theorem 6.79 (*First Trace Theorem*) *Let*

$$Q_+ = \{x = (x', x_N) \in \mathbb{R}^N \mid |x'| < 1, \ 0 < x_N < 1\},$$

$$Q_0 = \{x = (x', x_N) \in \mathbb{R}^N \mid |x'| < 1, \ x_N = 0\}.$$

The trace operator $T : W^{1,p}(Q_+) \to L^p(Q_0)$, an extension of the map $Tu(x', x_N) = u(x', 0)$, for $u \in C^\infty(\bar{Q}_+)$, is a continuous linear map of $W^{1,p}(Q_+)$ into $L^p(Q_0)$.

Proof.

(i) For $u \in C^\infty(\bar{Q}_+)$,

$$\|u\|_{L^p(Q_+)}^p = \int_{0 < x_N < 1} \int_{|x'| < 1} |u(x', x_N)|^p \, dx' dx_N$$

$$= \int_{|x'| < 1} |u(x', a)|^p \, dx' \text{ for some } a, \ 0 < a < 1.$$

Now

$$u(x', a) - u(x', 0) = \int_0^a \frac{\partial u}{\partial x_N}(x', x_N) dx_N$$

$$\leq \int_0^1 \left| \frac{\partial u}{\partial x_N} \right| dx_N$$

$$\leq \left(\int_0^1 \left| \frac{\partial u}{\partial x_N} \right|^p dx_N \right)^{1/p}.$$

Therefore,

$$|u(x', 0)| \leq |u(x', a)| + \left(\int_0^1 \left| \frac{\partial u}{\partial x_N} \right|^p dx_N \right)^{1/p},$$

or

$$|u(x', 0)| \leq c \left[|u(x', a)| + \left(\int_0^1 \left| \frac{\partial u}{\partial x_N} \right|^p dx_N \right)^{1/p} \right].$$

Thus,

$$\int_{|x'| < 1} |u(x', 0)|^p \, dx' \leq c \left[\int_{|x'| < 1} |u(x', a)|^p \, dx' + \int_{|x'| < 1} \int_0^1 \left| \frac{\partial u}{\partial x_N} \right|^p dx' dx_N \right]$$

$$\leq c \left[\|u\|_{L^p(Q_+)}^p + \|\nabla u\|_{L^p(Q_+)}^p \right],$$

or $\|u\|_{L^p(Q_0)} \leq c \|u\|_{W^{1,p}(Q_+)}$.

(ii) Because $C^\infty(\bar{Q}_+)$ is dense in $W^{1,p}(Q_+)$, the trace operator

$$Tu(x', x_N) = u(x', 0),$$

can be extended as a continuous linear operator of $W^{1,p}(Q_+)$ into $L^p(Q_0)$. ∎

Theorem 6.80 (*Second Trace Theorem*) *There is a constant $c > 0$, such that for $u \in C_c^1(\mathbb{R}^N)$, the operator $Tu(x', x_N) = u(x', 0)$ with $\|Tu\|_{L^p(\mathbb{R}^{N-1})} \leq c \|u\|_{W^{1,p}(\mathbb{R}_+^N)}$ can be extended as a continuous linear operator on $W^{1,p}(\mathbb{R}_+^N)$ into $L^p(\mathbb{R}^{N-1})$.*

Proof.

(*i*) Let $u \in C_c^1(\mathbb{R}^N)$, $G(t) = |t|^{p-1} t$. Then

$$G(u(x',0)) = -\int_0^\infty \frac{\partial}{\partial x_N} G(u(x',x_N)) dx_N$$

$$= -\int_0^\infty G'(u(x,x_N)) \frac{\partial}{\partial x_N}(x',x_N) dx_N,$$

or

$$|u(x',0)|^p \le p \int_0^\infty |u(x',x_N)|^{p-1} \left| \frac{\partial u}{\partial x_N}(x',x_N) \right| dx_N$$

$$\le c \int_0^\infty |u(x',x_N)|^p \, dx_N + \int_0^\infty \left| \frac{\partial u}{\partial x_N}(x',x_N) \right|^p.$$

Thus,

$$\int_{\mathbb{R}^{N-1}} |u(x',0)|^p \, dx' \le p \left[\int_{\mathbb{R}_+^N} |u(x',x_N)|^p + \int_{\mathbb{R}_+^N} \left| \frac{\partial u}{\partial x_N}(x',x_N) \right| \right]^p,$$

or

$$\left(\int_{\mathbb{R}^{N-1}} |u(x',0)|^p \, dx' \right)^{1/p} \le c \|u\|_{W^{1,p}(\mathbb{R}_+^N)}.$$

(*ii*) By the Forth Density Theorem 6.24,

$$\left(\int_{\mathbb{R}^{N-1}} |u(x',0)|^p \, dx' \right)^{1/p} \le c \|u\|_{W^{1,p}(\mathbb{R}_+^N)}$$

for $u \in W^{1,p}(\mathbb{R}_+^N)$.   ∎

Remark 6.81 *The kernel $KerT$ of the trace operator $T : W^{1,p}(\mathbb{R}_+^N) \to L^p(\mathbb{R}^{N-1})$ is*

$$KerT = W_0^{1,p}(\mathbb{R}_+^N).$$

By the local coordinates, we have:

Theorem 6.82 (*Third Trace Theorem*) *Let $\Omega \subset \mathbb{R}^N$ be a bounded open set of class C^1. The trace operator $T : W^{1,p}(\Omega) \to L^p(\partial\Omega)$ is a continuous linear operator.*

Let

$$Q_+ = \{x = (x',x_N) \in \mathbb{R}^N \mid |x'| < 1, \ 0 < x_N < 1\},$$

$$Q_0 = \{x = (x',0) \in \mathbb{R}^N \mid |x'| < 1\},$$

$$Q_- = \{x = (x',x_N) \in \mathbb{R}^N \mid |x'| < 1, \ -1 < x_N < 0\},$$

$$Q = \{x = (x',x_N) \in \mathbb{R}^N \mid |x'| < 1, \ |x_N| < 1\}.$$

Lemma 6.83 (*Matching Lemma*) *Let* $T : W^{1,p}(Q_+) \to L^p(Q_0)$, *and* $U : W^{1,p}(Q_-) \to L^p(Q_0)$ *be the trace operators. If* $u \in W^{1,p}(Q_-)$, $v \in W^{1,p}(Q_-)$ *with* $Tu = Uv$. *Define*

$$
w(x) = \begin{cases}
u(x) & \text{for } x \in Q_+, \\
v(x) & \text{for } x \in Q_-, \\
Tu(x) & \text{for } x \in Q_0.
\end{cases}
$$

Then $u \in W^{1,p}(Q)$.

Proof. Define

$$
\alpha(x', x_N) = v(x', -x_N) \text{ for } (x', x_N) \in Q_+.
$$

Then $\alpha \in W^{1,p}(Q_+)$ such that $Tu = T\alpha$ on Q_0, or $(u - \alpha) \in Ker\ T = W_0^{1,p}(Q_+)$. Take a sequence (u_n) in $C^\infty(\mathbb{R}^N)$ such that $u_n \to u$ in $W^{1,p}(Q_+)$, and a sequence (ζ_n) as before such that

$$
\zeta_n = 0 \text{ on } x_n \leq \frac{1}{n},
$$
$$
\psi_n = \zeta_n(\alpha - u) \to \alpha - u \text{ in } W^{1,p}(Q_+).
$$

Then $u_n + \psi_n \to \alpha$ in $W^{1,p}(Q_+)$. Define

$$
w_n(x', x_N) = \begin{cases}
u_n(x, x'_N) & \text{for } (x', x_N) \in Q_+ \cup Q_0, \\
u_n(x', -x_N) + \psi_n(x', -x_N) & \text{for } (x', x_N) \in Q_-.
\end{cases}
$$

Then $u \in W^{1,p}(Q)$. Moreover,

$$
\|w_n - w\|_{L^p(Q)} \leq \|u_n - u\|_{L^p(Q_+)} + \|(u_n + \psi_n) - \alpha\|_{L^p(Q_+)}
$$
$$
\to 0 \text{ as } n \to \infty,
$$
$$
\left\| \frac{\partial w_n}{\partial x_i} \right\|_{L^p(Q_+)} \leq \left\| \frac{\partial u_n}{\partial x_i} \right\|_{L^p(Q_+)} + \left\| \frac{\partial(u_n + \psi_n)}{\partial x_i} \right\|_{L^p(Q_+)}
$$
$$
\leq c \qquad \text{for } n = 1, 2, \cdots, \text{ for } i = 1, \cdots, N.
$$

Therefore there is a subsequence of (w_n), still denoted by (w_n), such that

$$
w_n \to w \quad \text{in } W^{1,p},
$$

or $w \in W^{1,p}(Q)$.   ∎

6.10   Lipschitz Functions

Definition 6.84 *Let* Ω *be a bounded open set in* $\mathbb{R}^N$, $u : \Omega \to \mathbb{R}$. *Assume* u *is Hölder-continuous with exponent* α, $0 < \alpha < 1$, *in* Ω, *and write* $u \in C^\alpha$ *if* $\|u\|_{C^\alpha} =$

$$\sup_{x,y\in\Omega,x\neq y} \frac{|u(x)-u(y)|}{|x-y|^{\alpha}} < \infty, \ then \ u \ is \ Lipschitz\text{-}continuous \ in \ \Omega \ (written \ u \in Lip(\Omega)),$$

if

$$\|u\|_{Lip} = \sup_{x,y\in\Omega,x\neq y} \frac{|u(x)-u(y)|}{|x-y|} < \infty,$$

$u \in C^1(\Omega)$ *if* $u, D^{\alpha}u \in C(\Omega), |\alpha| \leq 1.$

Example 6.85 *Define, for* $x \in I = (-1,1)$, $f(x) = |x|^{\alpha}$, $0 < \alpha < 1$, $g(x) = |x|$, $h(x) = |x|^{\beta}$, $1 < \beta$.

(i) $f \in C^{\theta}(I), 0 < \theta \leq \alpha;$

(ii) $g \in Lip(I);$

(iii) $h \in C^1(I).$

Proof. It is easy to see that $f, g, h \in C^{\infty}(I\backslash\{0\}).$

(i) For $0 < |x| < 1$,

$$\frac{|f(x) - f(0)|}{|x - 0|^{\theta}} = \frac{|x|^{\alpha}}{|x|^{\theta}} = |x|^{\alpha-\theta} \leq 1 \ \ for \ \ all \ \ 0 < |x| < 1.$$

If $-1 < x < 0 < y < 1$, then

$$\frac{|f(x) - f(y)|}{|x - y|^{\theta}} \leq \frac{|f(x) - f(0)|}{|x - y|^{\theta}} + \frac{|f(0) - f(y)|}{|x - y|^{\theta}}$$

$$\leq \frac{|f(x) - f(0)|}{|x|^{\theta}} + \frac{|f(0) - f(y)|}{|y|^{\theta}}$$

$$\leq 2.$$

Thus $f \in C^{\theta}(I)$.

(ii) Similar to the proof of (i).

(iii) $h(x) = |x|^{\beta}, \beta > 1$

$$h'(x) = \beta x^{\beta-1} \ \ for \ x > 0,$$

$$h'(x) = -\beta(-x)^{\beta-1} \ \ for \ x < 0,$$

$$\lim_{x\to 0} h'(x) = 0,$$

$$h'(0+) = \lim_{x\to 0^+} \frac{h(x) - h(0)}{x - 0} = \lim_{x\to 0^+} \frac{x^{\beta}}{x} = 0,$$

$$h'(0-) = \lim_{x\to 0^-} \frac{h(x) - h(0)}{x - 0} = \lim_{x\to 0^-} \frac{(-x)^{\beta}}{x} = 0.$$

Therefore $h \in C^1(I)$.   ∎

Theorem 6.86 *Let $\Omega \subset \mathbb{R}^N$ be a bounded open set.*

(i) *If u is absolutely continuous in each variable on segments in Ω for a.e. other variables, such that u, $D_i u \in L^p(\Omega)$, $i = 1, \cdots, N$, then $u \in W^{1,p}(\Omega)$;*

(ii) *If u is a Lipschitz-continuous function on Ω, then u is absolutely continuous in each variable on segments in Ω for a.e. other variables, such that u, $\frac{\partial u}{\partial x_i} \in L^\infty(\Omega)$ for $i = 1, \cdots, N$. In particular, $\mathrm{Lip}\,(\Omega) \subset W^{1,p}(\Omega)$, $1 \leq p \leq \infty$.*

Proof.

(i) Take a cube $I = [a_1, b_1] \times \cdots \times [a_N, b_N]$ such that $\Omega \subset I$. For $\varphi \in C_c^\infty(\Omega)$, then, for $i = 1, \cdots, N$

$$\int_\Omega u \frac{\partial \varphi}{\partial x_i} = \int_I u \frac{\partial \varphi}{\partial x_i} = -\int_I (D_i u)\varphi = -\int_\Omega (D_i u)\varphi$$

or $\frac{\partial u}{\partial x_i} = D_i u$, where $D_i u$ is the derivative in the classical sense, so $u \in W^{1,p}(\Omega)$.

(ii)

$$\sum_{i=1}^k |u(x_1, \cdots, x_{i-1}, \alpha_{ij}, x_{i+1}, \cdots, x_N) - u(x_1, \cdots, x_{i-1}, \beta_{ij}, x_{i+1}, \cdots, x_N)|$$

$$\leq c \sum_{i=1}^k |(x_1, \cdots, x_{i-1}, \alpha_{ij}, x_{i+1}, \cdots, x_N) - (x_1, \cdots, x_{i-1}, \beta_{ij}, x_{i+1}, \cdots, x_N)|$$

$$\leq c \sum_{i=1}^k |\alpha_{ij} - \beta_{ij}|.$$

Therefore, u is absolutely continuous on each variable on segments in Ω for a.e. other variables. Thus, there is $\frac{\partial u}{\partial x_i}$ a.e. on Ω for $i = 1, \cdots, N$. However,

$$|u(x) - u(x_0)| \leq c |x - x_0|.$$

Therefore

$$\left| \frac{\partial u}{\partial x_i}(x_0) \right| \leq c \text{ for } i = 1, \cdots, N \text{ for every } x_0 \in \Omega.$$

Now u, $\frac{\partial u}{\partial x_i} \in L^\infty(\Omega)$, or $u \in W^{1,\infty}(\Omega)$. ∎

Theorem 6.87 (*Rademacher Theorem*) *Let $\Omega \subset \mathbb{R}^N$ be a bounded open set, and u a Lipschitz-continuous function on Ω. Then u possesses a total differential almost everywhere on Ω.*

Proof.   Let Z_0 be the set of points where the derivatives $\frac{\partial u}{\partial x_i}$ fail to exist. By Theorem 6.86, the measure $|Z_0| = 0$. Let E be an everywhere dense denumerable set of points ζ in the unit sphere S^{N-1}. For each $\zeta \in E$, $x = x_\zeta(y)$ is a rotation with $\zeta = x_\zeta(e_1)$, and set

$$v(y) = u(x_\zeta(y)),$$
$$v(e_1) = u(\zeta).$$

The derivatives $\frac{\partial v}{\partial y_i}$ exist at $\{y \mid x_\zeta(y) \notin Z_0\}$. Actually, for each $\zeta \in E$, there is a set Z_ζ such that $\frac{\partial v}{\partial y_i}$ exists if $x_\zeta(y) \notin Z_\zeta$, and $|Z_\zeta| = 0$. If $x_0 \notin Z_0 \cup \left(\bigcup_{\zeta \in E} Z_\zeta \right) = A$, then there is $\frac{\partial v}{\partial \zeta}(x_0)$ $\forall \zeta \in E$. For $\alpha \in S^{N-1}$, $\epsilon > 0$, there is $\zeta \in E$ such that $|\zeta - \alpha| < \frac{\epsilon}{2(2c+1)}$, where c is the Lipschitz constant of u. Now for $x \notin A$,

$$\frac{\partial u}{\partial \zeta}(x) = \lim_{t \to 0} \frac{u(x + t\zeta) - u(x)}{t}.$$

Take t_0 such that $t_1, t_2 < t_0$, then

$$\left| \frac{u(x + t_1\zeta) - u(x)}{t_1} - \frac{u(x + t_2\zeta) - u(x)}{t_2} \right| < \frac{\epsilon}{2},$$

or

$$\left| \frac{u(x + t_1\zeta) - u(x)}{t_1} - \frac{u(x + t_2\zeta) - u(x)}{t_2} \right|$$
$$\leq \left| \frac{u(x + t_1\alpha) - u(x)}{t_1} - \frac{u(x + t_1\zeta) - u(x)}{t_2} \right|$$
$$+ \left| \frac{u(x + t_1\zeta) - u(x)}{t_1} - \frac{u(x + t_2\zeta) - u(x)}{t_2} \right|$$
$$+ \left| \frac{u(x + t_2\alpha) - u(x)}{t_2} - \frac{u(x + t_2\alpha) - u(x)}{t_2} \right|$$
$$\leq 2c|\alpha - \zeta| + \frac{\epsilon}{2} < \epsilon.$$

Thus there is $\frac{\partial u}{\partial \alpha}(x)$.   ∎

Theorem 6.88 *Let $\Omega \subset \mathbb{R}^N$ be a bounded open set,*

$$A^{1,\infty}(\Omega) = \overline{C^\infty(\mathbb{R}^N)}|_\Omega^{W^{1,\infty}(\Omega)}.$$

Then

$$A^{1,\infty}(\Omega) \subset Lip(\Omega) \subset W^{1,\infty}(\Omega).$$

Proof.

(i) For $u \in C^\infty(\mathbb{R}^N) \mid_\Omega$, x, $y \in \Omega$, then

$$u(y) - u(x) = \int_0^1 \frac{du}{dt}(x + t(y - x))dt$$
$$= \int_0^1 \nabla u(x + t(y - x)) \cdot (y - x)dt.$$

Then $|u(y) - u(x)| \le c\,|x - y|$, or u is Lipschitz-continuous.

(ii) For general $u \in A^{1,\infty}(\Omega)$. Take a sequence (u_n) in $C^\infty(\mathbb{R}^N) \mid_\Omega$ such that

$$u_n \to u \text{ in } W^{1,\infty}(\Omega).$$

Now
$$u_n(y) - u_n(x) = \int_0^1 \nabla u_n(x + t(y - x)) \cdot (y - x)dt.$$

Letting $n \to \infty$,

$$u(y) - u(x) = \int_0^1 \nabla u(x + t(y - x)) \cdot (y - x)dt.$$

Thus,
$$|u(y) - u(x)| \le c\,|y - x|\,.$$

Thus u is Lipschitz-continuous.

(iii) $Lip(\Omega) \subset W^{1,\infty}(\Omega)$ follows from Theorem 6.86.   ∎

Notes Most of the material in this chapter was adapted from Brezis [7], and Adams [1]. For related materials see Adams [2], Friedman [23], Giusti [25], and Mazja [34].

Chapter 7

OPERATOR THEORY

In this chapter we study compact, Fredholm, and monotone operators.

7.1 Compact Operators

Compact operators admit many nice properties.

Definition 7.1 *Let E be a Banach space and $\Omega \subset E$ a bounded set. Then $T : \bar{\Omega} \to E$ is compact if T is continuous on $\bar{\Omega}$ and $T(\bar{\Omega})$ is relatively compact in E.*

Let $Q(\bar{\Omega}, E)$ be the family of all compact operators $T : \bar{\Omega} \to E$, under the norm

$$\|f\|_\infty = \sup_{x \in \bar{\Omega}} \|f(x)\|.$$

Then $Q(\bar{\Omega}, E)$ forms a Banach space.

Definition 7.2 *Let $T : E \to E$ be a linear compact operator. $\mu \in \mathbb{R}$ is a characteristic value of T if there is $u \neq 0$ in E such that $u = \mu T u$. The characteristic values of T are the inverses of the eigenvalues of T.*

Proposition 7.3 *Let $T : E \to E$ be a compact linear operator. Then:*

(i) *The characteristic values of T are countable, and the only possible accumulation point is ∞. Thus, the characteristic values in $[0, 1]$ are finite numbers. The eigenvalues of T are countable, the only possible accumulation point is 0, and each nonzero eigenvalue has finite multiplicity;*

(ii) *Let μ be a characteristic value of T, $\mu \neq \infty$. The sequence*

$$Ker(\mu T - I) \subset Ker(\mu T - I)^2 \subset \cdots \subset Ker(\mu T - I)^k \subset \cdots$$

is stationary: there is $n_0 \in N$ such that

$$Ker(\mu T - I)^k = Ker(\mu T - I)^{n_0} \quad \text{for } k \geq n_0.$$

$Ker(\mu T - I)^{n_0}$ *is called the characteristic space associated with* μ. *This space* $Ker(\mu T - I)^{n_0}$ *is of finite dimension* m. *The number* m *is called the algebraic multiplicity of the characteristic value* μ. *This characteristic space is an invariant subspace of* T;

(*iii*) *If* μ *is not a characteristic value of* T, *then* $(\mu T - I)$ *admits a bounded inverse.*

Proof.   See Gilbarg-Trudinger [24, pp. 78-79].   ∎

Theorem 7.4 *Let* $T : B_r(0) \to E$ *be a compact operator and let* T *be Fréchet-differentiable at* 0. *Then* $T'(0) : E \to E$ *is a compact linear operator.*

Proof. $T \in Q(B_r(0), E)$. For ϵ, $0 < \epsilon \le 1$, define the operator $S_\epsilon : \overline{B_r(0)} \to E$ by $S_\epsilon(x) = \frac{T(\epsilon x) - T(0)}{\epsilon}$. Then S_ϵ is a compact operator: $S_\epsilon \in Q(B_r(0), E)$. Because T is Fréchet-differentiable at 0, we claim that

$$S_\epsilon \to T'(0) \quad \text{uniformly on } \overline{B_r(0)}, \text{ as } \epsilon \to 0.$$

In fact

$$\|T(y) - T(0) - T'(0)y\| \le \epsilon \|y\| \quad \text{for } \|y\| < \delta.$$

Take $y = \epsilon x$, $x \in B_r(0)$, ϵ small such that $\|\epsilon x\| < \delta$. From the above inequality we obtain

$$\left\| \frac{T(\epsilon x) - T(0)}{\epsilon} - T'(0)x \right\| < \epsilon \|x\| < \epsilon r.$$

For $\{x_n\} \subset B_r(0)$, there is $z \in E$ such that $\frac{T(\epsilon x) - T(0)}{\epsilon} \to z$. By the above inequality, $T'(0)x_n \to z$. $T'(0)[B_r(0)]$ is relatively compact, or $T'(0)$ is a compact linear operator.
                                                                                  ∎

Lemma 7.5 (*F. Riesz Lemma*) *Let* M *be a proper closed linear subspace of a normal linear space* X. *Then for each* θ, $0 < \theta < 1$, *there is* $x_0 \in X$ *such that* $\|x_0\| = 1$, $d(x_0, M) \ge \theta$.

Proof.   See Lemma 1.6.   ∎

Theorem 7.6 *Let* X *be a normed linear space. If the unit sphere* $\{x \in X \mid \|x\| = 1\}$ *is compact, then* X *has finite dimensions.*

Proof.   See Theorem 1.7.   ∎

For a linear operator A, denote the kernel of A by $Ker A$ and the range of A by $Range A$.

Theorem 7.7 *Let E be a Banach space, $K : E \to E$ a linear compact operator, $A = I - K$. Then $Range A$ is closed in E, and*

$$\dim Ker A = \dim \operatorname{coker} A < \infty,$$

where dim coker $A =$codim $Range A = \dim(E/Range A)$. In particular, either $Range A = X$, $Ker A = \{0\}$ or $Range A \subsetneq E$, and $Ker A \supsetneq \{0\}$.

Proof.   We divide the proof into the following steps.

(i) For $x \in E$, $\epsilon > 0$, there is $x_0 \in E$ such that

$$Ax = Ax_0,$$
$$\operatorname{dist}(x, Ker A) = \operatorname{dist}(x_0, Ker A),$$
$$\operatorname{dist}(x, Ker A) \leq \|x_0\| \leq \operatorname{dist}(x, Ker A) + \epsilon.$$

In fact, take $x_1 \in Ker A$ with $\|x - x_1\| < d(x, Ker A) + \epsilon$. Set $x_0 = x - x_1$, then

$$Ax_0 - Ax,$$
$$\operatorname{dist}(x_0, Ker A) = \inf_{y \in Ker A} \|x_0 - y\| = \inf_{y \in Ker A} \|x - x_1 - y\|$$
$$= \inf_{y \in Ker A} \|x - y\| = \operatorname{dist}(x, Ker A),$$
$$\operatorname{dist}(x, Ker A) = \inf_{y \in Ker A} \|x - y\| \leq \|x - x_1\|$$
$$= \|x_0\| \leq \operatorname{dist}(x, Ker A) + \epsilon.$$

(ii) There is a constant $c > 0$ with $d(x, Ker A) \leq c\|Ax\|$ for $x \in E$: if we suppose not, then there is a sequence $\{x_n\}$ in E such that

$$\operatorname{dist}(x_n, Ker A) = 1,$$

but

$$\lim_{n \to \infty} \|Ax_n\| = 0.$$

By (i) there is a sequence $\{y_n\}$ in E such that

$$Ay_n = Ax_n \quad \text{for all } n,$$
$$\operatorname{dist}(y_n, Ker A) = \operatorname{dist}(x_n, Ker A),$$
$$1 = \operatorname{dist}(x_n, Ker A) \leq \|y_n\| \leq \operatorname{dist}(x_n, Ker A) + 1 = 2.$$

Thus $1 \leq \|y_n\| \leq 2$, $\lim_{n \to \infty} \|Ay_n\| = 0$. Because K is compact, there are a subsequence of $\{y_n\}$, still denoted by $\{y_n\}$, and $z \in E$ such that

$$Ky_n \to z.$$

Now
$$y_n = Ay_n + Ky_n \to 0 + z = z$$
and
$$Ay_n \to Az, \quad \text{so } Az = 0.$$

We have $z \in Ker\,A$, which implies $1 = d(y_n, Ker\,A) \le \|y_n - z\| \to 0$, a contradiction.

(iii) $Range\,A$ is closed.
   Case 1. $Ker\,A = \{0\}$, and $\|x\| \le c\,\|Ax\|$ for $x \in E$: let $\{y_n\} \subset Range\,A$, $y_n \to y$ in E. Take $\{x_n\} \subset E$ with $Ax_n = y_n$, and

$$\|x_n - x_m\| \le c\,\|Ax_n - Ax_m\| = c\,\|y_n - y_m\| \to 0 \quad \text{as } n, m \to \infty.$$

   Choose $x \in E$, $x_n \to x$ in E. Then $Ax_n \to Ax$, or $Ax = y$.
   Case 2. $Ker\,A \underset{\ne}{\supset} \{0\}$. The map $\bar{A} : E/Ker\,A \to Range\,A$ is defined by

$$\bar{A}\bar{x} = Ax.$$

Now
$$\|\bar{x}\| = \text{dist}(x,\ Ker\,A) \le c\,\|Ax\| = c\,\|\bar{A}\bar{x}\| \quad \text{for } \bar{x} \in E/Ker\,A.$$

By case 1, $Range\,A$ is closed, and note that

$$Range(\bar{A}) = Range(A).$$

(iv) $\dim Ker\,A < \infty$: let $\{x_n\} \subset S = \{x \in Ker\,A \mid \|x\| = 1\}$. There are a subsequence, still denoted by $\{x_n\}$, and $y \in E$ such that

$$Kx_n \to y$$

i.e., $x_n = Ax_n + Kx_n = Kx_n \to y$. Thus $S \subset Ker\,A$ is compact. Then $\dim Ker\,A$ is finite by Theorem 7.6.

(v) $\text{codim}\,Range\,A < \infty$: note $A = I - K$, $A^* = I - K^*$ where K^* is compact and $\text{codim}\,Range\,A = \dim Ker(A^*) < \infty$ by (iv). That K^* is compact follows from (vi) below.

(vi) Let X, Y be normed linear spaces, and $K \in K(X, Y)$ a linear compact operator. Then $K^* \in K(Y^*, X^*)$. Recall that in a complete metric space, every totally bounded set is relatively compact. Because X^* is complete, it suffices to prove $K^*(W)$ is totally bounded in X^* for every bounded set W in Y^*. Set

$$B = \{x \in X \mid \|x\| \le 1\}.$$

$K(B)$ is relatively compact and hence totally bounded in Y. For $\epsilon > 0$, there are $x_1, \cdots, x_n \in B$ such that for $x \in B$, $\|Kx - Kx_i\| < \epsilon$ for some i. Define the map $T : Y^* \to \mathbb{R}^n$ by

$$Ty^* = (y^*Kx_1, \cdots, y^*Kx_n),$$

then T is of finite rank. If $W \subseteq Y^*$ is bounded, take $c > 0$ with

$$\|y^*\| \le c \quad \forall y^* \in W.$$

Then $T(W)$ is relatively compact. Assume there are $y_1^*, \cdots, y_m^* \in W$ such that for $y^* \in W$, $\|Ty^* - Ty_j^*\| < \epsilon$ for some j, i.e.,

$$\|y^*Kx_i - y_j^*Kx_i\| < \epsilon \quad \text{for } i = 1, \cdots, n.$$

$$\|y^*Kx - y_j^*Kx\|$$
$$\le \|K^*Kx - y^*Kx_i\| + \|y^*Kx_i - y_j^*Kx_i\| + \|y_j^*Kx_i - y_j^*Kx\|$$
$$\le \|y^*\| \, \|Kx - Kx_i\| + \epsilon + \|y_j^*\| \, \|Kx_i - Kx\| \le 2c\epsilon + \epsilon \quad \text{for } x \in B.$$

However,

$$y^*Kx = (Kx, y^*) = (x, K^*y^*).$$

Thus,

$$\|K^*y^* - K^*y_j^*\| \le (2c+1)\epsilon.$$

(vii) codim $Range A = 0$ if and only if dim $Ker A = 0$: because codim $Range A$ = dim $Ker(A^*)$, we assert that if codim $Range A = 0$ then dim $Ker A = 0$. Suppose $Ker(A^*) = \{0\}$. $Range A$ is closed, and $Range A = [(Range A)^\perp]^\perp = Ker(A^*)^\perp = \{0\}^\perp = E$. Suppose $Ker A \ne \{0\}$. Take $x_1 \ne 0$, $Ax_1 = 0$. Choose $x_n \in E$ with $Ax_{n+1} = x_n$, $n = 1, 2, 3, \cdots$. Note that A^n are bounded linear operators with $\|A^n\| \le \|A\|^n$, and Ker A^n are closed linear subspaces of E with

$$Ker(A) \subset Ker(A^2) \subset \cdots \subset Ker(A^n) \subset \cdots.$$

Now

$$A^n x_n = A^{n-1}(Ax_n) = A^{n-1}x_{n-1} = \cdots = Ax_1 = 0,$$
$$A^{n-1}x_n = A^{n-2}x_{n-1} = \cdots = Ax_2 = x_1 \ne 0.$$

Thus, $x_n \in Ker(A^n)\backslash Ker(A^{n-1})$, and hence, $Ker(A^{n-1}) \subsetneq Ker(A^n)$. That is, $Ker(A) \subsetneq Ker(A^2) \subsetneq \cdots \subsetneq Ker(A^n) \subsetneq \cdots$. By the F. Riesz Lemma 7.5, choose $z_n \in Ker(A^n)$, $\|z_n\| = 1$, $d(z_n, Ker(A^{n-1})) \ge \dfrac{1}{2}$. For $n > m$

$$\|Kz_n - Kz_m\| = \|z_n - Az_n - z_m + Az_m\|$$
$$= \|z_n - (z_m - Az_m + Az_n)\|$$
$$\ge \frac{1}{2},$$

because $A^{n-1}(z_m - Az_m + Az_n) = 0$. This contradicts the assumption that K is compact. Therefore $Ker(A) = \{0\}$.

$(viii)$ $\dim Ker A = \operatorname{codim} Range A$: suppose

$$Ker A =< x_1, \cdots, x_n >, \quad N(A^*) =< f_1, \cdots, f_m >,$$

we claim that there is $f_0 \in E^*$, $x_0 \in E$ such that

$$\begin{cases} f_0(x_j) = 0 & \text{for } 1 \le j < n, \\ f_0(x_n) \ne 0, \end{cases}$$

and

$$\begin{cases} f_i(x_0) = 0 & \text{for } 1 \le i < m, \\ f_m(x_0) \ne 0. \end{cases}$$

In fact, set $M =< x_1, \cdots, x_{n-1} >$, $d(x_n, M) > 0$. The Hahn-Banach Theorem implies that there is $f_0 \in E^*$ such that

$$\|f_0\| = 1,$$
$$f_0(x_n) = d(x_n, M) > 0,$$
$$f_0(M) = 0.$$

We then find x_0 by induction. Suppose the relationship holds for $\ell - 1 \ge 1$. Choose z_i, $i = 1, \cdots, \ell - 1$, with

$$f_j(z_i) = \delta_{ij}.$$

For $x \in E$

$$f_j(x - \sum_{k=1}^{l-1} f_k(x)z_k) = f_j(x) - f_j(x) = 0 \text{ for } 1 \le j < \ell$$

i.e.,

$$x - \sum_{k=1}^{l-1} f_k(x)z_k \in< f_1, \cdots, f_{\ell-1} >^{\perp}.$$

If $f_\ell = 0$ on $< f_1, \cdots, f_{\ell-1} >^{\perp}$, then

$$f_\ell(x - \sum_{k=1}^{l-1} f_k(x)z_k) = 0,$$

i.e.,

$$f_\ell = \sum_{k=1}^{l-1} f_\ell(z_k)f_k$$

contradicts the assumption that $f_1, \cdots, f_\ell$ are linearly independent. Thus $x_0 \in <$ $f_1, \cdots,$ there is $f_{\ell-1} >^\perp$, with $f_\ell(x_0) \neq 0$. Define the map $K_0 : E \to E$ by

$$K_0(x) = f_0(x)x_0 \quad \text{for } x \in E.$$

Then

$$(x, K_0^* x^*) = (K_0 x, x^*) = (f_0(x)x_0, x^*) = f_0(x)(x_0, x^*) = x^*(x_0)f_0(x).$$

Thus,

$$K_0^* x^* = x^*(x_0)f_0 \quad \text{for } x^* \in E^*.$$

(ix) Set $A_1 = A - K_0$, and we have $\dim Ker(A_1) = n - 1$, $\dim Ker(A_1^*) = m - 1$: suppose $x \in Ker(A_1)$. Then

$$A_1 x = 0, \quad \text{or } Ax = K_0 x = f_0(x)x_0.$$

However, $f_m(x_0) \neq 0$, so $x_0 \notin Ker(A^*)^\perp = Range A$. We conclude that $f_0(x) = 0$, and $Ax = 0$, or $x \in Ker(A)$. Set

$$x = \sum_{j=1}^{n} \alpha_j x_j.$$

Now

$$0 = f_0(x) = \sum_{j=1}^{n} \alpha_j f_0(x_j) = \alpha_n f_0(x_n).$$

We have $\alpha_n = 0$ because $f_0(x_n) \neq 0$, or

$$x = \sum_{j=1}^{n-1} \alpha_j x_j.$$

Conversely, if $x = \sum_{j=1}^{n-1} \alpha_j x_j \in Ker(A)$, then

$$f_0(x) = \sum_{j=1}^{n-1} \alpha_j f_0(x_j) = 0.$$

Thus $Ax = 0$, and $K_0 x = f_0(x)x_0 = 0$. We conclude that $A_1 x = Ax - K_0 x = 0$, or $x \in Ker(A_1)$. Therefore $Ker(A_1) = n - 1$. Next, suppose $f \in Ker(A_1^*)$, then

$$A^* f = K_0^* f = f(x_0)f_0.$$

Because $f_0(x_n) \neq 0$, we have $f_0 \notin Ker(A)^\perp : Range(A^*)$, or $f(x_0) = 0$. Thus $A^*f = 0$. If $f = \sum_{j=1}^{m} \beta_j f_j$, then

$$0 = f(x_0) = \sum_{j=1}^{m} \beta_j f_j(x_0) = \beta_m f_m(x_0).$$

$\beta_m = 0,$ because $f_m(x_0) \neq 0$. Thus $f = \sum_{j=1}^{m-1} \beta_j f_j$. Conversely, if $f = \sum_{j=1}^{m-1} \beta_i f_j \in N(A^*)$, then $A^*f = 0$, and $K_0^* f = f(x_0)f_0 = 0$. Thus, $A_1^* f = 0$, and we have $\dim \ N(A_1^*) = m - 1$.

(x) $m = n$: because $\dim Ker(A_1) = n - 1$, $\dim Ker(A_1^*) = m - 1$. However, $A_1 = I - (K + K_0)$, where $K + K_0$ is a linear compact operator. Continuing this process, we reach

$$\hat{A} = I - \hat{K} \quad \text{where } \hat{K} \text{ is a linear compact operator,}$$

and either $\dim Ker(\hat{A}) = 0$ or $\dim Ker(\hat{A}^*) = 0$. (vi) implies

$$\dim Ker(\hat{A}) = \dim Ker(\hat{A}^*) = 0. \qquad \blacksquare$$

Theorem 7.7 can be stated as the following well-known result.

Theorem 7.8 (*Fredholm Alternative Theorem*) *Let E be a normed linear space, and $T : E \to E$ a compact linear operator. Then either:*

(i) *the homogeneous equation*
$$x - Tx = 0$$
has a nontrivial solution $x \in E$; or

(ii) *for each $y \in E$ the equation*
$$x - Tx = y$$
has a uniquely determined solution $x \in E$. Furthermore, in case (ii), there is the operator $(I - T)^{-1}$ and is bounded.

Lemma 7.9 *Let N be a finite-dimensional subspace of a normed linear space X. Then there is a closed linear subspace M such that*

$$X = N \oplus M.$$

Proof.   See Lemma 1.9. $\blacksquare$

Lemma 7.10 *Let N be a closed linear subspace of a normed linear space X, of* codim $N = \dim N^{\perp} = n$. *Then there is an n-dimensional subspace M of X such that $X = N \oplus M$.*

Proof. See Lemma 1.10. ∎

Remark 7.11 *Let E be a B-space, $K : E \to E$ a compact linear operator, and $A = I - K$. Proposition 7.3 implies Range A is closed with codim Range $A < \infty$. By Lemma 7.10 there is a closed linear subspace M such that*

$$E = Range A \oplus M.$$

Consider the projection $P : E \to M$. We obtain

$$M \cong E/Range A.$$

We have

$$\dim M = \dim E/Range A = \text{codim } Range A = \dim(Range A)^{\perp}$$
$$= \dim N(A^*) = \dim \text{coker } A.$$

Lemma 7.12 *Let $K \subset E$ be a compact set. For $\epsilon > 0$ there is a finite-dimensional subspace F_ϵ of E, $g_\epsilon \in C(K, F_\epsilon)$ such that $\|x - g_\epsilon(x)\| < \epsilon$ for $x \in K$.*

Proof. See Lemma 9.63. ∎

Lemma 7.13 *Let $T \in Q(\Omega, E)$, and $f = I - T$. Then:*

(i) *f is closed: that is, f maps closed sets to closed sets;*

(ii) *f is proper: that is, f^{-1} maps compact sets to compact sets.*

Proof.

(i) Let $A \subset \bar{\Omega}$ be closed, and let $\{u_n\} \subset A$ be such that $f(u_n) \to z$ in E. Note that $\{u_n\}$ is bounded because $\bar{\Omega}$ is bounded. Because T is compact, there is a subsequence, still denoted by $\{u_n\}$, and $w \in E$ such that

$$T(u_n) \to w,$$

so

$$u_n = f(u_n) + T(u_n) \to z + w.$$

Thus, $z + w \in A$, and $f(u_n) \to f(z + w) = z$, or $z \in f(A)$.

(ii) Let $K \subset E$ be compact, and $\{u_n\} \subset f^{-1}(K)$. Then $\{f(u_n)\} \subset K$. There is a subsequence, still denoted by $f(u_n)$, $y \in K$ such that

$$f(u_n) \to y.$$

Because T is compact, there is a subsequence, still denoted by $\{u_n\}$, and $z \in K$ such that
$$T(u_n) \to z,$$
so $u_n = f(u_n) + T(u_n) \to y + z$, or $f^{-1}(K)$ is compact.   ∎

Proposition 7.14 *Let $t \to S(\cdot, t)$ be a continuous map of $[0,1]$ into $Q(\bar{\Omega}, E)$. Then $S \in Q(\bar{\Omega} \times [0,1], E)$.*

Proof.   Let $\{x_n\} \subset \bar{\Omega}$, $\{t_n\} \subset [0,1]$, $t \in [0,1]$, $t_n \to t$. Because $S(\cdot, t) \in Q(\bar{\Omega}, E)$, there is a subsequence, still denoted by $\{x_n\}$, and $z \in E$ such that

$$S(x_n, t) \to z \quad \text{as } n \to \infty.$$

Now

$$\begin{aligned}
|z - S(x_n, t_n)| &\leq |z - S(x_n, t)| + |S(x_n, t) - S(x_n, t_n)| \\
&\leq |z - S(x_n, t)| + \|S(\cdot, t) - S(\cdot, t_n)\|_\infty \\
&\to 0 \quad \text{as } n \to \infty.
\end{aligned}$$

Thus $S(x_n, t_n) \to z$, and it follows $S \in Q(\bar{\Omega} \times [0,1], E)$.   ∎

7.2   Fredholm Operators

Definition 7.15 *Let E, F be Banach spaces. A linear operator $A : E \to F$ is called Fredholm if $\dim Ker(A) < \infty$, $RangeA$ is closed, and $codim\ RangeA < \infty$.*

$\dim Ker(A) - codim RangeA$ *is called the index of A.*

Theorem 7.16 *Let E, F be B-spaces, $A : E \to F$ a Fredholm operator, and*

$$E = KerA \oplus E_0, F = F_0 \oplus RangeA.$$

Then there is a bounded linear operator $A_0 : F \to E$ such that:

(i) $Ker(A_0) = F_0$;

(ii) $Range(A_0) = E_0$;

(iii) $A_0 A = I$ *on* E_0;

(iv) $AA_0 = I$ on $Range A$;

(v) $A_0 A = I - T_1$ on E where T_1 is of finite rank, $Range(T_1) = Ker A$;

(vi) $AA_0 = I - T_2$ on F where T_2 is of finite rank, $Range(T_2) = F_0$.

Proof.   Setting $\hat{A} = A\big|_{E_0}$, then $\hat{A} : E_0 \to Range A$ is a bijective bounded linear operator. Consider the projections $P : F \to F_0$, $I - P : F \to Range(A)$. Set $A_0 = \hat{A}^{-1}(I - P)$ and $T_1 = I - AA_0$ and we obtain the proofs.   ∎

Denote by $\Phi(E, F)$ the family of all Fredholm operators.

Theorem 7.17 Let $A \in L(E, F)$. If there are A_1, $A_2 \in L(F, E)$, $K_1 \in K(E, E)$, $K_2 \in K(F, F)$ such that

$$A_1 A = I - K_1 \quad \text{on } E,$$
$$AA_2 = I - K_2 \quad \text{on } F,$$

then $A \in \Phi(E, F)$ is a Fredholm operator.

Proof.   Because $Ker(A) \subseteq Ker(A_1 A) = Ker(I - K_1)$, we have $\dim Ker(A) \le \dim Ker(I - K_1) < \infty$. Because $AA_2 = I - K_2$, we have

$$A_2^* A^* = I - K_2^*.$$

Hence,

$$Ker(A^*) \subseteq Ker(A_2^* A^*) \subseteq Ker(I - K_2^*),$$

therefore

$$\text{codim } Range A = \dim Ker(A^*) \le \dim \ Ker(I - K_2^*) < \infty.$$

Note that $Range A$ is closed; this follows because $\text{Range}(I - K_2) \subseteq Range A$, and codim $Range(I - K_2)$ is finite.   ∎

Theorem 7.18 If $A \in \Phi(E, F)$ and $B \in \Phi(F, G)$, then $BA \in \Phi(E, G)$, and

$$\text{index}(BA) = \text{index } B + \text{index } A.$$

Proof.

(i) By Theorem 7.16, 7.17, there are $A_0 \in \Phi(F, E)$, $B_0 \in \Phi(G, F)$, $T_1 \in K(E, E)$, T_2, $T_3 \in K(F, F)$, $T_4 \in K(G, G)$ such that

$$A_0 A = I - T_1 \quad \text{on } E,$$
$$AA_0 = I - T_2 \quad \text{on } F,$$
$$B_0 B = I - T_3 \quad \text{on } F,$$
$$BB_0 = I - T_4 \quad \text{on } G.x$$

Then

$$A_0 B_0 BA = A_0(I - T_3)A$$
$$= A_0 A - A_0 T_3 A$$
$$= I - T_1 - A_0 T_3 A$$
$$= I - T_5 \qquad \text{on } E,$$
$$BAA_0 B_0 = B(I - T_2)B_0$$
$$= I - T_4 - BT_2 B_0$$
$$= I - T_6 \quad \text{on } G,$$

where $T_5 \in K(E, E)$, $T_6 \in K(G, G)$. By Theorem 7.17, $BA \in \Phi(E, G)$.

(ii) index(BA) = index B+index A : denote $\alpha(A) = \dim Ker A$ and $\beta(A) = \text{codim} \, Range A$. Set $F_1 = Range A \cap Ker(B)$. Now $\dim F_1 \leq \dim Ker(B) < \infty$. Choose F_2, F_3, F_4 from finite-dimensional subspaces such that

$$Range A = F_1 \oplus F_2, \quad Ker(B) = F_1 \oplus F_3, \quad F_0 = F_3 \oplus F_4.$$

Clearly, $F_3 \cap Range A = \{0\}$. Let $d_i = \dim \ F_i$ for i. Now

$$F = Range A \oplus F_0 = Range A \oplus F_3 \oplus F_4$$
$$= F_1 \oplus F_2 \oplus F_3 \oplus F_4.$$

Because $Ker(A) \subset Ker(BA)$, Range$(BA) \subset Range(B)$. Take E_1, G_4 such that

$$Ker(BA) = Ker(A) \oplus E_1,$$
$$Range(B) = Range(BA) \oplus G_4.$$

We claim that $A : E_1 \to F_1$, and $B : F_4 \to G_4$ are bijective linear, and hence,

$$\dim E_1 = \dim F_1 = d_1,$$
$$\dim G_4 = \dim F_4 = d_4.$$

In fact,

(a) $A : E_1 \to F_1$ is bijective: $A(E_1) \subset Range A$, $Ker(BA) = Ker(A) \oplus E_1$, so $E_1 \subset Ker(BA)$, or $A(E_1) \subset Ker(B)$. Therefore $A(E_1) \subset Range(A) \cap Ker(B) = F_1$. Conversely, if $0 \neq y \in F_1$, take $x \in E$ such that $Ax = y$, $By = 0$, so $BAx = 0$. We have $x \in Ker(BA) = Ker(A) \oplus E_1$. Write $x = x_0 + x_1$ where $x_0 \in Ker(A)$, $x_1 \in E_1$. $y = Ax = Ax_1$, $BAx_1 = By = 0$. Thus $y \in AE_1$, or $AE_1 = F_1$.

(b) $B : F_4 \to G_4$ is bijective: $Range(B) = Range(BA) \oplus G_4$, or $B(F) = BA(E) \oplus G_4$. However, $F_1 = Range(A) \cap Ker(B)$, $A(E) = Range A = F_1 \oplus F_2$, $N(B) = F_1 \oplus F_3$, $F = F_1 \oplus F_2 \oplus F_3 \oplus F_4$. Thus $B(F) = B(F_2) \oplus B(F_4)$, but $BA(E) = B(F_1) \oplus B(F_2) = B(F_2)$, so $B(F) = BA(E) \oplus B(F_4)$. We conclude that $BF_4 = G_4$. Clearly, $B : F_4 \to G_4$ is 1-1.

Now

(α) $Ker(BA) = Ker(A) \oplus E_1 \Rightarrow \alpha(BA) = \alpha(A) + d_1$;

(β) $G = Range(B) \oplus M = Range(BA) \oplus G_4 \oplus M \Rightarrow \dim M = \beta(B)$, and $\dim M + d_4 = \beta(BA) \Rightarrow \beta(BA) = \beta(B) + d_4$;

(γ) $Ker(B) = F_1 \oplus F_3 \Rightarrow \alpha(B) = d_1 + d_3$;

(δ) $F = RangeA \oplus F_3 \oplus F_4 \Rightarrow \beta(A) + d_3 + d_4$.

By (α) — (δ)

$$
\begin{aligned}
i(BA) &= \alpha(BA) - \beta(BA) \\
&= \alpha(A) + d_1 - \beta(B) - d_4 \\
&= \alpha(A) - \beta(B) + d_1 - d_4 \\
&= \alpha(A) - \beta(B) + \alpha(B) - \beta(A) \\
&= \alpha(A) - \beta(A) + \alpha(B) - \beta(B) \\
&= i(A) + i(B).
\end{aligned}
$$
■

Lemma 7.19 *Suppose $A \in \Phi(E, F)$, then there exist $A_0 \in L(F, E)$, $F_1 \in K(E, E)$ and $F_2 \in K(F, F)$ such that*

$$
\begin{aligned}
A_0 A &= I - F_1 \quad \text{on } E, \\
A A_0 &= I - F_2 \quad \text{on } F.
\end{aligned}
$$

Then $A_0 \in \Phi(F, E)$, and $i(A_0) = -i(A)$.

Proof. By Theorem 7.17, $A_0 \in \Phi(F, E)$, and then by Theorem 7.18,

$$
i(A_0) + i(A) = i(A_0 A) = i(I - F_1) = 0,
$$

so

$$
i(A_0) = -i(A).
$$
■

Theorem 7.20 *If $A \in \Phi(E, F)$ and $K \in K(E, F)$, then $A + K \in \Phi(E, F)$, and $i(A + K) = i(A)$.*

Proof. By Theorem 7.16, there are $A_0 \in \Phi(F, E)$ and $T_1 \in K(E, E)$, $T_2 \in K(F, F)$ such that

$$
\begin{aligned}
A_0 A &= I - T_1 \quad \text{on } E, \\
A A_0 &= I - T_2 \quad \text{on } F.
\end{aligned}
$$

Hence,

$$A_0(A + K) = A_0 A + A_0 K$$
$$= I - T_1 + A_0 K$$
$$= I - K_1 \qquad \text{on } E,$$
$$(A + K)A_0 = AA_0 + KA_0$$
$$= I - T_2 + KA_0$$
$$= I - K_2 \quad \text{on } F,$$

where $K_1 = T_1 - A_0 K \in K(E, E)$, $K_2 = T_2 - KA_0 \in K(F, F)$. Hence, $A + K \in \Phi(E, F)$. By Lemma 7.19, $i(A_0) = -i(A + K)$, $i(A_0) = -i(A)$, so $i(A) = i(A + K)$. ∎

Theorem 7.21 *Let $A \in \Phi(E, F)$. Then there is $\eta > 0$ such that for any $T \in L(E, F)$ with $\|T\| < \eta$, we have $A + T \in \Phi(E, F)$, $i(A + T) = i(A)$, and $\alpha(A + T) \leq \alpha(A)$.*

Proof.

(i) There are $A_0 \in \Phi(F, E)$, $T_1 \in K(E, E)$ and $T_2 \in K(F, F)$ such that

$$A_0 A = I - T_1 \quad \text{on } E,$$
$$AA_0 = I - T_2 \quad \text{on } F.$$

Then
$$A_0(A + T) = A_0 A + A_0 T = I - T_1 + A_0 T \quad \text{on } E.$$

Take $\eta = \|A_0\|^{-1}$. Then

$$\|A_0 T\| \leq \|A_0\| \|T\| < 1, \quad \|TA_0\| < 1, \quad \text{if } \|T\| < \eta.$$

Therefore, $I + A_0 T$ and $I + TA_0$ have bounded inverses, and

$$(I + A_0 T)^{-1} A_0 (A + T) = (I + A_0 T)^{-1}(I - T_1 + A_0 T)$$
$$= I - (I + A_0 T)^{-1} T_1 \qquad \text{on } E,$$
$$(A + T)A_0(I + TA_0)^{-1} = (I - T_2 + TA_0)(I + TA_0)^{-1}$$
$$= I - T_2(I + TA_0)^{-1} \quad \text{on } F.$$

We conclude that $A + T \in \Phi(E, F)$.

(ii) $i((I + A_0 T)^{-1}) + i(A_0) + i(A + T) = 0$: $I + A_0 T$ is bijective, so

$$i((I + A_0 T)^{-1}) = \dim Ker((I + A_0 T)^{-1}) - \text{codim} \, Range((I + A_0 T)^{-1}) = 0.$$

Thus,
$$i(A) = -i(A_0) = i(A + T).$$

(iii) By Theorem 7.16, $A_0 A = I$ on E_0 where $E = Ker A \oplus E_0$ and

$$A_0(A + T) = A_0 A + A_0 T = I + A_0 T \quad \text{on } E_0.$$

Because $\|A_0 T\| < 1$, $A_0(A+T)$ is 1-1 on E_0. We claim that $(A+T) \cap E_0 = \{0\}$. In fact, $x \in Ker(A+T) \cap E_0$, then $x \in E_0$, $(A+T)x = 0$ or $x \in E_0$, $A_0(A+T)x = 0$. Therefore $x = 0$.

We claim that $\alpha(A + T) = \dim Ker(A + T) \leq \dim Ker A = \alpha(A)$. Thus $\lim_{T \to 0} \alpha(A+T) \leq \alpha(A)$, that is, α is upper semi-continuous. In fact, it suffices to prove that if $E = Ker A \oplus E_0$, $Ker A$ is finite-dimensional, and $M \subset E$ a linear subspace, $M \cap E_0 = \{0\}$, then $\dim M \leq \dim Ker A$. Suppose $\dim Ker A < n$. For $x_1, \cdots, x_n$ in M, $x_k = x_{k_0} + x_{k_1} \in E_0 \oplus Ker A$, $1 \leq k \leq n$, there are $\alpha_1, \cdots, \alpha_n$, which are not all zero with

$$\sum_{k=1}^{n} \alpha_k x_{k_1} = 0.$$

Hence, $\sum_{k=1}^{n} \alpha_k x_k = \sum_{k=1}^{n} \alpha_k x_{k_0} \in M \cap E_0 = \{0\}$. Therefore, $\sum_{k=1}^{n} \alpha_k x_k = 0$, that is, $\dim M < n$. Thus, $\dim M \leq \dim N$.   ∎

Theorem 7.22 *If $K \in K(E, E)$ and $A = I - K$, then there is $n \geq 1$ such that $Ker(A^n) = Ker(A^k)$ for $k \geq n$.*

Proof.   Suppose $Ker(A^k) = Ker(A^{k+1})$, and $j \geq k$. If $x \in Ker(A^{j+1})$ then $A^{j-k}x \in Ker(A^{k+1}) = Ker(A^k)$, or $x \in Ker(A^j)$. Thus, $Ker(A^j) = Ker(A^{j+1})$ for $j \geq k$. Suppose $Ker(A^k) \subsetneq Ker(A^{k+1})$ for $k \geq 1$. By the F. Riesz Lemma 7.5, there is $z_k \in Ker(A^{k+1})$ such that

$$\|z_k\| = 1, \quad \mathrm{d}(z_k, \, Ker(A^k)) \geq \frac{1}{2}.$$

For $j > k$,

$$\begin{aligned}
\|Kz_j - Kz_k\| &= \|z_j - Az_j - z_k + Az_k\| \\
&= \|z_j - (z_k - Az_k + Az_j)\| \\
&\geq \frac{1}{2} \qquad\qquad \text{because } A^j(z_k - Az_k + Az_j) = 0.
\end{aligned}$$

Thus $\{Kz_k\}$ contains no convergent subsequence. This contradicts the assumption that K is a compact operator.   ∎

7.3 Monotone Operators in Hilbert Spaces

Definition 7.23 *Let H be a Hilbert space, $M \subset H$ a subset, and $A : M \to M$. A is monotone if*

$$(Ax - Ay, x - y) \geq 0 \quad \text{for } x, y \in M.$$

A is nonexpansive if

$$\|Ax - Ay\| \leq \|x - y\| \quad \text{for } x, y \in M.$$

A is expansive if

$$\|Ax - Ay\| \geq \|x - y\|.$$

Theorem 7.24 *If we let H be a Hilbert space, $M \subset H$ a subset, and $A : M \to M$, then A is monotone if and only if $I + \lambda A$ is expansive for $\lambda > 0$.*

Proof. " $\Rightarrow$ " Suppose A is monotone. Then, for $\lambda > 0$,

$$
\begin{aligned}
&((I + \lambda A)x - (I + \lambda A)y, x - y) \\
&= (x + \lambda Ax - y - \lambda Ay, x - y) \\
&= (x - y, x - y) + \lambda(Ax - Ay, x - y) \\
&\geq \|x - y\|^2,
\end{aligned}
$$

i.e.,

$$
\begin{aligned}
&\|(I + \lambda A)x - (Ix - \lambda A)y\| \\
&= \sup_{x' \neq y'} \frac{|(x + \lambda Ax - y - \lambda Ay,\ x' - y')|}{\|x' - y'\|} \\
&\geq \frac{|(x + \lambda Ax - y - \lambda Ay,\ x - y)|}{\|x - y\|} \qquad x \neq y \\
&\geq \|x - y\|.
\end{aligned}
$$

" $\Leftarrow$ " Suppose $\|(x + \lambda Ax) - (y + \lambda Ay)\| \geq \|x - y\|$, for $\lambda > 0$. Then

$$\|(x - y) + \lambda(Ax - Ay)\|^2 \geq \|x - y\|^2,$$

or

$$\|x - y\|^2 + 2\lambda(Ax - Ay, x - y) + \lambda^2 \|Ax - Ay\|^2 \geq \|x - y\|^2,$$

that is,

$$\lambda \|Ax - Ay\| + 2(Ax - Ay, x - y) \geq 0.$$

Letting $\lambda \to 0$, we obtain

$$(Ax - Ay, x - y) \geq 0. \qquad \blacksquare$$

Lemma 7.25 (*Minty Lemma*) *Let H be a Hilbert space, $\Omega \subset H$ a convex set, and let $A : \Omega \to H$ be monotone and continuous on every segment in Ω. For fixed $u \in \Omega$, $z \in H$. The following are equivalent:*

(i) $(Au - z, v - u) \geq 0$ for $v \in \Omega$;

(ii) $(Av - z, v - u) \geq 0$ for $v \in \Omega$.

Note that if $u \in \overset{\circ}{\Omega}$, then (i) implies $Au = z$.

Proof. $(i) \Rightarrow (ii)$. Suppose $(Au - z, v - u) \geq 0$ for $v \in \Omega$. Then

$$(Av - z, v - u) - (Au - z, v - u) = (Av - Au, v - u) \geq 0.$$

Thus,

$$(Av - z, v - u) \geq (Au - z, v - u) \geq 0 \quad \forall v \in \Omega.$$

$(ii) \Rightarrow (i)$. Suppose $(Av - z, v - u) \geq 0$ for $v \in \Omega$. For any $w \in \Omega$, $0 \leq t < 1$, set

$$v = tu + (1 - t)w,$$

or

$$v - u = (1 - t)(w - u).$$

Now

$$
\begin{aligned}
0 \leq (Av - z, v - u) &= (Av - z, (1 - t)(w - u)) \\
&= (1 - t)(Av - z, w - u)
\end{aligned}
$$

so

$$(Av - z, w - u) \geq 0.$$

Letting $t \to 1$, we have

$$(A(\lim_{t \to 1} v) - z, w - u) \geq 0$$

or

$$(Au - z, w - u) \geq 0. \qquad \blacksquare$$

Theorem 7.26 *Let H be a Hilbert space and $B \subset H$ a bounded closed convex set, and let $f : B \to B$ be nonexpansive. Then f has a fixed point in B. Furthermore, the set of fixed points of f is a convex set.*

Proof. We may assume $0 \in \overset{\circ}{B}$. For $0 < \lambda < 1$ with the Contraction Mapping Theorem applied to λf, there is a unique $x_\lambda \in B$ with

$$\lambda f(x_\lambda) = x_\lambda.$$

Set $A = I - f$, $A_\lambda = I - \lambda f$. We have:

(i) $A_\lambda x_\lambda = 0$;

(ii) $\|A_\lambda v - Av\| = \|v - \lambda f(v) - v + f(v)\| = (1-\lambda)\,\|f(v)\| \to 0$ for $v \in B$ as $\lambda \to 1$;

(iii) A, A_λ are monotone. To show A_λ is monotone:

$$(A_\lambda u - A_\lambda v, u - v) = (u - \lambda f(u) - v + \lambda f(v),\, u - v)$$
$$= -\lambda(f(u) - f(v), u - v) + \|u - v\|^2$$
$$\geq 0.$$

Similarly, A is monotone. Because B is bounded, there is a subsequence, still denoted by $\{x_\lambda\}$, and $u \in B$ such that

$$x_\lambda \to u \quad \text{weakly as } \lambda \to 1.$$

For any $v \in B$,

$$(A_\lambda v, v - x_\lambda) = (A_\lambda v - A_\lambda x_\lambda, v - x_\lambda) \geq 0, \quad \text{because } A_\lambda x_\lambda = 0.$$
$$|(A_\lambda v, v - x_\lambda) - (Av, v - u)|$$
$$\leq |(A_\lambda v, v - x_\lambda) - (Av, v - x_\lambda)| + |(Av, v - x_\lambda) - (Av, (v - u)|$$
$$= |\lambda - 1|\,|(f(v), v - x_\lambda)| + |(Av, u - x_\lambda)|$$
$$\to 0 \qquad \text{as } \lambda \to 1.$$

Hence $(Av, v - u) \geq 0$. By the Minty Lemma,

$$(Au, v - u) \geq 0 \quad \text{for } v \in B.$$

That is,

$$(u - f(u), v - u) \geq 0 \quad \text{for } v \in B.$$

If we replace $v = f(u)$ in the above inequality, we obtain

$$-(u - f(u), u - f(u)) \geq 0.$$

Hence,

$$f(u) = u.$$

Now if $u \in B$, then

$$u = f(u) \quad \text{if and only if} \quad Au = 0$$
$$\text{if and only if} \quad (Au, v - u) \geq 0 \quad \text{for} \quad v \in B$$
$$\text{if and only if} \quad (Av, v - u) \geq 0 \quad \text{for} \quad v \in B.$$

Let $u_1 = f(u_1)$, $u_2 = f(u_2)$, then, for $0 \le t \le 1$,

$$(Av, v - [tu_1 + (1 - t)u_2])$$
$$= t(Av, v - u_1) + (1 - t)(Av, v - u_2) \ge 0.$$

Therefore $tu_1 + (1 - t)u_2$ is a fixed point of f. We conclude that the set of all fixed points of f forms a convex set.   ∎

Remark 7.27 *Theorem 7.26 is false for general Banach spaces. Consider the Banach space*

$$\ell_0 = \{x = (x_1, x_2, \cdots) \mid x_i \to 0 \quad \text{as } i \to \infty\}$$

under the norm $\|x\| = \max_i |x_i|$.

Let B be the unit ball in ℓ_0, $f : B \to B$ defined by

$$f(x_1, x_2, \cdots) = (1, x_1, x_2, \cdots).$$

For $x = (x_1, x_2, \cdots)$, $y = (y_1, y_2, \cdots)$,

$$\|f(x) - f(y)\| = \|(0, x_1 - y_1, x_2 - y_2, \cdots)\| = \|x - y\|,$$

and therefore f is nonexpansive. Let $x = (x_1, x_2, \cdots)$ with $f(x) = x$:

$$(1, x_1, x_2, \cdots) = (x_1, x_2, \cdots),$$

and we conclude that $1 = x_1 = x_2 = \cdots = \cdots$, $x = (1, 1, \cdots) \notin \ell_0$.

Remark 7.28 *In elementary calculus, if* $f : (a, b) \to \mathbb{R}$ *is of* C^2, *then* f *is convex if and only if* $f'' \ge 0$ *if and only if* f' *is increasing if and only if* $(f'(x) - f'(y))(x - y) \ge 0$.

We have the following analogous result in Hilbert space.

Theorem 7.29 *Let* H *be a Hilbert space,* $\Omega \subset H$ *an open convex set, and* $f : \Omega \to \mathbb{R}$ *be smooth. For* $x \in \Omega$, *there is* $z(x) \in H$ *such that*

$$f'(x)y = (y, z(x)) \text{ for } y \in H$$

or $f'(x) = z(x)$. *Then* f *is convex on* Ω *if and only if* $z(x) = f'(x)$ *is monotone on* Ω.

Proof. "$\Longrightarrow$" Suppose f is convex on Ω. For $0 < \lambda_1 < \lambda_2$, that is, $0 < \frac{\lambda_1}{\lambda_2} < 1$, and $w = u + \lambda_2 v$,

$$f(u + \lambda_1 v) = f(u + \frac{\lambda_1}{\lambda_2} \lambda_2 v)$$
$$= f(u + \frac{\lambda_1}{\lambda_2}(w - u))$$
$$= f((1 - \frac{\lambda_1}{\lambda_2})u + \frac{\lambda_1}{\lambda_2}w)$$
$$\leq (1 - \frac{\lambda_1}{\lambda_2})f(u) + \frac{\lambda_1}{\lambda_2}f(w).$$

Thus,

$$\lambda_2 f(u + \lambda_1 v) \leq \lambda_2 f(u) - \lambda_1 f(u) + \lambda_1 f(u + \lambda_2 v),$$

or

$$\lambda_2 f(u + \lambda_1 v) - \lambda_1 f(u + \lambda_2 v) - \lambda_2 f(u) + \lambda_1 f(u) \leq 0.$$

Thus,

$$\frac{f(u + \lambda_1 v) - f(u)}{\lambda_1} - \frac{f(u + \lambda_2 v) - f(u)}{\lambda_2} \leq 0,$$

so $\dfrac{f(u + \lambda v) - f(u)}{\lambda}$ increases as λ increases, but

$$\lim_{\lambda \to 0} \frac{f(u + \lambda v) - f(u)}{\lambda} = f'(u)v.$$

Thus, for $\lambda > 0$,

$$f'(u)v \leq \frac{f(u + \lambda v) - f(u)}{\lambda}.$$

Let $\lambda = 1$, $w = u + v$,

$$f(w) - f(u) \geq f'(u)(w - u) = (w - u, z(u))$$

or

$$f(u) + (w - u, z(u)) \leq f(w).$$

Set $u = u_1 \in \Omega$, $w = u_2 \in \Omega$, to obtain

$$f(u_1) + (u_2 - u_1, z(u_1)) \leq f(u_2). \tag{7.1}$$

We change the roles of u_1, u_2, to obtain

$$f(u_2) + (u_1 - u_2, z(u_2)) \leq f(u_1). \tag{7.2}$$

Add (7.1) and (7.2) to obtain

$$(u_2 - u_1, z(u_2) - z(u_1)) \geq 0,$$

and so $z(u)$ is monotone on Ω.

"$\Longleftarrow$" Suppose $f'(x)$ is monotone on Ω. Fix $u, v \in \Omega$, and define the map $\varphi : [0, 1] \to \mathbb{R}$ by

$$\varphi(\lambda) = f(u + \lambda(v - u)).$$

Then

$$\varphi'(\lambda) = f'(u + \lambda(v - u))(v - u)$$
$$= (v - u, z(u + \lambda(v - u))).$$

If $0 \leq \lambda_1 < \lambda_2$, then

$$\varphi'(\lambda_2) - \varphi'(\lambda_1) = (v - u, z(u + \lambda_2(v - u))) - (v - u, z(u + \lambda_1(v - u)))$$
$$= (v - u, z(u + \lambda_2(v - u)) - z(u + \lambda_1(v - u)))$$
$$= \frac{1}{\lambda_2 - \lambda_1}([u + \lambda_2(v - u)] - [u - \lambda_1(v - u)],$$
$$z(u + \lambda_2(v - u)) - z(u + \lambda_1(v - u))) \geq 0.$$

Thus φ' is increasing, or φ is convex. In particular,

$$\varphi(\lambda) = \varphi((1 - \lambda)0 + \lambda \cdot 1) \leq (1 - \lambda)\varphi(0) + \lambda\varphi(1),$$

that is,

$$f(u + \lambda(v - u)) \leq (1 - \lambda)f(u) + \lambda f(v).$$

We conclude that f is convex. ∎

Theorem 7.30 *Let H be a Hilbert space, $\Omega \subset H$ a closed bounded convex set, and $f : \Omega \to \mathbb{R}$ smooth convex. Then the following are equivalent:*

(i) $f(u) = \min\limits_{v \in \Omega} f(v);$

(ii) $(f'(u), v - u) \geq 0$ *for* $v \in \Omega;$

(iii) $(f'(v), v - u) \geq 0$ *for* $v \in \Omega.$

Proof. *(i)* $\Rightarrow$ *(ii)*. Suppose $f(u) = \min\limits_{v \in \Omega} f(v)$. For $v \in \Omega$, $\lambda \in (0, \delta)$, we obtain

$$f(u) \leq f((1 - \lambda)u + \lambda v) = f(u + \lambda(v - u)).$$

Then

$$\lim_{\lambda \to 0^+} \frac{f(u + \lambda(v - u)) - f(u)}{\lambda} \geq 0,$$

and hence,
$$(f'(u), (v - u)) \geq 0.$$

$(ii) \Rightarrow (i)$. Suppose $(f'(u), v - u) \geq 0 \ \forall v \in \Omega$. Note that

$$\frac{f((1 - \lambda)u + \lambda v) - f(u)}{\lambda} \leq \frac{(1 - \lambda)f(u) + \lambda f(v) - f(u)}{\lambda}$$
$$= f(v) - f(u).$$

Letting $\lambda \to 0^+$

$$0 \leq (f'(u), v - u)) \leq f(v) - f(u),$$

or

$$f(u) \leq f(v) \quad \forall v \in \Omega.$$

Hence,

$$f(u) = \min_{v \in \Omega} f(v).$$

$(ii) \Rightarrow (iii)$. Suppose $(f'(u), v - u) \geq 0 \ \forall v \in \Omega$. Because f' is monotone,

$$(f'(v) - f'(u), v - u) \geq 0 \qquad \text{for } v \in \Omega.$$

Adding these two inequalities, we obtain

$$(f'(v), v - u) \geq 0 \quad \forall v \in \Omega.$$

$(iii) \Rightarrow (ii)$. Suppose $(f'(v), v - u) \geq 0 \ \forall v \in \Omega$. Fix $w \in \Omega$ and set $v = (1 - \lambda)u + \lambda w$ for some $\lambda \in (0, 1)$. Now

$$0 \leq (f'(v), v - u) = \lambda(f'((1 - \lambda)u + \lambda w), w - u)$$

or

$$(f'((1 - \lambda)u + \lambda w), w - u)) \geq 0.$$

Let $\lambda \to 0$ to obtain

$$(f'(u), w - u)) \geq 0 \quad \forall w \in \Omega. \qquad \blacksquare$$

Remark 7.31 *Theorem 8.5 asserts that if E is a reflexive Banach space, $K \subset E$ a closed convex set, $f : K \to \mathbb{R}$ is convex and lower semicontinuous, $\lim_{\|x\| \to \infty} f(x) = \infty$, then f achieves its minimum on K. We have an analogous result as follows.*

Theorem 7.32 *Let H be a Hilbert space, $B \subset H$ the closed unit ball, and let $A : B \to H$ be monotone and continuous on $B \cap M$, M any finite-dimensional subspace of H. Then:*

(i) There is $x_0 \in B$ such that $(Ax_0, y - x_0) \geq 0 \ \forall \, y \in B$. Moreover, the set of such x_0 is convex;

(ii) If for $x \in \partial B$, Ax never points opposite to x : i.e. $x + \lambda Ax \neq 0 \ \forall \, \lambda \geq 0$, $\|x\| = 1$, then $Ax_0 = 0$.

Proof.

(i) For $y \in B$, set

$$S(y) = \{x \in B \mid (Ay, y - x) \geq 0\}.$$

Then

(a) $S(y) \subset B$.

(b) $S(y)$ is convex: $x_1, x_2 \in S(y)$, $t \in [0, 1]$,

$$(Ay, y - [(1 - t)x_1 + tx_2]) = (1 - t)(Ay, y - x_1) + t(Ay, y - x_2) \geq 0.$$

(c) $S(y)$ is closed: for $(x_n) \subset S(y)$, $x_n \to x$, then

$$(Ay, y - x) = \lim_{n \to \infty} (Ay, y - x_n) \geq 0.$$

(d) $\{S(y)\}_{y \in B}$ have the finite intersection property: for $y_1, \cdots, y_n \in B$, let $M = \langle y_1, \cdots, y_n \rangle$. Then $M \cap B$ is a compact convex set in M. Recall the Hartman-Stampacchia Theorem 8.8: if $K \subset \mathbb{R}^N$ is compact convex and $A : K \to \mathbb{R}^N$ is continuous, then there is $u \in K$ such that $(Au, v - u) \geq 0$ for $v \in K$. By that theorem, we obtain $x \in M \cap B$ such that

$$(Ax, y - x) \geq 0 \quad \text{for } y \in M \cap B.$$

Because A is monotone: $(Ay - Ax, y - x) \geq 0$, we have

$$(Ay, y - x) \geq 0 \quad \forall \, y \in M \cap B.$$

Thus, $x \in S(y)$ for $y \in M \cap B$. In particular, $\{S(y)\}_{y \in B}$ has the finite intersection property. Because $S(y) \subset B$ is closed convex for $y \in b$, $S(y)$ is weakly closed. If $\bigcap_{y \in B} S(y) = \emptyset$, then $B = \bigcap_{y \in B} S(y)^c$, where $S(y)^c$ is weakly open. However, B is weakly compact, so

$$B = S(y_1)^c \cup \cdots \cup S(y_m)^c,$$

i.e., $S(y_1) \cap \cdots \cap S(y_m) = \emptyset$, a contradiction. We have $\bigcap_{y \in B} S(y) \neq \emptyset$. Take $x_0 \in B$ such that $x_0 \in S(y)$ for $y \in B$, i.e., $(Ay, y - x_0) \geq 0 \ \forall \, y \in B$. If we set

$$K = \{x_0 \in B \mid (Ax_0, y - x_0) \geq 0 \ \forall \, y \in B\},$$

then

$$K = \{x_0 \in B \mid (Ay, y - x_0) \geq 0 \ \forall \, y \in B\}.$$

Therefore the set K of solutions x_0 is convex.

(ii) If $x_0 \in \overset{\circ}{B}$ with $(Ax_0, y - x_0) \geq 0 \; \forall y \in B$, then the set $\{y - x_0 \mid y \in B\}$ contains a ball $B(0, r)$. We claim that $Ax_0 = 0$. In fact, $(Ax_0, t) \geq 0, \; \forall t \in B(0, r)$, i.e., $(Ax_0 \, t) = 0 \; \forall t \in B(0, r)$. For $y \in H$, set $y = \frac{1}{s} sy$ such that $sy \in B(0, r)$, then $(Ax_0, y) = \frac{1}{s}(Ax_0, sy) = 0 \; \forall y \in H$, so $Ax_0 = 0$. If $x_0 \in \partial B$. Suppose $Ax_0 \neq 0$, $(Ax_0, y - x) \geq 0 \; \forall y \in B$. By the Hahn-Banach Theorem, there is a supported affine hyperplane through x_0. There is $\lambda > 0$ such that $x_0 + \lambda Ax_0 = 0$, contradicting $x_0 + \lambda Ax_0 \neq 0 \; \forall \lambda \geq 0$. Therefore $Ax_0 = 0$. ∎

Corollary 7.33 *Let H be a Hilbert space, $A : H \to H$ satisfying:*

(i) *A is monotone;*

(ii) *A is continuous on finite-dimensional subspaces of H;*

(iii) *$\dfrac{(Ax, x)}{\|x\|} \to \infty$ as $\|x\| \to \infty$ uniformly.*

Then A maps H onto H.

Proof. Fix $y \in H$, and set $B : H \to H$ defined by

$$Bx = Ax - y \quad \text{for } x \in H.$$

Then

(i) B is monotone:

$$(Bx_1 - Bx_2, x_1 - x_2) = (Ax_1 - y - Ax_2 + y, x_1 - x_2)$$
$$= (Ax_1 - Ax_2, x_1 - x_2) \geq 0.$$

(ii) B is continuous on finite-dimensional subspaces of H.

(iii) $\dfrac{(Bx, x)}{\|x\|} = \dfrac{(Ax - y, x)}{\|x\|} = \dfrac{(Ax, x)}{\|x\|} - \dfrac{(y, x)}{\|x\|} \to \infty$ as $\|x\| \to \infty$.

From (iii), there is $r > 0$ such that

$$\|x\| \geq r \text{ implies that } \frac{(Bx, x)}{\|x\|} \geq 1, \quad \text{or } (Bx, x) \geq \|x\|.$$

For $\|x\| = r$, if $Bx = \delta x$, $\delta < 0$, then

$$\delta \|x\|^2 = \delta(x, x) = (\delta x, x) = (Bx, x) \geq \|x\|,$$

i.e., $\delta \|x\| \geq r$ contradicts $\delta < 0$, $r > 0$. Hence we have, for $\|x\| = r$, $Bx \neq \delta x, \forall \delta < 0$ or $x + \lambda Bx \neq 0$ for $\lambda > 0$. Theorem 7.32 implies there is x such that

$$Bx = 0,$$

i.e.,

$$Ax = y.$$ ∎

Remark 7.34 *Note that*

$$(Ax, x) \geq \|x\|^2$$

$$\Rightarrow \frac{(Ax, x)}{\|x\|} \to \infty \quad \text{as } \|x\| \to \infty$$

$$\Rightarrow \|Ax\| \to \infty \quad \text{as } \|x\| \to \infty.$$

We have the following strong result:

Corollary 7.35 *Let H be a Hilbert space, $A : H \to H$ satisfying:*

(i) *A is monotone;*

(ii) *A is continuous on finite-dimensional subspaces of H;*

(iii) *$\|Ax\| \to \infty$ as $\|x\| \to \infty$ uniformly.*

Then A maps H onto H.

Proof. For $\epsilon > 0$, define the map $A_\epsilon : H \to H$ by

$$A_\epsilon = A + \epsilon I.$$

Then

(i) A_ϵ is monotone:

$$(A_\epsilon x - A_\epsilon y, x - y) = (Ax + \epsilon x - Ay - \epsilon y, x - y)$$
$$= (Ax - Ay, x - y) + \epsilon(x - y, x - y) \geq 0 \quad \text{for} \quad x, y \in H.$$

(ii) A_ϵ is continuous on finite-dimensional subspaces of H.

(iii) $\dfrac{(A_\epsilon x, x)}{\|x\|} \to \infty$ as $\|x\| \to \infty$: note that

$$(Ax - A0, x - 0) \geq 0 \quad \text{or} \quad (Ax, x) \geq (A0, x),$$

and

$$\frac{(A0, x)}{\|x\|} \leq \frac{\|A0\| \|x\|}{\|x\|} = \|A0\| \quad \forall x \in H.$$

Thus,

$$\frac{(A_\epsilon x, x)}{\|x\|} x = \frac{(Ax + \epsilon x, x)}{\|x\|} = \frac{(Ax, x)}{\|x\|} + \epsilon \|x\| \geq \frac{(A0, x)}{\|x\|} + \epsilon \|x\|$$
$$\geq \epsilon \|x\| - \|A0\| \to \infty \quad \text{as } \|x\| \to \infty.$$

Corollary 7.33 implies that $A_\epsilon : H \to H$ is onto. Fixing $y \in H$, then for some $x_\epsilon \in H$, $A_\epsilon x_\epsilon = y$. We have

$$\|y\| \geq \frac{(y, x_\epsilon)}{\|x_\epsilon\|} = \frac{(A_\epsilon x_\epsilon, x_\epsilon)}{\|x_\epsilon\|} = \frac{(Ax_\epsilon + \epsilon x_\epsilon, x_\epsilon)}{\|x_\epsilon\|}$$

$$= \frac{(Ax_\epsilon, x_\epsilon)}{\|x_\epsilon\|} + \epsilon \|x_\epsilon\| \geq \frac{(A0, x_\epsilon)}{\|x_\epsilon\|} + \epsilon \|x_\epsilon\|$$

$$\geq \epsilon \|x_\epsilon\| - \|A0\| \quad \text{because} \quad \frac{|(A0, x_\epsilon)|}{\|x_\epsilon\|} \leq \|A0\|,$$

i.e., $\epsilon \|x_\epsilon\| \leq \|A0\| + \|y\| = K$, independent of ϵ. Recall that $Ax_\epsilon + \epsilon x_\epsilon = y$, or $Ax_\epsilon = y - \epsilon x_\epsilon$. Thus, $\|Ax_\epsilon\| \leq \|y\| + \epsilon \|x_\epsilon\| \leq c$, independent of ϵ. By assumption (iii), $\|x_\epsilon\| \leq d$, independent of ϵ. By the Banach-Alaoglu-Bourbaki Theorem, there is a subsequence, still denoted by $\{x_\epsilon\}$, and $x \in H$ such that

$$x_\epsilon \rightharpoonup x \quad \text{weakly.}$$

Because $Ax_\epsilon + \epsilon x_\epsilon = y$, $\|x_\epsilon\| \leq d$, we have $Ax_\epsilon \to y$. Letting $\epsilon \to 0$ in $(Ax_\epsilon - av, x_\epsilon - v) \geq 0 \; \forall v \in H$, we obtain

$$(y - Av, x - v) \geq 0 \quad \text{for } v \in H,$$

i.e.,

$$(Av - y, v - x) \geq 0 \quad \text{for } v \in H.$$

The Minty Lemma implies that

$$(Ax - y, v - x) \geq 0 \quad \text{for } v \in H.$$

As before, because the set $\{v - x \mid v \in H\}$ contains a ball $B(x, \delta)$, we obtain

$$Ax = y,$$

i.e., A is onto. ∎

Notes It is quite common to use monotone operators to prove the existence of solutions of differential equations. For related material see Brezis [8], Gilbarg-Trudinger [24], Nirenberg [36], Schechter [43], and Zeidler [54].

Chapter 8

VARIATIONAL METHODS

In this chapter we study various variational methods.

8.1 Stampacchia Theorem and Lax-Milgram Theorem

Theorem 8.1 (*Riesz-Fréchet Representation Theorem*) *Let H be a Hilbert space with dual H'. Assume $\varphi \in H'$, then there is a unique $f \in H$ such that $\varphi(v) = (f, v)$ for all $v \in H$, and $\|f\|_H = \|\varphi\|_{H'}$.*

Proof.   Method 1. For $g \in H$, consider the map $Tg : H \to \mathbb{R}$ defined by

$$Tg(v) = (g, v) \quad \text{for } v \in H.$$

Because $|Tg(v)| = |(g, v)| \leq \|g\| \, \|v\|$, and

$$\|g\| = \frac{|Tg(g)|}{\|g\|} \leq \|Tg\|_{H'},$$

we obtain $\|Tg\|_{H'} = \|g\|_H$. In particular, $Tg \in H'$ and $T : H \to H'$ is a linear isometry. Furthermore $T(H)$ is dense in H'. In fact, because H is reflexive, if $h \in H'' = H$ such that $h(Tg) = (g, h) = 0$ for all $g \in H$, then $h = 0$, but $T(H)$ is closed in H', so $T(H) = H'$. In particular, given $\varphi \in H'$, there is a unique $f \in H$ such that $\varphi = Tf$. Therefore $\varphi(v) = Tf(v) = (f, v)$ for $v \in H$, and $\|f\|_H = \|\varphi\|_{H'}$.

Method 2. In the proof by this method, we do not require the reflection of H. Given $\varphi \in H'$, let $M = \varphi^{-1}(0)$. M is a closed subspace of H. If $M = H$, take $f = 0$. Suppose $M \subsetneq H$, take $g \notin M$, $|g| = 1$, $(g, M) = 0$. For example, for any $g_0 \notin M$, take $g = \dfrac{g_0 - P_M g_0}{|g_0 - P_M g_0|}$, then $g \notin M$, $|g| = 1$, and $(g, M) = 0$. Now for any $v \in H$, let $\lambda = \dfrac{\varphi(v)}{\varphi(g)}$, $w = v - \lambda g$. We have

$$v = \lambda g + w, \; \varphi(w) = 0, \; \text{and } 0 = (g, w) = (g, v - \lambda g).$$

Thus $\lambda = (g, v)$. If we define $f = \varphi(g)g$, then

$$\varphi(v) = \lambda \varphi(g) = (g, v)\varphi(g)$$
$$= (f, v) \quad \text{for all } v \in H,$$

211

and $\|f\|_H = \|\varphi\|_{H'}$ as proved in Method 1.   ∎

We recall some results in Functional Analysis. Let E be a Banach space with dual E'.

Theorem 8.2 *Let $C \subset E$ be convex. Then C is $\sigma(E, E')$-closed if and only if C is strongly closed.*

Proof.   By the Hahn-Banach Theorem 1.1.   ∎

Theorem 8.3 *Suppose $\varphi : E \to (-\infty, \infty]$ is convex, and lower semicontinuous with respect to the strong topology. Then φ is weakly lower semicontinuous. In particular, if $x_n \rightharpoonup x$ weakly, then*

$$\varphi(x) \leq \lim_{n \to \infty} \inf \varphi(x_n).$$

Proof.   See Theorem 1.17.   ∎

Theorem 8.4 *Let E be a reflexive Banach space, and let $K \subset E$ be any convex, bounded, and closed set. Then K is weakly compact.*

Proof.   See Theorem 1.18.   ∎

Now we come to the useful classical existence result of extremes:

Theorem 8.5 *Let E be a reflexive Banach space, and $A \subset E$ a nonempty closed, convex set. Assume that $\varphi : A \to (-\infty, \infty]$ is a convex, lower semicontinuous function, $\varphi \not\equiv \infty$ such that φ is coercive:*

$$\lim_{\substack{x \in A, \\ \|x\| \to \infty}} \varphi(x) = \infty. \tag{8.1}$$

Then φ attains its minimum on A: there is $x_0 \in A$ such that

$$\varphi(x_0) = \min_{x \in A} \varphi(x).$$

If A is bounded, then the condition (8.1) will be omitted.

Proof.   Take $a \in A$ with $\lambda = \varphi(a) < \infty$, and consider

$$A_\lambda = \{x \in A \mid \varphi(x) \leq \lambda\}.$$

A_λ is closed, because $A_\lambda = \varphi^{-1}((-\infty, \lambda])$, and φ is lower semicontinuous. A_λ is also convex because, for $x, y \in A_\lambda$, $t \in [0, 1]$,

$$\varphi[(1 - t)(x) + t(y)] \leq (1 - t)\varphi(x) + t\varphi(y) \leq \lambda.$$

By (8.1), A_λ is bounded. Theorem 8.4 implies that A_λ is weakly compact, and Theorem 8.3 implies that φ is weakly lower semicontinuous. Let $\alpha = \inf\{\varphi(x) \mid x \in A_\lambda\}$. Take a sequence $(x_n) \subset A_\lambda$, $x_0 \in A_\lambda$ such that $\varphi(x) \to \alpha$ and $x_n \rightharpoonup x_0$ weakly. By the weak lower semicontinuity of φ, we have $\varphi(x_0) \leq \varphi(x_n)$ for $n = 1, 2, \cdots$, and consequently $\varphi(x_0) \leq \alpha$, and $\varphi(x_0) \leq \varphi(x)$ for $x \in A$. Therefore $\varphi(x_0) = \min\limits_{x \in A} \varphi(x)$.

■

Theorem 8.6 (*Projection on a closed convex set*) *Let H be a Hilbert space, and $K \subset H$ a nonempty closed convex set. For $f \in H$ there is unique $u \in K$ such that*

$$|f - u| = \min_{v \in K} |f - v|, \tag{8.2}$$

and u can be characterized by

$$\begin{cases} u \in K, \\ (f - u, v - u) \leq 0 & \text{for } v \in K. \end{cases} \tag{8.3}$$

Moreover, if we denote $u = P_K f$, then

$$|P_K f_1 - P_K f_2| \leq |f_1 - f_2| \quad \text{for } f_1, f_2 \in H.$$

Proof.

(i)   Existence: we give here two different proofs:
  Method 1. Consider the function $\varphi : K \to \mathbb{R}$ defined by

$$\varphi(v) = |f - v| \quad \text{for } v \in K.$$

φ is convex: for $t \in [0, 1]$, u, $v \in K$

$$\begin{aligned}
\varphi((1 - t)u + tv) &= |(1 - t)f - (1 - t)u + tf - tv| \\
&\leq (1 - t)|f - u| + t|f - v| \\
&= (1 - t)\varphi(u) + t\varphi(v).
\end{aligned}$$

φ is continuous: for u, $v \in K$

$$\begin{aligned}
|\varphi(u) - \varphi(v)| &= ||f - u| - |f - v|| \\
&\leq |u - v|.
\end{aligned}$$

Moreover, $\lim\limits_{\substack{x \in k, \\ \|x\| \to \infty}} \varphi(x) = \infty$. The existence then follows from Theorem 8.5.

  Method 2. Given $f \in H$, let $d = \inf\limits_{v \in k} |f - v|$. Choose a minimizing sequence (v_n) in K such that $d_n = |f - v_n| \to d$ as $n \to \infty$. (v_n) is Cauchy: the Parallelogram Theorem implies

$$|f - v_n + f - v_m|^2 + |f - v_n - f + v_m|^2 = 2|f - v_n|^2 + 2|f - v_m|^2$$

or

$$\left| f - \frac{v_n + v_m}{2} \right|^2 + \left| \frac{v_n - v_m}{2} \right|^2 = \frac{d_n^2 + d_m^2}{2},$$

but $\left| f - \dfrac{v_n + v_m}{2} \right| \geq d$, because $\dfrac{v_n + v_m}{2} \in K$. Hence,

$$\left| \frac{v_n - v_m}{2} \right|^2 \leq \frac{1}{2}(d_n^2 + d_m^2) - d^2,$$

and consequently, $\lim\limits_{n,m \to \infty} |v_n - v_m| = 0$. Assume there is $u \in K$ such that $v_n \to u$ as $n \to \infty$ or $d = |f - u|$.

(ii) Equivalence of (8.2) and (8.3):
(8.2) $\Rightarrow$ (8.3). Given $f \in H$, take $u \in K$ with

$$|f - u| = \min_{v \in k} |f - v|.$$

For any $v \in K$, let $w = (1 - t)u + tv$, $t \in (0, 1)$. Now,

$$|f - u| \leq |f - w| = |f - (1 - t)u - tv| = |f - u - t(v - u)|.$$

Then

$$|f - u|^2 \leq |f - u|^2 - 2t(f - u, v - u) + t^2 |v - u|^2,$$

or

$$2(f - u, v - u) \leq t |v - u|^2.$$

Letting $t \to 0$, we conclude that $(f - u, v - u) \leq 0$.
(8.3) $\Rightarrow$ (8.2). Suppose $(f - u, v - u) \leq 0$ for all $v \in K$. Because

$$|f - v|^2 = |(f - u) - (v - u)|^2$$
$$= |f - u|^2 - 2(f - u, v - u) + |v - u|^2,$$

we have $|f - u| \leq |f - v|$ for $v \in K$, or $|f - u| = \min\limits_{v \in k} |f - v|$.

(iii) Uniqueness: Let $u_1, u_2 \in K$ such that, for $v \in K$,

$$(f - u_1, v - u_1) \leq 0,$$
$$(f - u_2, v - u_2) \leq 0.$$

Substituting u_2, u_1 for v in the first and second inequalities, respectively, we have

$$(f - u_1, u_2 - u_1) \leq 0, \tag{8.4}$$
$$(f - u_2, u_1 - u_2) \leq 0,$$

or
$$(u_2 - f, u_2 - u_1) \leq 0. \tag{8.5}$$

Add (8.4), (8.5) to obtain $(u_2 - u_1, u_2 - u_1) \leq 0$, or $u_2 = u_1$.

(iv) Nonexpansion: let $u_1 = P_K f_1$, $u_2 = P_K f_2$, for $v \in K$.

$$(f_1 - u_1, v - u_1) \leq 0 \tag{8.6}$$
$$(f_2 - u_2, v - u_2) \leq 0 \tag{8.7}$$

Substituting u_2, u_1 for v in (8.6), (8.7), respectively, we have

$$(-f_1 + u_1, u_1 - u_2) \leq 0,$$
$$(f_2 - u_2, u_1 - u_2) \leq 0.$$

Add these two inequalities to obtain

$$(f_2 - f_1, u_1 - u_2) + |u_1 - u_2|^2 \leq 0,$$

then

$$|u_1 - u_2|^2 \leq (f_1 - f_2, u_1 - u_2)$$
$$\leq |f_1 - f_2| |u_1 - u_2|.$$

Thus, $|u_1 - u_2| \leq |f_1 - f_2|$.   ∎

Corollary 8.7 *Let H be a Hilbert space and $M \subset H$ a closed linear subspace. Then P_M is a linear operator. Moreover, $u = P_M f$ is characterized by*

$$\begin{cases} u \in M \\ (f - u, v) = 0 \qquad \text{for } v \in M. \end{cases}$$

Proof. Clearly P_M is a linear operator. By Theorem 8.6, $u = P_M f$ is characterized by

$$u \in M,$$
$$(f - u, w - u) \leq 0 \quad \text{for } w \in M.$$

Because M is a linear subspace, for $v \in M$, there are w_1, $w_2 \in M$ such that

$$v = w_1 - u = u - w_2.$$

Therefore
$$(f - u, v) \geq 0, \text{ and } (f - u, v) \leq 0.$$
Thus $(f - u, v) = 0$ for $v \in M$.   ∎

Let H be a Hilbert space. A bilinear form $a : H \times H \to \mathbb{R}$ is continuous if there is a constant $c_1 > 0$ such that

$$|a(u, v)| \leq c_1 |u| |v| \quad \text{for } u, v \in H.$$

$a(u, v)$ is coercive if there is a constant $c_2 > 0$ such that

$$a(u, u) \geq c_2 |u|^2 \quad \text{for } u \in H.$$

Theorem 8.8 (*Stampacchia Theorem*) *Let H be a Hilbert space, $a(u, v)$ a continuous coercive bilinear form on H, and K a nonempty closed convex set in H. For $\varphi \in H'$, there is a unique $u \in K$ such that*

$$a(u, v - u) \geq \varphi(v - u) \quad \text{for } v \in K.$$

Moreover, if $a(u, v)$ is symmetric, then

$$\begin{cases} u \in K, \\ \dfrac{1}{2} a(u, u) - \varphi(u) & = \min_{v \in k} \left\{ \dfrac{1}{2} a(v, v) - \varphi(v) \right\}. \end{cases}$$

Proof.   Apply the Riesz-Fréchet Representation Theorem 8.1 to obtain an $f \in H$ such that

$$\varphi(v) = (f, v) \quad \text{for } v \in H.$$

For any fixed $u \in H$, the function

$$v \to a(u, v)$$

is a continuous linear functional on H. Apply the Riesz-Fréchet Representation Theorem 8.1 again to ensure that there is $Au \in H$ such that

$$a(u, v) = (Au, v) \quad \text{for } v \in H.$$

Clearly, $A : H \to H$ is linear satisfying for $u \in H$,

$$|Au| \leq c_1 |u|,$$
$$(Au, u) \geq c_2 |u|^2,$$

for some constants c_1, c_2 with $c_1 > c_2 > 0$. Note that, for $\rho > 0$,

$$\rho(Au, v - u) \geq \rho(f, v - u) \quad \text{for } v \in k$$
$$\text{if and only if} (\rho f - \rho Au + u - u, v - u) \leq 0 \quad \text{for } v \in k$$
$$\text{if and only if} u = P_K(\rho f - \rho Au + u).$$

Consider the map $S : K \to K$ defined by

$$Sv = P_K(\rho f - \rho Av + v).$$

Then, for $x, y \in K$,

$$|Sx - Sy| = |P_K(\rho f - \rho Ax + x) - P_K(\rho f - \rho Ay + y)|$$
$$\leq |(x - y) - \rho (Ax - Ay)|$$

or

$$|Sx - Sy|^2 \leq |x - y|^2 - 2\rho(Ax - Ay, x - y) + \rho^2 |Ax - Ay|^2$$
$$\leq (1 - 2\rho c_2 + \rho^2 c_1^2) |x - y|^2.$$

Let $0 < \rho < (2c_2)/c_1^2$, $k^2 = 1 - 2\rho c_2 + \rho^2 c_1^2$, with $k > 0$. Then $0 < k < 1$, and

$$|Sx - Sy| \leq k |x - y|.$$

By the Banach Fixed Point Theorem 2.1, there is a unique $u \in K$ such that $u = Su$: $u = P_K(\rho f - \rho Au + u)$, or $\rho(Au, v - u) \geq \rho(f, v - u)$ for $v \in K$. We therefore have

$$a(u, v - u) = (Au, v - u)$$
$$\geq (f, v - u)$$
$$= \varphi(v - u) \text{ for } v \in K.$$

Suppose $a(u, v)$ is symmetric. Then $a(u, v)$ is a scalar product in H associated with the norm $a(u, u)^{1/2}$, which is equivalent to the norm $(u, u)^{1/2}$. By the Riesz-Fréchet Representation Theorem 8.1, there is $g \in H$ such that

$$\varphi(v) = a(g, v) \text{ for } v \in H.$$

Recall that

$$a(u, v - u) \geq \varphi(v - u) \text{ for } v \in K,$$

or $a(g, v - u) = \varphi(v - u) \leq a(u, v - u)$, thus $a(g - u, v - u) \leq 0$. Theorem 8.6 implies

$$a(g - u, g - u) = \min_{v \in k} a(g - v, g - v)$$

or

$$a(u, u) - 2a(g, u) = \min_{v \in k} [a(v, v) - 2a(g, v)],$$

so

$$\frac{1}{2}a(u, u) - \varphi(u) = \min_{v \in k} \left[\frac{1}{2}a(v, v) - \varphi(v) \right]. \qquad \blacksquare$$

Theorem 8.9 *If $a(u,v)$ is a bilinear form on a Hilbert space H with $a(v,v) \geq 0$ for $v \in H$, then $v \to a(v,v)$ is a convex function.*

Proof. Assume $u,\, v \in H$, $\alpha,\, \beta \geq 0$, and $\alpha + \beta = 1$. Set $f(v) = a(v,v)$. Note that

$$0 \leq a(u-v, u-v) = a(u,u) + a(v,v) - a(u,v) - a(v,u),$$

so

$$a(u,v) + a(v,u) \leq a(u,u) + a(v,v).$$

Now

$$
\begin{aligned}
f(\alpha u + \beta v) &= a(\alpha u + \beta v, \alpha u + \beta v) \\
&= \alpha^2 a(u,u) + \alpha\beta a(u,v) + \alpha\beta a(v,u) + \beta^2 a(v,v) \\
&\leq \alpha^2 a(u,u) + \alpha\beta a(u,u) + \alpha\beta a(v,v) + \beta^2 a(v,v) \\
&= \alpha a(u,u) + \beta a(v,v) \\
&= \alpha f(u) + \beta f(v).
\end{aligned}
$$
∎

Theorem 8.10 (*Lax-Milgram Theorem*) *Let $a(u,v)$ be a bilinear continuous coercive form on H. Then for $\varphi \in H'$, there is a unique $u \in H$ such that*

$$a(u,v) = \varphi(v) \quad \text{for } v \in H.$$

Moreover, if a is symmetric, then u is characterized by

$$
\left\{
\begin{array}{l}
u \in H, \\
\dfrac{1}{2} a(u,u) - \varphi(u) \;\; = \min\limits_{v \in H} \{ \dfrac{1}{2} a(v,v) - \varphi(v) \}.
\end{array}
\right.
$$

Proof.   Method 1. The Stampacchia Theorem 8.8 implies that there is a unique $u \in H$ such that

$$
\begin{aligned}
a(u,w) = a(u, w+u-u) &\geq \varphi(w+u-u) \\
&= \varphi(w).
\end{aligned}
$$

Similarly,

$$a(u,-w) \geq \varphi(-w).$$

Thus, $a(u,w) = \varphi(w)$ for $w \in H$. The last part of the theorem follows from Theorem 8.8.

Method 2. The Riesz-Fréchet Representation Theorem 8.1 implies that there is $f \in H$ such that

$$\varphi(v) = (f,v) \text{ for } v \in H.$$

For fixed $u \in H$, the map $v \to a(u, v)$ is a continuous linear functional. By the Riesz-Fréchet Representation Theorem 8.1, there is $Au \in H$ such that

$$a(u, v) = (Au, v).$$

Clearly, $A : H \to H$ is linear and for $v \in H$ satisfies

$$|Av| \leq c\,|v|$$
$$(Av, v) \geq \alpha\,|v|^2 \quad (\text{ which implies } |Av| \geq \alpha\,|v|).$$

(i) Range A is closed: let $(v_n = Au_n)$ be a Cauchy sequence in Range A, then

$$\alpha\,|u_n - u_m| \leq |Au_n - Au_m|$$
$$\leq c\,|v_n - v_m| \to 0 \quad \text{as } n,\ m \to \infty.$$

Therefore (u_n) is Cauchy, so define $u \in H$ with $u_n \to u$. Then

$$Au_n \to Au.$$

(ii) Range A is dense in H : If there is $w \in H$ such that

$$a(t,\ w) = (At, w) = 0 \quad \text{for } t \in H,$$

take $t = w$, then $\alpha\,|w|^2 \leq a(w, w) = 0$. Then $w = 0$. By (i), (ii) Range$A = H$, take $u \in H$ such that $Au = f$, and we obtain

$$a(u, v) = (Au, v) = (f, v) = \varphi(v) \quad \text{for } v \in H.$$

The last part of the theorem follows from Theorem 8.8. ∎

8.2 Ljusternik-Schnirelman Constrained Theorems

Let E be a real Banach space. $\sum = \sum(E)$ consists of all closed sets A in E, symmetric with respect to 0, $0 \notin A$, for example, the sets $\partial B_1(0)$, $\{-x, x\}$ for $x \neq 0$.

Definition 8.11 *Assume $A \in \sum$. We call the genus of A, in symbols $\gamma(A)$, the least integer $n \in N \cup \{0\}$ such that there is $f \in C(A, \mathbb{R}^n \backslash \{0\})$, where f is odd. Set $\gamma(A) = 0$, and if there is no such n, set $\gamma(A) = \infty$.*

Example 8.12

(i) *Let $x \in \mathbb{R}^N$, $r < \|x\|$, and $A = \overline{B_r(x)} \cup \overline{B_r(-x)}$. Then $\gamma(A) = 1$. In fact, define*

$$f(y) = \begin{cases} 1 & \text{on } \overline{B_r(x)}, \\ -1 & \text{on } \overline{B_r(-x)}. \end{cases}$$

Then $f \in C(A, \mathbb{R} \backslash \{0\})$, and f is odd;

(ii) *Assume $x_1, \cdots, x_m \in \mathbb{R}^N$, $A = \{x_1, \cdots, x_m, -x_1, \cdots, -x_m\}$, $0 \notin A$. Define*

$$f(x) = \begin{cases} 1 & x \in \{x_1, \cdots, x_m\}, \\ -1 & x \in \{-x_1, \cdots, -x_m\}. \end{cases}$$

Then $f \in C(A, \mathbb{R}\backslash\{0\})$ with f odd. Thus $\gamma(A) = 1$.

Proposition 8.13 *Given $A, B \in \sum$:*

(i) *If there is an odd map $f : C(A, B)$, then $\gamma(A) \le \gamma(B)$;*

(ii) *If $A \subset B$ then $\gamma(A) \le \gamma(B)$;*

(iii) *If there is an odd homeomorphism $h \in C(A, B)$, then $\gamma(A) = \gamma(B)$;*

(iv) *$r(A \cup B) \le \gamma(A) + \gamma(B)$;*

(v) *If $\gamma(B) < \infty$, then $\gamma(\bar{A} - \bar{B}) \ge \gamma(A) - \gamma(B)$;*

(vi) *If A is compact, then $\gamma(A) < \infty$. Moreover, there is $\delta > 0$ such that $\gamma(N_\delta(A)) = \gamma(A)$;*

(vii) *If $\gamma(A) > k$, V is a k-dimensional subspace of E, and $V^\perp$ is an algebraically and topologically complementary subspace of V, then $A \cap V^\perp \ne \emptyset$.*

Proof.   We may assume $\gamma(A) < \infty$ and $\gamma(B) < \infty$.

(i) Suppose $\gamma(B) = n$, and $\varphi \in C(B, \mathbb{R}^n\backslash\{0\})$ odd, then the map $\varphi \circ f \in C(A, \mathbb{R}^n\backslash\{0\})$ is odd.

(ii) The identity map is odd.

(iii) Trivial.

(iv) Let $\gamma(A) = n$, $\gamma(B) = m$. Suppose $\varphi \in C(A, \mathbb{R}^n\backslash\{0\})$, and $\psi \in C(B, \mathbb{R}^m\backslash\{0\})$ are odd. The Tietze Extension Theorem extends φ, ψ to $\bar{\varphi} \in C(E, \mathbb{R}^n)$, $\bar{\psi} \in C(E, \mathbb{R}^m)$, respectively. If necessary, replace $\bar{\varphi}(x)$ by $\frac{1}{2}(\bar{\varphi}(x) - \bar{\varphi}(-x))$, and $\bar{\psi}(x)$ by $\frac{1}{2}(\bar{\psi}(x) - \bar{\psi}(-x))$; we may assume that $\bar{\varphi}$ and $\bar{\psi}$ are odd. Take $f = (\bar{\varphi}, \bar{\psi})$. It follows that $f \in C(A \cup B, \mathbb{R}^{n+m}\backslash\{0\})$ is odd, so $\gamma(A \cup B) \le \gamma(A) + \gamma(B)$.

(v) Note $A \subset \overline{(A - B)} \cup B$.

(vi) For $x \in E$ and $0 < r < \|x\|$, $D = \overline{B_r(x)} \cup \overline{B_r(-x)}$. We have $\overline{B_r(x)} \cap \overline{B_r(-x)} = \emptyset$. The map $f : D \to \mathbb{R}\backslash\{0\}$ defined by

$$f(y) = \begin{cases} \|y\| & \text{for } y \in \overline{B_r(x)}, \\ -\|y\| & \text{for } y \in \overline{B_r(-x)}, \end{cases}$$

is continuous and odd. Therefore $\gamma(D) = 1$. A is compact and hence can be covered by finitely many such D, so $\gamma(A) < \infty$. Suppose $\gamma(A) = n$, and $\varphi \in C(A, \mathbb{R}^n \backslash \{0\})$ is odd. Extend φ oddly to $\bar{\varphi} \in C(E, \mathbb{R})$. Then there is $\delta > 0$ such that $\bar{\varphi}(N_\delta(A)) \neq 0$. Hence, $\gamma(N_\delta(A)) \leq \gamma(A)$, but clearly $\gamma(A) \leq \gamma(N_\delta(A))$.

(vii) Let $E = V \oplus V^\perp$, and $P : E \to V$ be the projection of E onto V. If $A \cap V^\perp = \emptyset$, then $p \in C(A, V \backslash \{0\})$ is odd. Hence, $\gamma(A) \leq k$, a contradiction.   ∎

Lemma 8.14 *Let $\Omega \subset \mathbb{R}^N$ be a bounded open set, symmetric with respect to 0, $0 \in \Omega$. Then $\gamma(\partial\Omega) = N$. In particular, $\gamma(S^{N-1}) = N$, also $\gamma(A) = N$ if $A \in \sum$ is homeomorphic to S^{N-1} by an odd homeomorphism.*

Proof. Because the identity map $I \in C(\partial\Omega, \mathbb{R}^N \backslash \{0\})$, I is odd, we have $\gamma(\partial\Omega) \leq N$. If $\gamma(\partial\Omega) = j < N$: then $f \in C(\partial\Omega, \mathbb{R}^j \backslash \{0\})$, f is odd. This contradicts Corollary 9.47. Thus $\gamma(\partial\Omega) = N$.   ∎

Theorem 8.15 *Let $\Omega \subset \mathbb{R}^N$ be a bounded open set, symmetric with respect to 0, $0 \in \Omega$, $N > M$, and $f \in C(\partial\Omega, \mathbb{R}^M)$. Let $A = \{x \in \partial\Omega \mid f(x) = 0\}$, $B = \{x \in \partial\Omega \mid f(x) = f(-x)\}$. Then:*

(i) *If f is odd, then $\gamma(A) \geq N - M$;*

(ii) $\gamma(B) \geq N - M$.

Proof.

(i) Note that, $A \neq \emptyset$. By Proposition 8.13 (vi), there is $\delta > 0$ such that $\gamma(N_\delta(A)) = \gamma(A)$. Let $Z_\epsilon = \{x \in \partial\Omega \mid |f(x)| \leq \epsilon\}$. We claim that there is $\epsilon_0 > 0$ such that $N_\delta(A) \supset Z_\epsilon$ for $0 < \epsilon < \epsilon_0$. Otherwise there is a sequence $\{\epsilon_n\}$, $\epsilon_n \downarrow 0$, $x_n \in Z_{\epsilon_n}$ such that $x_n \notin N_\delta(A)$. Then $\bar{x} \notin \text{int}(N_\delta(A))$, and there is a subsequence, still denoted by $\{x_n\}$, such that $x_n \to \bar{x}$. Then $\bar{x} \notin A$. However, $|f(\bar{x})| = 0$, or $\bar{x} \in A$, a contradiction. Now, for ϵ, $0 < \epsilon < \epsilon_0$, $A \subset Z_\epsilon \subset N_\delta(A)$, or $\gamma(A) \leq \gamma(Z_\epsilon) \leq \gamma(N_\delta(A))$. We have $\gamma(Z_\epsilon) = \gamma(A)$. Set

$$W_\eta = \{x \in \partial\Omega \mid |f(x)| \geq \eta\}$$

and $\rho(x) = \dfrac{x}{|x|}$ for $x \neq 0$. Now $g = \rho \circ f \in C(w_\eta, S^{M-1})$, g is odd, so, by Proposition 8.13 (i) and (v),

$$\gamma(W_\eta) \leq \gamma(S^{M-1}) = M,$$
$$\gamma\left(\overline{\partial\Omega \backslash W_\eta}\right) \geq \gamma(\partial\Omega) - \gamma(W_\eta).$$

Thus $\gamma(\partial\Omega \backslash W_\eta) \geq N - M$. Because $\overline{\partial\Omega \backslash W_\eta} = 2\eta$, set $\eta = \epsilon$ for some ϵ, $0 < \epsilon < \epsilon_0$, so that $\gamma(A) = \gamma(Z_\epsilon) \geq N - M$.

(ii) Note that $B \neq \emptyset$. Let $g(x) = f(x) - f(-x)$, then apply (i) to obtain the result. ∎

Lemma 8.16 *There are N antipodal closed sets $\{B_1, \cdots, B_N\}$ such that the set*

$$S^{N-1} : B_i = C_i \cup (-C_i), \ C_i \cap (-C_i) = \emptyset, \ i = 1, \cdots, N, \ B_i \subset S^{N-1}, \ S^{N-1} = \bigcup_{i=1}^{N} B_i$$

is convex.

Proof. By induction on N,

$N = 1 : S^0 = \{1\} \cup \{-1\}$.

$N = 2 : S^1 = B_1 \cup B_2, \ B_1 = \{(x, y) \in S^1 \mid |x| \geq \frac{1}{2}\}$,

$$B_2 = \{(x, y) \in S^1 \mid |y| \geq \frac{1}{2}\}.$$

Suppose cases $1, 2, \cdots, N$ are correct, and let $S^{N-1} = \bigcup_{i=1}^{N} B_i'$, where $\{B_1', \cdots, B_N'\}$ are antipodal closed sets, $B_i' = C_i' \cup (-C_i')$, $C_i' \cap (-C_i') = \emptyset$. Denote $x = (x', x_{N+1}) \in \mathbb{R}^{N+1}$, where $x' = (x_1, \cdots, x_N)$. Identify $\mathbb{R}^N$ with the hyperplane $\{x \in \mathbb{R}^{N+1} \mid x_{N+1} = 0\}$. Set

$$C_{N+1} = \{(x', x_{N+1}) \in S^N \mid x_{N+1} \geq \frac{1}{4}\}.$$

$$C_i = \{(x', x_{N+1}) \in S^N \mid x_{N+1} \leq \frac{1}{2}, \ \frac{x'}{\sqrt{1 - x_{N+1}^2}} \in C_i'\}, \ i = 1, \cdots, N.$$

Then the $\{C_i\}$ are closed sets in S^N, $C_i \cap (-C_i) = \emptyset$. Set $B_i = C_i \cup (-C_i)$, $1 \leq i \leq N+1$. In fact, C_i is the intersection of a sectorial cone with the x_{N+1}-axis and C_i' and S^N. Note that if $(x', x_{N+1}) \in S^N$, then $|x'|^2 + x_{N+1}^2 = 1$, or $|x'| = \sqrt{1 - x_{N+1}^2}$. Thus, $\frac{x'}{\sqrt{1 - x_{N+1}^2}} = \frac{x'}{|x'|} \in S^{N-1}$. Now, for $(x', x_{N+1}) \in S^N$, if $x_{N+1} \geq \frac{1}{4}$ then $(x', x_{N+1}) \in C_{N+1} \cup (-C_{N+1})$. If $|x_{N+1}| \leq \frac{1}{2}$ then $\frac{x'}{\sqrt{1 - x_{N+1}^2}} = \frac{x'}{|x'|} \in S^{N-1}$, or $\frac{x'}{|x'|} \in C_i' \cup (-C_i')$ for some i, $1 \leq i \leq N$. Therefore $\{B_i\}_{1 \leq i \leq N+1}$ covers S^N, also

$$(x', x_{N+1}) \in C_i \text{ if and only if } \frac{x'}{|x'|} \in C_i'$$

$$\text{if and only if } \frac{x'}{|x'|} \notin (-C_i')$$

$$\text{if and only if } \frac{-x'}{|-x'|} \notin C_i'$$

$$\text{if and only if } (x', x_{N+1}) \notin -C_i.$$

Therefore $C_i \cap (-C_i) = \emptyset$ and the theorem follows. ∎

Lemma 8.17 *Let $A \in \sum$. Then $\gamma(A) = n$ if and only if n is the least integer such that there are n sets $A_1, \cdots, A_n \in \sum$ such that $\gamma(A_i) = 1$, $1 \leq i \leq n$, $A \subset \bigcup\limits_{i=1}^{n} A_i$.*

Proof.   Let $A \in \sum$, $\gamma(A) = j$. We require $D_1, \cdots, D_j \in \sum$, $\gamma(D_i) = 1$, $1 \leq i \leq j$, $A \subset \bigcup\limits_{i=1}^{j} D_i$. In fact, there is $f \in C(A, \mathbb{R}^j \backslash \{0\})$, and f is odd. By Lemma 8.16, there are antipodal closed sets $B_i = C_i \cup (-C_i) \subset S^{j-1}$, $C_i \cap (-C_i) = \emptyset$, $i = 1, \cdots, j$, $S^{j-1} = \bigcup\limits_{i=1}^{j} B_i$. Let $p(x) = \frac{x}{|x|}$ for $x \in \mathbb{R}^j \backslash \{0\}$, then $\rho \circ f \in C(A, S^{j-1})$. Set

$$D_i = f^{-1} \circ \rho^{-1}(B_i) = [f^{-1} \circ \rho^{-1}(C_i)] \cup [f^{-1} \circ \rho^{-1}(-C_i)].$$

Then the D_i are closed, $D_i \in \sum$, $A \subset \bigcup\limits_{i=1}^{j} D_i$, and the D_i are the composition of two closed disjoint sets, so $\gamma(D_i) \leq 1$. Therefore

$$A \subset \bigcup_{i=1}^{j} D_i \text{ implies that } \gamma(A) = j \leq \sum_{i=1}^{j} \gamma(D_i).$$

Thus $\gamma(D_i) = 1$, $i = 1, \cdots, j$. If there is at least n such that $A \subset \bigcup\limits_{i=1}^{n} A_i$, then $\gamma(A) \leq n$. However, if $\gamma(A) = j < n$, then by the previous part, there are j such coverings, and this contradicts the least n. Thus $\gamma(A) = n$. Suppose $\gamma(A) = n$. Then there is $\{A_1, \cdots, A_n\} \subset \sum$, such that $A \subset \bigcup\limits_{i=1}^{n} A_i$, $\gamma(A_i) = 1$, $j \leq i \leq n$. On the other hand, we cannot have $m < n$, $A \subset \bigcup\limits_{i=1}^{m} B_i$, $B_i \in \sum$, $\gamma(B_i) = 1$, $1 \leq i \leq m$. Otherwise $\gamma(A) \leq m < n$, a contradiction.   ∎

Definition 8.18

(i) *Let E be a topological space and $A \subset E$ a closed set. We say that A is of the (Ljusternik-Schnirelman) category, in symbols $\mathrm{cat}_E A = 1$, if the canonical injection $A \to E$ is homotopic to a constant function: that is, there is a continuous function $H(x, t) \in C(A \times [0, 1], E)$ such that, for $x \in A$.*

$$H(x, 0) = x,$$
$$H(x, 1) = x_0.$$

That is, A can be deformed continuously to a point.

(ii) *We say A is of category n, in symbols, $\mathrm{cat}_E A = n$, if there are closed sets $A_1, \cdots, A_n$ in E such that*

$$A \subset \bigcup_{i=1}^{n} A_i, \ \mathrm{cat}_E A_i = 1, \ 1 \leq i \leq n,$$

and n is the least integer with such properties.

The properties of category are similar to those of genus.

Definition 8.19 *Let E be a real Banach space, $U \subseteq E$ an open set, and $f \in C^1(U, \mathbb{R})$. Then $v \in E$ is a pseudogradient vector for f at $u \in U$ if the following are satisfied:*

$$\begin{cases} \|v\| \le 2 \, \|f'(u)\| \\ (f'(u), v) \ge \|f'(u)\|^2 \end{cases} \tag{8.8}$$

The pseudogradient vector is not unique in general, and a convex combination of pseudogradient vectors for f at u is a pseudogradient vector for f at u. Assume

$$f \in C^1(E, \mathbb{R}), \quad \widetilde{E} = \{u \in E \mid f'(u) \ne 0\},$$

then $v : \widetilde{E} \to E$ is called a pseudogradient vector field on $\widetilde{E}$ if $v(x)$ is locally Lipschitz-continuous, and $v(x)$ is a pseudogradient vector for f at $x \in \widetilde{E}$.

Lemma 8.20 *Let $f \in C^1(E, \mathbb{R})$. There is a pseudogradient vector field for f on $\widetilde{E}$.*

Proof.   Fix $u \in \widetilde{E}$. Because $0 < \|f'(u)\| = \sup\limits_{\substack{w \in E, \\ \|W\|=1}} (f'(u), w)$, we have $w \in E$, $\|w\| = 1$, and $(f'(u), w) > \frac{2}{3} \|f'(u)\|$. Setting $z = \frac{3}{2} \|f'(u)\| \, w$, we have

$$\|z\| = \frac{3}{2} \|f'(u)\| < 2 \, \|f'(u)\|,$$

$$(f'(u), \frac{3}{2} \|f'(u)\| \, w) > \|f'(u)\|^2.$$

Thus, z is a pseudogradient vector for f at u. The continuity of f' implies that there is an open neighborhood N_u of u such that z is a pseudogradient vector for f at each $v \in N_u$. Each such N_u covers $\widetilde{E}$. Take its locally finite refinement $\{N_{u_i}\}$, and set

$$\rho_i(x) = \text{dist}(x, N_{u_i}^c).$$

Then ρ_i is a Lipschitz-continuous function such that

$$\rho_i(N_{u_i}^c) = 0.$$

Set

$$\beta_i(x) = \frac{\rho_i(x)}{\sum\limits_{r} \rho_r(x)}, \quad \text{for } x \in \widetilde{E}.$$

Then the β_i are locally Lipschitz-continuous. Set

$$v(x) = \sum_i z_i \beta_i(x),$$

where z_i is a pseudogradient vector for f at u_i. Note that $\sum_i \beta_i(x) = 1$, and for each $x \in \widetilde{E}$, $v(x)$ is a finite convex combination of pseudogradient vectors for f at x. Thus $v(x)$ is a pseudogradient vector field for f on $\widetilde{E}$. ∎

Corollary 8.21 *In Lemma 8.20, if f is even, then $v(x)$ can be chosen to be odd.*

Proof. If f is even, replace v by

$$w(x) = \frac{1}{2}(v(x) - v(-x)).$$

Then w is odd and locally Lipschitz-continuous. Moreover,

$$\|w(x)\| \le \frac{1}{2}(\|v(x)\| + \|v(-x)\|)$$
$$\le \frac{1}{2}[2\|f'(x)\| + 2\|f'(x)\|]$$
$$= 2\|f'(x)\|,$$
$$(f'(x), w(x)) = \frac{1}{2}[(f'(x), v(x)) + (f'(x), -v(-x)]$$
$$= \frac{1}{2}[(f'(x), v(x)) + (f'(-x), v(-x)]$$
$$\ge \frac{1}{2}[\|f'(x)\|^2 + \|f'(-x)\|^2]$$
$$= \|f'(x)\|^2. ∎$$

Definition 8.22 *Let E be a real Banach space, $f \in C^1(E, \mathbb{R})$. We say that f satisfies the Palais-Smale condition, or (PS) condition, if for any sequence $\{x_n\}$ in E with $\{f(x_n)\}$ bounded, and $f'(x_n) \to 0$ strongly in E', then there is a convergent subsequence of $\{x_n\}$ in E.*

Theorem 8.23 *(Deformation Theorem) Let E be a real Banach space, and $f \in C^1(E, \mathbb{R})$ satisfies the (PS) condition. For $c \in \mathbb{R}$, let N be a neighborhood of*

$$K_c = \{x \in E \mid f(x) = c, \; f'(x) = 0\},$$

and set $A_d = \{x \in E \mid f(x) \le d\}$, for each $d \in \mathbb{R}$. Then there is $\eta(t, x) \equiv \eta_t(x)$ in $C([0, 1] \times E, E)$, and there are constants $0 < \epsilon < \bar{\epsilon}$, such that:

(i) $\eta_0(x) = x$ for $x \in E$;

(ii) $\eta_t(x) = x$ for $t \in [0, 1]$, for $x \notin f^{-1}[c - \bar{\epsilon}, c + \bar{\epsilon}]$;

(iii) For $t \in [0, 1]$, η_t is a homeomorphism of E onto E;

(iv) $f(\eta_t(x)) \leq f(x)$ for $x \in E$, $t \in [0, 1]$;

(v) $\eta_1(A_{c+\epsilon} \backslash N) \subset A_{c-\epsilon}$;

(vi) If $K_c = \emptyset$ then $\eta_1(A_{c+\epsilon}) \subset A_{c-\epsilon}$;

(vii) If f is even, then $\eta_t(x)$ is odd in x.

Proof.   We may assume $K_c \neq \emptyset$, otherwise choose $M_\delta = \emptyset$ below. Clearly K_c is closed. The (PS) condition implies that K_c is compact: every sequence has a convergent subsequence. N is a neighborhood of K_c, and we may choose a small $\delta > 0$ with $M_\delta = \text{Int } N_\delta(K_c) \subset N$, where $N_\delta(K_c) = \{x \in E \mid \text{dist}(x, K_c) \leq \delta\}$. Thus in the proof of (v) we may replace N by M_δ. There are constants b, $\bar{\epsilon} > 0$ depending on δ such that

$$\|f'(x)\| \geq b \quad \text{for } x \in A_{c+\bar{\epsilon}} \backslash A_{c-\bar{\epsilon}} \backslash M_{\delta/8}. \tag{8.9}$$

Otherwise take the sequences

$$b_n \searrow 0, \ \epsilon_n \searrow 0,$$
$$x_n \in A_{c+\epsilon_n} \backslash A_{c-\epsilon_n} \backslash M_{\delta/8},$$
$$\|f'(x_n)\| < b_n.$$

Apply the (PS) condition to show that there is a subsequence of (x_n) converging to x, and hence $f(x) = c$, $f'(x) = 0$, $x \notin M_{\delta/8}$, but this contradicts $K_c \subseteq M_{\delta/8}$. Because (8.9) remains valid if $\bar{\epsilon}$ is decreased, we can assume

$$0 < \bar{\epsilon} < \min\{\frac{b\delta}{32}, \frac{b^2}{8}, \frac{1}{8}\}. \tag{8.10}$$

For $0 < \epsilon < \bar{\epsilon}$, set

$$A = \{x \in E \mid f(x) \geq c + \bar{\epsilon}, \text{ or } f(x) \leq c - \bar{\epsilon}\},$$
$$B = \{x \in E \mid c - \epsilon \leq f(x) \leq c + \epsilon\}.$$

Then $A \cap B = \emptyset$. The function $g : E \to \mathbb{R}$ defined by

$$g(x) = \text{d}(x, A)[\text{d}(x, A) + \text{d}(x, B)]^{-1}$$

is locally Lipschitz-continuous, satisfying

$$\begin{cases} g = 0 & \text{on } A, \\ g = 1 & \text{on } B, \\ 0 \leq g(x) \leq 1. \end{cases}$$

Similarly, there is a locally Lipschitz-continuous function $\bar{g} : E \to \mathbb{R}$ such that

$$\begin{cases} \bar{g} = 0 & \text{on } M_{\delta/8}, \\ \bar{g} = 1 & \text{on } E \backslash M_{\delta/4}, \\ 0 \leq \bar{g}(x) \leq 1. \end{cases}$$

Note if f is even, then A, B, M_δ are symmetric with respect to 0, and g and $\bar{g}$ are even. Set

$$h(s) = \begin{cases} 1 & \text{if } 0 \leq s \leq 1, \\ \dfrac{1}{S} & \text{if } s \geq 1, \end{cases}$$

h is Lipschitz-continuous. Apply Lemma 8.20 to $f \in C^1(E, \mathbb{R})$ to obtain a pseudo-gradient vector field $v(x)$ for f on $\widetilde{E}$. Then set

$$V(x) = \begin{cases} -g(x)\bar{g}(x)h(\|v(x)\|)v(x) & x \in \widetilde{E}, \\ 0 & \text{otherwise.} \end{cases}$$

Then $V(x)$ is a locally Lipschitz-continuous vector field on E, and $0 \leq \|V(x)\| \leq 1$, V is odd if f is even. Consider the Cauchy problem:

$$\begin{cases} \dfrac{d\eta}{dt} = V(\eta(t, x)), \\ \eta(0, x) = x. \end{cases} \tag{8.11}$$

The basic Existence-Uniqueness Theorem for Ordinary Differential Equations implies that for each $x \in E$ there is a unique solution $\eta(t, x)$ of (8.11) on $t \in (t^-(x), t^+(x))$, a maximal interval depending on x. That V is bounded implies $t^\pm(x) = \pm\infty$ for $x \in E$. Otherwise, for example, $t^+(x) < \infty$, letting $t_n \nearrow t^+(x)$. Integrating (8.11), we obtain

$$\|\eta(t_{n+1}, x) - \eta(t_n, x)\| = \left\| \int_{t_n}^{t_{n+1}} V(\eta(t), x)dt \right\|$$
$$\leq \|t_{n+1} - t_n\|,$$

so $\eta(t_n, x) \to \bar{x}$ as $n \to \infty$. We solve

$$\begin{cases} \dfrac{d\eta}{dt} = V(\eta(t, x)), \\ \eta(t^+(x), x) = \bar{x}, \end{cases}$$

to obtain the contradiction of the maximality of $t^+(x)$. The continuous dependence of solutions of equation (8.11) on the initial data x implies $\eta(t, x) = \eta_t(x) \in C([0, 1] \times E, E)$ satisfying:

(i) $\eta_0(x) = x$ for $x \in E$.

(ii) Fix $x \notin f^{-1}[c - \bar{\epsilon}, c + \bar{\epsilon}]$. Then $x \in A$. Therefore $g(x) = 0$ or $V(x) = 0$. Note that $\eta(t, x) = x$ for all t is a solution of

$$\begin{cases} \dfrac{d\eta_t(x)}{dt} = V(\eta_t(x)), \\ \eta(0, x) = x. \end{cases}$$

By the uniqueness of the solution, we obtain

$$\eta(t, x) = x \quad \text{for } t \in [0, 1].$$

(iii) Note the semigroup property for solutions of (8.11):

$$\eta_{t_1 + t_2} = \eta_{t_1} \eta_{t_2}, \text{ and } \eta_0(x) = x.$$

If x_1, x_2 satisfy $\eta_t(x_1) = \eta_t(x_2)$, then $x_1 = \eta_{-t}\eta_t(x_1) = \eta_{-t}\eta_t(x_2) = x_2$. For $x \in E$, $x = \eta_0(x) = \eta_t \eta_{-t}(x)$. Thus, η_t is a homeomorphism of E onto E.

(iv) This is trivial if $\eta_t(x) = x$. If $\eta_t(x) \neq x$, then $V(\eta_t(x))$ is defined, and

$$\begin{aligned} \frac{d}{dt} f(\eta_t(x)) &= (f'(\eta_t(x)), \frac{d}{dt}\eta_t(x)) \\ &= (f'(\eta_t(x)), V(\eta_t(x))) \\ &= (f'(\eta_t(x)), -g(\eta_t(x))\bar{g}(\eta_t(x))h(\|v(\eta_t(x))\|)v(\eta_t(x)) \qquad (8.12) \\ &= -g(\eta_t(x))\bar{g}(\eta_t(x))h(\|v(\eta_t(x))\|)(f'(\eta_t(x), v(\eta_t(x))) \\ &\leq 0. \end{aligned}$$

Thus, $f(\eta_t(x))$ is decreasing as a function of t, i.e., $f(\eta_t(x)) \leq f(\eta_0(x)) = f(x)$ for $x \in E$, $t \in [0, 1]$.

(v) To show that $\eta_1(A_{c+\epsilon} \backslash M_\delta) \subset A_{c-\epsilon}$. If $x \in A_{c-\epsilon}$, then

$$f(\eta_t(x)) \leq f(x) \leq c - \epsilon \quad \text{for } t \in [0, 1],$$

so $\eta_t(x) \in A_{c-\epsilon}$. Thus, it is only necessary to prove that $x \in Y = A_{c+\epsilon} \backslash (A_{c-\epsilon} \cup M_\delta)$ implies that $\eta_1(x) \in A_{c-\epsilon}$. Let $x \in Y$. The inequality (8.12) implies that

$$\frac{df(\eta_t(x))}{dt} \leq 0.$$

Because $g = 0$ on $A_{c-\bar{\epsilon}}$, the orbit $\eta_t(x)$ cannot enter the subset $A_{c-\bar{\epsilon}}$ of $A_{c-\epsilon}$ as in the proof of (ii). Now

$$\begin{aligned} f(\eta_0(x)) &= f(x) \leq c + \epsilon \leq c + \bar{\epsilon}, \\ \eta_t(x) &\notin A_{c-\bar{\epsilon}} \Rightarrow f(\eta_t(x)) \geq c - \bar{\epsilon}, \end{aligned}$$

so

$$0 \leq f(\eta_0(x)) - f(\eta_t(x)) \leq c + \bar{\epsilon} - (c - \bar{\epsilon}) = 2\bar{\epsilon} \quad \text{for } t \in [0, 1].$$

Suppose $x \in Y$, and $\eta_s(x) \in Z = A_{c+\epsilon} \backslash (A_{c-\epsilon} \cup M_{\delta/2})$ for $s \in [0, T]$ and $\eta_s \notin Z$ for $s > T$. This will certainly be the case for small T by the continuity of $\eta(t, x)$. Suppose $0 < T < 1$. For such s, $\eta(s, x) \in \tilde{E}$ because by (8.9)

$$\|f'(\eta_s(x))\| \geq b,$$

and by definitions

$$g(\eta_s(x)) = \bar{g}(\eta_s(x)) = 1 \quad \text{for } s \in [0, T].$$

Now

$$2\bar{\epsilon} \geq f(\eta_0(x)) - f(\eta_T(x)) = \int_0^T -\frac{df(\eta_s(x))}{ds} ds$$

$$= \int_0^T h(\|v(\eta_s(x))\|)(f'(\eta_s(x), v(\eta_s(x))))ds$$

$$\geq \int_0^T h\,\|(v(\eta_s(x))\|)\,\|f'(\eta_s(x))\|^2 \geq b \int_0^T h(\|v(\eta_s(x))\|)\,\|f'(\eta_s(x)\|\,ds$$

$$\geq \frac{b}{2} \int_0^T h(\|v(\eta_s(x))\|)\,\|v(\eta_s(x))\|\,ds \geq \frac{b}{2} \left\| \int_0^T h(\|v(\eta_s(x))\|)v(\eta_s(x))ds \right\|$$

$$= \frac{b}{2} \left\| \int_0^T v(\eta_s(x)) \right\| = \frac{b}{2} \left\| \int_0^T \frac{d\eta_s(x)}{ds} dx \right\| = \frac{b}{2} \|\eta_T(x) - x\|,$$

and consequently,

$$\|\eta_T(x) - x\| \leq \frac{4\bar{\epsilon}}{b} \leq \frac{4}{b} \cdot \frac{b\delta}{32} = \frac{\delta}{8}.$$

There is $\eta > 0$ such that $T < t < T + \eta$ implies that $\|\eta_t(x) - \eta_T(x)\| < \frac{\partial}{8}$. Thus, for such t, $\|\eta_t(x) - x\| \leq \|\eta_t(x) - \eta_T(x)\| + \|\eta_T(x) - x\| < \frac{\delta}{4}$, and we have $\eta_t(x) \notin M_{\delta/2}$, or $\eta_t(x) \in Z$, contradicting the maximality of T. Hence $T = 1$. Thus, the orbit $\eta(t, x)$ cannot leave Z and enter $M_{\delta/2}$. Consequently, the only way that $\eta(t, x)$ can leave Z is to enter $A_{c-\epsilon}$. We claim this occurs for some $t \in (0, 1)$ thereby proving (v). If not, $\eta(t, x) \in Z$ for all $t \in (0, 1)$

$$\frac{df(\eta_t(x))}{dt} = -g(\eta_t(x))\bar{g}(\eta_t(x))h(\|v(\eta_t(x))\|)(f'(\eta_t(x), v(\eta_t(x))))$$

$$\leq -h(\|v(\eta_t(x))\|)\,\|f'(\eta_t(x))\|^2.$$

If $\|v(\eta_t(x))\| \leq 1$, then $h(\|v(\eta_t(x))\|) = 1$, and

$$\frac{df(\eta_t(x))}{dt} \leq -\|f'(\eta_t(x))\|^2 \leq -b^2.$$

If $\|v(\eta_t(v))\| > 1$, then $h(\|v(\eta_t(x))\|) = \|v(\eta_t(x))\|^{-1}$, and

$$\frac{df(\eta_t(x))}{dt} \leq -\frac{\|f'(\eta_t(x))\|^2}{\|v(\eta_t(x))\|} \leq -\frac{\|v(\eta_t(x))\|}{4} \leq -\frac{1}{4}.$$

These imply that

$$\frac{df(\eta_t(x))}{dt} \leq -\min(b^2, \frac{1}{4}).$$

Integrating both sides, we obtain

$$\min(b^2, \frac{1}{4}) \leq f(\eta_0(x)) - f(\eta(x)) \leq 2\bar{\epsilon},$$

which violates (8.10). We conclude that $\eta_{t_0}(x) \notin Z$ for some $t_0 \in [0,1]$, and then

$$\eta_{t_0}(x) \in A_{c-\epsilon}. \text{ By (8.11), } \eta_1(x) \in A_{c-\epsilon}.$$

(*vi*) If $K_c = \emptyset$, take $N = \emptyset$.

(*vii*) Discussed above.   ∎

Remark 8.24 *Although Theorem 8.23 is set in the framework of a real Banach space, the ideas can be applied more widely. For example, suppose E is a real Hilbert space and $S_r = \partial B_r$. Let $\widetilde{f} = f \mid_{S_r}$. Then $\widetilde{f}'(x) = f'(x) - r^{-2}(f'(x), x)x$. Set*

$$\widetilde{K}_c = \{x \in S_r \mid f(x) = c, \ f'(x) = r^{-2}(f'(x), x)x\}.$$

Then, provided that $\widetilde{f}$ satisfies the (PS)condition and $A_{c+\epsilon}$, N, V, and so on are replaced by their appropriate relativizations to S_r, there is a $\widetilde{\eta} \in C([0,1] \times S_r, S_r)$ having the properties stated in Theorem 8.23 and only minor modifications are required for the proof.

We commence our study of the Ljusternik-Schnirelman theorems: the free case. From linear algebra we obtain:

Theorem 8.25 *Let L be an $N \times N$ symmetric matrix, $f(x) = \frac{1}{2}(Lx, x)$, then $f'(x) = Lx$. There are n eigenvalues λ_i, $i = 1, \cdots, N$, and eigenvectors u_i such that*

$$f'(u_i) = \lambda_i u_i.$$

Surprisingly, we have:

Theorem 8.26 *Assume $f \in C^1(\mathbb{R}^N, \mathbb{R})$ with f even, and $S_r = \{x \in \mathbb{R}^N \mid \|x\| = r\}$ for $r > 0$. Then for each $r > 0$, $\widetilde{f} = f \mid_{S_r}$ possesses at least n distinct pairs of critical points x :*

$$f'(x) = \lambda x, \quad \lambda = r^{-2}(f'(x), x).$$

Proof.   For convenience we take $r = 1 : S_1 = S^{N-1}$. The proof consists of three steps:

(i) Minimax characterization of critical values of $\widetilde{f}$: let $\gamma_k = \{A \subset S^{N-1} \mid \gamma(A) \geq k\}$. By Lemma 8.14, $\gamma_k \neq \emptyset$, $1 \leq k \leq N$. Define

$$b_k = \inf_{A \in \gamma_k} \max_{x \in A} f(x)$$

If $\gamma(A) > k$, there is $B \subset A$ such that $\gamma(B) = k$. Hence, $\max\limits_{B} f \leq \max\limits_{A} f$, so b_k can be characterized as

$$b_k = \inf_{\substack{B \in \gamma_k, \\ \gamma(B)=k}} \max_{x \in k} f(x).$$

Because b_{k+1} is an infimum over a smaller class of sets than b_k, $b_1 \leq b_2 \leq \cdots \leq b_N$. Next note that for $x \in S^{N-1}$, $A = \{x, -x\} \in \gamma_1$. Hence $b_1 = \min\limits_{S^{N-1}} f(x)$. Moreover, $b_N = \max\limits_{S^{N-1}} f(x)$. To verify this, it suffices to show that S^{N-1} is the only set in γ_N. Suppose $A \subsetneq S^{N-1}$, the coordinate system can be assumed oriented so that $(0, \cdots, 0, \pm 1) \notin A$. Define $p(x_1, \cdots, x_N) = (x_1, \cdots, x_{N-1}, 0)$. Therefore $p \in C(A, \mathbb{R}^{N-1} - \{0\})$, and is odd, so $\gamma(A) \leq N - 1$. Set $\widetilde{A}_c = \{x \in S^{N-1} \mid f(x) \leq c\}$. Then

$$b_k = \inf\{r \in \mathbb{R} \mid \gamma(\widetilde{A}_r) \geq k\}. \tag{8.13}$$

Indeed, let r_k denote the right hand side of (8.13). Suppose $\gamma(\widetilde{A}_r) \geq k$. Then $\widetilde{A}_r \in \gamma_k$ and $b_k \leq \max\limits_{\widetilde{A}_K} f = r$. Because this is valid for all such r, $b_k \leq r_k$. If $b_k < r_k$, there is an $A \in \gamma_k$ such that $\max\limits_{A} f = s < r_k$. Hence, $A \subset \widetilde{A}_s$ and by Proposition 8.13 (ii), $\gamma(\widetilde{A}_s) \geq \gamma(A) \geq k$, but then $r_k \leq s$, a contradiction. Another set of critical values of f is given by

$$c_j = \sup_{\theta \in \gamma_j} \min_{x \in \theta} f(x) 1 \leq j \leq N.$$

This follows from the b_j case on replacing f by $-f$. It is easy to see that $c_n = b_1 \leq \cdots \leq c_1 = b_N$. However, simple examples show that in general $c_j \neq b_{N+1-j}$ for $j \neq 1, N$.

(ii) b_k is a critical value of $\widetilde{f}$: because S^{N-1} is compact, it is trivial to verify that $\widetilde{f}$ satisfies the (PS) condition. By Remark 8.24, Theorem 8.23 is applicable here. If b_k is not a critical value of $\widetilde{f}$, by Theorem 8.23 (vi), $\epsilon > 0$, there is $\eta_1 \in C(E, E)$ such that $A = \eta_1(\widetilde{A}_{b_k+\epsilon}) \subset \widetilde{A}_{b_k-\epsilon}$. Because (8.13) shows $\gamma(\widetilde{A}_{b_k+\epsilon}) \geq k$ and η_1 is odd, by Proposition 8.13 (i), $\gamma(A) \geq k$. Hence $A \in \gamma_k$, but

$$b_k \leq \max_A f \leq \max_{\widetilde{A}_{b_k-\epsilon}} f \leq b_k - \epsilon,$$

a contradiction. Thus b_k is a critical value of $\widetilde{f}$.   ∎

A multiplicity lemma.

Lemma 8.27 *If $b_k = \cdots = b_{k+p-1} \equiv b$ and $\widetilde{k}_b = \{x \in S^{N-1} \mid \widetilde{f}(x) = b,\ f'(x) = 0\}$, then $\gamma(\widetilde{k}_b) \geq p$.*

Proof.   If $r(\widetilde{k}_b) \leq p - 1$, by Proposition 8.13 (vi) there is a $\delta > 0$ such that $r(N_\delta(\widetilde{k}_b)) \leq p - 1$. By Theorem 8.23 (v) with $N = int\ N_\delta(\widetilde{k}_b)$, there is $\epsilon > 0$ such that $\eta_1(\widetilde{A}_{b+\epsilon} - N) \equiv Q \subset \widetilde{A}_{b-\epsilon}$. Because $b = b_{k+p-1}$, (8.13) shows that $\gamma(\widetilde{A}_{b+\epsilon}) \geq k + p - 1$. Therefore by Proposition 8.13 (v) and (i), $\gamma(\widetilde{A}_{b+\epsilon} - N) = \gamma(\overline{\widetilde{A}_{b+\epsilon} - N}) \geq (k + p - 1) - (p - 1) = k$ and $\gamma(Q) \geq k$. Hence $Q \in \gamma_k$ and

$$b_k \leq \max_Q f \leq \max_{\widetilde{A}_{b_k-\epsilon}} \leq b_k - \epsilon,$$

which is a contradiction.   ∎

Remark 8.28 *If $p > 1$ in Lemma 8.27, $\widetilde{k}_b$ contains infinitely many critical points because a finite set has genus 1.*

Theorem 8.29 *Suppose E is a real infinite-dimensional Hilbert space and $f \in C^1(E, \mathbb{R})$ is even, then $f \mid_{S_r}$ possesses infinitely many distinct pairs of critical points.*

Proof.   For $u \in E$, $f'(u) \in E' \cong E$. At a critical point of $\widetilde{f} = f \mid_{S_r}$ we have

$$(f'(u), x) = \lambda(u, x) \text{ for all } x \in E.$$

Then let

$$b_k = \inf f_{A \in \gamma_k}(x) \quad 1 \leq k < \infty$$
$$= \inf\{r \in \mathbb{R} \mid \gamma(\widetilde{A}_r) \geq k\}.$$ ∎

Krasnoselskii gave conditions on f under which the above hypotheses are satisfied.

Lemma 8.30 *Let E be a real infinite-dimensional separable Hilbert space, $f \in C^1(E, \mathbb{R})$ be even with $f(0) = 0$, and $f(x) < 0$ if $x \neq 0$. Suppose further that f' maps weakly convergent to strongly convergent sequences and $f'(x) \neq 0$ if $x \neq 0$. Then f satisfies the hypotheses of Theorem 8.26 for all $r > 0$.*

Proof.   Again for convenience we take $r = 1$ and $S_1 = S$. Because f' maps weakly convergent to strongly convergent sequences, f' is compact. Therefore f is weakly continuous. If $\inf\limits_{S} f = -\infty$, there is a sequence $(u_n) \subset S$ such that $f(u_n) < -n$. Because (u_n) is bounded and E is a Hilbert space, (u_n) possesses a weakly convergent subsequence $u_{n_k} \to u$. Therefore $f(u_{n_k}) \to f(u) = -\infty$, which is impossible because f is continuous. Hence, f is bounded from below. Because $f < 0$ on S, to complete the proof it suffices to verify the $(PS)^-$ condition. Let us suppose there is a sequence (u_n) and constants K, $\alpha > 0$ such that $-K \leq f(u_n) \leq -\alpha < 0$ and

$$f \mid f'_S(u_n) = f'(u_n) - (f'(u_n), u_n)u_n \to 0. \tag{8.14}$$

As above, (u_n) possesses a weakly convergent subsequence $u_{n_k} \to u$ and $f(u_{n_k}) \to f(u) \leq -\alpha$. Therefore $u \neq 0$ and $f'(u) \neq 0$. Moreover, $f'(u_{n_k}) \to f'(u)$ and $(f'(u_{n_k}), u_{n_k}) \to (f'(u), u)$. Thus, taking the inner product of (8.14) with $f'(u_{n_k})$, we show

$$\|f'(u_n)\|^2 - (f'(u_n), u_n)^2 \to 0.$$

Thus,

$$0 \neq \|f'(u)\|^2 = (f'(u), u)^2.$$

Thus, $|(f'(u_{n_k}), u_{n_k})| \geq c > 0$ for large n_k, and dividing by this quantity in (8.14) yields the convergence of u_{n_k} : $(f'(u), u))[1 - \|u_n\|^2] \to 0$. Hence the $(PS)^-$ condition. $\blacksquare$

Remark 8.31

(*i*) *Krasnoselskii proved that if E is not separable, there is no f satisfying the above hypothesis.*

(*ii*) *The $(PS)^-$ condition implies that each critical value b_k is of "finite multiplicity": $\gamma(K_{b_k}) < \infty$. Moreover, $b_k \to 0$ as $k \to \infty$, as was shown by Krasnoselskii.*

(*iii*) *Another set of critical values for $f \mid_S$ under the hypotheses of Lemma 8.30 can be obtained by defining*

$$\widetilde{b}_k = \inf_{A \in \widetilde{\gamma}_k} \max_{u \in A} f(u),$$

where $\widetilde{\gamma}_k = \{A \subset S \mid A$ is compact and $r(A) \geq k\}$. Clearly, $\widetilde{b}_k \geq b_k$.

8.3 Ljusternik-Schnirelman Theorems

Let E be a real Banach space and let $\bar{\gamma}_k = \{A \subset E \mid \gamma(A) \geq k\}$.

Theorem 8.32 *Let $f \in C^1(E, \mathbb{R})$ with f even and $f(0) = 0$. If f satisfies $(PS)^-$ condition and*

$$-\infty < b_k \equiv \inf_{A \in \bar{\gamma}} \sup_{x \in A} < 0,$$

then b_k is a critical value of f. Moreover, if $-\infty < b_k = \cdots = b_{k+p-1} \equiv b$, then $\gamma(k_b) \geq p$.

Proof.   Similar to Theorem 8.36.   ∎

Lemma 8.33 (*Mountain Pass Lemma*) *Let E be a real Banach space, $f \in C^1(E, \mathbb{R})$ satisfying the (PS) condition. Define the ball $B_p = \{x \in E \mid \|x\| < \rho\}$, and the sphere $S_\rho = \{x \in E \mid \|x\| = \rho\}$. Suppose there is $\alpha > 0$ such that*

$$f(0) = 0,$$
$$f \geq \alpha \text{ on } S_\rho,$$
$$f > 0 \text{ on } B_\rho/\{0\},$$
$$f(e) = 0 \text{ for some } 0 \neq e \in E.$$

Then

$$c = \inf_{g \in \Gamma} \max_{y \in [0,1]} f(g(y))$$

is a critical value of f with $0 < \alpha \leq c < \infty$, where $\Gamma = \{g \in C([0,1], E) \mid g(0) = 0, \ g(1) = e\}$.

Proof.   The hypothesis implies $c \geq \alpha$. Suppose c is not a critical value of f, then $K_c = \{x \in E \mid f(x) = c, \ f'(x) = 0\} = \emptyset$. The Deformation Theorem 8.23 implies $c > \epsilon > 0$, there is $\eta_1 \in C(E, E)$ such that

$$\eta_1(A_{c+\epsilon}) \subset A_{c-\epsilon}$$

choose $g \in \Gamma$ with $c - \epsilon \leq c \leq \max_{x \in [0,1]} f(g(x)) \leq c + \epsilon$. By Theorem 8.23 (ii), $\eta_1(0) = 0$, $\eta_1(e) = e$. Note that $\eta_1 \circ g \in \Gamma$, and

$$c \leq \max_{x \in [0,1]} f(\eta_1 \circ g(x)) \leq c - \epsilon, \quad \text{a contradiction.}$$ ∎

Let E be an infinite-dimensional Banach space, $f \in C^1(E, \mathbb{R})$. We list some properties:
(fi) There are $f(0) = 0$ and ρ and $\alpha > 0$ with $f > 0$ in $B_\rho \setminus \{0\}$, $f \geq \alpha$ on ∂B_ρ;
(fii) There is an $e \in E$, $e \neq 0$, such that $f(e) = 0$;

$(fiii)$ f is even;

(fiv) If X is a finite-dimensional subspace of E, $X \cap \hat{A}_0$ is bounded where

$$\hat{A}_c = \{x \in E \mid f(x) \geq c\}$$

Let Γ_* denote the set of homeomorphisms h of E onto E with $h(0) = 0$ and $h(B_1) \subset \hat{A}_0$. Set $\Gamma^* = \{h \in \Gamma_* \mid h$ is odd$\}$ and

$$\Gamma_m = \{K \subset E \mid K \text{ is compact, symmetric with respect to } 0,$$
$$\text{and for all } h \in \Gamma^*, r(K \cap h(S)) \geq m, \text{ where } S = S_1.\}.$$

Observe that if $h \in \Gamma^*$, $h(S) \subset E - \{0\}$ and is closed and symmetric. Therefore $K \cap h(S) \in \sum(E)$.

Lemma 8.34 *Let f satisfy (fi) and (fiv). Then:*

(i) $\Gamma_m \neq \emptyset$;

(ii) $\Gamma_{m+1} \subset \Gamma_m$;

(iii) *If $K \in \Gamma_m$ and $Y \in \sum(E)$ with $\gamma(Y) \leq r < m$, then $\overline{K - Y} \in \Gamma_{m-r}$;*

(iv) *If φ is an odd homeomorphism of E onto E with $\varphi^{-1}(\hat{A}_0) \subset \hat{A}_0$, then $\varphi : \Gamma_m \to \Gamma_m$.*

Proof.

(i) Let X be an m-dimensional subspace of E. If $K_R = X \cap B_R$, K_R is compact and symmetric. By (fiv), if R is sufficiently large, $K_R \supset X \cap \hat{A}_0$. If $h \in \Gamma^*$, $h(B_1) \subset \hat{A}_0$, so $K_R \supset X \cap h(B_1)$ and $K_R \cap h(S) = X \cap h(S)$. Moreover, $h(0) = 0$ and $h(B_1)$ is a neighborhood of 0 in E, h being a homeomorphism. Therefore $X \cap h(B_1)$ is a symmetric bounded neighborhood of 0 in X with its boundary contained in $X \cap h(S)$. Because X is isomorphic to $\mathbb{R}^m$, it follows from Proposition 8.13 (ii), (iii) and Theorem 8.23 that

$$r(K_R \cap h(S)) = r(X \cap h(S)) \geq r(\partial(X \cap h(B_1))) = m. \qquad (8.15)$$

Because the right hand side of (8.15) can have at most genus m, actually we have equality in (8.15).

(ii) Trivial.

(iii) $\overline{K - Y}$ is compact and symmetric. If $h \in \Gamma^*$, then $\overline{K - Y} \cap h(s) = \overline{(K \cap h(s)) - Y}$, so by Proposition 8.13 (v),

$$\overline{K - Y} \cap h(s) \geq \gamma(K \cap h(S)) - \gamma(Y) \geq m - r.$$

(iv) Let $K \in \Gamma_m$. Then $\varphi(K)$ is compact and symmetric and by Proposition 8.13 (iii),

$$\gamma(\varphi(K) \cap h(S)) = \gamma(K \cap \varphi^{-1} \circ h(S)) \quad \text{for } h \in \Gamma^*. \tag{8.16}$$

Because $h(S) \subset \hat{A}_0$, the hypotheses on φ imply $\varphi^{-1} \circ h \in \Gamma^*$. Hence, the right hand side of (8.16) is not less than m and $\varphi : \Gamma_m \to \Gamma_m$. ∎

Theorem 8.35 *Assume $f \in C^1(E, \mathbb{R})$ and satisfies (f_i), (f_{iii}), (f_{iv}), and the $(PS)^+$ condition. For $m \in \mathbb{N}$, define $b_m = \inf\limits_{K \in \Gamma_m} \max\limits_{u \in K} f(u)$. Then:*

(i) b_m is a critical value of f with $0 < \alpha \leq b_m \leq b_{m+1}$;

(ii) If $b_m = \cdots = b_{m+r-1} \equiv b$, $\gamma(K_b) \geq r$;

(iii) $b_m \to \infty$ as $m \to \infty$.

Proof. Because $h(u) = \rho u \in \Gamma^*$, $K \cap h(S) = K \cap S_\rho \neq \emptyset$ for all $K \in \Gamma_m$. Therefore, by (f_i), $\max\limits_K f \geq \alpha$ for all $K \in \Gamma_m$, so $b_m \geq \alpha$. Because $\Gamma_{m+1} \subset \Gamma_m$, $b_{m+1} \geq b_m$. To show that b_m is a critical value of f, it suffices to prove the stronger multiplicity assertion (ii). Suppose $\gamma(K_b) < r$. By Proposition 8.13 (vi), there is a $\delta > 0$ such that $\gamma(N_\delta(K_b)) < r$. Let $N = int N_\delta(K_b)$. Applying Theorem 8.23 (iii), (vi), (vii) shows that there is an $\epsilon \in (0, \alpha)$ and an odd homeomorphism η_1 of E onto E such that $\eta_1(A_{b+\epsilon} - N) \subset A_{b-\epsilon}$. Choose $K \in \Gamma_{m+r-1}$ such that $\max\limits_K f(u) \leq b + \epsilon$. By Lemma 8.35 (iii), $K - N = K - N_\delta(K_b) \equiv Q \in \Gamma_m$. Moreover, $\eta_1^{-1}(\hat{A}_0) \subset \hat{A}_0$ by Theorem 8.23 (iv). Hence, $\eta_1(Q) \in \Gamma_m$ by Lemma 8.34 (iv). However, $b \leq \max\limits_{\eta_1(Q)} f(u) \leq b - \epsilon$, a contradiction. Lastly, to verify (iii), observe that by (i), either (a) there is an $n \in \mathbb{N}$ such that $b_m = b_n$ for $m \geq n$, (b) (b_m) has a finite limit point, or (c) $b_m \to \infty$ as $m \to \infty$. If (a) is true, by (ii), $\gamma(K_{b_n}) = \infty$. However, the $(PS)^+$ condition implies K_{b_n} is compact, and therefore, by Proposition 8.13 (vi), $\gamma(K_{b_n}) < \infty$. To exclude (b), suppose $b_m \to \bar{b} < \infty$. Let $K = \{u \in E \mid b_1 \leq f(u) \leq \bar{b} \text{ and } f'(u) = 0\}$. Again, by the $(PS)^+$ condition, K is compact. Suppose $\gamma(K) = j$. By Proposition 8.13 (vi), there is a $\delta > 0$ such that $\gamma(N_\delta(K)) = j$. Let $N = int N_\delta(K)$. By Theorem 8.23, there is an $\epsilon > 0$ and an odd homeomorphism η_1 of E onto E such that $\eta_1(A_{\bar{b}+\epsilon} - N) \subset A_{\bar{b}-\epsilon}$. Let m be the smallest integer such that $b_m > \bar{b} - \epsilon$. We can assume ϵ is small enough so that $m \geq 1$. Choose $K \in \Gamma_{m+j}$ so that $\max\limits_K f \leq \bar{b} + \epsilon$. By Lemma 8.34 (iii), (iv), $Q = K - N = \overline{K - N_\delta(K)} \in \Gamma_m$ and $\eta_1(Q) \in \Gamma_m$. Because $Q \subset A_{\bar{b}+\epsilon} - N$, $\max\limits_{\eta_1(Q)} f \leq \bar{b} - \epsilon < b_m$, a contradiction. ∎

8.4 Concentration-Compactness principle

First we state a result.

Theorem 8.36 (*Helly Theorem*) *For $T > 0$, given a sequence $\{f_n(t)\}$ of uniformly bounded variation in $[0, T]$, either there is a uniformly bounded subsequence $\{f_{n_k}(t)\}$ converging everywhere to a function $f(t)$ of bounded variation in $[0, T]$, or else $\{|f_n(t)|\}$ diverges uniformly to ∞ in $[0, T]$, as $n \to \infty$.*

Proof. See Torchinsky [50, p. 39]. ∎

Now we are in a position to understand the concentration lemma.

Lemma 8.37 *Let $(\rho_n)_{n \geq 1}$ be a sequence in $L^1(\mathbb{R}^N)$ satisfying*

$$\rho_n \geq 0 \quad \text{in } \mathbb{R}^N, \quad \int_{\mathbb{R}^N} \rho_n = \lambda,$$

where $\lambda > 0$ is fixed. Then there is a subsequence $(\rho_{n_k})_{k \geq 1}$, satisfying one of the following three possibilities:

(i) *(compactness) There is $y_k \in \mathbb{R}^N$ such that $\rho_{n_k}(\cdot + y_k)$ is tight: for $\epsilon > 0$, there is some $R > 0$ such that $\displaystyle\int_{y_k + B_R} \rho_{n_k} \geq \lambda - \epsilon$;*

(ii) *(vanishing)*

$$\lim_{k \to \infty} \sup_{y \in \mathbb{R}^N} \int_{y + B_R} \rho_{n_k} = 0 \text{ for every } R > 0;$$

(iii) *(dichotomy) There is $\alpha \in (0, \lambda)$ such that for $\epsilon > 0$, there are $k_0 \geq 1$ and $\{\rho_{k,1}\}$, $\{\rho_{k,2}\} \subset L_+^1(\mathbb{R}^N)$ satisfying for $k \geq k_0$:*

$$\|\rho_{n_k} - (\rho_{k,1} + \rho_{k,2})\|_{L^1} < \epsilon, \quad \left| \int_{\mathbb{R}^N} \rho_{k,1} - \alpha \right| < \epsilon,$$

$$\left| \int_{\mathbb{R}^N} \rho_{n,2} - (\lambda - \alpha) \right| < \epsilon,$$

$$\text{dist} \, (\text{supp} \, \rho_{k,1}, \text{supp} \, \rho_{k,2}) \to \infty \quad \text{as } k \to \infty.$$

Proof. Consider the concentration function of a measure:

$$Q_n(t) = \sup_{y \in \mathbb{R}^N} \int_{y + B_t} \rho_n(x) dx.$$

$(Q_n(t))_n$ is a sequence of nondecreasing, nonnegative, uniformly bounded functions on $\mathbb{R}_+$, and

$$\lim_{t \to \infty} Q_n(t) = \lambda.$$

By Theorem 8.36, there is a subsequence $\{Q_{n_k}\}$ of $\{Q_n\}$ and a nondecreasing nonnegative function Q such that $0 \leq Q \leq \lambda$ satisfying

$$Q_{n_k}(t) \to Q(t), \text{for each } t \geq 0, \text{ as } k \to \infty.$$

Set
$$\alpha = Q(t).$$

Then $\alpha \in [0, \lambda]$. We consider the following cases:

(i) If $\alpha = \lambda$, then
$$\lim_{t \to \infty} Q(t) = \lambda.$$

For $\dfrac{\lambda}{2} < \mu < \lambda$ there is $R_1 = R_1(\mu)$ such that $Q(R_1) = \lim_{k \to \infty} Q_{n_k}(R_1) > \mu$. There is $N_0 > 0$ such that $Q_{n_k}(R) > \mu$ for $k \geq N_0$. Moreover, there is $R_2(\mu)$ such that
$$Q_{n_k}(R_2) > \mu \quad \text{for } k = 1, \cdots, N_0 - 1.$$

Take $R(\mu) = R_1(\mu) + R_2(\mu)$, for all $k \geq 1$. We have
$$Q_{n_k}(R) = \sup_{y \in \mathbb{R}^N} \int_{y+B_R} \rho_{n_k}(x) dx > \mu$$

Take $y_k(\mu) \in \mathbb{R}^N$ such that:
$$\int_{y_k+B_R} \rho_{n_k}(\xi) d\xi > \mu.$$

We claim that
$$\left| y_k(\mu) - y_k\left(\frac{\lambda}{2}\right) \right| \leq R\left(\frac{\lambda}{2}\right) + R(\mu) \quad \text{for all } \mu \geq \frac{\lambda}{2}.$$

Otherwise, if
$$\left| y_k(\mu) - y_k\left(\frac{\lambda}{2}\right) \right| > R\left(\frac{\lambda}{2}\right) + R(\mu),$$

then
$$\left(y_k(\mu) + B_{R(\mu)}\right) \cap \left(y_k\left(\frac{\lambda}{2}\right) + B_{R\left(\frac{\lambda}{2}\right)}\right) = \emptyset,$$

and
$$\int_{\mathbb{R}^N} \rho_{n_k}(x) dx \geq \int_{y_k(\mu)+B_{R(\mu)}} \rho_{n_k}(x) dx + \int_{y_k\left(\frac{\lambda}{2}\right)+B_{R\left(\frac{\lambda}{2}\right)}} \rho_{n_k}(x) dx$$
$$> \mu + \frac{\lambda}{2} > \lambda,$$

a contradiction. Setting $R'(\mu) = R\left(\frac{\lambda}{2}\right) + 2R(\mu)$, then
$$y_k(\mu) + B_{R(\mu)} \subset y_k\left(\frac{\lambda}{2}\right) + B_{R'(\mu)},$$

and hence,

$$\int_{y_k(\frac{\lambda}{2})+B'_R(\mu)} \rho_{n_k}(\xi)d\xi \geq \int_{y_k(\mu)+B_R(\mu)} \rho_{n_k}(\xi)d\xi > \mu.$$

This concludes (i).

(ii) If $\alpha = 0$, then $\lim_{k\to\infty N} \sup_{y\in\mathbb{R}^N} \int_{y+B_R} \rho_{n_k}(x)dx = 0$ for each $R > 0$; that is,

$$\lim_{k\to\infty} Q_{n_k}(R) = 0.$$

Otherwise, suppose

$$\lim_{k\to\infty} Q_{n_k}(R) = Q(R) > 0.$$

Then $0 = \lim_{k\to\infty} Q(t) \geq Q(R) > 0$, a contradiction. Thus, (ii) holds.

(iii) Suppose $\alpha = \lim_{t\to\infty} Q(t)$, $\alpha \in (0, \lambda)$. For $\epsilon > 0$ choose $R > 0$ such that $R' \geq R$ implies

$$\alpha - \epsilon < Q(R') < \alpha + \epsilon.$$

We have

(a) There is $k_0 \in N$ such that $k \geq k_0$, which implies

$$\alpha - \epsilon < Q_{n_k}(R) < \alpha + \epsilon.$$

Then there are $y_k \in \mathbb{R}^N$ such that

$$\alpha - \epsilon < \int_{y_k+B_R} \rho_{n_k}(\xi)d\xi < \alpha + \epsilon.$$

(b) There is a sequence $R_k > R$, $R_k \to \infty$ as $k \to \infty$ such that

$$\alpha - \epsilon < Q_{n_k}(R_k) < \alpha + \epsilon.$$

Note that

$$\alpha - \epsilon < \int_{y_k+B_R} \rho_{n_k}(\xi)d\xi \leq \int_{y_k+B_{R_k}} \rho_{n_k}(\xi)d\xi \leq Q_{n_k}(R_k) < \alpha + \epsilon.$$

Set

$$\rho_{k,1} = \rho_{n_k}\chi_{(y_k+B_R)}, \quad \rho_{k,2} = \rho_{n_k}\chi_{(y_k+B_{R_k})^c}.$$

Then by (a)

$$\int_{\mathbb{R}^N} |\rho_{n_k} - \rho_{k,1} - \rho_{k,2}| = \int_{R \le |x - y_k| \le R_k} \rho_{n_k}$$

$$= \int_{|x - y_k| \le R_k} \rho_{n_k} - \int_{|x - y_k| < R} \rho_{n_k}$$

$$\le Q_{n_k}(R_k) - Q_{n_k}(R) + 2\epsilon$$

$$\le (\alpha + \epsilon) - (\alpha - \epsilon) + 2\epsilon$$

$$= 4\epsilon.$$

Moreover, by (b)

$$\left| \int_{\mathbb{R}^N} \rho_{k,1} - \alpha \right| = \left| \int_{y_k B_k} \rho_{n_k} - \alpha \right| < \epsilon.$$

$$\left| \int_{\mathbb{R}^N} \rho_{k,2} - (\lambda - \alpha) \right| = \left| \int_{\mathbb{R}^N} \rho_{n_k} - \int_{y_k + B_{R_k}} \rho_{n_k} - \lambda + \alpha \right|$$

$$= \left| \int_{y_k + B_{R_k}} \rho_{n_k} - \alpha \right| < \epsilon.$$

$$\text{dist}(\text{supp}\, \rho_{k,1}, \text{supp}\, \rho_{k,2}) \to \infty \quad \text{as } k \to \infty. \qquad \blacksquare$$

We assume that

$$H = H^1(\mathbb{R}^N),$$

$$E(u) = \int_{\mathbb{R}^N} |\nabla u|^2 + a(x)u^2,$$

$$J(u) = \int_{\mathbb{R}^N} b(x)u^p,$$

$$E^\infty(v) = \int_{\mathbb{R}^N} (|\nabla u|^2 + \bar{a}u^2),$$

$$J^\infty(u) = \int_{\mathbb{R}^N} \bar{b}u^p,$$

where $0 < \alpha_1 \le a(x) \le \alpha_2, 0 < \beta_1 \le b(x) \le \beta_2, \lim_{|x| \to \infty} a(x) = \bar{a}, \lim_{|x| \to \infty} b(x) = \bar{b}.$

$$I_\lambda = \inf\{E(u) \mid u \in H, J(u) = \lambda\}, \ \lambda > 0. \qquad (M_\lambda)$$

$$I_\lambda^\infty = \inf\{E^\infty(n) \mid u \in H, J^\infty(n)\}, \ \lambda > 0. \qquad (M_\lambda^\infty)$$

Lemma 8.38 $I_\lambda \leq I_\alpha + I_{\lambda-\alpha}^\infty$ for $\alpha \in [0, \lambda)$, where $I_0 = 0$.

Proof.   For $\epsilon > 0$, let u_ϵ and v_ϵ be of compact support such that

$$\begin{cases} I_\alpha \leq E(u_\epsilon) \leq I_\alpha + \epsilon, & J(u_\epsilon) = \alpha. \\ I_{\lambda-\alpha}^\infty \leq E^\infty(v_\epsilon) \leq I_{\lambda-\alpha}^\infty + \epsilon, & J^\infty(v_\epsilon) = \lambda - \alpha. \end{cases}$$

Let $v_\epsilon^n = v_\epsilon(\cdot + n\chi)$, where given $\chi \in \mathbb{R}^N$, $|\chi| = 1$. For n large enough, the distance between the supports of u_ϵ and v_ϵ^n is strictly positive and goes to ∞, and we recall that

$$E(u_\epsilon + v_\epsilon^n) = E(u_\epsilon) + E^\infty(v_\epsilon^n) + o(1) \quad \text{as } n \to \infty$$
$$J(u_\epsilon + v_\epsilon^n) = J(u_\epsilon) + J^\infty(v_\epsilon^n) = \lambda \quad \text{for large } n.$$

Because E^∞, J^∞ are translation-invariant, and we finally obtain:

$$\begin{aligned} I_\lambda \leq E(u_\epsilon + v_\epsilon^n) &= E(u_\epsilon) + E^\infty(v_\epsilon^n) + o(1) \\ &= E(u_\epsilon) + E^\infty(v_\epsilon) + o(1) \\ &\leq I_\alpha + I_{\lambda-\alpha}^\infty + 2\epsilon. \end{aligned}$$

We conclude that

$$I_\lambda \leq I_\alpha + I_{\lambda-\alpha}^\infty. \qquad \blacksquare$$

Lemma 8.39 Let $1 < p \leq \infty$, $1 \leq q < \infty$ with $q \neq \frac{Np}{N-p}$ if $p < N$. Assume that $\{u_n\}$ is bounded in $L^q(\mathbb{R}^N)$ and $\{\nabla u_n\}$ is bounded in $L^p(\mathbb{R}^N)$ and

$$\lim_{n\to\infty} \sup_{y\in\mathbb{R}^N} \int_{y+B_R} |u_n|^q \, dx = 0 \quad \text{for some } R > 0.$$

Then $u_n \to 0$ in $L^\alpha(\mathbb{R}^N)$, for α between q and $p^* = \frac{Np}{N-p}$.

Proof.   Suppose $\{u_n\}$ is bounded in $L^\infty(\mathbb{R}^N)$. Then for $\beta > \min\{q, p^* = \frac{Np}{N-p}\}$, or $\beta > q$ if $p \geq N$, we have for $\beta > q$,

$$\int_{y+B_R} |u_n|^\beta \leq \int_{y+B_R} |u_n|^q \, |u_n|^{\beta-q} \leq c \int_{y+B_R} |u_n|^q.$$

For β between q and p^*, $\beta = (1-t)q + tq^*$, $t \in (0,1)$,

$$\int_{y+B_R} |u_n|^\beta = \int_{y+B_R} |u_n|^{(1-t)q} |u_n|^{tp^*}$$

$$\leq \left(\int_{y+B_R} |u_n|^q \right)^{1-t} \left(\int_{y+B_R} |u_n|^{p^*} \right)^t$$

$$\leq c \left(\int_{y+B_R} |u_n|^q \right)^{1-t} \|\nabla u_n\|_{L^p(\mathbb{R}^N)}^{tp^*}$$

$$\leq c \left(\int_{y+B_R} |u_n|^q \right)^{1-t},$$

we have

$$\lim_{n\to\infty} \sup_{y\in\mathbb{R}^N} \int_{y+B_R} |u_n|^\beta = 0 \quad \text{for } \beta > \min\{q, p^*\}, \text{ or } \beta > q \text{ if } p \geq N.$$

Take $\bar{q}$ such that $\bar{q} > q$, and $\infty > (\bar{q}-1)p' > q$. By the Hölder inequality,

$$\int_{y+B_R} |u_n|^{\bar{q}-1} |\nabla u_n| \leq \left(\int_{y+B_R} |u_n|^{(\bar{q}-1)p'} \right)^{1/p'} \left(\int_{y+B_R} |\nabla u_n|^p \right)^{1/p}$$

$$\leq c \left(\int_{y+B_R} |u_n|^{(\bar{q}-1)p'} \right)^{1/p'}.$$

Therefore

$$\lim_{n\to\infty} \sup_{y\in\mathbb{R}^N} \int_{y+B_R} |u_n|^{\bar{q}-1} |\nabla u_n| = 0.$$

If $\gamma \in (1, 1^* = \frac{N}{N-1})$, then there is $C_0 > 0$, independent of y, such that

$$\left(\int_{y+B_R} |u_n|^{\bar{q}r} \right)^{1/r} = \left[\int_{y+B_R} (|\nabla u_n|^{\bar{q}})^r \right]^{1/r}$$

$$\leq \left(\int_{y+B_R} |u_n|^{\bar{q}} + |\nabla(|u_n|^{\bar{q}})| \right)$$

$$= \left[\int_{y+B_R} (|\nabla u_n|^{\bar{q}} + \bar{q} |u_n|^{\bar{q}-1} |\nabla u_n|) \right],$$

so

$$\int_{y+B_R} |u_n|^{\bar{q}r} \le \left[\int_{y+B_R} (|\nabla u_n|^{\bar{q}} + \bar{q}|u_n|^{\bar{q}-1}|\nabla u_n|) \right]^r$$

$$= \left[\int_{y+B_R} (|\nabla u_n|^{\bar{q}} + \bar{q}|u_n|^{\bar{q}-1}|\nabla u_n|) \right]^{r-1} \cdot$$

$$\left[\int_{y+B_R} (|\nabla u_n|^{\bar{q}} + \bar{q}|u_n|^{\bar{q}-1}|\nabla u_n|) \right]$$

$$= \epsilon_n^{r-1} \int_{y+B_R} (|\nabla u_n|^{\bar{q}} + \bar{q}|u_n|^{\bar{q}-1}|\nabla u_n|),$$

where $\epsilon_n \to 0$ as $n \to \infty$. Cover the unit cube $I = [0,1] \times [0,1] \times \cdots \times [0,1]$ by m_1 balls $B(x_i, R)$. Then $\{y_j + B(x_i, R)\}$ cover $\mathbb{R}^N$, where y_j has integer coordinates, $j = 1, 2, \cdots, i = 1, \cdots, m_1$. Each unit cube $y_j + I$ is covered by at most m_2 balls $y_{j_k} + B(x_{i_k}, R)$, and each ball $y_j + B(x_i, R)$ can only intersect at most m_3 cubes $y_k + I$. We obtain

$$\int_{\mathbb{R}^N} |u_n|^{\bar{q}r} = \sum_{j=1}^{\infty} \int_{y_j+I} |u_n|^{\bar{q}r} \le \sum_{j=1}^{\infty} m_2 \int_{y_{j_k}+B(x_{i_k},R)} |u_n|^{\bar{q}r}$$

$$\le \sum_{j=1}^{\infty} m_2 \epsilon_n^{r-1} \int_{y_{j_k}+B(x_{i_k},R)} (|u_n|^{\bar{q}} + \bar{q}|u_n|^{\bar{q}-1}|\nabla u_n|)$$

$$\le \sum_{j=1}^{\infty} m_2 m_3' \epsilon_n^{r-1} \int_{y_{j_k}+I} (|u_n|^{\bar{q}} + \bar{q}|u_n|^{\bar{q}-1}|\nabla u_n|)$$

$$\le c\epsilon_n^{r-1} \int_{\mathbb{R}^N} (|u_n|^{\bar{q}} + \bar{q}|u_n|^{\bar{q}-1}|\nabla u_n|)$$

$$\le c\epsilon_n^{r+1} \qquad \text{because } \{u_n\} \text{ is bounded in } L^{\infty}(\mathbb{R}^N),$$

because α lies between q and p^*. Let $s = \bar{q}r > q$, $s > p^*$, then $\alpha = (1-t)a + ts$, $a = q$ or p^*,

$$\int_{\mathbb{R}^N} |u_n|^{\alpha} = \int_{\mathbb{R}^N} |u_n|^{(1-t)a} |u_n|^{ts}$$

$$\le \left(\int_{\mathbb{R}^N} |u_n|^{a} \right)^{1-t} \left(\int_{\mathbb{R}^N} |u_n|^s \right)^t$$

$$\le c \left(\int_{\mathbb{R}^N} |u_n|^s \right)^t$$

$$= o(1) \quad \text{as } n \to \infty.$$

Therefore $u_n \to 0$ in $L^{\alpha}(\mathbb{R}^N)$ for α lying between q and $p^* = \frac{Mp}{N-p}$ as $n \to \infty$. In the general case, let $c > 0$, $v_n = \min\{|u_n|, c\}$. Then by the above proof, for α between

q and p^*, $v_n \to 0$ in $L^\alpha(\mathbb{R}^N)$ as $n \to \infty$. Take $\beta > \alpha$, then β lies between q and $p^* = \frac{Mp}{N-p}$ as $n \to \infty$.

$$
\begin{aligned}
\int_{\mathbb{R}^N} |u_n|^\alpha &= \int_{\mathbb{R}} |v_n|^\alpha + \int_{\mathbb{R}^N} |u_n|^\alpha \chi_{\{|u_n| \geq c\}} \\
&\leq \int_{\mathbb{R}^N} |v_n|^\alpha + \int_{\mathbb{R}^N} |u_n|^\beta |u_n|^{\alpha-\beta} \chi_{\{|u_n| \geq c\}} \\
&\leq \int_{\mathbb{R}^N} |v_n|^\alpha + \frac{1}{c^{\beta-\alpha}} \int_{\mathbb{R}^N} |u_n|^\beta \\
&\leq \int_{\mathbb{R}^N} |v_n|^\alpha + \frac{k}{c^{\beta-\alpha}}.
\end{aligned}
$$

Thus

$$
\varlimsup_{n \to \infty} \int_{\mathbb{R}^N} |u_n|^\alpha \leq \frac{K}{c^{\beta-\alpha}} \quad \text{for all } c > 0.
$$

Letting $c \to \infty$, we obtain

$$
\lim_{n \to \infty} \int_{\mathbb{R}^N} |u_n|^\alpha = 0.
$$
∎

Lemma 8.40 *Let h be a real-valued function on $[0, \lambda]$ with $\lambda > 0$ satisfying $h(\theta\alpha) < \theta h(\alpha)$ for every $\alpha \in (0, \lambda)$, $\theta \in (1, \frac{\lambda}{\alpha}]$. Then we have:*

$$
h(\lambda) < h(\alpha) + h(\lambda - \alpha) \text{ for every } \alpha \in (0, \lambda).
$$

Proof. We divide the proof into two steps.

(i) If $\alpha \geq \lambda - \alpha$ (or $\alpha \geq \frac{\lambda}{2}$) then

$$
h(\lambda) < \frac{\lambda}{2} h(\alpha) = h(\alpha) + \frac{\lambda-\alpha}{\alpha} h(\alpha) = h(\alpha) + \frac{\lambda-\alpha}{\alpha} h(\frac{\lambda-\alpha}{\lambda-\alpha}\alpha)
$$
$$
\begin{cases}
= h(\alpha) + h(\lambda - \alpha) & \text{if } \lambda - \alpha < \alpha \\
< h(\alpha) + h(\lambda - \alpha) & \text{if } \lambda - \alpha < \alpha.
\end{cases}
$$

(ii) if $\alpha < \lambda - \alpha = \beta$ then $\alpha < \frac{\lambda}{2}$, $\lambda - \beta = \lambda - (\lambda - \alpha) = \alpha$ and $\beta = \lambda - \alpha > \lambda - \frac{\lambda}{2} = \frac{\lambda}{2}$. By (i), $h(\lambda) < h(\beta) + h(\lambda - \beta) = h(\lambda - \alpha) + h(\alpha)$. ∎

Theorem 8.41 (*Concentration-Compactness Principle*) *For each fixed $\lambda > 0$, all minimizing sequences of the problem*

$$
I_\lambda = \inf\{E(u) \mid u \in H, J(u) = \lambda\} \quad (M_\lambda)
$$

are relatively compact in H if and only if

$$
I_\lambda < I_\alpha + I_{\lambda-\alpha}^\infty \text{for each } \alpha \in [0, \lambda).
$$

Proof. We divide the proof into several steps.

(I) If $I_\lambda = I_\alpha + I_{\lambda-\alpha}^\infty$ for some $\alpha \in [0, \lambda)$, then there is a minimizing sequence that is not relatively compact.

(a) The case $\alpha \in (0, \lambda)$: Let $\{u_n\}$, $\{v_n\}$ be minimizing sequences with compact supports of problems (M_α), $(M_{\lambda-\alpha}^\infty)$, respectively: that is,

$$E(u_n) \to I_\alpha, \quad E^\infty(\tilde{v}_n) = E^\infty(v_n) \to E_{\lambda-\alpha}^\infty \quad \text{as } n \to \infty,$$
$$J(u_n) = \alpha \quad , \quad J^\infty(\tilde{v}_n) = J^\infty(v_n) = \lambda - \alpha \quad \text{for } n = 1, 2, \cdots.$$

where $\tilde{v}_n(x) = v_n(x + \xi_n)$, $\xi_n \in \mathbb{R}^N$. Choosing $|\xi_n|$ large enough, we may assume

$$\text{dist}(\text{supp} u_n, \ \text{supp} \tilde{v}_n) \to \infty \quad \text{as } n \to \infty.$$

Let $w_n = u_n + \tilde{v}_n$ for $n = 1, 2, \cdots$. Note that $\{c_n w_n\}$ is a minimizing sequence of the problem (M_λ) for some sequence $c_n \to 1$:

$$E(w_n) = E(u_n) + E^\infty(\tilde{v}_n) + o(1)$$
$$\to I_\alpha + I_{\lambda-\alpha}^\infty = I_\lambda \qquad\qquad \text{as } n \to \infty.$$
$$J(w_n) = J(u_n) + J^\infty(\tilde{v}_n) + (1) \to \lambda \qquad\qquad \text{as } n \to \infty.$$

Let $J(w_n) = a_n$, $E(w_n) = b_n$, $a_n \to \lambda$. Let $\alpha_n = \dfrac{\lambda}{a_n}$, then $\alpha_n \to 1$. Now

$$J(\alpha_n{}^{1/p} w_n) = \int b(x)(\alpha_n{}^{1/p} w_n)^p = \alpha_n a_n = \lambda,$$
$$E(\alpha_n{}^{1/p} w_n) = (\alpha_n)^{2/p}(|\nabla w_n|^2 + a(x)w_n{}^2) = \alpha_n{}^{2/p} b_n \to I_\lambda.$$

However, $\{w_n\}$, and in particular $\{c_n w_n\}$, is not relatively compact in H. Otherwise, suppose $w_n \to w$ strongly in H : this implies that $\{u_n\}$ and $\{\tilde{v}_n\}$ are Cauchy in H. In fact, for large n we have

$$\int |\nabla(w_n - w_m)|^2 + \int a\,|w_n - w_n|^2$$
$$= \int |\nabla(u_n - u_m)|^2 + \int |\nabla(\tilde{v}_n - \tilde{v}_m)|^2 + \int a\,|u_n - u_m|^2 + \int a\,|\tilde{v}_n - \tilde{v}_m|^2.$$

Therefore there are $u, v \in H$ such that

$$u_n \to u \quad \text{strongly in } H,$$
$$\tilde{v}_n \to v \quad \text{strongly in } H.$$

Clearly $v = 0$, and then $w = u$. Now, because $\|w_n - w\|_{H^1} \to 0$, we have

$$\int b\,|w_n - w|^2 \to 0, \int b\,|w_n|^{2^*} \le c \text{ for all } n.$$

Let $p = (1-t)2 + t \cdot 2^*$,

$$\int b(x) |w_n - w|^p = \int b^{1-t} |w_n - w|^{(1-t)2} b^t) |w_n - w|^{t2^*}$$

$$\leq \left(\int b |w_n - w|^2\right)^{1-t} \left(\int b |w_n - w|^{2^*}\right)^t$$

$$\leq c \left(\int b |w_n - w|^2\right)^{1-t}$$

$$\to 0 \quad \text{as } n \to \infty.$$

or

$$J(w) = \lim_{n \to \infty} J(w_n) = \lambda,$$

$$J(w) = J(u) = \lim_{n \to \infty} J(u_n) = \alpha,$$

a contradiction.

(b) The case $\alpha = 0 : I_\lambda = I_\lambda^\infty$: Let $\{v_n\}$ be a minimizing sequence of the problem (M_λ^∞) : that is,

$$E^\infty(v_n) = E^\infty(\tilde{v}_n) \to I_\lambda^\infty, \quad J^\infty(v_n) = J^\infty(\tilde{v}_n) = \lambda,$$

where $\tilde{v}_n(x) = v_n(x + \xi_n)$, $\xi_n \in \mathbb{R}^N$. If we let

$$w_n = \tilde{v}_n \quad \text{for } n = 1, 2, \cdots,$$

then

$$J(w_n) = J^\infty(\tilde{v}_n) = \lambda,$$

$$E(w_n) = E^\infty(\tilde{v}_n) \to I_\lambda^\infty = I_\lambda \quad \text{as } n \to \infty.$$

Therefore $[w_n]$ is a minimizing sequence of the problem (M_λ). However, $\{w_n\}$ is not relatively compact in H. Otherwise, suppose $w_n \to w$ strongly in H. Because $\tilde{v}_n \to 0$ in H, then $w = 0$. Now

$$J(0) = 0,$$

$$J(0) = J(w) = \lim_{n \to \infty} J(\tilde{v}_n) = \lambda,$$

a contradiction.

(II) If $I_\lambda < I_\alpha + I_{\lambda-\alpha}^\infty$ for each $\alpha \in [0, \lambda]$, then every minimizing sequence is relatively compact: let $\{u_n\}$ be a minimizing sequence of the problem (M_λ) :

$$J(u_n) = \lambda,$$

$$E(u_n) \to I_\lambda \quad \text{as } n \to \infty.$$

Apply the Concentration Lemma 237 with $\rho_n(x) = b(x)u_n^p$, $\int \rho_n = J(u_n) = \lambda$. There is a subsequence $\{\rho_n\}$ that satisfies one of the following three possibilities:

(i) there is $y_k \in \mathbb{R}^N$ such that for $\epsilon > 0$ there is some $R > 0$ such that

$$\int_{y_k + B_R} \rho_k \geq \lambda - \epsilon.$$

(ii) $\displaystyle\lim_{k \to \infty} \sup_{y \in \mathbb{R}^N} \int_{y + B_R} \rho_n = 0$ for every $R > 0$.

(iii) For some $\alpha \in (0, \lambda)$ such that for $\epsilon > 0$ there are $k_0 \geq 1$, $\{\rho_{k,1}\}$ and $\{\rho_{k,2}\} \subset L_+^1(\mathbb{R}^N)$ such that for $k \geq k_0$

$$\|\rho_k - (\rho_{k,1} + \rho_{k,2})\|_{L^1} < \epsilon,$$

$$\left| \int_{\mathbb{R}^N} \rho_{k,1} - \alpha \right| < \epsilon,$$

$$\left| \int_{\mathbb{R}^N} \rho_{k,2} - (\lambda - \alpha) \right| < \epsilon,$$

$$\text{dist}(\text{supp}\rho_{k,1}, \text{supp}\rho_{k,2}) \to \infty \quad \text{as } k \to \infty$$

(a) Case (ii) cannot occur. By Lemma 8.39, if

$$\lim_{\epsilon \to \infty} \sup_{y \in \mathbb{R}^N} \int_{y + B_R} |u_n|^p = 0 \quad \text{as } n \to \infty, \text{ for each } R > 0,$$

we then have

$$u_n \to 0 \text{ in } L^\alpha(\mathbb{R}^N) \text{ for } p < \alpha < \frac{2N}{N-2}.$$

However, $\{u_n\}$ is bounded in $L^2(\mathbb{R}^N)$. Let $2 < p < \alpha$, $p = (1-t)2 + t\alpha$, $t \in (0, 1)$. Then

$$\int_{\mathbb{R}^N} |u_n|^p = \int_{\mathbb{R}^N} |u_n|^{(1-t)2} |u_n|^{t\alpha}$$

$$\leq \left(\int_{\mathbb{R}^N} |u_n|^2 \right)^{1-t} \left(\int |u_n|^\alpha \right)^t$$

$$\leq c \left(\int |u_n|^\alpha \right)^t \to 0 \text{ as } n \to \infty.$$

We conclude that

$$\lambda = J(u_n) = \int_{\mathbb{R}^N} b(x) |u_n|^p \to 0,$$

a contradiction.

(b) If $I_\lambda < I_\alpha + I_{\lambda-\alpha}^\infty$ for $\alpha \in [0, \lambda)$, then case (iii) cannot occur: as before, let

$$Q_n(t) = \sup_{y \in \mathbb{R}^N} \int_{y+B_t} \rho_n \ ,$$

$$Q_n(t) \to Q(t) \text{ for each } t \geq 0 \ ,$$

$$\lim_{t \to \infty} Q(t) = \alpha \in (0, \lambda).$$

For $\epsilon > 0$, choose $R_0 > 0$ such that $Q(R_0) \geq \alpha - \epsilon$. We have

$$Q_n(R_0) = \int_{y_n+B_{R_0}} |u_n|^p \geq \alpha - 2\epsilon \quad \text{for some } y_n \in \mathbb{R}^N.$$

In addition, as $n \to \infty$, there is $R_n \to \infty$, such that

$$Q_n(R_n) \leq \alpha + \epsilon.$$

Let $\xi, \varphi \in C_b^\infty$ satisfying $0 \leq \xi, \varphi \leq 1$

$$\xi = 0 \text{ on } B_2^c, \qquad\qquad \xi = 1 \text{ on } B_1,$$
$$\varphi \equiv 0 \text{ on } B_1, \qquad\qquad \varphi \equiv 1 \text{ on } B_2^c.$$

We denote

$$\xi_n(x) = \xi\left(\frac{x - y_n}{R_1}\right), \quad \varphi_n(x) = \varphi\left(\frac{x - y_n}{R_n}\right),$$

where R_1 is determined as shown below. For R_1, and n large enough we have

$$\left| \int_{\mathbb{R}^N} \xi_n^2 |\nabla u_n|^2 - \int_{\mathbb{R}^N} |\nabla(\xi_n u_n)|^2 \right| \leq \epsilon,$$

$$\left| \int_{\mathbb{R}^N} \varphi_n^2 |\nabla u_n|^2 - \int_{\mathbb{R}^N} |\nabla(\varphi_n u_n)|^2 \right| \leq \epsilon,$$

where we note that $\nabla\xi_n(x) = \frac{1}{R_1}\nabla\xi_n\left(\frac{x - y_n}{R_1}\right)$ and $\nabla\varphi_n(x) =$

$\frac{1}{R_n}\nabla\varphi_n\left(\frac{x - y_n}{R_n}\right)$. If we set $u_n^1 = \xi_n u_n$, $u_n^2 = \varphi_n u_n$, for n large enough,

$$\left\| u_n - (u_n^1 + u_n^2) \right\|_{L^p} \leq 3\epsilon,$$

$$\left| \lambda - \int_{\mathbb{R}^N} (b(x)|u_n^1|^p + b(x)|u_n^2|^p) \right| \leq c\epsilon,$$

$$\int_{\mathbb{R}^N} |\nabla u_n|^2 \geq \int_{\mathbb{R}^N} |\nabla u_n^1|^2 + \int_{\mathbb{R}^N} |u_n^2|^2 - 2\epsilon,$$

$$\int_{\mathbb{R}^N} a(x)|\nabla u_n|^2 \geq \int_{\mathbb{R}^N} a(x)|\nabla u_n^1|^2 + \int_{\mathbb{R}^N} a(x)|u_n^2|^2,$$

or

$$E(u_n) \geq E(u_n^1) + E(u_n^2) - 2\epsilon,$$

$$\|u_n - (u_n^1 + u_n^2)\|_{L^\alpha} \leq \delta_\alpha(\epsilon), \quad \text{where } \delta_\alpha(\epsilon) \to 0 \text{ as } \epsilon \to 0, p < \alpha < \frac{2N}{N-2}.$$

We may assume that

$$\int_{\mathbb{R}^N} b(x) |u_n^1|^P \to \lambda_1, \quad \int_{\mathbb{R}^N} b(x) |u_n^2|^P \to \lambda_2 \quad \text{as } n \to \infty,$$

and thus, $|\lambda - (\lambda_1 + \lambda_2)| \leq c\epsilon$. Suppose $\lambda_1 \leq 0$, we may assume that there is $v > 0$ such that

$$\int_{\mathbb{R}^N} |u_n^1|^2 + a(x) |u_n^1|^2 \geq v \|u_n^1\|_{H^1} \quad \text{for } n = 1, 2, \cdots.$$

From

$$\int_{y_n + B_{R_0}} |u_n|^P \geq \alpha - 2\epsilon,$$

we have

$$\|u_n^1\|_{H^1} \geq c \|u_n\|_{L^p} \geq c > 0,$$

so

$$\int_{\mathbb{R}^N} |u_n^1|^2 + a(x) |u_n^1|^2 \geq \delta > 0.$$

Letting $n \to \infty$ we have

$$I_\lambda \geq \liminf_{n \to \infty} \int (|\nabla u_n^1|^2 + a |u_n^1|^2) + \liminf_{n \to \infty} \int (|\nabla u_n^2|^2 + a |u_n^2|^2)$$

$$\geq \delta + I_{\lambda_2}.$$

However, from $|\lambda - (\lambda_1 + \lambda_2)| \leq c\epsilon$, we obtain $\lambda - \lambda_1 - \lambda_2 \leq c\epsilon$ or $\lambda \leq \lambda_2 + c\epsilon$ for all $\epsilon > 0$. Thus, $I_\lambda \leq I_{\lambda_2 + c\epsilon}$. In fact, if $a \leq b$ then $I_a \leq I_b$, for take $u \in H^1(\mathbb{R}^N)$ such that $J(u) = b$, $E(u) \leq I_b + \epsilon$, then we can find $\theta \in (0, 1]$ such that $J(\theta u) = a$. Therefore

$$I_a \leq E(\theta u) = \theta^2 E(u) < I_b + \epsilon \quad \text{for each } \epsilon > 0.$$

Thus $I_a \leq I_b$. Moreover, I_μ is continuous on $\mu \in (o, \lambda]$. Fix $\alpha \in (0, \lambda]$. Then

$$I_\alpha = \inf\{E(u) \mid u \in H^1(\mathbb{R}^N), J(u) = \alpha\}.$$

$$I_{\theta\alpha} = \inf\{E(\theta^{1/p} u) \mid u \in H^1(\mathbb{R}^N), J(u) = \alpha\}$$

$$= \theta^{2/p} \inf\{E(u) \mid u \in H^1(\mathbb{R}^N), J(u) = \alpha\}$$

$$= \theta^{2/p} I_\alpha.$$

Therefore $\lim_{\theta\to\infty} I_{\theta\alpha} = I_\alpha$. Letting $\epsilon \to 0$ in the inequality $I_\lambda \le I_{\lambda_2+c\epsilon}$, we have $I_\lambda \le I_{\lambda_2}$. Hence, we conclude that

$$I_\lambda \ge \delta + I_{\lambda_2} \ge \delta + I_\lambda,$$

a contradiction. We have $\lambda_1 > 0$. Similarly, we have $\lambda_2 > 0$. Suppose $|y_n| \to \infty$. Because $\operatorname{supp} u_n^1 \subset y_n + B_{R_1}$, we have

$$\left| \int_{\mathbb{R}^N} (|\nabla u_n^1|^2 + a(x)|u_n^1|^2) - \int_{\mathbb{R}^N} (|\nabla u_n^1|^2 + \bar{a}|u_n^1|^2) \right| \to 0 \text{ as } n \to \infty, \text{ and}$$

$$\int_{\mathbb{R}^N} (b(x)|u_n^1|^p + \bar{b}|u_n^1|^p) \to 0 \quad \text{as } n \to \infty,$$

or

$$J(u_n^1) = \int_{\mathbb{R}^N} \bar{b}|u_n^1|^p \to \lambda_1.$$

Now, for $\epsilon > 0$ there is some $N > 0$ such that $n \ge N$ implies

$$\lambda_1 - \epsilon < J(u_n^1) = a_n < \lambda_1 + \epsilon,$$

and we have

$$I_{\lambda_1-\epsilon} \le I_{a_n} \le E(u_n^1) \le I_{\lambda_1+\epsilon}.$$

Letting $n \to \infty$,

$$I_{\lambda_1} = I_{\lambda_1}^\infty.$$

Suppose $\{|y_n|\}$ is bounded. Because

$$\operatorname{supp} u_n^2 \subset \mathbb{R}^N \backslash (y_n + B_{R_n}), \text{ and } R_n \to \infty,$$

we have

$$\left| \int_{\mathbb{R}^N} (|\nabla u_n^2|^2 + a(x)|u_n^2|^2) - \int_{\mathbb{R}^N} (|\nabla u_n^2|^2 + \bar{a}|u_n^2|^2) \right| \to 0 \quad \text{as } n \to \infty,$$

$$\int_{\mathbb{R}^N} \bar{b}|u_n^2|^p \to \lambda_2.$$

Similarly we have

$$I_{\lambda_2} = I_{\lambda_2}^\infty.$$

In both cases, we have

$$I_\lambda \ge I_{\lambda_1}^\infty + I_{\lambda_2} \text{ or } I_\lambda \ge I_{\lambda_2}^\infty + I_{\lambda_1},$$
$$|\lambda - (\lambda_1 + \lambda_2)| \le c\epsilon.$$

Letting $\epsilon \to 0$ we have

$$I_\lambda \ge I_{\lambda_1}^\infty + I_{\lambda-\lambda_1},$$

a contradiction. Therefore we deduce that there is $y_n \in \mathbb{R}^N$ such that for $\epsilon > 0$ there is some $R > 0$ such that

$$\int_{B_R} b(x) |u_n(x + y_n)|^p \geq \lambda - \epsilon.$$

(c) If $I_\lambda < I_\lambda^\infty$, then $\{y_n\}$ is bounded: suppose on the contrary $\{y_n\}$ is unbounded, say $|y_n| \to \infty$. Note that for $\epsilon > 0$, there is $R_\epsilon > 0$ such that for $\tilde{u}_n(x) = u_n(x + y_n)$,

$$\int_{B_{R_\epsilon}^c} b(x) |\tilde{u}_n|^p < \epsilon \quad \text{for each } n.$$

Then there is $N > 0$ such that $n \geq N$ implies, for fixed $R > 0$,

$$y_n + B_R \subset B_{R_\epsilon}^c.$$

This asserts that

$$\int_{B_R} |\tilde{u}_n|^p \to 0,$$

so

$$\left| \int_{\mathbb{R}^N} b(x) |\tilde{u}_n|^p - \int_{\mathbb{R}^N} \bar{b} |\tilde{u}_n|^p \right|$$

$$\leq \|b(x) - \bar{b}\|_\infty \cdot \int_{B_R} |\tilde{u}_n|^p + \left(\int_{B_R^c} |\tilde{u}_n|^p \right) \cdot \|b(x) - \bar{b}\|_{L^\infty(B_R^c)}.$$

Letting $n \to \infty$ and then $R \to \infty$, we obtain

$$\lim_{n \to \infty} \left| \int_{\mathbb{R}^N} b(x) |\tilde{u}_n|^p - \int_{\mathbb{R}^N} \bar{b} |\tilde{u}_n|^p \right| = 0.$$

This asserts that $I_\lambda \geq I_\lambda^\infty$, a contradiction.

(d) Conclusion: now $\{y_n\}$ is bounded. Assume $y_n \to y_0$ and that there is $R_1 > 0$ such that $y_n \in B(0, R_1)$ for each $n \in N$. For given $\epsilon > 0$, there is $R_2 > 0$ such that

$$\int_{B(0,R_2)} b(x) |u_n(x + y_n)|^p \, dx \geq \lambda - \epsilon.$$

Take $R = R_1 + R_2$, then

$$\int_{B(0,R)} b(x) |u_n(x)|^p \, dx \geq \int_{B(0,R_2)} b(x) |u_n(x + y_n)|^p \, dx \geq \lambda - \epsilon.$$

There is $u \in L^p(\mathbb{R}^N)$ such that

$$u_n \to u \text{ weakly in } H^1(\mathbb{R}^N) \text{ a.e. on } \mathbb{R}^N, \text{ and strongly in } L_{loc}^p(\mathbb{R}^N),$$

so

$$\int_{\mathbb{R}^N} b(x)\,|u|^p = \lambda.$$

However,

$$E(u) \leq \varliminf_{n \to \infty} E(u_n) = I_\lambda.$$

Thus,

$$E(u) = \min\{E(w) \mid w \in H^1(\mathbb{R}^N),\ J(w) = \lambda\}.$$

We have $E(u_n) \to E(u)$. Let $v_m = u_m - u$. Then $E(u_n) = E(u) + E(v_n) + o(1)$, so $\|u_n - u\|_{H^1} \leq cE(v_n) = E(u_n) - E(u) + o(1) = o(1)$, and u is a minimizer of (M_λ). ∎

In general, let H be a function space on $\mathbb{R}^N$, and let J, E be functionals defined on H of the type:

$$E(u) = \int_{\mathbb{R}^N} e(x,\,Au(x))dx,$$

$$J(u) = \int_{\mathbb{R}^N} j(x,\,Bu(x))dx,$$

where $e(x,\,p)$, $j(x,\,q)$ are real-valued functions defined on $\mathbb{R}^N \times \mathbb{R}^m$, $\mathbb{R}^N \times \mathbb{R}^n$, respectively, and j is nonnegative. A, B are operators (possibly nonlinear) from H into E, F (function spaces defined on $\mathbb{R}^N$ with values in $\mathbb{R}^m$, $\mathbb{R}^n$), which commute with translations of $\mathbb{R}^N$. Assume $J(0) = 0$. We consider the following minimization problems:

$$I_1 = \inf\{E(u) \mid u \in H, J(u) = 1\}. \qquad\qquad (M)$$
$$I_\lambda = \inf\{E(u) \mid u \in H, J(u) = \lambda\},\ \lambda > 0. \qquad\qquad (M_\lambda).$$

We assume

$$j(x,q) \to j^\infty(q),\ e(x,p) \to e^\infty(p) \quad \text{as } |x| \to \infty, \text{ for all } p, q \in \mathbb{R}^m,\ \mathbb{R}^n.$$

Consider the problem at infinity:

$$I_\lambda^\infty = \inf\{E^\infty(n) \mid u \in H,\ J^\infty(n)\},\ \lambda > 0, \qquad\qquad (M_\lambda^\infty),$$

where

$$E^\infty(u) = \int_{\mathbb{R}^N} e^\infty(Au(x))dx,$$

$$J^\infty(u) = \int_{\mathbb{R}^N} j^\infty(Bu(x))dx.$$

We assume that, for $\lambda > 0$,

(H1) $K_\lambda = \{u \in H \mid J(u) = \lambda\} \neq \emptyset$;

(H2) $I_\lambda > -\infty$;

(H3) The minimizing sequences for (M_λ), (M_λ^∞) are bounded in H;

(H4) E^∞, J^∞ are translation invariant;

(H5) If u, v have compact support, $v_n(x) = v(x + ne)$ for some $e \neq 0$, then

$$J(u + v_n) = J(u) + J(v_n) \quad \text{for large } n \text{ and}$$
$$E(u + v_n) = \{E(u) + E^\infty(v_n)\} + 0(1) \quad \text{as } n \to \infty;$$

(H6) The vanishing property of Lemma 8.37 (ii) implies that $I_\lambda \geq 0$;

(H7) $\lim_{\sigma \to 0} E^\infty(\sigma^{-N/2} u(\frac{\cdot}{\sigma})) < 0$.

Theorem 8.42 (*Concentration-Compactness Principle*)

(i) *If e, j depend on x, for each fixed $\lambda > 0$, all minimizing sequences of the problem*

$$I_\lambda = \inf\{E(u) \mid u \in H, J(u) = \lambda\} \qquad (M_\lambda)$$

are relatively compact in H if and only if

$$I_\lambda < I_\alpha + I_{\lambda-\alpha}^\infty \text{ for each } \alpha \in [0, \lambda);$$

(ii) *If e, j do not depend on x, for each fixed $\lambda > 0$, all minimizing sequences of the problem*

$$I_\lambda = \inf\{E(u) \mid u \in H, J(u) = \lambda\} \qquad (M_\lambda)$$

are relatively compact up to a translation in H if and only if

$$I_\lambda < I_\alpha + I_{\lambda-\alpha} \text{for each } \alpha \in (0, \lambda)$$

(Recall that in this case, $I_{\lambda-\alpha} = I_{\lambda-\alpha}^\infty$ for each $\alpha(\in (0, \lambda))$.)

Lemma 8.43 *Let μ, ν be two bounded nonnegative measures on $\mathbb{R}^N$ satisfying for some constant $c \geq 0$:*

$$\left(\int_{\mathbb{R}^N} |\varphi|^q \, d\nu\right)^{1/q} \leq c \left(\int_{\mathbb{R}^N} |\varphi|^p \, d\mu\right)^{1/p}, \text{for each } \varphi \in D(\mathbb{R}^N), \qquad (8.17)$$

where $1 \leq p < q \leq +\infty$. Then there exist an at most countable set J, families $(x_j)_{j \in J}$ of distinct points in $\mathbb{R}^N$, and $(\nu_j)_{j \in J}$ in $(0, \infty)$ such that:

$$\nu = \sum_{j \in J} \nu_j \delta_{x_j}, \quad \mu \geq c^{-P} \sum_{j \in J} \nu_j^{p/q} \delta_{x_j}.$$

In particular: $\sum_{j \in J} \nu_j^{p/q} < \infty$. If in addition $\nu(\mathbb{R}^N)^{1/q} \geq c\mu(\mathbb{R}^N)^{1/p}$ reduces to a single point and $\nu = \gamma \delta_{x_0} = \gamma^{-p/q} c^p \mu$, for some $x_0 \in \mathbb{R}^N$ and for some $\gamma \geq 0$.

Proof. We first remark that (8.17) holds by density for all φ bounded measurable. Therefore we see that in particular ν is absolutely continuous with respect to μ i.e., $\nu = f\mu$ where $f \in L_+^1(\mu)$. Because

$$\nu(A) \leq c\mu(A)^{q/p}, \quad \text{for every Borel } A \text{ in } \mathbb{R}^N.$$

We have in fact $f \in L_+^\infty(\mu)$. Next, if $\mu = g\nu + \sigma$ where $g \in L_+^1(\nu)$, σ is a bounded nonnegative measure such that if $K = \text{supp}\,\sigma$, $\nu(K) = 0$. Considering $\widetilde{\mu} = 1_K\mu$ and taking φ in (8.17) of the form $1_K\varphi$ where φ is bounded measurable, we see that without loss of generality we may assume that $\sigma = 0$. We next denote by $\nu_k = g^\alpha \chi_{\{g \leq k\}}\nu$, where $\alpha = q/(q-p)$. We will prove that ν_k is given by a finite number of Dirac masses; this will prove that $\nu\chi_{\{g \leq k\}}$ is a finite number of Dirac masses for all $k < \infty$, and letting $k \to \infty$, the claim on ν will be proved (because $\nu(\{g = +\infty\}) = 0$). To prove our claim on ν_k, we take φ in (8.17) of the form:

$$\varphi = g^{1(q-p)}\chi_{\{g \leq k\}}\psi,$$

where ψ is an arbitrary bounded measurable function. We thus obtain for all φ :

$$\left(\int_{\mathbb{R}^N} |\varphi|^q \, d\nu_k\right)^{1/q} \leq c \left(\int_{\mathbb{R}^N} |\varphi|^p \, d\mu_k\right)^{1/p},$$

(indeed, $g^{p/(q-p)}\chi_{\{g \leq k\}}\mu = g^{q/(q-p)}\chi_{\{g \leq k\}}\nu$). This reversed Hölder inequality now yields our claim on ν_k : a short proof of this standard statement is the following. For any Borel set A, the above inequality gives:

$$\nu_k(A)^{1/q} \leq c\nu_k(A)^{1/p}.$$

Therefore either $\nu_k(A) = 0$, or $\nu_k(A) \geq \delta = C^{-p/(q-p)} > 0$. Because for each $x \in \mathbb{R}^N$, $\nu_k(\{x\}) = \lim_{\epsilon \downarrow 0} \nu_k(B(x, \epsilon))$, we have for all $x \in \mathbb{R}^N$:

$$\text{either } \nu_k(\{x\}) \geq \delta, \text{ or there is } \epsilon > 0, \, \nu_k(B(x, \epsilon)) = 0.$$

Thus, there are a finite number of distinct points x_j in $\mathbb{R}^N$ such that:

$$\nu_k(\{x_j\}) \geq \delta \qquad \text{for each} \quad i \leq j \leq m,$$
$$\nu_k(B(x, \epsilon)) = 0 \qquad \text{for some} \quad \epsilon = \epsilon(x) > 0 \text{ for each } x \notin \{x_j \mid 1 \leq j \leq m\}.$$

Let K be any compact set in $0 = \{x \mid x \neq x_j \text{ for all } 1 \leq j \leq m\}$; we have by a finite covering of K by balls $B(x, \epsilon(x)) : \nu_k(K) = 0$, therefore $\nu_k(0) = 0$; and our claim is proved. At this point, we have proved the representation of ν and by (8.17) we have

$$\mu(\{x_j\}) \geq c^{-p}\nu(\{x_j\})^{p/q}.$$

Finally, if $\nu(\mathbb{R}^N)^{1/q} \geq c\mu(\mathbb{R}^N)^{1/p}$, taking $\varphi \equiv 1$ in $(*)$ we see that $\nu(\mathbb{R}^N)^{1/q} = C\mu(\mathbb{R}^N)^{1/p}$; and using the Hölder inequality we find for all $\varphi \in D(\mathbb{R}^N)$:

$$\left(\int_{\mathbb{R}^N} |\varphi|^q \, d\nu \right)^{1/q} \leq c\mu(R^N)^\theta \left(\int_{\mathbb{R}^N} |\varphi|^p \, d\mu \right)^{1/p},$$

where $\theta = (q - p)/(pq)$. Observing that

$$\nu(\mathbb{R}^N) = c^q \mu(\mathbb{R}^N)^{q/p} = \{c\mu(\mathbb{R}^N)^\theta\}^q \mu(\mathbb{R}^N),$$

we deduce from the above inequality: $\nu = \{c\mu(\mathbb{R}^N)^\theta\}^q \mu$. Therefore we have for all $\varphi \in D(\mathbb{R}^N)$:

$$\left(\int_{\mathbb{R}^N} |\varphi|^q \, d\nu \right)^{1/q} \leq \nu(\mathbb{R}^N)^{-\theta} \left(\int_{\mathbb{R}^N} |\varphi|^p \, d\nu \right)^{1/p}.$$

The above proof has to this point shown that: $\nu = \sum_{i=1}^m \nu_i \delta_{x_i}$, where $m \geq 1$, $(x_i)_i$ are m distinct points in $\mathbb{R}^N$ and $\nu_i > 0$. We choose $\varphi \in D(\mathbb{R}^N)$ such that $\varphi(x_i) = \alpha_i > 0$; thus we find for all $\alpha_i > 0$:

$$\left(\sum_{i=1}^m \alpha_i^q \nu_i \right)^{1/q} \left(\sum_{i=1}^m \nu_i \right)^{(q-p)/pq} \leq \left(\sum_{i=1}^m \alpha_i^p \nu_i \right)^{1/p}.$$

This is possible if and only if $m = 1$.   ■

Remark 8.44 *Lemma 8.43 is of course valid in an arbitrary measure space and the various conclusions hold, provided one replaces points in $\mathbb{R}^N$ by atoms.*

Lemma 8.45 (*Second Concentration-Compactness Lemma*) *Let $(u_n)_n$ be a bounded sequence in $D^{m,p}$ converging weakly to some u and such that $|D^m u_n|^p$ converges weakly to μ and $|u_n|^q$ converges tightly to ν, where μ, ν are bounded nonnegative measures on $\mathbb{R}^N$. Then we have:*

(i) *There exist some at most countable set J and two families $(x_j)_{j \in J}$ of distinct points in $\mathbb{R}^N$, $(\nu_j)_{j \in J}$ in $(0, \infty)$ such that:*

$$\nu = |u|^q + \sum_{j \in J} \nu_j \delta_{x_j}. \tag{8.18}$$

(ii) *In addition, we have*

$$\mu \geq |D^m u|^p + \sum_{j \in J} \mu_j \delta_{x_j}, \tag{8.19}$$

for some $\mu_j > 0$ satisfying:

$$\nu_j^{p/q} \le \mu_j/I, \quad \text{for all } j, \tag{8.20}$$

hence, $\sum\limits_{j \in J} \nu_j^{p/q} < \infty.$

(iii) *If $v \in D^{m,p}(\mathbb{R}^N)$, and $|D^m(u_n + v)|^p$ converges weakly to some measure $\widetilde{u}$, then $\widetilde{u} - \mu \in L^1(\mathbb{R}^N)$; and therefore:*

$$\widetilde{u} \ge |D^m(u + v)|^p + \sum_{j \in J} \mu_j \delta_{x_j}.$$

(iv) *If $u \equiv 0$ and $\left(\int d\mu \right) \le \left(\int d\mu \right)^{p/q}$, then J is a singleton and*

$$\nu = \gamma \delta_{x_0} = \mu(I\gamma^{p/q})^{-1} \quad \text{for some } \gamma > 0, \ x_0 \in \mathbb{R}^N.$$

Proof. We first treat the case when $u \equiv 0$. The goal is to obtain some reversed Hölder inequality between ν and μ, which will give the various relations contained in Lemma 8.43. Let $\varphi \in D(\mathbb{R}^N)$. Then by the Sobolev inequalities we have

$$\left(\int_{\mathbb{R}^N} |\varphi|^q \, |u_n|^q \, dx \right)^{1/q} I^{1/p} \le \left(\int_{\mathbb{R}^N} |D^m(\varphi \, u_n)|^p \, dx \right)^{1/p}. \tag{8.21}$$

Then the left-hand-side member of (8.21) goes to $\left(\int_{\mathbb{R}^N} |\varphi|^q \, d\nu \right)^{1/q} I^{1/p}$ as n goes to ∞. Now the right-hand-side member is estimated as follows:

$$\left| \left(\int_{\mathbb{R}^N} |D^m(\varphi \, u_n)|^p \, dx \right)^{1/p} - \left(\int_{\mathbb{R}^N} |\varphi|^p \, |D^m u_n|^p \, dx \right)^{1/p} \right|$$
$$\le c \sum_{j=0}^{m-1} \left(\int_{\mathbb{R}^N} |D^{m-j}\varphi|^p \, |D^j u_n|^p \, dx \right)^{1/p}.$$

Using the fact that φ has compact support and the Rellich Theorem, we see that this bound goes to 0 as n goes to ∞. Therefore, passing to the limit in (8.21), we obtain for all $\varphi \in D(\mathbb{R}^N)$:

$$\left(\int_{\mathbb{R}^N} |\varphi|^q \, d\nu \right)^{1/q} \le I^{-\frac{1}{p}} \left(\int_{\mathbb{R}^N} |\varphi|^p \, d\mu \right)^{1/q}. \tag{8.22}$$

The lemma is also proved when $u \equiv 0$, by the application of Lemma 8.43. We now consider the general case of a weak limit u not necessarily 0. Of course (8.21) still holds, and if we denote by $v_n = u_n - u$, the Brezis-Lieb Lemma yields for all $\varphi \in D(\mathbb{R}^N)$:

$$\int_{\mathbb{R}^N} |\varphi|^q |u_n|^q \, dx - \int_{\mathbb{R}^N} |\varphi|^q |v_n|^q \, dx \to \int_{\mathbb{R}^N} |\varphi|^q |u|^q \, dx.$$

Clearly, however, v_n is bounded in $D^{m,p}$ and $|v_n|^q$ is tight; therefore, applying what we proved above we obtain the representation (8.18) of ν. Next, passing to the limit in (8.21) and using the Rellich Theorem as before, we find for all $\varphi \in D(\mathbb{R}^N)$:

$$\left(\int_{\mathbb{R}^N} |\varphi|^q \, d\nu \right)^{1/q} I^{1/p} \leq \|\varphi\|_{L^p(\mu)} + C \sum_{i=0}^{m-1} \left(\int_{\mathbb{R}^N} |D^{m-i}\varphi|^p |D^i u|^p \, dx \right)^{1/p}.$$

If φ satisfies $0 \leq \varphi \leq 1$, $\varphi(0) = 1$, Supp $\varphi = B(0, 1)$, $\varphi \in D(\mathbb{R}^N)$; we apply the above inequality with $\varphi(\frac{x-x_j}{\epsilon})$ for $\epsilon > 0$ and where j is fixed in J. We obtain

$$v_j^{1/q} I^{1/p} \leq \mu(B(x_j, \epsilon))^{1/p}$$

$$+ C \sum_{j=1}^{m-1} \left(\int_{B(x_j,\epsilon)} \epsilon^{-p(m-i)} \left| D^{m-i}\varphi \left(\frac{x-x_j}{\epsilon} \right) \right|^p |D^i u|^p \, dx \right)^{1/p}.$$

Now we may estimate each term of the sum by Hölder inequalities recalling that $D^i u \in L^{q_i}(\mathbb{R}^N)$ (by Sobolev inequalities):

$$\epsilon^{-p(m-i)} \int_{B(x_j,\epsilon)} \left| D^{m-i}\varphi \left(\frac{x-x_j}{\epsilon} \right) \right|^p |D^i u|^p \, dx$$

$$\leq \left(\int_{B(x_j,\epsilon)} |D^i u|^{q_i} \, dx \right)^{p/q_i} \epsilon^{-p(m-i)} \left(\int_{\mathbb{R}^N} \left| D^{m-i}\varphi \left(\frac{x}{\epsilon} \right) \right|^{p_i} \, dx \right)^{(q_i-p)/q_i},$$

where $p_i = q_i p(q_i - p)^{-1}$, $(q_i - p)/q_i = (m - i)p/N$. Hence, we have:

$$v_j^{1/q} I^{1/p} \leq \mu(B(x_j, \epsilon))^{1/p} + C \sum_{i=1}^{m-1} \left(\int_{B(x_j,\epsilon)} |D^i u|^{q_i} \, dx \right)^{p/q_i}.$$

This implies that $\mu(\{x_j\}) > 0$, and

$$\mu \geq v_j^{p/q} I \, \delta_{x_j}, \quad \text{for each } j \in J$$

and thus, $\mu \geq \sum_{j \in J} I \, v_j^{p/q} \delta_{x_j} = \mu_1$. Because by weak convergence we also have $\mu \geq |D^m u|^p$ and because $|D^m u|^p$ and μ_1 are orthogonal, (8.19) and (8.20) are proved.

Finally, to prove (iii), we simply observe that for all $\varphi \in C_b(\mathbb{R}^N)$, $\varphi \geq 0$:

$$\left| \left(\int_{\mathbb{R}^N} \varphi \, |D^m(u_n + v)|^p \, dx \right)^{1/p} - \left(\int_{\mathbb{R}^N} \varphi \, |D^m u_n|^p \, dx \right)^{1/p} \right|$$
$$\leq \left(\int_{\mathbb{R}^N} \varphi \, |D^m v|^p \, dx \right)^{1/p}.$$

Passing to the limit in n, we find:

$$\left| \left(\int_{\mathbb{R}^N} \varphi \, d\widetilde{u} \right)^{1/p} - \left(\int_{\mathbb{R}^N} \varphi \, d\mu \right)^{1/p} \right| \leq \left(\int_{\mathbb{R}^N} \varphi \, h \, dx \right)^{1/p}$$

where $h \in L^1_+(\mathbb{R}^N)$. This shows that the singular parts of $\widetilde{u}$ and μ are the same; and we conclude our analysis. ∎

8.5   Ekeland Variational Principle

Theorem 8.46 (*Ekeland Variational Principle*) *Let X be a complete metric space, $U : X \to \mathbb{R} \cup \{+\infty\}$ a proper, nonnegative, and lower semicontinuous function. Let $\epsilon > 0$ and $x_\epsilon \in X$ be given such that*

$$U(x_\epsilon) \leq \epsilon + \inf U.$$

Then for any $k > 0$, there is some point $y_\epsilon \in X$ such that

$$U(y_\epsilon) \leq U(x_\epsilon),$$
$$d(x_\epsilon, y_\epsilon) \leq \frac{1}{k},$$
$$U(x) > U(y_\epsilon) - k\epsilon d(x, y_\epsilon) \quad \text{for all } x \neq y_\epsilon.$$

Proof.   We associate with U the dynamic system defined by

$$G(x) = \{y \mid U(y) + d(x, y) \leq U(x)\}.$$

Because U is lower semicontinuous, $G(x)$ is a closed subset of X. Let the forward cone $C(x)$ be the set of all points in X that can be reached in a finite number of steps if we start at x. It has the obvious properties

$$x \in \mathcal{C}(x), \tag{8.23}$$
$$\text{if } y \in \mathcal{C}(x), \text{ and } z \in \mathcal{C}(y), \text{ then } z \in \mathcal{C}(x). \tag{8.24}$$

Indeed, if $y \in C(x)$, there is a motion $(x_n)_{n=0}^\infty : x_{n+1} \in G(x_n)$ for all n such that $x_0 = x$, $x_k = y$. Because $z \in C(y)$, there is a sequence $(y_n)_{n=0}^\infty : y_{n+1} \in G(y_n)$ for all

n such that $y_0 = y$, $y_\ell = z$. Set

$$z_0 = x_0 = x, \cdots, \; z_k = x_k = y,$$
$$z_{k+1} = y_1, \quad z_{k+\ell} = z, \cdots.$$

We obtain a sequence $(z_n)_{n=0}^\infty$ such that

$$z_0 = x_0 = x, \; z_{k+\ell} = z, \; z_{n+1} \in G(z_n).$$

Therefore $(z_n)_{n=0}^\infty$ is a motion, and $z \in C(x)$.

(i) $G(x) = C(x)$ for all x.
   If $U(x) = +\infty$, then $G(x) = C(x) = X$.
   If $U(x) < \infty$, let $y \in G(x)$, and $z \in G(y)$, and we have

$$U(y) + d(x,y) \leq U(x),$$
$$U(z) + d(y,z) \leq U(y).$$

Because $U(x)$ is finite, so are $U(y)$ and $U(z)$. Adding these, we obtain

$$U(z) + d(x,z) \leq U(x),$$

so $z \in G(x)$. Therefore $G(y) \subset G(x)$ whenever $y \in G(x)$. It follows by induction that $C(x) \subset G(x)$. Because the converse inclusion holds by definition, the two sets coincide.

(ii) If $U(x) < \infty$, there is a motion $(x_n)_{n=0}^\infty$ and a point $\bar{x} \in X$ such that

$$x_0 = x \text{ and } \{\bar{x}\} = \bigcap_{n=0}^\infty C(x_n).$$

Indeed, for any y with $U(y) < \infty$, define

$$V(y) = \inf\{U(z) \mid z \in C(y)\}.$$

Because U is positive and $C(y)$ is nonempty, $V(y)$ is some real number. For $y \in C(x)$, there is a motion $(x_n)_{n=0}^\infty : x_0 = x$, $x_k = y$.

$$U(y) + d(y, x_{k-1}) \leq U(x_{k-1}),$$
$$U(x_{k-1}) + d(x_{k-1}, x_{k-2}) \leq U(x_{k-2}),$$
$$\vdots$$
$$U(x_1) + d(x_1, x) \leq U(x).$$

Summing, we obtain

$$U(y) + d(y, x) \leq U(x),$$

or

$$d(y, x) \leq U(x) - U(y)$$
$$\leq U(x) - V(x).$$

Thus,

$$\text{diam}\, \mathcal{C}(x) \leq 2(U(x) - V(x)).$$

Consider a sequence (not a motion) $(y_n)_{n=0}^{\infty}$ defined by

$$y_{n+1} \in \mathcal{C}(y_n), \quad U(y_{n+1}) \leq V(y_n) + 2^{-n}.$$

By (8.23), $C(y_{n+1}) \subset C(y_n)$, and consequently,

$$V(y_{n+1}) \geq V(y_n).$$

Now,

$$V(y_{n+1}) \leq U(y_{n+1}) \leq V(y_n) + 2^{-n} \leq V(y_{n+1}) + 2^{-n}.$$

Hence,

$$U(y_{n+1}) - V(y_{n+1}) \leq 2^{-n}, \quad \text{or diam}\mathcal{C}(y_{n+1}) \leq 2^{1-n}.$$

It follows that $C(y_n)$ is a nested (decreasing) sequence of closed sets in the complete metric space. Their diameters go to zero, so there is $\bar{x} \in X$ such that

$$\{\bar{x}\} = \bigcap_{n=0}^{\infty} \mathcal{C}(y_n).$$

Because $y_{n+1} \in C(y_n)$, there is a motion $(x_k^{(n)})_{k=0}^{\infty}$ such that $x_0^{(n)} = y_n$, $x_{k_n}^{(n)} = y_{n+1}$. Define a motion $(x_k)_{k=0}^{\infty}$ by

$$x_0 = x_0^{(0)} = y_0 = x, \quad \cdots, \quad x_{k_0} = x_{k_0}^{(0)} = y_1 = x_0^{(1)},$$
$$x_{k_0+1} = x_1^{(1)}, \cdots, \quad x_{k_0+k_1} = x_{k_1}^{(1)} = y_2,$$
$$\cdots$$

Clearly for any k, we can find n and m such that

$$x_k \in \mathcal{C}(y_n) \quad \text{and} \quad y_m \in \mathcal{C}(x_k).$$

Hence,

$$\mathcal{C}(y_m) \subset \mathcal{C}(x_k) \subset \mathcal{C}(y_n).$$

Therefore,

$$\{\bar{x}\} = \bigcap_{n=0}^{\infty} \mathcal{C}(y_n) = \bigcap_{n=0}^{\infty} \mathcal{C}(x_n).$$

(iii) If $U(x) < \infty$, then there is $\bar{x} \in C(x)$ such that $G(\bar{x}) = \{\bar{x}\}$. By (ii) there is a motion $\{x_n\}_{n=0}^{\infty}$ such that

$$x_0 = x, \text{ and } \{\bar{x}\} = \bigcap_{n=0}^{\infty} C(x_n).$$

Because $\bar{x} \in C(x_n)$, we have $C(\bar{x}) \subset C(x_n)$ for all n. By (i), $G(\bar{x}) = C(\bar{x})$, so $G(\bar{x}) \subset C(x_n)$ for all n, or

$$\{\bar{x}\} \subset G(\bar{x}) \subset \bigcap_{n=0}^{\infty} C(x_n) = \{\bar{x}\},$$

and we have

$$G(\bar{x}) = \{\bar{x}\}.$$

(iv) If $U(x_0) < \infty$, there is $y \in X$ such that

$$\begin{cases} U(y) + d(x_0, y) \leq U(x_0), \\ U(y) < U(x) + d(x, y) \quad \text{for every} \quad x \neq y. \end{cases}$$

Because $U(x_0) < \infty$, by (iii) we have

$$y \in C(x_0), \text{ and } G(y) = \{y\}.$$

By (i), $G(x_0) = C(x_0)$. Now

$$y \in G(x_0) \text{ implies } U(y) + d(x_0, y) \leq U(x_0)$$
$$G(y) = \{y\} \text{ implies } x \in G(y) = \{y\},$$

or

$$U(x) + d(y, x) \leq U(y) \quad \text{if and only if} \quad y = x,$$

that is, if $x \neq y$, then

$$U(y) < U(x) + d(y, x).$$

(v) For $\epsilon > 0$, $x_\epsilon \in X$ such that $U(x_\epsilon) \leq \epsilon + \inf U$. Then for any $k > 0$, there is some point $y_\epsilon \in X$ such that

$$U(y_\epsilon) \leq U(x_\epsilon),$$
$$d(x_\epsilon, y_\epsilon) \leq \frac{1}{k},$$
$$U(x) > U(y_\epsilon) - k\epsilon d(x, y_\epsilon) \quad \text{for all } x \neq y_\epsilon.$$

In fact, we can replace d by $k\epsilon d$, the function U by $U - \inf U$, and the point x_0 by some x_ϵ satisfying

$$U(x_\epsilon) \leq \epsilon + \inf U.$$

Define y_ϵ as a solution provided by (iv)

$$\begin{cases} U(y_\epsilon) + k\epsilon d(x_\epsilon, y_\epsilon) \leq U(x_\epsilon), \\ U(y_\epsilon) < U(x) + k\epsilon d(x, y_\epsilon) \quad \text{for every } x \neq y_\epsilon. \end{cases}$$

Therefore

$$U(y_\epsilon) \leq U(x_\epsilon),$$
$$U(x) > U(y_\epsilon) - k\epsilon d(x, y_\epsilon),$$
$$k\epsilon d(x_\epsilon, y_\epsilon) \leq U(y_\epsilon) - \inf U + k\epsilon d(x_\epsilon, y_\epsilon) \leq U(x_\epsilon) - \inf U \leq \epsilon,$$

or

$$d(x_\epsilon, y_\epsilon) \leq \frac{1}{k}. \qquad \blacksquare$$

The two most important cases for k are $k = 1$, and $k = \epsilon^{-1/2}$ as follows.

Corollary 8.47 *Let X be a complete metric space, and $U : X \to \mathbb{R} \cup \{+\infty\}$ a proper lower semicontinuous function bounded below. Then for any $\epsilon > 0$, there exist some points y_ϵ where*

$$U(y_\epsilon) \leq \epsilon + \inf U,$$
$$U(x) > U(y_\epsilon) - \epsilon d(x, y_\epsilon) \text{ for all } x \neq y_\epsilon.$$

Corollary 8.48 *Let X be a complete metric space, and let $U : X \to \mathbb{R} \cup \{+\infty\}$ be a lower semicontinuous function bounded below. Let $\epsilon > 0$ and $x_\epsilon \in X$ be such that*

$$U(x_\epsilon) \leq \epsilon + \inf U.$$

Then there are some points y_ϵ, at which

$$U(y_\epsilon) \leq U(x_\epsilon),$$
$$d(x_\epsilon, y_\epsilon) \leq \sqrt{\epsilon},$$
$$U(x) \geq U(y_\epsilon) - \sqrt{\epsilon} d(x, y_\epsilon) \text{ for all } y \neq x_\epsilon.$$

Theorem 8.49 *Assume X is a Banach space and $U : X \to \mathbb{R}$ is Gâteaux-differentiable at $y_\epsilon \in X$ satisfying*

$$U(x) > U(y_\epsilon) - k\epsilon d(x, y_\epsilon) \quad \text{for all } x \neq y_\epsilon,$$

then

$$\|U'(y_\epsilon)\|_{X^*} \leq k\epsilon.$$

Proof. Taking $y \in X$, $\|y\| = 1$, and setting $x = y_\epsilon + ty$ with $t > 0$, then

$$\frac{1}{t}[U(y_\epsilon + ty) - U(y_\epsilon)] > -k\epsilon \text{ for all } t > 0.$$

Letting $t \to 0$ we obtain

$$< U'(y_\epsilon), y > \geq -k\epsilon \quad \text{for all } y \in X, \ \|y\| = 1.$$

Because $-y$ is also a unit vector, we have

$$- < U'(y_\epsilon), y > \geq -k\epsilon \quad \text{for all } y \in X, \ \|y\| = 1.$$

From both impossibilities, it follows that

$$\| < U'(y_\epsilon), y > \| \leq k\epsilon \quad \text{for all } y \in X, \ \|y\| = 1.$$

Therefore

$$\|U'(y_\epsilon)\|_{X^*} \leq k\epsilon. \qquad \blacksquare$$

Corollary 8.50 *Assume X is a Banach space, and that $U : X \to \mathbb{R}$ is lower semi-continuous, Gâteaux-differentiable, and bounded below. Let $\epsilon > 0$, and $x_\epsilon \in X$ be given, with*

$$U(x_\epsilon) \leq \epsilon + \inf U.$$

Then there are some points y_ϵ at which

$$U(y_\epsilon) \leq U(x_\epsilon),$$
$$\|x_\epsilon - y_\epsilon\| \leq \sqrt{\epsilon},$$
$$\|U'(y_\epsilon)\|_{X^*} \leq \sqrt{\epsilon}.$$

Corollary 8.51 *Assume X is a Banach space, and that $U : X \to \mathbb{R}$ is lower semi-continuous, Gâteaux-differentiable, and bounded from below. Then there is a sequence y_n such that when $n \to \infty$,*

$$U(y_n) \to \inf U,$$
$$U'(y_n) \to 0 \quad \text{in } X^*.$$

8.6 Best Sobolev Constant

The best Sobolev constant S plays an important role: for a smooth bounded domain Ω in $\mathbb{R}^N$, it is defined by

$$S_\Omega = \inf_{\substack{\phi \in H_0^1(\Omega), \\ \phi \neq 0}} \frac{\displaystyle\int |\nabla \Phi|^2}{\|\Phi\|_{L^{2^*}}^2}, \quad \text{where } \frac{1}{2^*} = \frac{1}{2} - \frac{1}{N}.$$

By the Sobolev Embedding Theorem, $S_\Omega > 0$. Here are some more facts about S_Ω.

Lemma 8.52 S_Ω *is independent of* Ω.

Proof. For $k > 0$, $u \neq 0$, set $u_k(x) = u(kx)$. Then

$$\frac{\partial}{\partial x_i} u_k(x) = k \frac{\partial u}{\partial y_i}(kx), \ y = kx,$$

and we have

$$\|\nabla u_k\|_{L^2(k\Omega)} = k^{(2-N)/2} \|\nabla u\|_{L^2(\Omega)},$$
$$\|u_k\|_{L^{2^*}(k\Omega)} = k^{(2-N)/2} \|u\|_{L^{2^*}(\Omega)},$$

or

$$\frac{\|\nabla u_k\|^2_{L^2(k\Omega)}}{\|u_k\|^2_{L^{2^*}(k\Omega)}} = \frac{\|\nabla u\|^2_{L^2(\Omega)}}{\|u\|^2_{L^{2^*}(\Omega)}}.$$

Hence, we conclude that

$$S_\Omega = \inf_{u \in H_0^1(\Omega), u \neq 0} \frac{\|\nabla \overline{u}\|^2_{L^2(\Omega)}}{\|\overline{u}\|^2_{L^{2^*}(\Omega)}} = \inf_{u \in H_0^1(k\Omega), u \neq 0} \frac{\|\nabla u\|^2_{L^2(k\Omega)}}{\|u\|^2_{L^{2^*}(k\Omega)}} = S_{k\Omega}.$$

Similarly for $x_0 \in \Omega$, let $\overline{u}(x) = u(x - x_0)$, and we have

$$\frac{\|\nabla u\|^2_{L^2(\Omega-x)}}{\|u\|^2_{L^{2^*}(\Omega-x)}} = \frac{\|\nabla u\|^2_{L^2(\Omega)}}{\|u\|^2_{L^{2^*}(\Omega)}}, \quad \text{or } S_\Omega = S_{\Omega-x_0}.$$

For any $\Omega_1, \Omega_2 \subset \mathbb{R}^N$, we have $\Omega_2 \subset k_1(\Omega_1 - x_1)$, $\Omega_1 \subset k_2(\Omega_2 - x_2)$, or

$$S_{\Omega_1} = \inf_{u \in H_0^1(\Omega_1), u \neq 0} \frac{\|\nabla u\|^2_{L^2}}{\|u\|^2_{L^{2^*}}} \geq \inf_{w \in H_0^1(k_2(\Omega_2-x_2)), u \neq 0} \frac{\|\nabla w\|^2_{L^2}}{\|u\|^2_{L^{2^*}}} = S_{\Omega_2},$$

$$S_{\Omega_2} = \inf_{u \in H_0^1(\Omega_2), u \neq 0} \frac{\|\nabla u\|^2_{L^2}}{\|u\|^2_{L^{2^*}}} \geq \inf_{w \in H_0^1(k_1(\Omega_1-x_1)), u \neq 0} \frac{\|\nabla w\|^2_{L^2}}{\|u\|^2_{L^{2^*}}} = S_{\Omega_1}.$$

We conclude that $S_{\Omega_1} = S_{\Omega_2}$.   ∎

Lemma 8.53 S_Ω *is not achieved in any bounded domain* Ω.

Proof. Suppose S_Ω was attainedb y $\qquad\qquad u \in H_0^1(\Omega)$. We may assume $u \geq 0$ in Ω, and $\|u\|_{L^{2^*}} = 1$, and satisfies

$$S_\Omega = \inf_{\substack{w \in H_0^1(\Omega), \\ w \neq 0}} \frac{\|\nabla w\|^2_{L^2}}{\|w\|^2_{L^{2^*}}} = \|\nabla u\|^2_{L^2}.$$

Fix a ball B containing Ω, and set

$$\widetilde{u}(x) = \begin{cases} u(x) & if \ x \in \Omega, \\ 0 & if \ x \notin \Omega. \end{cases}$$

Then S_Ω is also achieved by $\widetilde{u} \in H_0^1(B)$. By the Lagrange Multiplier Theorem, we obtain

$$-\Delta \widetilde{u} = \lambda \widetilde{u}^p \text{ in } B, \text{ where } \lambda > 0, \ p + 1 = 2^* \text{ or } p = \frac{N+2}{N-2}.$$

Let $v = \lambda^{1/(1-p)} \widetilde{u}$. Then

$$\begin{cases} -\triangle v = v^p & in \ B, \\ v = 0 & on \ \partial B, \end{cases}$$

which contradicts the Pohozaev Theorem (see Struwe [47, p.171]).   ∎

Lemma 8.54 *When* $\Omega = \mathbb{R}^N$, *let*

$$S = \inf \left\{ \frac{\displaystyle\int_{\mathbb{R}^N} |\nabla u|^2}{\left[\displaystyle\int_{\mathbb{R}^N} |u|^{2^*} \right]^{2/2^*}} \ \Big| \ u \in D^{1,2}\left(\mathbb{R}^N\right), \ u \not\equiv 0 \right\}.$$

Then S is achieved by the functions:

$$U_{\delta,y}(x) = \frac{[N(N-2)\delta]^{(N-2)/4}}{(\delta + |x-y|^2)^{(N-2)/2}}, \quad for \ \delta > 0, \ y \in \mathbb{R}^N.$$

Proof. See Talenti [48].   ∎

Lemma 8.55 *Let* $\Omega \subset \mathbb{R}^N$ *be a domain. Then* $S_\Omega = S$.

Proof.

(i) Suppose $\Omega = B = B_1(0)$. Because $H_0^1(B) \subset H_0^1(\mathbb{R}^N)$, we have $S_B \geq S$. We claim that $S_B \leq S$. In fact, let $\varphi \in C^\infty(\mathbb{R}^N)$, be a radial function such that $0 \leq \varphi \leq 1$,

$$\varphi(r) = \begin{cases} 0 & if \ 0 \leq r \leq 1/2, \\ 1 & if \ r \geq 1. \end{cases}$$

We know that $S = A^{2/N}$, where U_{a,x_1} as in Lemma 8.54 satisfying

$$S = \frac{\displaystyle\int |\nabla U_{a,x_0}|^2}{\left(\displaystyle\int |U_{a,x_1}|^{2^*} \right)^{2/2^*}}, \quad \int |\nabla U_{a,x_1}|^2 = \int |U_{a,x_0}|^{2^*} = A.$$

Let
$$u_a(x) = \varphi(x) U_{a,x_0}(x).$$
Then $u_a \in H_0^1(\Omega)$. Note that

$$\int_B |\nabla u_a|^2 = \int_{B_{1/a}} |\nabla U|^2 \to \int_{\mathbb{R}^N} |\nabla U|^2 \quad \text{as } a \to 0,$$

$$\int_B |u_a|^{2^*} = \int_{B_{1/a}} |U|^{2^*} \to \int_{\mathbb{R}^N} |U|^{2^*},$$

$$\int_B |\nabla u_a|^2 = \int_{B \setminus B_{1/2}} [|\nabla \varphi|^2 |u_a|^2 + |\nabla u_a|^2 |\varphi|^2] + \int_{B_{1/2}} |\nabla u_a|^2.$$

Note that $u_a \to 0$ on $B_{1/2}^c$ as $a \to 0$, so

$$\int_{B_{1/2}} |\nabla u_a|^2 \to \int_{\mathbb{R}^N} |\nabla U|^2.$$

Thus, for $\epsilon > 0$, there is $\delta > 0$ such that $a \le \delta$ and we have

$$\int_B |\nabla u_a|^2 < A + \epsilon.$$

Similarly, for $\epsilon > 0$, there is $\delta > 0$ such that $a \le \delta$ and we have

$$\int_B |u_a|^{2^*} = \int_{B \setminus B_{1/2}} |\varphi U_a|^{2^*} + \int_{B_{1/2}} |\varphi U_a|^{2^*}$$

$$\ge \int_{B_{1/2}} |U_a|^{2^*} \ge \left(\int_{\mathbb{R}^N} |U|^{2^*} \right) - \epsilon = A - \epsilon,$$

so we have

$$S_B \le \frac{\int_B |\nabla u_a|^2}{\left(\int_B |u_a|^{2^*} \right)^{2/2^*}} \le \frac{A + \epsilon}{(A - \epsilon)^{2/2^*}}.$$

Letting $\epsilon \to 0$,

$$S_B \le \frac{A}{A^{2/2^*}} = S.$$

This asserts that $S_B = S$.

(ii) Suppose $\Omega \subset \mathbb{R}^N$ is a domain. Choose $B(x_0, r) \subset \Omega \subset \mathbb{R}^N$. Then we have by (i) and Lemma 8.52,

$$S_B = S_{B(x_0,1)} \ge S_\Omega \ge S \ge S_B.$$

Therefore $S_\Omega = S$.   ■

Theorem 8.56 (*Lieb Theorem*) *Let* $\Omega \subset \mathbb{R}^N$ *be bounded, and*

$$S_\lambda = \inf_{\substack{u \in H_0^1(\Omega), \\ \|u\|_{L^{2^*}} = 1}} (\|\nabla u\|_{L^2}^2 - \lambda \|u\|_{L^2}^2), \quad \lambda > 0.$$

If $S_\lambda < S$, *then* S_λ *is achieved.*

Proof. Let $(u_j) \subset H_0^1(\Omega)$ be a minimizing sequence for S_λ; that is,

$$\|u_j\|_{L^{2^*}} = 1,$$

$$\|\nabla u_j\|_{L^2}^2 - \lambda \|u_j\|_{L^2}^2 = S_\lambda + o(1), \quad \text{as } j \to \infty. \tag{8.25}$$

Because

$$\|u_j\|_{L^2} \le c \|u_j\|_{L^{2^*}} = c \quad \text{for each } j,$$

$$\|\nabla u_j\|_{L^2}^2 = S_\lambda + \lambda \|u_j\|_{L^2}^2 + o(1) \le c \quad \text{for each } j.$$

Now, $\{u_j\}$ is bounded in $H_0^1(\Omega)$. There is a subsequence, still denoted by u_j such that, for some $u \in H_0^1(\Omega)$,

$$\begin{cases} u_j \rightharpoonup u & \text{weakly in } H_0^1(\Omega), \\ u_j \rightharpoonup u & a.e. \text{ on } \Omega, \text{ strongly in } L^2(\Omega), \\ \|u\|_{L^{2^*}} \le 1. \end{cases}$$

Recall that $S = S_0 = \inf\limits_{\substack{u \in H_0^1(\Omega), \\ \|u\|_{L^{2^*}} = 1}} \|\nabla u\|_{L^2}^2$. For each j, $\|u_j\|_{L^{2^*}} = 1$, we have $\|\nabla u_j\|_{L^2}^2 \ge S$. With this inequality and the assumption we obtain

$$0 < S - S_\lambda \le \lambda \|u_j\|_{L^2}^2 + o(1).$$

Letting $j \to \infty$, we obtain

$$0 < S - S_\lambda \le \lambda \|u\|_{L^2}^2.$$

Hence, we have $u \not\equiv 0$. Set $v_j = u_j - u$, and we have

$$\begin{cases} v_j \rightharpoonup 0 & \text{weakly in } H_0^1(\Omega), \\ v_j \to 0 & a.e. \text{ on } \Omega, \end{cases}$$

and

$$\nabla u_j = \nabla v_j + \nabla u.$$

Now,

$$\begin{aligned} \|\nabla u_j\|_{L^2}^2 &= \|\nabla v_j + \nabla u\|_{L^2}^2 \\ &= \|\nabla v_j\|_{L^2}^2 + \|\nabla u\|_{L^2}^2 + 2\int (\nabla v_j)(\nabla u) \\ &= \|\nabla v_j\|_{L^2}^2 + \|\nabla u\|_{L^2}^2 + o(1). \end{aligned}$$

Substitute this equality into (8.25) to obtain

$$\|\nabla v_j\|_{L^2}^2 + \|\nabla u\|_{L^2}^2 - \lambda \|u\|_{L^2}^2 = S_\lambda + o(1). \tag{8.26}$$

By the Brezis-Lieb Lemma 1.45,

$$\|u + v_j\|_{L^{2^*}}^{2^*} = \|u\|_{L^{2^*}}^{2^*} + \|v_j\|_{L^{2^*}}^{2^*} + o(1).$$

Now,

$$1 = \|u_j\|_{L^{2^*}} = \|u + v_j\|_{L^{2^*}} \quad \text{implies that} \quad \|u\|_{L^{2^*}} \le 1, \ \|v_j\|_{L^{2^*}} \le 1,$$

so

$$\|u\|_{L^{2^*}}^2 \ge \|u\|_{L^{2^*}}^{2^*}, \quad \|v_j\|_{L^{2^*}}^2 \ge \|v_j\|_{L^{2^*}}^{2^*},$$

or

$$1 \le \|u\|_{L^{2^*}}^2 + \|v_j\|_{L^{2^*}}^2 + o(1).$$

By the definition of S, $S \le \dfrac{\|\nabla v_j\|_{L^{2^*}}^2}{\|v_j\|_{L^{2^*}}^2}$ or $\|v_j\|_{L^{2^*}}^2 \le \dfrac{1}{S} \|\nabla v_j\|_{L^2}^2$. We conclude that

$$1 \le \|u\|_{L^{2^*}}^2 + \frac{1}{S} \|\nabla v_j\|_{L^2}^2 + o(1). \tag{8.27}$$

We claim that

$$\|\nabla u\|_{L^2}^2 - \lambda \|u\|_{L^2}^2 \le S_\lambda \|u\|_{L^{2^*}}^2,$$

which will imply

$$S_\lambda = \frac{\|\nabla u\|_{L^2}^2 - \lambda \|u\|_{L^2}^2}{\|u\|_{L^{2^*}}^2}, \quad u \not\equiv 0. \qquad \blacksquare$$

Theorem 8.57 $S_\lambda > 0$ *if and only if* $\lambda < \lambda_1$, *where* λ_1 *is the first eigenvalue of* $-\triangle$ *in the Dirichlet problem.*

Proof. In fact, suppose $S_\lambda > 0$, and choose $\varphi > 0$ in $H_0^1(\Omega)$ with $\triangle \varphi + \lambda_1 \varphi = 0$, or

$$\|\nabla \varphi\|_{L^2}^2 - \lambda \|\varphi\|_{L^2}^2 = \int (-\triangle \varphi)\varphi - \lambda \varphi^2 = (\lambda_1 - \lambda) \int \varphi^2.$$

Then

$$\frac{\|\nabla \varphi\|_{L^2}^2 - \lambda \|\varphi\|_{L^2}^2}{\|\varphi\|_{L^{2^*}}^2} = (\lambda_1 - \lambda)\frac{\|\varphi\|_{L^2}^2}{\|\varphi\|_{L^{2^*}}^2} \ge S_\lambda > 0, \quad \text{or } \lambda_1 > \lambda.$$

On the other hand, if $\lambda < \lambda_1$, choose α such that $\lambda < \alpha < \lambda_1 = \inf \dfrac{\|\nabla w\|_{L^2}^2}{\|w\|_{L^2}}$. We have

$$\|\nabla w\|_{L^2}^2 > \alpha \|w\|_{L^2}^2 = [\lambda + (\alpha - \lambda)] \|w\|_{L^2}^2.$$

Thus,
$$\|\nabla w\|_{L^2}^2 - \lambda \|w\|_{L^2}^2 \geq \delta \|w\|_{L^2}^2 \text{ where } \delta = \alpha - \lambda > 0.$$
Assume $\{u_n\}$ and $u \not\equiv 0$, as in the beginning of the proof.
$$\|\nabla u_j\|_{L^2}^2 - \lambda \|u_j\|_{L^2}^2 = S_\lambda + o(1),$$
$$u_j \to u \quad \text{in } L^2(\Omega),$$
$$\delta \|u_j\|_{L^2}^2 \leq \|\nabla u_j\|_{L^2}^2 - \lambda \|u_j\|_{L^2}^2 .$$
Letting $j \to \infty$ we have $\delta \|u\|_{L^2}^2 \leq S_\lambda$, or $S_\lambda > 0$.

(i) Suppose $0 < S_\lambda < S$ (i.e., $\lambda < \lambda_1$). By the inequality (8.27)
$$1 \leq \|u\|_{L^{2^*}}^2 + \frac{1}{S} \|\nabla u_j\|_{L^2}^2 + o(1),$$
we have
$$S_\lambda \leq S_\lambda \|u\|_{L^{2^*}}^2 + \frac{S_\lambda}{S} \|\nabla u_j\|_{L^2}^2 + o(1)$$
$$\leq S_\lambda \|u\|_{L^{2^*}}^2 + \|\nabla v_j\|_{L^2}^2 + o(1).$$
From (8.26),
$$\|\nabla u\|_{L^2}^2 + \|\nabla v_j\|_{L^2}^2 - \lambda \|u\|_{L^2}^2 = S_\lambda + o(1)$$
$$\leq S_\lambda \|u\|_{L^{2^*}}^2 + \|\nabla v_j\|_{L^2}^2 + o(1),$$
or
$$\|\nabla u\|_{L^2}^2 - \lambda \|u\|_{L^2}^2 \leq S_\lambda \|u\|_{L^{2^*}}^2 .$$

(ii) Suppose $S_\lambda \leq 0$ (i.e., $\lambda \geq \lambda_1$). Now $\|u\|_{L^{2^*}} \leq 1$ implies that $S_\lambda \leq S_\lambda \|u\|_{L^{2^*}}^2$. By (8.26), we have
$$\|\nabla u\|_{L^2}^2 - \lambda \|u\|_{L^2}^2 \leq \|\nabla u\|_{L^2}^2 + \|\nabla v_j\|_{L^2}^2 - \lambda \|u\|_{L^2}^2$$
$$= S_\lambda + o(1)$$
$$\leq S_\lambda \|u\|_{L^{2^*}}^2 + o(1),$$
or $\|\nabla u\|_{L^2}^2 - \lambda \|u\|_{L^2}^2 \leq S_\lambda \|u\|_{L^{2^*}}^2 .$ $\blacksquare$

8.7   Palais-Smale Sequences

Theorem 8.58 *If*
$$\alpha = \min_{\|u\| \to \infty} \inf F(u) \quad \text{is finite,}$$
then there is a sequence (u_n) in X such that $\|u_n\| \to \infty$, $F(u_n) \to \alpha$, and $\|F'(u_n)\| \to 0$.

Proof. Set, for $r \geq 0$,

$$m(r) = \inf_{\|u\| \geq r} F(u).$$

Clearly, m is a nondecreasing function and $\lim_{r \to \infty} m(r) = \alpha$. Then for any positive $\epsilon < \dfrac{1}{2}$, we have

$$\alpha - \epsilon^2 \leq m(r) \quad \text{for } r \geq \bar{r}.$$

We may take $\bar{r} \geq \epsilon^{-1}$. Choose z_0 with $\|z_0\| \geq 2\bar{r}$ such that

$$F(z_0) < m(2\bar{r}) + \epsilon^2 \leq \alpha + \epsilon^2.$$

Applying Ekeland's Principle in $\{\|x\| \geq \bar{r}\}$, we find some z, $\|z\| \geq \bar{r}$ satisfying

$$F(x) - F(z) + \epsilon \|x - z\| \geq 0 \quad \text{provided } \|x\| \geq \bar{r},$$
$$\alpha - \epsilon^2 \leq m(\bar{r}) \leq F(z) \leq F(z_0) - \epsilon \|z - z_0\|.$$

It follows that $\|z - z_0\| \leq 2\epsilon$. Hence, $\|z\| > \bar{r}$ and we may conclude that $\|F'(z)\| \leq \epsilon$.
∎

Corollary 8.59 *If F is bounded below and satisfies the (PS) condition, then F is coercive: $F(u) \to \infty$ as $\|u\| \to \infty$.*

Proof. Suppose $\alpha = \min \inf_{\|u\| \to \infty} F(u)$ is finite. Then by Theorem 8.58, there is a $(PS)_\alpha -$ sequence that contains no convergent subsequences.
∎

Theorem 8.60 *If F is bounded below,*

$$\alpha = \inf_{u \in X} F(u)$$

and $\{x_n\}$ is a minimizing sequence, then there is a minimizing $(PS)_\alpha$ sequence $\{y_n\}$ such that $\|x_n - y_n\| < 1/n$.

Theorem 8.61 *Assume F is bounded below and satisfies the (PS) condition. Then every minimizing sequence has a convergent subsequence.*

Proof. Let (x_n) be a minimizing sequence. For a subsequence, still denoted (x_n), we may assume that $F(x_n) \leq \inf F + 1/n^2$. By Ekeland's Principle, there is y_n in X such that

$$F(y) - F(y_n) + (1/n)\|y - y_n\| \geq 0 \quad \text{for each } y \in X,$$
$$F(y_n) \leq F(x_n) - (1/n)\|x_n - y_n\|.$$

Thus, $\|F'(y_n)\| \leq 1/n$, $F(y_n) \leq \inf F + (1/n^2)$ and $\|x_n - y_n\| \leq 1/n$. By the (PS) condition, the sequence (y_n) has a convergent subsequence (y_{n_k}) and (x_{n_k}) also converges.
∎

Lemma 8.62 *Let N be a metric space and let $f : N \to X^*$ be a continuous map. Then, given $\epsilon > 0$, there is a locally Lipschitz map $v : N \to X$ such that for all $\xi \in N$*

$$\|v(\xi)\| \leq 1, \quad < f(\xi), v(\xi) > 1 \geq \|f(\xi)\| - \epsilon.$$

Let $F : X \to \mathbb{R}$ be a C^1 function satisfying the following (MPL)-condition: there is an open neighborhood U of 0 and some point $u_0 \notin \bar{U}$ such that

$$F(0), \ F(u_0) < c_0 \leq F(u) \quad \text{for each } u \in \partial U.$$

Consider the family A of all continuous paths joining 0 to u_0, and set

$$c := \inf_{p \in A} \max_{u \in p} F(u). \tag{8.28}$$

Lemma 8.63 *(Standard MPL) Under the (MPL)-condition and with c given by (8.28), there is a sequence (u_n) in X such that*

$$F(u_n) \to c \text{ and } \|F'(u_n)\| \to 0.$$

In addition, if F satisfies the $(PS)_c$ condition with c given by (8.28), then c is a critical value.

In connection with this well-known (MPL) we would like to call attention to the following two forms. As before, F is a real C^1 function on a Banach space X. Let K be a compact metric space and let K^* be a nonempty closed subset $\neq K$. Let

$$A = \{p \in C(K; X); \ p = p^* \text{ on } K^*\},$$

where p^* is a fixed continuous map on K. Define

$$c = \inf_{p \in A} \max_{\xi \in K} F(p(\xi)),$$

so that

$$c \geq \max_{\xi \in K^*} F(p^*(\xi)).$$

Theorem 8.64 *Assume that for every $p \in A$, $\max_{\xi \in K} F(p(\xi))$ is attained at some point in $K \backslash K^*$. Then there is a sequence (u_n) in X such that*

$$F(u_n) \to c \text{ and } \|F'(u_n)\| \to 0.$$

If in addition F satisfies the $(PS)_c$ condition, then c is a critical value. Moreover, if (p_n) is any sequence in A such that

$$\max_{\xi \in K} F(p_n(\xi)) \to C,$$

then there is a sequence (ξ_n) in K such that $F(p_n(\xi_n)) \to c$ and $\|F'(p_n)(\xi_n))\| \to 0$. The standard (MPL) is clearly a special case of this with $K = [0,1]$, $K^* = \{0,1\}$ and with $p^*(t) = t u_0$.

Proof. For $\xi \in K$, set

$$d(\xi) = \min\{\text{dist}(\xi, K^*), 1\},$$

and consider for any fixed $\epsilon > 0$, and $p \in A$,

$$G(p, \xi) = F(p(\xi)) + \epsilon d(\xi).$$

Set

$$\psi_\epsilon(p) = \max_{\xi \in K} G(p(\xi), \xi). \tag{8.29}$$

$$c_\epsilon = \inf_{p \in A} \psi_\epsilon(p).$$

Clearly, $c \leq c_\epsilon \leq c + \epsilon$. For $M = A$ (equipped with the usual metric), we see easily that $\psi_\epsilon(p)$ is continuous on A. By Ekeland's Principle, there is $p \in A$ such that

$$\psi_\epsilon(x) - \psi_\epsilon(p) + \epsilon d(p, x) \geq 0 \quad \text{for each } x \in P, \tag{8.30}$$

$$c \leq c_\epsilon \leq \psi_\epsilon(p) \leq c_\epsilon + \epsilon \leq c + 2\epsilon.$$

By our main hypothesis

$$\psi_\epsilon(p) > \max_{\xi \in K^*} F(p(\xi)). \tag{8.31}$$

Set

$$B_\epsilon(p) = \{\xi \in K; \ G(p(\xi), \ \xi) = \psi_\epsilon(p)\}.$$

We shall prove that there is some $\xi_0 \in B_\epsilon(p)$ such that

$$\|F'(p(\xi_0))\| \leq 2\epsilon.$$

The conclusion of the first part of the theorem then follows by choosing $\epsilon = 1/n$ and $u_n = p(\xi_0)$. Applying Lemma 8.62 with $N = K$ and $f(\xi) = F'(p(\xi))$, we obtain a continuous map $v : K \to X$ such that for all $\xi \in K$,

$$\|v(\xi)\| \leq 1, \quad < F'(p(\xi)), v(\xi) >\geq \|F'(p(\xi))\| - \epsilon. \tag{8.32}$$

By (8.31), $B_\epsilon(p) \subset K \backslash K^*$. Thus, there is a continuous nonnegative function $\alpha(\xi) \leq 1$ on K that equals 1 on $B_\epsilon(p)$, and vanishes on K^*. We shall take for q, in (8.29), small variations of the path p :

$$q_h(\xi) = p(\xi) - hw(\xi) \quad \text{for } 0 < h \text{ small},$$

and $w(\xi) = \alpha(\xi)v(\xi)$. In what follows, $\epsilon > 0$ is fixed while we let $h \to 0$. Observe that

$$\psi_\epsilon(q_h) = \max_{\xi \in K} G(q_h(\xi), \xi)$$

is attained at some point $\xi_h \in K$. For a suitable sequence $h_n \to 0$, ξ_{h_n} converges to some ξ_0 that belongs to $B_\epsilon(p)$. By (8.31), with $q = q_h$, and by (8.30), we obtain

$$F(p(\xi_h) - hw(\xi_h)) + \epsilon d(\xi_h) - \psi_\epsilon(p) + \epsilon h \geq 0. \tag{8.33}$$

On the other hand, it is easy to check that as $h \to 0$,

$$F(p(\xi_h) - hw(\xi_h)) = F(p(\xi_n)) - <F'(p(\xi_h)), hw(\xi_h)> +o(h). \tag{8.34}$$

Combining (8.33) and (8.34), we see that

$$-h < F'(p(\xi_h)), w(\xi_h) > +\epsilon h + o(h) \geq 0$$

(note that $F(p(\xi_h)) + \epsilon d(\xi_h) \leq \psi_\epsilon(P)$). Hence,

$$< F'(p(\xi_h)), w(\xi_h) > \leq \epsilon + o(1).$$

As $h \to 0$ we find

$$< F'(p(\xi_0)), v(\xi_0) > \leq \epsilon, \tag{8.35}$$

which by (8.34) yields (8.35). The last assertion is established by constructing first, via Ekeland's Principle, a sequence (q_n) in A such that

$$\psi_{\epsilon_n^2}(q) - \psi_{\epsilon_n^2}(q_n) + \epsilon_n \mathrm{d}(q, q_n) \geq 0 \quad \text{for each } q \in \mathcal{A}.$$
$$\psi_{\epsilon_n^2}(q) \leq \psi_{\epsilon_n^2}(q_n) - \epsilon_n \mathrm{d}(p_n, q_n).$$

Here, (ϵ_n) is a sequence of positive numbers, $\epsilon_n \to 0$, such that $\max_{\xi \in K} F(p_n(\xi)) \leq c + \epsilon_n^2$. It follows that $d(p_n, q_n) \leq 2\epsilon_n$. The preceding argument (applied with q_n in place of p) leads to the existence of some $\xi_n \in K$ such that

$$F(q_n(\xi_n)) = c + O(\epsilon_n^2), \quad \|F'(q_n(\xi_n))\| \leq 2\epsilon_n.$$

This is the desired sequence (ξ_n). Indeed, by the $(PS)_c$ condition, a subsequence of $q_n(\xi_n)$ converges to a critical point and the corresponding subsequence of $p_n(\xi_n)$ converges to the same limit. A standard argument shows that for the full sequence, $F(p_n(\xi_n)) \to c$ and $\|F'(p_n(\xi_n))\| \to 0$. ∎

We consider a function $F \in C^1$ in X and set

$$F_a = \{u \in X; \ F(u) \leq a\},$$
$$K_a = \text{set of critical points of F where } F = a.$$

Theorem 8.65 (*Deformation Theorem*) *Let $c \in \mathbb{R}$. For any given $\delta < \frac{1}{8}$, there is a continuous deformation $\eta : [0,1] \times X \to X$ such that:*

$$\eta(0, u) = u \quad \text{for each } u \in X; \tag{8.36}$$

$$\eta(t, \cdot) \text{ is a homeomorphism of } X \text{ onto } X, \quad \text{for each } t \in [0, 1]; \tag{8.37}$$

$$\eta(t, u) = u \text{ for each } t \in [0, 1] \quad \text{if } |F(u) - c| \geq 2\delta \text{ or if } \|F'(u)\| \leq \sqrt{\delta}; \tag{8.38}$$

$$0 \leq F(u) - F(\eta(t, u)) \leq 4\delta \quad \text{for each } u \in X, \text{ for each } t \in [0, 1]; \tag{8.39}$$

$$\|\eta(t, u) - u\| \leq 16\sqrt{\delta} \quad \text{for each } u \in X, \text{ for each } t \in [0, 1]. \tag{8.40}$$

If $u \in F_{c+\delta}$, then either

$$
\begin{aligned}
&\text{(i) } \eta(1, u) \in F_{c-\delta} \text{ or} \\
&\text{(ii) for some } t_1 \in [0, 1], \text{ we have } \|F'(\eta(t_1, u))\| < 2\sqrt{\delta}.
\end{aligned}
\tag{8.41}
$$

More generally, let $\tau \in [0, 1]$ and assume that for all $t \in [0, \tau]$, $\eta(t, u)$ belongs to the set

$$\widetilde{N} = \{v \in X \mid |F(v) - c| \leq \delta \text{ and } \|F'(v)\| \geq 2\sqrt{\delta}\}. \tag{8.42}$$

Then $F(\eta(v, u)) \leq F(u) - \tau/4$.

Proof. In addition to the set $\widetilde{N}$ in (8.42) we shall make use of the set

$$N = \{u \in X \mid |F(u) - c| < 2\delta \text{ and } \|F'(u)\| > \sqrt{\delta}\}.$$

Because $\widetilde{N}$ and N^c are disjoint closed sets, there is a locally Lipschitz nonnegative function $g \leq 1$ satisfying

$$g = \begin{cases} 1 & \text{on } \widetilde{N}, \\ 0 & \text{outside } N. \end{cases}$$

For example,

$$g(u) = \frac{\text{dist}(u, N^c)}{\text{dist}(u, N^c) + \text{dist}(u, \widetilde{N})}.$$

Consider the vector field

$$V(u) = \begin{cases} -g(u)\dfrac{v(u)}{\|v(u)\|^2} & \text{on } N, \\ 0 & \text{outside } N, \end{cases}$$

where v is a pseudogradient defined on $\{F'(u) \neq 0\}$. Clearly V is locally Lipschitz on X and $\|V(u)\| \leq 1\sqrt{\delta}$ for all $u \in X$. Consider the flow $\eta(t) = \eta(t, u)$ defined by

$$\frac{d\eta}{dt} = V(\eta), \quad \eta\mid_{t=0} = u.$$

Clearly, η is defined for $t \in [0, 1]$ and satisfies (8.36)–(8.37) and

$$\frac{d}{dt} F(\eta(t)) \le -\frac{1}{4} g(\eta(t)) \text{ for each } u \in X, \text{ for each } t \in [0, 1].$$

In particular, we have

$$\int_0^t g(\eta(s)) ds \le 4(F(u) - F(\eta(t))). \tag{8.43}$$

(8.39): if $|F(u) - c| \ge 2\delta$, then $\eta(t, u) = u$, for all $t \in [0,1]$, and the conclusion is obvious. Hence, we may assume that $|F(u) - c| < 2\delta$. If $F(\eta(1)) \ge c - 2\delta$, the proof is finished. Suppose that $F(\eta(1)) < c - 2\delta$. Then there is $t_1 \in [0, 1]$ such that $F(\eta(t_1)) = c - 2\delta$ and because $\eta(t) = \eta(t_1)$ for $t \ge t_1$, it follows that $F(u) - F(\eta(1)) = F(u) \quad F(\eta(t_1)) \le c + 2\delta - (c - 2\delta) = 4\delta$.
(8.40):

$$\|\eta(t) - u\| \le \int_0^t \|\frac{d\eta}{dt}(s)\| ds = \int_0^t \|V(\eta(s))\| ds < \int_{[0, t] \cap \{s | \eta(s) \in N\}} \frac{g(\eta(s)) ds}{\|v(\eta(s))\|}$$

$$\le \int_{[0, t] \cap \{s | \eta(s) \in N\}} \frac{g(\eta(s))}{\|F'(\eta(s))\|} \le \frac{1}{\sqrt{\delta}} \int_0^t g(\eta(s)) ds$$

$$\le \frac{4}{\sqrt{\delta}} (F(u) - F(\eta(t))) \le 16\sqrt{\delta} \quad \text{by (8.42) and (8.38)}.$$

(8.42): if $\eta(t) \in \widetilde{N}$ for $0 \le t \le \tau$, then $g(\eta(t)) = 1$ and the assertion of (8.42) follows from (8.43).   ∎

Theorem 8.66 *Assume the conditions of Theorem 8.58 and that there is a closed set $\mathcal{B}$ in X, disjoint from $p * (K*)$, on which*

$$F \ge c \quad \text{(defined in (8.28))} \tag{8.44}$$

and such that

$$\text{for each } p \in \mathcal{A} \quad p(K) \text{ intersects } \mathcal{B}.$$

Then there is a sequence (u_n) in X satisfying

$$F(u_n) \to c, \|F'(u_n)\| \to 0 \text{ and } \text{dist}(u_n, \mathcal{B}) \to 0.$$

Proof.   For any given $\delta > 0$, we shall show that there is a point $\hat{u}$ such that $c \le F(\hat{u}) < c + \delta$, $\|F'(\hat{u})\| < 2\sqrt{\delta}$ and $\text{dist}(\hat{u}, \mathcal{B}) \le 32\sqrt{\delta}$. Letting $\delta \to 0$ through a sequence δ_n, the corresponding $\hat{u}_n$ have the desired properties. We take $\delta < \frac{1}{8}$ so that $32\sqrt{\delta} < \text{dist}(\mathcal{B}, p^*(K^*))$. Let η be a deformation in Theorem 8.65. Let $p \in \mathcal{A}$ be such that $\max_{\xi \in K} F(p(\xi)) < c + \delta$. Let $0 \le \zeta(v) \le 1$ be a continuous function on X, which equals 1 if $\text{dist}(v, \mathcal{B}) \leq 16\sqrt{\delta}$, and vanishes if $\text{dist}(v, \mathcal{B}) \geq 32\sqrt{\delta}$. Consider the "path"

$$q(\xi) = \eta(\zeta(p(\xi)), p(\xi)).$$

Clearly, $q \in A$. Let $\widetilde{u} \in q(K) \cap \mathcal{B}$. Therefore $\widetilde{u} = \eta(\zeta(p(\widetilde{\xi})), p(\widetilde{\xi}))$ for some $\widetilde{\xi} \in K$. Set $u = p(\widetilde{\xi})$. By property (8.40)

$$\|\eta(t, p(\widetilde{\xi})) - p(\widetilde{\xi})\| \leq 16\sqrt{\delta} \quad \text{for every } t \in [0, 1],$$

and so $\zeta(p(\widetilde{\xi})) = 1$. Hence, $\widetilde{u} = \eta(1, u)$, and $c \leq F(\eta(t, p(\widetilde{\xi}))) < c + \delta$, for all $t \in [0, 1]$. Alternative (8.40) in Theorem 8.65 must therefore hold, and hence, for some $t_1 \in [0, 1]$, $\hat{u} = \eta(t_1, p(\widetilde{\xi}))$ satisfies

$$\|F'(\hat{u})\| < 2\sqrt{\delta}.$$

Furthermore, $\|\hat{u} - \widetilde{u}\| \leq 32\sqrt{\delta}$, by (8.40). ∎

8.8   Global Compactness

Definition 8.67 *Let H be a Hilbert space and let $f : H \to \mathbb{R}$ be a function of class C^1.*

(i) *f satisfies the (PS)−condition if for every sequence $\{u_n\}$ in H that satisfies the properties $\{f(u_n)\}$ is bounded, and $\|f'(u_n)\| \to 0$ as $n \to \infty$, then $\{u_n\}$ contains a convergent subsequence;*

(ii) *for $c \in \mathbb{R}$, a sequence $\{u_n\}$ in H is called a $(PS)_c$ sequence for f if $f(u_n) \to c$, $\|f'(u_n)\|_{H'} \to 0$, as $n \to \infty$;*

(iii) *For $c \in \mathbb{R}$, f has the $(PS)_c$ condition if every $(PS)_c$ sequence contains a convergent subsequence.*

Consider $F : H_0^1(\Omega) \to \mathbb{R}$ such that

$$F(u) = \frac{1}{2} \int_\Omega |\nabla u|^2 - \frac{1}{2^*} \int_\Omega |u|^{2^*} \quad \text{for } u \in H_0^1(\Omega),$$

where $2^* = \dfrac{2N}{N-2}$, $\Omega \subset \mathbb{R}^N$ is a bounded domain, $N \geq 3$. Then we have:

(i) F is of C^1;

(ii) If $F'(u) = 0$ for $u \in H_0^1(\Omega)$, then u is a weak solution of the equation

$$\begin{cases} -\triangle u = u^{(N+2)/(N-2)} & \text{in } \Omega, \\ u \in H_0^1(\Omega). \end{cases}$$

Theorem 8.68 *The $(PS)_c$ condition of F fails at the levels $c = k\sum$, where $\sum = \frac{1}{N}S^{N/2}$, $k = 1, 2, \cdots$.*

Proof. Set

$$U(x) = \frac{1}{(1 + |x|^2)^{\frac{N-2}{2}}}.$$

$$U_{\epsilon,x_0}(x) = \frac{1}{\epsilon^{(N-2)/2}}U(\frac{x - x_0}{\epsilon}) = \frac{\epsilon^{(N-2)/2}}{(\epsilon^2 + |x - x_0|^2)^{\frac{N-2}{2}}}, \quad \epsilon > 0, \ x_0 \in \mathbb{R}^N.$$

Recall that if we set

$$a^{(N+2)/(N-2)} = N(N - 2),$$

then

$$-\triangle(aU_{\epsilon,x_0}) = (aU_{\epsilon,x_0})^{(N+2)/(N-2)} \quad \text{in } \mathbb{R}^N.$$

Fix a point of concentration $\bar{x} \in \Omega$, a speed of concentration $\epsilon_n > 0$, $\epsilon_n \to 0$, a function $\zeta \in C_0^\infty(\Omega)$ such that $\zeta \geq 0$ and $\zeta = a$ on $B(\bar{x}, p) \subset \Omega$. Consider, for $n = 1, 2, \cdots$,

$$u_n(x) = \zeta(x)U_{\epsilon_n}(x - \bar{x}) = \zeta(x)U_{\epsilon_n,\bar{x}}(x) = \zeta(x)U_n(x).$$

(i) $F(u_n) \to \sum$: for $\epsilon > 0$ and $x_0 \in \mathbb{R}^N$, set

$$V(x) = U_{\epsilon,x_0}(x).$$

Then

$$S = \frac{\int_{\mathbb{R}^N} |\nabla(aV)|^2}{\|aV\|_{L^{2^*}}^2},$$

and we have

$$\int_{\mathbb{R}^N} |\nabla(aV)|^2 = S\|aV\|_{L^{2^*}}^2,$$

but,

$$\int_{\mathbb{R}^N} |\nabla(aV)|^2 = \int_{\mathbb{R}^N} (-\nabla(aV))^2(aV) = \int (aV)^{2^*-1}(aV)$$

$$= \int (aV)^{2^*} = \|aV\|_{L^{2^*}}^{2^*}.$$

Thus,

$$\|aV\|_{L^{2^*}}^{2^*} = S\|aV\|_{L^{2^*}}^2,$$

or

$$\int_{\mathbb{R}^N} |\nabla(aV)|^2 = \|aV\|_{L^{2^*}}^{2^*} = S^{2^*/(2^*-2)} = S^{N/2},$$

so

$$F(aV) = \frac{1}{2} \int_{\mathbb{R}^N} |\nabla(aV)|^2 - \frac{1}{2^*} \int_{\mathbb{R}^N} |(aV)|^{2^*} = (\frac{1}{2} - \frac{1}{2^*}) S^{N/2} = \frac{1}{N} S^{N/2} = \sum .$$

We claim that

(a) $\nabla U_n \to 0$ in $L^2(\mathbb{R}^N \backslash B(\bar{x}, \rho))$, where $U_n(x) = U_{\epsilon_n, \overline{X}}(x)$,

$$\int_{|x-\bar{x}| \geq \rho} |\nabla U_n(x)|^2 \, dx = \epsilon_n^{2-N} \int_{|x-\bar{x}| \geq \rho} \left| \nabla U(\frac{x-\bar{x}}{\epsilon_n}) \right|^2 dx$$

$$\leq \epsilon_n^{2-N} \epsilon_n^N \int_{|y| \geq (\rho/\epsilon_n)} |\nabla U(y)|^2 \, dy$$

$$\leq \epsilon_n^2 \int_{|y| \geq 1} |\nabla U(y)|^2 \, dy \quad \text{for large } n$$

$$\to 0 \quad \text{as } n \to \infty,$$

where $|\nabla U(y)|^2 = (2-N)^2 (1+|y|^2)^{-N} |y|^2 \leq c(1+|y|^2)^{-N+1} \in L^1(B_1(0)^c)$.

(b) Let $K \subset (\Omega \backslash \{\bar{x}\})$ be compact, or in general let $K \subset (\mathbb{R}^N \backslash \{\bar{x}\})$ be bounded, $\text{dist}(\bar{x}, K) > 0$. Then $u_n(x) \to 0$ and $U_n(x) \to 0$ uniformly on K as $n \to \infty$,

$$|u_n(x)| = |\zeta(x) U_n(x)| \leq |\zeta(x)| \frac{\epsilon_n^{(N-2)/2}}{(\epsilon_n^2 + |x-\bar{x}^2|)^{\frac{N-2}{2}}}$$

$$\leq c_1 \frac{\epsilon_n^{(N-2)/2}}{(\epsilon_n^2 + \alpha^2)^{\frac{N-2}{2}}} \to 0 \quad \text{as } n \to \infty.$$

(c) $\frac{1}{2} \int_\Omega |\nabla u_n|^2 - \frac{1}{2} \int_{\mathbb{R}^N} |\nabla(aU_n)|^2 = o(1)$ as $n \to \infty$: Let $B = B(\bar{x}, \rho)$. Then

$$\frac{1}{2} \int_\Omega |\nabla u_n|^2 = \frac{1}{2} \int_\Omega |\nabla(\zeta(x) U_n(x)|^2 \, dx$$

$$= \frac{1}{2} \int_B |\nabla(aU_n)|^2 + \frac{1}{2} \int_{\Omega \backslash B} |\nabla(\zeta(x) U_n|^2$$

$$= \frac{1}{2} \int_B |\nabla(aU_n)|^2 + \frac{1}{2} \int_{\mathbb{R}^N \backslash B} |\nabla(aU_n)|^2 - \frac{1}{2} \int_{\mathbb{R}^N \backslash B} |\nabla(aU_n)|^2$$

$$+ \frac{1}{2} \int_{\Omega \backslash B} |\zeta \nabla(U_n)|^2 + \frac{1}{2} \int_{\Omega \backslash B} |(\nabla \xi) U_n|^2 .$$

Therefore

$$\frac{1}{2} \int_\Omega |\nabla u_n|^2 - \frac{1}{2} \int_{\mathbb{R}^N} |\nabla(aU_n)|^2$$

$$= \frac{-1}{2} \int_{\mathbb{R}^N \backslash B} |\nabla(aU_n)|^2 + \frac{1}{2} \int_{\Omega \backslash B} |\zeta \nabla(U_n)|^2 + \frac{1}{2} \int_{\Omega \backslash B} |(\nabla \zeta) U_n|^2 = o(1)$$

(d) $\dfrac{1}{2^*}\displaystyle\int_\Omega |u_n|^{2^*} - \dfrac{1}{2^*}\displaystyle\int_{\mathbb{R}^N} |aU_n|^{2^*} = o(1)$ as $n \to \infty$.

$$\frac{1}{2^*}\int_\Omega |u_n|^{2^*} = \frac{1}{2^*}\int_B |u_n|^{2^*} + \frac{1}{2^*}\int_{\Omega\backslash B} |u_n|^{2^*}$$

$$= \frac{1}{2^*}\int_B |aU_n|^{2^*} + \frac{1}{2^*}\int_{\mathbb{R}^N\backslash B} |aU_n|^{2^*} - \frac{1}{2^*}\int_{\mathbb{R}^N\backslash B} |aU_n|^{2^*}$$

$$+ \frac{1}{2^*}\int_{\Omega\backslash B} |u_n|^{2^*}$$

$$= \frac{1}{2^*}\int_{\mathbb{R}^N} |aU_n|^{2^*} - \frac{1}{2^*}\int_{\mathbb{R}^N\backslash B} |aU_n|^{2^*} + \frac{1}{2^*}\int_{\Omega\backslash B} |u_n|^{2^*},$$

or

$$\frac{1}{2^*}\int_\Omega |u_n|^{2^*} - \frac{1}{2^*}\int_{\mathbb{R}^N} |aU_n|^{2^*} = -\frac{1}{2^*}\int_{\mathbb{R}^N\backslash B} |aU_n|^2 + \frac{1}{2^*}\int_{\Omega\backslash B} |u_n|^{2^*}.$$

By (b), $\dfrac{1}{2^*}\int_{\Omega\backslash B} |u_n|^{2^*} = o(1)$ as $n \to \infty$,

$$\frac{1}{2^*}\int_{\mathbb{R}^N\backslash B} |aU_n|^{2^*} = \frac{1}{2^*}\int_{|x-\bar{x}|\geq\rho} |aU_n|^{2^*}$$

$$= \frac{a^{2^*}}{2^*}\int_{|x-\bar{x}|\geq\rho} \left|\epsilon_n^{(2-N)/2}U\left(\frac{x-\bar{x}}{\epsilon_n}\right)\right|^{2^*} dx$$

$$= \frac{a^{2^*}}{2^*}\int_{|y|\geq(\rho/\epsilon_n)} |U(y)|^{2^*} dy$$

$$\to 0 \quad \text{as } n \to \infty, \text{ because } U \in L^{2^*}(\mathbb{R}^N).$$

Thus,

$$\frac{1}{2^*}\int_\Omega |u_n|^{2^*} - \frac{1}{2^*}\int_{\mathbb{R}^N} |aU_n|^{2^*} = o(1) \quad \text{as } n \to \infty,$$

or

$$F(u_n) - \sum = \frac{1}{2}\int_\Omega |\nabla u_n|^2 - \frac{1}{2^*}\int_\Omega |u_n|^{2^*} - \frac{1}{2}\int_{\mathbb{R}^N} |\nabla(aU_n)|^2$$

$$- \frac{1}{2^*}\int_{\mathbb{R}^N} |aU_n|^{2^*}$$

$$= o(1) \quad \text{as } n \to \infty.$$

That is, $F(u_n) \to \sum$ as $n \to \infty$.

(ii) $F \left\| '(u_n) \right\|_{H^{-1}(\Omega)} \to 0$ as $n \to \infty$.

$$\|F'(u_n)\|_{H^{-1}(\Omega)} = \sup_{v \in H_0^1(\Omega), \|v\|=1} |F'(u_n)v|$$

$$= \sup_{v \in H_0^1(\Omega), \|v\|=1} \left| \int_\Omega (\nabla u_n \nabla v - |u_n|^{2^*-1} v) \right|,$$

but $-\triangle(aU_n) = (aU_n)^{2^*-1}$, or $\int_{\mathbb{R}^N} \nabla(aU_n)\nabla v = \int_{\mathbb{R}^N} (aU_n)^{2^*-1}v$, for $v \in H_0^1(\Omega)$.

(a) $\int_\Omega \nabla u_n \nabla v = \int_{\mathbb{R}^N} \nabla(aU_n)\nabla v + o(1)$ as $n \to \infty$: In fact,

$$\int_\Omega \nabla u_n \nabla v = \int_B \nabla(aU_n)\nabla v + \int_{\Omega\backslash B} \nabla(\zeta U_n)\nabla v$$

$$+ \int_{\mathbb{R}^N\backslash B} \nabla(aU_n)\nabla v - \int_{\mathbb{R}^N\backslash B} \nabla(aU_n)\nabla v$$

$$= \int_{\mathbb{R}^N} \nabla(aU_n)\nabla v + \int_{\Omega\backslash B} \nabla(\zeta U_n)\nabla v - \int_{\mathbb{R}^N\backslash B} \nabla(aU_n)\nabla v.$$

By (a) of (i),

$$\left| \int_{\mathbb{R}^N\backslash B} \nabla(aU_n)\nabla v \right| \le \left(\int_{\mathbb{R}^N\backslash B} |\nabla(aU_n)|^2 \right)^{1/2} \left(\int_{\mathbb{R}^N\backslash B} |\nabla v|^2 \right)^{1/2}$$

$$\le c \left(\int_{\mathbb{R}^N\backslash B} |\nabla(aU_n)|^2 \right)^{1/2} \to 0 \quad \text{as } n \to \infty.$$

$$\left| \int_{\Omega\backslash B} \nabla(\zeta U_n)\nabla v \right| \le \left(\int_{\Omega\backslash B} |\nabla(\zeta U_n)|^2 \right)^{1/2} \left(\int_{\Omega\backslash B} |\nabla v|^2 \right)^{1/2}$$

$$\le c \left[\left(\int_{\Omega\backslash B} |\nabla\zeta|^2 |U_n|^2 \right)^{1/2} + \left(\int_{\Omega\backslash B} |\zeta|^2 |\nabla U_n|^2 \right)^{1/2} \right]$$

$$\to 0 \quad \text{as } n \to \infty \text{ because } \zeta \in C_c^\infty(\Omega),$$

so

$$\int_\Omega \nabla u_n v = \int_{\mathbb{R}^N} \nabla(aU_n)\nabla v + o(1) \quad \text{as } n \to \infty.$$

(b) $\int_\Omega |u_n|^{2^*-1} v = \int_{\mathbb{R}^N} |aU_n|^{2^*-1} v + o(1)$ as $n \to \infty$.

$$\int_\Omega |u_n|^{2^*-1} v = \int_B |aU_n|^{2^*-1} v + \int_{\Omega \backslash B} |u_n|^{2^*-1} v$$

$$+ \int_{\mathbb{R}^N \backslash B} |aU_n|^{2^*-1} v - \int_{\mathbb{R}^N \backslash B} |aU_n|^{2^*-1} v$$

$$= \int_{\mathbb{R}^N} |aU_n|^{2^*-1} v + \int_{\Omega \backslash B} |u_n|^{2^*-1} v - \int_{\mathbb{R}^N \backslash B} |aU_n|^{2^*-1} v.$$

However, by (b) of (i), and (a) of (ii),

$$\left| \int_{\Omega \backslash B} |u_n|^{2^*-1} v \right| \le \left(\int_{\Omega \backslash B} |u_n|^{(2^*-1)\cdot 2} \right)^{1/2} \left(\int_{\Omega \backslash B} |v|^2 \right)^{1/2}$$

$$\le c \left(\int_{\Omega \backslash B} |u_n|^{(2^*-1)\cdot 2} \right)^{1/2} \to 0 \quad \text{as } n \to \infty.$$

$$\left| \int_{\mathbb{R}^N \backslash B} |aU_n|^{2^*-1} v \right| \le \left(\int_{\mathbb{R}^N \backslash B} |aU_n|^{(2^*-1)p} \right)^{1/p} \left(\int_{\mathbb{R}^N \backslash B} |v|^{p'} \right)^{1/p'}$$

$$\le c \left(\int_{\mathbb{R}^N \backslash B} |aU_n|^{2^*} \right)^{1/p} \to 0 \quad \text{as } n \to \infty,$$

where $p = \frac{2^*}{2^*-1} = \frac{2N}{N+2}$, $p' = \frac{2N}{N-2} = 2^*$. Thus,

$$\int_\Omega |u_n|^{2^*-1} v = \int_{\mathbb{R}^N} |aU_{\epsilon_n}|^{2^*-1} v + o(1) \quad \text{as } n \to \infty,$$

or as $n \to \infty$

$$\|F'(u_n)\|_{H^{-1}} = \sup_{\substack{v \in H_0^1(\Omega) \\ \|v\|=1}} |\int_\Omega (\nabla u_n \nabla v - |u_n|^{2^*-1} v)|$$

$$= \sup_{\substack{v \in H_0^1(\Omega) \\ \|v\|=1}} |\int_{\mathbb{R}^N} (\nabla(aU_n)\nabla v - |aU_n|^{2^*-1} v)| + o(1) = o(1) \ .$$

(iii) $u_n \to 0$ in $L^2(\Omega)$.

$$\int_\Omega |u_n|^2 = \int_\Omega |\zeta(x)|^2 \left| \frac{1}{\epsilon_n^{(N-2)/2}} U\left(\frac{x - \bar{x}}{\epsilon_n}\right) \right|^2 dx = \int_\Omega |\zeta(x)|^2 \, \epsilon_n^{2-N} \left| U\left(\frac{x-\bar{x}}{\epsilon_n}\right) \right|^2 dx$$

$$= \int_{\Omega - \{\bar{x}\}} |\zeta(x + \bar{x})|^2 \, \epsilon_n^{2-N} \left| U\left(\frac{x}{\epsilon_n}\right) \right|^2 dx \le c \int_{B(0,r)} \epsilon_n^{2-N} \left| U\left(\frac{x}{\epsilon_n}\right) \right|^2 dx$$

$$= c \int_{B(0,r/\epsilon_n)} \epsilon_n^{2-N} \epsilon_n^N \, |U(y)|^2 \, dy = c\epsilon_n^2 \int_{B(0,r/\epsilon_n)} \frac{1}{\left(1+|y|^2\right)^{N-2}} dy.$$

$$= c\epsilon_n^2 \int_0^{r/\epsilon_n} \frac{t^{N-1}}{(1+t^2)^{N-2}} dt \le c\epsilon_n^2 \int_0^1 t^{N-1} dt + c\epsilon_n^2 \int_1^{r/\epsilon_n} t^{N-1} t^{4-2N} dt$$

$$= c\epsilon_n^2 + c\epsilon_n^{N-2} \to 0 \text{ as } n \to \infty.$$

(iv) $\{u_n\}$ is not relatively compact in $H_0^1(\Omega)$: suppose there are a subsequence $\{u_{n_k}\}$ of $\{u_n\}$, and $u \in H_0^1(\Omega)$ such that

$$u_{n_k} \to u \quad \text{in } H_0^1(\Omega) \quad \text{as } k \to \infty,$$

and we have $u_{n_k} \to u$ in $L^2(\Omega)$. Then by (iii), $u = 0$, or $u_{n_k} \to 0$ in $H_0^1(\Omega)$. Thus,

$$F(u_{n_k}) \to 0.$$

By (i), $\sum = 0$, a contradiction.   ∎

We may superimpose k such gadgets. To do so, fix k distinct points of concentration $\bar{x}_1, \bar{x}_2, \cdots, \bar{x}_k$ in Ω, and k speeds of concentration $\epsilon_{1,n}, \epsilon_{2,n}, \cdots, \epsilon_{k,n}$ (any positive sequences tending to 0). Then the sequence of functions

$$u_n(x) = \sum_{i=1}^k \zeta_i(x) U_{\epsilon_{i,n}}(x - \bar{x}_i), \tag{8.45}$$

where $\zeta_i \in C_c^\infty(\Omega)$, $\zeta_i \equiv a$ near $\bar{x}_i$, is not relatively compact in $H_0^1(\Omega)$, but satisfies

$$F(u_n) \to k \sum.$$
$$F'(u_n) \to 0.$$

It is a striking fact that the $(PS)_c$ condition fails only at the levels $k\sum$, and that the formula (8.45) provides a good representation of the "critical points at infinity" for the levels $k\sum$.

Theorem 8.69 *For any $c \in \mathbb{R}$, a $(PS)_c$ sequence $\{u_n\}$ of F is bounded in $H_0^1(\Omega)$.*

Proof.   Note that

$$\frac{1}{2} \int_\Omega |\nabla u_n|^2 - \frac{1}{2^*} \int_\Omega |u_n|^{2^*} = c + o(1). \tag{8.46}$$

$$\triangle u_n = |u_n|^{2^*-1} + \phi_n \quad \text{with } \|\phi_n\|_{H^{-1}} \to 0. \tag{8.47}$$

Multiply (8.47) with u_n and then integrate to obtain

$$\int_\Omega |\nabla u_n|^2 = \int_\Omega |u_n|^{2^*} + \int_\Omega \phi_n u_n. \tag{8.48}$$

By (8.46) and (8.48), we have

$$(\frac{1}{2} - \frac{1}{2^*}) \int_\Omega |\nabla u_n|^2 = c + \frac{1}{2^*} \int_\Omega \phi_n u_n + o(1)$$
$$\leq c_1 + \|\phi_n\|_{H^{-1}} \|u_n\|_{H^1}$$
$$\leq c_1 + c_2 \|u_n\|_{H^1},$$

or

$$\|u_n\|_{H^1}^2 \leq c_3 + c_4 \|u_n\|_{H^1}.$$

Thus,

$$(\|u_n\|_{H^1} - \frac{c_4}{2})^2 \leq c_5,$$

or

$$\|u_n\|_{H^1} \leq c \quad \text{for each } n. \qquad \blacksquare$$

Remark 8.70 *By the Sobolev Embedding Theorem, or the above proof,*
$\{\|u_n\|_{L^{2^*}}\}$ *is bounded.*

Theorem 8.71 *If we assume that the only critical point of F is 0, then the $(PS)_c$ condition of F fails precisely at the levels $c = k\sum$, $k = 1, 2, \cdots$.*

Proof.   It suffices to prove that if $c < \sum$ or $k\sum < c < (k+1)\sum$, the $(PS)_c$ condition holds.

(i) The $(PS)_c$ condition holds if $c < \sum$: let $\{u_n\}$ be a sequence in $H_0^1(\Omega)$ such that

$$F(u_n) \to c, \quad \|F'(u_n)\|_{H^{-1}} \to 0 \quad \text{as } n \to \infty.$$

By Theorem 8.69, $\{\|u_n\|_{H^1}\}$ and $\{\|u_n\|_{L^{2^*}}\}$ are bounded such that

$$\frac{1}{2} \int_\Omega |\nabla u_n|^2 - \frac{1}{2^*} \int_\Omega |u_n|^{2^*} = c + o(1),$$

$$\int_\Omega |\nabla u_n|^2 = \int_\Omega |u_n|^{2^*} + o(1),$$

and we have

$$(\frac{1}{2} - \frac{1}{2^*}) \int_\Omega |\nabla u_n|^2 = c + o(1).$$

Note that

$$\frac{1}{2} - \frac{1}{2^*} = \frac{1}{2} - \frac{N-2}{2N} = \frac{2}{2N} = \frac{1}{N}.$$

We have

$$\int_\Omega |\nabla u_n|^2 = Nc + o(1),$$

$$\int_\Omega |\nabla u_n|^{2^*} = Nc + o(1).$$

Sobolev's inequality implies that

$$\int_\Omega |\nabla u_n|^2 \geq S \|u_n\|_{L^{2^*}}^2 .$$

Thus,

$$Nc \geq S(Nc)^{2/2^*},$$

which implies that either $c = 0$ or

$$(Nc)^{1-2/2^*} = (Nc)^{2/N} \geq S,$$

or

$$c \geq \frac{1}{N} S^{N/2} = \sum,$$

by assumption $c < \sum$, so that $c = 0$. We conclude that

$$\int_\Omega |\nabla u_n|^2 = Nc + o(1) = o(1).$$

Thus, $u_n \to 0$ in $H_0^1(\Omega)$, or the $(PS)_c$ condition holds when $c < \sum$.

(ii) The $(PS)_c$ condition holds for c, $k \sum < c < (k+1) \sum$, $k = 1, 2, \cdots$. We claim that no sequence $\{u_n\}$ exists in $H_0^1(\Omega)$ such that $F(u_n) \to c$, $\|F'(u_n)\|_{H^{-1}} \to 0$, and $k \sum < c < (k+1) \sum$, $k = 1, 2, \cdots$. Suppose by contradiction that $\{u_n\}$ is such a sequence. Again we have

$$\frac{1}{2} \int_\Omega |\nabla u_n|^2 - \frac{1}{2^*} \int_\Omega |u_n|^{2^*} = c + o(1).$$

$$-\triangle u_n = |u_n|^{2^*-1} + \phi_n, \ \|\phi_n\|_{H^{-1}} \to 0, \ \|\phi_n\|_{H^{-1}} \to 0.$$

By Theorem 8.69, $\{u_n\}$ is bounded in $H_0^1(\Omega)$. There is a subsequence $\{u_{n_k}\}$ of $\{u_n\}$, and $u \in H_0^1(\Omega)$ such that

$$u_{n_k} \rightharpoonup u \quad \text{weakly in } H_0^1(\Omega),$$

$$u_{n_k} \to u \quad \text{strongly in } L^{2^*-1}(\Omega).$$

Multiply (8.47) by $\varphi \in H_0^1(\Omega)$ and then integrate it:

$$\int_\Omega (-\triangle u_{n_k}) \varphi = \int_\Omega |u_{n_k}|^{2^*-1} \varphi + \int_\Omega \phi_{n_k} \varphi,$$

or

$$\int_\Omega u_{n_k}(-\triangle\varphi) = \int_\Omega |u_{n_k}|^{2^*-1}\varphi + \int_\Omega \phi_{n_k}\varphi.$$

Letting $k \to \infty$, we obtain

$$\int_\Omega u(-\triangle\varphi) = \int_\Omega |u|^{2^*-1}\varphi,$$

or

$$\int_\Omega [(-\triangle u) - |u|^{2^*-1}]\varphi = 0, \quad \text{for } \varphi \in H_0^1(\Omega).$$

This implies that

$$F'(u) = -\triangle u - |u|^{2^*-1} = 0,$$

where

$$F(u) = \frac{1}{2}\int_\Omega |\nabla u|^2 - \frac{1}{2^*}\int_\Omega |u|^{2^*}.$$

Recall that we assume that $u = 0$ is the only critical point of f, so set $u = 0$ and then

$$u_{n_k} \rightharpoonup 0 \quad \text{weakly in } H_0^1(\Omega).$$

It is clear that $u_{n_k} \to 0$ strongly in $H_0^1(\Omega)$; otherwise, by (i), $c = 0$, which contradicts the assumption that $k \sum < c$ for some $k \geq 1$. Write $\{u_n\}$ instead of $\{u_{n_k}\}$, and take a sequence $\{x_n\}$ in Ω, and a sequence $\epsilon_n \to 0$ such that

$$\int_{B(x_n, \epsilon_n)} |\nabla u_n|^2 = v \quad \text{with } v > 0 \text{ is small enough.}$$

The singular behavior of $\{u_n\}$ is analyzed by a bottom-up technique. Namely, let u_n be extended by 0 outside Ω, and set

$$\bar{u}_n(x) = \epsilon_n^{(N-2)/2} u_n(\epsilon_n x + x_n) \quad \text{for } x \in \mathbb{R}^N,$$

so that $\bar{u}_n$ satisfies

$$\int_{B(0,1)} |\nabla \bar{u}_n|^2 = v, \quad \int_{\mathbb{R}^N} |\nabla \bar{u}_n|^2 = c,$$

and

$$-\triangle \bar{u}_n = |\bar{u}_n|^{2^*-1} + \overline{\phi}_n \quad \text{on } \Omega_n = \frac{\Omega - x_n}{\epsilon_n},$$

where $\overline{\phi}_n(x) = \epsilon_n^{(N-2)/2}\phi_n(\epsilon_n x + x_n)$ and $\|\phi_n\|_{H^{-1}} \to 0$ as $n \to 0$ as $n \to \infty$. Passing to the limit on the sequence $\{\bar{u}_n\}$ (or rather a subsequence) one can show that, for some $w \in H^1(\Omega)$

$$\bar{u}_n \to w \quad \text{strongly in } L_{\ell oc}^{2^*-1}(\mathbb{R}^N),$$
$$\triangle \bar{u}_n \to \triangle w \quad \text{strongly in } L_{\ell oc}^{2^*-1}(\mathbb{R}^N).$$

Thus,

$$\int_{B(0,1)} |\triangle w|^2 = v.$$

It follows that

$$\begin{cases} -\triangle w = |w|^{2^*-1} & \text{on } \mathbb{R}^N, \\ w \neq 0. \end{cases}$$

Therefore w is an old friend! We remove this first singularity by letting

$$v_n(x) = u_n(x) - \frac{1}{\epsilon_n^{(N-2)/2}} w\left(\frac{x - x_n}{\epsilon_n}\right) \quad \text{for } x \in \Omega.$$

Recall that

$$\frac{1}{2}\int_{\mathbb{R}^N} |\nabla w|^2 - \frac{1}{2}\int_{\mathbb{R}^N} |w|^{2^*} = \sum.$$

We can prove that $\{v_n\}$ satisfies

$$F(v_n) \to c - \sum, \quad \text{and } F'(v_n) \to 0.$$

(see Theorem 8.73 below for more detail.) In other words, the sequence $\{v_n\}$ is like the sequence $\{u_n\}$, except that c is replaced by $c - \sum$. Iterating this construction k times, we obtain a sequence $\{v_n^k\}$ such that

$$F(v_n^k) \to c - k\sum < \sum, \quad \text{and } F'(v_n^k) \to 0.$$

Note that each singularity contributes the same amount of energy, namely $\sum$. By part (i), $c - k\sum = 0$, a contradiction because $k\sum < c$. ∎

Let

$$F(u) = \frac{1}{2}\int_{\Omega} (|\nabla u|^2 - \lambda u^2) - \frac{1}{2^*}\int_{\Omega} |u|^{2^*},$$

$$F^* = \frac{1}{2}\int_{\Omega} |\nabla u|^2 - \frac{1}{2^*}\int_{\Omega} |u|^{2^*}.$$

Lemma 8.72 Let $\{u_n\}$ in $H_0^1(\Omega)$ satisfy

$$-\triangle u_n = |u_n|^{2^*-2} + \phi_n \quad \text{in } \Omega,$$

$$\|\phi_n\|_{H^{-1}} \to 0,$$

$$0 < \alpha \leq \int_{\Omega} |\nabla u_n|^2 \leq c_0 \quad \text{for } n = 1, 2, \cdots.$$

Then the following exist:

(i) *a nonconstant solution w of $\triangle w = |w|^{2^*-1} w$ on $\mathbb{R}^N$;*

(ii) *a sequence $\{x_n\}$ in Ω;*

(iii) *a sequence $\{\epsilon_n > 0\}$ with $\lim_{n\to\infty} \epsilon_n = 0$, such that (for some subsequence still denoted by u_n)*

$$\bar{u}_n(x) = \epsilon_n^{(N-2)/2} u_n(\epsilon_n x + x_n) \to w \quad \text{a.e. for } x \in \mathbb{R}^N,$$

$$\nabla \bar{u}_n \rightharpoonup \nabla w \quad \text{weakly in } L^2(\mathbb{R}^N),$$

$$\frac{1}{\epsilon_n} \text{dist}(x_n, \partial\Omega) \to \infty,$$

$$\|\bar{u}_n\|^2_{H^1} = \|\bar{u}_n - w\|^2_{H^1} + \|w\|^2_{H^1} + o(1),$$
$$F(\bar{u}_n) = F^*(\bar{u}_n - w) + F(w) + o(1),$$
$$F'(\bar{u}_n) = (F^*)'(\bar{u}_n - w) + F'(w) + o(1).$$

Proof.   We introduce the concentration functions:

$$Q_n(t) = \sup_{x \in \mathbb{R}^N} \int_{x+t\Omega} |\nabla u_n|^2 \quad \text{for } t \geq 0.$$

We claim that each $Q_n(t)$ is continuous, nondecreasing in t, and satisfies

$$Q_n(0) = 0,$$
$$Q_n(1) = Q_n(\infty) = \int_\Omega |\nabla u_n|^2 \geq \alpha.$$

In fact, for the continuity of $Q_n(t)$: fix $t_0 \geq 0$. For $\epsilon > 0$, there is $x_0 \in \mathbb{R}^N$ such that

$$Q_n(t_0) - \epsilon < \int_{x_0+t_0\Omega} |\nabla u_n|^2.$$

Now, for $t > 0$,

$$Q_n(t_0) - Q_n(t) \leq \int_{x_0+t_0\Omega} |\nabla u_n|^2 + \epsilon - \int_{x_0+t\Omega} |\nabla u_n|^2,$$

but

$$\int_{x_0+t\Omega} |\nabla u_n|^2 = \int_{x_0+t_0\Omega} |\nabla u_n|^2 + o(1) \quad \text{as } t \to t_0.$$

Thus,

$$Q_n(t) \to Q_n(t_0) \quad \text{as } t \to t_0.$$

Fix a constant a with

$$0 < a < \min\{\frac{1}{4c_0}, \alpha\}.$$

For each n, there is some $x_n \in \bar{\Omega}$, $0 < \epsilon_n < 1$ such that

$$Q_n(\epsilon_n) = \int_{x_n + \epsilon_n \Omega} |\nabla u_n|^2 = a.$$

Set

$$\bar{u}_n(x) = \epsilon_n^{(N-2)/2} u_n(\epsilon_n x + x_n) \quad \text{for } n = 1, 2, \cdots.$$

Then

$$\int_{\mathbb{R}^N} |\nabla \bar{u}_n|^2 = \int_{\mathbb{R}^N} |\nabla u_n|^2 \leq c.$$

Therefore, $\{u_n\}$ is in $H^1_{loc}(\mathbb{R}^N)$ with

$$\|\bar{u}_n\|_{D^{1,2}(\mathbb{R}^N)} \leq c \quad \text{for } n = 1, 2, \cdots.$$

By the Rellich-Kondrakov Theorem, there is $w \in D^{1,2}(\mathbb{R}^N)$ such that

$$\bar{u}_n \to w \quad \text{strongly in } L^2_{loc}(\mathbb{R}^N), \text{ a.e. on } \mathbb{R}^N,$$
$$\nabla \bar{u}_n \rightharpoonup \nabla w \quad \text{weakly in } L^2(\mathbb{R}^N).$$

We may assume that $x_n \to x_0$, $\Omega_n = \dfrac{1}{\epsilon_n}(\Omega - x_n)$, so that $\Omega_n \to D$. We now distinguish several cases:

Case (i) $\epsilon_n \to \ell > 0$.

Case (ii) $\epsilon_n \to 0$, and $\frac{1}{\epsilon_n} \text{dist}(x_n, \partial\Omega) \to m < \infty$, so that D is a half-plane.

Case (iii) $\epsilon_n \to 0$ and $\frac{1}{\epsilon_n} \text{dist}(x_n, \partial\Omega) \to \infty$, so that $D = \mathbb{R}^N$.

We claim that cases (i) and (ii) cannot occur. Let $\theta_n \in H^1_0(\Omega)$ be the solution of

$$-\triangle \theta_n = \phi_n \quad \text{in } \Omega.$$

Then

$$\|\theta_n\|^2_{H^1(\Omega)} = \int_\Omega |\nabla \theta_n|^2 = -\int_\Omega (\nabla \theta_n)\theta_n = \int \theta_n \phi_n \leq \|\theta_n\|_{H^1}\|\phi_n\|_{H^1}.$$

Thus,

$$\|\theta_n\|_{H^1(\Omega)} \leq \|\phi_n\|_{H^1(\Omega)} \to 0.$$

From

$$-\triangle u_n = |u_n|^{2^*-2} u_n + \phi_n \quad \text{in } \Omega,$$

we obtain

$$-\triangle(u_n - \theta_n) = |u_n|^{2^*-2} u_n \quad \text{in } \Omega.$$

Let

$$\bar{\theta}_n(x) = \epsilon_n^{(N-2)/2} \theta_n(\epsilon_n x + x_n).$$

Then
$$-\triangle(\bar{u}_n - \bar{\theta}_n) = |\bar{u}_n|^{2^*-2}\,\bar{u}_n \quad \text{on } \Omega_n.$$

Note that
$$\int_\Omega |\nabla\bar{\theta}_n|^2 = \int_\Omega |\nabla\theta_n|^2 = o(1).$$

If we multiply by $\varphi \in H_0^1(\Omega_n)$ and integrate with the equation
$$-\triangle(\bar{u}_n - \bar{\theta}_n) = |\bar{u}_n|^{2^*-2}\,\bar{u}_n \quad \text{in } \Omega,$$

then we obtain
$$\int_{\Omega_n} \nabla\bar{\theta}_n \nabla\varphi - \int_{\Omega_n} \nabla\bar{\theta}_n \nabla\varphi = \int_{\Omega_n} |\bar{u}_n|^{2^*-2}\,\bar{u}_n\varphi.$$

Letting $n \to \infty$, we obtain
$$\int_D \nabla w \nabla\varphi = \int_D |w|^{2^*-2}\,w\varphi,$$

or
$$\begin{cases} -\triangle w = |w|^{2^*-2}\,w & \text{in } D, \\ w = 0 & \text{on } \partial D. \end{cases}$$

Suppose case (i) holds. Then by the Pohozaev identity,
$$w = 0 \quad \text{in } D.$$

Thus,
$$\bar{u} \rightharpoonup 0 \quad \text{weakly in } H_0^1(\Omega),$$
$$\bar{u}_n \to 0 \quad \text{strongly in } L^2(\Omega).$$

We claim that
$$\nabla\bar{u} \to 0 \quad \text{strongly in } L^2_{\ell oc}(\mathbb{R}^N).$$

It suffices to prove that
$$\int \zeta^2 |\nabla\bar{u}|^2 = o(1),$$

where $\zeta \in C_c^\infty(\mathbb{R}^N)$ with $\mathrm{supp}\zeta \subset x+\Omega$ for some $x \in \mathbb{R}^N$. If we multiply the equation
$$-\triangle(\bar{u}_n - \bar{\theta}_n) = |\bar{u}_n|^{2^*-2}\,\bar{u}_n$$

by $\zeta^2\bar{u}_n$, we find
$$\int \nabla\bar{u}_n \nabla(\zeta^2\bar{u}_n) - \int \nabla\bar{\theta}_n \nabla(\zeta^2\bar{u}_n) = \int \zeta^2 |\bar{u}_n|^{2^*}. \tag{8.49}$$

Note that

$$[c]ll \left|\nabla(\zeta^2 \bar{u}_n)\right|^2 = \left|\zeta \nabla \bar{u}_n + (\nabla \zeta)\bar{u}_n\right|^2$$
$$= \zeta^2 \left|\nabla \bar{u}_n\right|^2 + 2\zeta(\nabla \zeta)(\nabla \bar{u}_n)\bar{u}_n + \left|\nabla \zeta\right|^2 \bar{u}_n^2, \tag{8.50}$$

$$\nabla \bar{u}_n \nabla(\zeta^2 \bar{u}_n) = \nabla \bar{u}_n[2\zeta(\nabla \zeta)\bar{u}_n + \zeta^2(\nabla \bar{u}_n)]$$
$$= 2\zeta(\nabla \zeta)(\nabla \bar{u}_n)\bar{u}_n + \zeta^2 \left|\nabla \bar{u}_n\right|^2.$$

These equalities imply that

$$\left|\nabla(\zeta \bar{u}_n)\right|^2 = \nabla \bar{u}_n \nabla(\zeta^2 \bar{u}_n) + \left|\nabla \zeta\right|^2 \bar{u}_n^2. \tag{8.51}$$

We can then write (8.49) as follows:

$$\int \left|\nabla(\zeta \bar{u}_n)\right|^2 - \int \left|\nabla \zeta\right|^2 \bar{u}_n - \int \nabla \bar{\theta}_n \nabla(\zeta^2 \bar{u}_n) = \int \zeta^2 \left|\bar{u}_n\right|^{2^*}. \tag{8.52}$$

Note that

$$\int \left|\nabla \zeta\right|^2 \bar{u}_n^2 + \int \nabla \bar{\theta}_n \nabla(\zeta^2 \bar{u}_n) = o(1),$$

and

$$\int \zeta^2 \left|\bar{u}_n\right|^{2^*} = \int \left|\zeta^2 \bar{u}_n\right|^2 \left|\bar{u}_n\right|^{2^*-2}$$
$$\leq \left(\int \left|\zeta^2 \bar{u}_n\right|^{2^*}\right)^{2/2^*} \left(\int \left|\bar{u}_n\right|^{2^*}\right)^{(2^*-2)/2^*}$$
$$\leq \left(\int \left|\nabla(\zeta \bar{u}_n)\right|^2\right) \left(\int \left|\nabla \bar{u}_n\right|^2\right)^{2/(N-2)} \quad \text{by the Sobolev inequality}$$
$$\leq a^{2/(N-2)} \left(\int \left|\nabla(\zeta \bar{u}_n)\right|^2\right).$$

By (8.52)

$$\int \left|\nabla(\zeta \bar{u}_n)\right|^2 = \int \zeta^2 \left|\bar{u}_n\right|^{2^*} + o(1) \leq a^{2/(N-2)} \left(\int \left|\nabla(\zeta \bar{u}_n)\right|^2\right) + o(1).$$

Thus we obtain

$$\left(1 - a^{2/(N-2)}\right) \int \left|\nabla(\zeta \bar{u}_n)\right|^2 = o(1),$$

but a is small, so

$$\int \left|\nabla(\zeta \bar{u}_n)\right|^2 = o(1).$$

By (8.49)

$$\int \zeta^2 |\nabla \bar{u}_n|^2 = o(1).$$

However,

$$\int_\Omega |\nabla \zeta u_n|^2 = \int |\nabla \bar{u}_n|^2 = a > 0.$$

This contradiction excludes case (i).

Suppose we have case (ii). Similarly to the above proof and by the Pohozaev Theorem, if $D = \mathbb{R}^N_+$, and

$$\begin{cases} -\triangle w = |w|^{2^*-2} w & \text{in } D, \\ w = 0 & \text{on } \partial D, \end{cases}$$

then $w = 0$ in D. We thus obtain

$$\bar{u}_n \rightharpoonup 0 \quad \text{weakly in } H^1_0(\mathbb{R}^N),$$
$$\bar{u}_n \to 0 \quad \text{strongly in } L^2(\mathbb{R}^N_+).$$

As in case (i), we can prove that

$$\nabla \bar{u}_n \to 0 \quad \text{strongly in } L^2_{\ell oc}(\mathbb{R}^N_+).$$

This contradicts

$$\int_\Omega |\nabla \bar{u}_n|^2 = \int_{\Omega_n} |\nabla u_n|^2 = a > 0.$$

Therefore case (ii) is excluded.

Hence, the only case that can occur is case (iii).

$$\epsilon_n \to 0, \quad \frac{1}{\epsilon_n} \operatorname{dist}(x_n, \partial\Omega) \to \infty,$$

so that $D = \mathbb{R}^N$. To conclude the proof of Lemma 8.72, it only remains to show that w is not a constant. We claim that

$$\nabla \bar{u}_n \to \nabla w \quad \text{strongly in } L^2_{\ell oc}(\mathbb{R}^N).$$

This result together with

$$\int_\Omega |\nabla \bar{u}_n|^2 = a > 0$$

implies

$$\int_\Omega |\nabla w|^2 = a > 0.$$

Therefore w is not a constant. In fact,

$$-\triangle w = |w|^{2^*-1} \quad \text{in } \mathbb{R}^N,$$

so $w = 0$, or

$$w(x) = \frac{[N(N-2)\lambda^2]^{(N-2)/4}}{[[\lambda^2 + |x - x_0|^2]^{(N-2)/2}},$$

which is a bounded function on $\mathbb{R}^N$. We require

$$\nabla \bar{u}_n \to \nabla w \quad \text{in } L^2(B_R), \quad \text{for each } R > 0.$$

If we let $v_n = \bar{u}_n - w$, then

$$\nabla v_n \rightharpoonup 0 \quad \text{weakly in } \mathcal{D}^{1,2}.$$

Moreover,

$$
\begin{aligned}
-\triangle v_n &= -\triangle u_n + \triangle w \\
&= |u_n|^{2^*-2} u_n + \phi_n - |w|^{2^*-2} w \\
&= |v_n|^{2^*-2} v_n + [|u_n|^{2*-2} u_n - |v_n|^{2^*-2} v_n + \phi_n - |w|^{2^*-2} w].
\end{aligned}
$$

Let

$$
\begin{aligned}
-\triangle h_n &= [|u_n|^{2^*-2} u_n - |v_n|^{2^*-2} + \phi_n - |w|^{2^*-2} w] \\
&= [|v_n + w|^{2^*-2} (v_n + w) - |v_n|^{2^*-2} v_n + \phi_n - |w|^{2^*-2} w].
\end{aligned}
$$

Multiply by h_n on both sides and then integrate to obtain

$$
\begin{aligned}
\int_{B_R} |\nabla h_n|^2 &= \int_{B_R} h_n dx \int_0^1 \frac{d}{dt} |w + t v_n|^{2^*-1} dt + \int \phi_n h_n - \int |v_n|^{2^*-2} v_n h_n \\
&\leq (2^* - 1) \int_{B_R} |w + tv|^{2^*-2} v_n h_n + o(1) \\
&\leq c \int_{B_R} \left(|w|^{2^*-2} v_n h_n + |v_n|^{2^*-1} h_n \right) + o(1) \\
&= o(1) \quad \text{as } n \to \infty.
\end{aligned}
$$

Thus,

$$h_n \to 0 \quad \text{strongly in } H_0^1,$$

or

$$-\triangle v_n = |v_n|^{2^*-2} v_n + \eta_n,$$
$$\eta_n \to 0 \quad \text{strongly in } H^{-1}.$$

As in the proof of case (i),

$$\nabla v_n \to 0 \quad \text{strongly in } L^2_{\ell oc}(\mathbb{R}^N).$$

Moreover, by the Brezis-Lieb Lemma 1.45, we have

$$\|u_n\|_{H^1}^2 = \|\bar{u}_n - w\|_{H^1}^2 + \|w\|_{H^1}^2 + o(1),$$
$$F(\bar{u}_n) = F^*(\bar{u}_n - w) + F(w) + o(1),$$
$$F'(\bar{u}_n) = (F^*)'(\bar{u}_n - w) + F'(w) + o(1).$$ ∎

Theorem 8.73 *Let $\{u_n\}$ be a $(PS)_c$ sequence for F. Then there are $k \in N$, a sequence of k points $\{y_n^j\} \subset \mathbb{R}^N$, a sequence of k positive numbers $\{\epsilon_m^j\}$, $1 \le j \le k$, and a sequence of $k+1$ functions $\{u_n^j\} \subset H_0^1(\Omega)$, $0 \le j \le k$, such that for some subsequence, still denoted by $\{u_n\}$,*

$$u_n(x) = u_n^0(x) + \sum \epsilon_n^{j\,-(N-2)/2} u_n^j \left(\frac{x - y_n^j}{\epsilon_n^j}\right),$$
$$u_n^0(x) \to u^0(x) \quad \text{strongly } H_0^1(\Omega),$$
$$u_n^j(x) \to u^j(x) \quad \text{strongly } H^1(\mathbb{R}^N), \quad 1 \le j \le k,$$

where

$$- \triangle u^\circ - \lambda u^\circ = |u^\circ|^{2^*-2} u^\circ \quad \text{in } \Omega,$$
$$- \triangle u^j = |u^j|^{2^*-2} u^j \quad \text{in } \mathbb{R}^N, \quad 1 \le j \le k,$$
$$\|u_m\|_{H^1}^2 = \sum_{j=1}^{k} \|u^j\|_{H^1}^2 + o(1),$$
$$F(u_m) = F(u^\circ) + \sum_{j=1}^{k} F^*(u^j) + o(1).$$

Proof. Because $\{u_n\}$ is a $(PS)_c$ sequence for F, $\{u_n\}$ is bounded in $H_0^1(\Omega)$. Assume

$$u_n \rightharpoonup u^\circ \quad \text{weakly in } H_0^1(\Omega).$$

Then

$$F'(u_n) \to F'(u^\circ) = 0,$$

or

$$-\triangle u^\circ - \lambda u^\circ = |u^\circ|^{2^*-2} u^\circ \quad \text{in } \Omega.$$

Put

$$g_n^1 = u_n - u^\circ.$$

Using the Brezis-Lieb Lemma 1.45, we have

$$\left\|g_n^1\right\|^2 = \|u_n\|^2 - \|u_0\|^2 + o(1),$$
$$F(g_n^1) = F(u_n) - F(u^\circ) + o(1),$$
$$F(g_n^1) = F'(u_n) - F'(u_0) + o(1).$$

Then $\{g_n^1\}$ is a $(PS)_c$ sequence for F, and

$$g_n^1 \rightharpoonup 0 \quad \text{weakly in } H_0^1(\Omega).$$

If $g_n^1 \to 0$ strongly in $H_0^1(\Omega)$, then we are done. Suppose $g_n^1 \not\to 0$ strongly in $H_0^1(\Omega)$. By Lemma 8.72, there is a sequence of points $\{x_n^1\}$, and a sequence of positive numbers $\{\eta_m^1\}$ such that

$$v_n^1(x) = (\eta_n^1)^{(N-2)/2} g_n^1(\eta_n^1 x + x_m^1)$$

converges weakly in $H_0^1(\mathbb{R}^N)$ to a solution u^1 of the equation

$$-\triangle u = |u|^{2^*-2} u \quad \text{in } \mathbb{R}^N.$$
$$F^*(v_n^1 - u^1) = F(g_m^1) - F^*(u^1) + o(1)$$
$$= F(u_m) - F(u^\circ) - F^*(u^1) + o(1),$$
$$\left\| v_n^1 - u^1 \right\|^2 = \left\| g_n^1 \right\|^2 - \left\| u^1 \right\|^2 + o(1)$$
$$= \left\| u_n \right\|^2 - \left\| u^\circ \right\|^2 - \left\| u^1 \right\|^2 + o(1).$$

Iterating this procedure we obtain a $(PS)_c$ sequence of functions:

$$g_n^j = v_n^{j-1} - u^{j-1}, \ g_n^j \rightharpoonup 0 \quad \text{weakly in } H_0^1(\Omega),$$
$$g_n^j = (\eta_n^j)^{(N-2)/2} g_n^j(\eta_n^j x + x_m^j) \rightharpoonup u^j \quad \text{weakly in } H_0^1(\Omega),$$

where $v_n^0 = u_n$ for $n = 1, 2, \cdots$,

$$[c]llF^*(v_n^j) = F^*(g_n^j) = F^*(v_m^{j-1} - u^{j-1}) + o(1)$$
$$= F^*(v_m^{j-1}) - F^*(u^{j-1}) + o(1) \tag{8.53}$$
$$= F(u_m) - F(u^\circ) - \sum_{i=1}^{j-1} F^*(u^i) + o(1),$$

$$\left\| v_m^j \right\|^2 = \left\| g_m^j \right\|^2 = \left\| v_m^{j-1} - u^{j-1} \right\|^2 + o(1)$$
$$= \left\| v_m^{j-1} \right\|^2 - \left\| u^{j-1} \right\|^2 + o(1)$$
$$= \left\| u_m \right\|^2 - \left\| u^\circ \right\|^2 - \sum_{i=1}^{j-1} \left\| u^j \right\| + o(1).$$

On the other hand, by the definition of S, we have

$$\left\| u^j \right\|_{L^2}^2 \geq S \left\| u^j \right\|_{L^{2^*}}^2 .$$

By the estimate

$$
\begin{aligned}
0 &= (F^{*\prime}(u^j),\, u^j) \\
&= \left\| \nabla u^j \right\|_{L^2}^2 - \left\| u^j \right\|_{L^{2^*}}^{2^*} \\
&= \left\| \nabla u^j \right\|_{L^2}^2 \left(1 - S^{-N/(N-2)} \left\| \nabla u^j \right\|_{L^2}^{-4/(N-2)} \right),
\end{aligned}
$$

so

$$
\left\| \nabla u^j \right\|_{L^{2^*}}^{2^*} \geq S^{N/2}.
$$

This inequality with (8.53) and the bound of $\{u_n\}$ tells us that the iteration must terminate at some index $k > 0$. Finally, put

$$
u_m^k = v_m^k(x) = (\eta_m^k)^{(N-2)/2} g_m^k(\eta_m^k x + x_m^k),
$$

$$
u_m^{k-1}(x) = v_m^{k-1}(x) - \frac{1}{(\eta_m^k)^{(N-2)/2}} u_m^k \left(\frac{x - x_m^k}{\eta_m^k} \right),
$$

$$
u_m^{k-2}(x) = v_m^{k-2}(x) - \frac{1}{(\eta_m^{k-1})^{(N-2)/2}} v_m^{k-1} \left(\frac{x - x_m^{k-1}}{\eta_m^{k-1}} \right)
$$

$$
= v_m^{k-2}(x) - \frac{1}{(\eta_m^{k-1})^{(N-2)/2}} u_m^{k-1} \left(\frac{x - x_m^{k-1}}{\eta^{k-1}} \right)_m
$$

$$
- \frac{1}{(\eta_m^k \eta_m^{k-1})^{(N-2)/2}} u_m^k \left(\frac{x - (x_m^{k-1} + \eta_m^{k-1} x_m^k)}{\eta_m^k \eta_m^{k-1}} \right),
$$

$$
u_m^{k-3}(x) = v_m^{k-3}(x) - \frac{1}{\eta_m^{k-2}} v_m^{k-2} \left(\frac{x - x_m^{k-2}}{\eta_m^{k-2}} \right)
$$

$$
= v_m^{k-3}(x) - \frac{1}{\eta_m^{k-2}} u_m^{k-2} \left(\frac{x - x_m^{k-2}}{\eta_m^{k-2}} \right)
$$

$$
= \frac{1}{\left(\eta_m^{k-2} \eta_m^{k-1} \right)^{(N-2)/2}} u_m^{k-1} \left(\frac{x - (x_m^{k-2} + \eta_m^{k-2} x_m^{k-1})}{\eta_m^{k-2} \eta_m^{k-1}} \right)
$$

$$
- \frac{1}{\left(\eta_m^{k-2} \eta_m^{k-1} \eta_m^k \right)^{(N-2)/2}} u_m^{k-1} \left(\frac{x - (x_m^{k-2} + \eta_m^{k-2} x_m^{k-1} + \eta_m^{k-2} \eta_m^{k-1} x_m^k)}{\eta_m^{k-2} \eta_m^{k-1} \eta_m^k} \right)
$$

$$
\cdots,
$$

$$
u_m^\circ(x) = u_m(x) - \sum_{j=1}^k \frac{1}{\left(\epsilon_m^j \right)^{(N-2)/2}} u_m^j \left(\frac{x - y_m^j}{\epsilon_m^j} \right),
$$

where

$$\epsilon_m^1 = \eta_m^1, \quad \epsilon_m^j = \prod_{i=1}^{j} \eta_m^i, \quad 2 \le j \le k,$$

$$y_m^1 = x_m^1, \quad y_m^j = y_m^1 + \sum_{i=2}^{j} (\epsilon_m^{i-1}) x_m^i, \quad 2 \le j \le k. \qquad \blacksquare$$

Notes The material in this Chapter is from Ambrosetti-Rabinowitz [3], Bahri-Coron [5], Brezis [7], [9], Brezis-Nirenberg [10], Lions [30], [31], [32], [33], Struwe [45], Zeidler [55], Talenti [48], and Torchinsky [50].

Chapter 9

TOPOLOGICAL DEGREE THEORY

In this chapter we present various topological degree properties.

9.1 Finite Dimensional Degree Theory

Let $\Omega \subset \mathbb{R}^N$ be a bounded open set. For $f \in C^1(\bar{\Omega}, \mathbb{R}^N)$, $f = (f_1, \cdots, f_N)$, set

$$S = \{x \in \Omega \mid J_f(x) = 0\},$$

where $J_f(x)$ is the Jacobian determinant of $x \in \bar{\Omega}$:

$$J_f(x) = \det \begin{pmatrix} \dfrac{\partial f_1}{\partial x_1}(x) & \cdots & \dfrac{\partial f_N}{\partial x_1}(x) \\ \vdots & & \vdots \\ \dfrac{\partial f_1}{\partial x_N}(x) & \cdots & \dfrac{\partial f_N}{\partial x_N}(x). \end{pmatrix}.$$

Fix $b \in \mathbb{R}^N$, $b \notin f(\partial \Omega) \cup f(S)$. If $x \in \Omega \cap f^{-1}(b)$, then by the Inversion Function Theorem, f is a diffeomorphism of a neighborhood of x onto a neighborhood of b. Therefore, $f^{-1}(b)$ is discrete and compact: $f^{-1}(b) = \{x_1, \cdots, x_k\}$ with $J_f(x_i) \neq 0$ for $i = 1, \cdots, k$.

Definition 9.1 *For $f \in C^1(\bar{\Omega}, \mathbb{R}^N)$ and $b \notin f(\partial \Omega) \cup f(S)$, the (topological) degree of f with respect to Ω and b is the integer*

$$\mathrm{d}(f, \Omega, b) = \sum_{x \in f^{-1}(b)} \mathrm{sgn}\, J_f(x), \text{ where } \mathrm{sgn}\, t = \begin{cases} 1 & \text{for } t > 0, \\ -1 & \text{for } t < 0. \end{cases}$$

Definition 9.2 *(The Heinz Integral of Degree) Assume $f \in C^1(\bar{\Omega}, \mathbb{R}^N)$, $b \notin f(\partial \Omega) \cup f(S)$, and $f^{-1}(b) = \{x_1, \cdots, x_k\}$ with $J_f(x_i) \neq 0$ for $1 \leq i \leq k$. Take a neighborhood O_i of x_i, such that $f : O_i \to f(O_i)$ is a diffeomorphism. Choose a neighborhood U_i of x_i such that $U_i \subset O_i$ and $\{U_i\}$ are disjoint to each other, $J_f(x) \neq 0$ for each $x \in U_i$, and $J_f(x)$ has constant sign on each U_i. Take $\epsilon_1 > 0$ so small that $B(b, \epsilon_1) \subset \bigcap_{i=1}^{k} f(U_i)$, and take $M_i = U_i \cap f^{-1}(B(b, \epsilon_1))$. Then the M_i are neighborhoods* *of x_i, disjoint from each other, J_f has constant sign on M_i, and $f : M_i \to B(b, \epsilon_1)$ is a diffeomorphism. Consider the compact set $D = \bar{\Omega} \backslash \bigcup_{i=1}^{k} M_i$. Note that $f(D)$ is compact when $b \notin f(D)$. Take $\epsilon_2 > 0$ such that $B(b, \epsilon_2) \cap f(D) = \emptyset$, and consequently $f^{-1}(B(b, \epsilon_2)) \cap D = \emptyset$. Thus, $f^{-1}(B(b, \epsilon_2)) \subset \bigcup_{i=1}^{k} M_i$. Let $\epsilon = \min(\epsilon_1, \epsilon_2)$, $N_i \subset M_i$ such that $f^{-1}(B(b, \epsilon)) = N_1 \cup \cdots \cup N_k$. For such $\epsilon > 0$, take $f_\epsilon \in C(\mathbb{R}^N, \mathbb{R})$ such that $\mathrm{supp}\, f_\epsilon \subset B(b, \epsilon)$, $\int_{\mathbb{R}^N} f_\epsilon(x) dx = 1$. Then*

$$\int_{\bar{\Omega}} f_\epsilon(f(x)) J_f(x)\, dx = \int_{\{x \in \Omega \,|\, |f(x)-b| < \epsilon\}} f_\epsilon(f(x)) J_f(x)\, dx$$

$$= \sum_{i=1}^{k} \int_{N_i} f_\epsilon(f(x)) J_f(x)\, dx$$

$$= \sum_{i=1}^{k} \int_{N_i} f_\epsilon(f(x))\, \mathrm{sgn}\, J_f(x)\, |J_f(x)|\, dx$$

$$= \sum_{i=1}^{k} \mathrm{sgn}\, J_f(x_i) \int_{B(b,\epsilon)} f_\epsilon(y)\, dy$$

$$= \sum_{i=1}^{k} \mathrm{sgn}(J_f(x_i))$$

$$= \mathrm{d}(f, \Omega, b).$$

We have, for small ϵ,

$$\mathrm{d}(f, \Omega, b) = \int_{\bar{\Omega}} f_\epsilon(f(x)) J_f(x) dx.$$

The integral $\int_{\bar{\Omega}} f_\epsilon(f(x)) J_f(x) dx$ is called the Heinz integral of degree f with respect to Ω and b.

We now study the degree properties of functions $f \in C^1(\bar{\Omega}, \mathbb{R}^N)$. We give a contraction result.

Lemma 9.3 *Let E be a Banach space, and $B_r(0)$ a ball in E with center 0 and radius r. Let $f : B_r(0) \to E$ satisfy $f(x) = x + g(x)$, where g is an α-contraction, $0 \le \alpha < 1$, and $g(0) = 0$. Then:*

(i) $f(B_r(0)) \supset B_{r(1-\alpha)}(0)$;

(ii) f is injective on $B_r(0)$.

Proof.

(i) Fix $x_0 \in B_{r(1-\alpha)}(0)$: that is, $|x_0| < r(1-\alpha)$. Take r_0, $0 < r_0 < r$ such that $|x_0| = r_0(1-\alpha)$. Define $h(x) = -g(x) + x_0$. For $x \in \overline{B_{r_0}(0)}$, we have

$$|h(x)| \le \alpha |x| + |x_0| \le \alpha r_0 + (1-\alpha)r_0 = r_0.$$

Thus, $h : \overline{B_{r_0}(0)} \to \overline{B_{r_0}(0)}$. Moreover, h is an α-contraction because g is an α-contraction. Thus, h admits a fixed point $x_1 \in B_{r_0}(0)$:

$$x_1 = h(x_1) = -g(x_1) + x_0.$$

Thus,

$$f(x_1) = g(x_1) + x_1 = x_0.$$

We conclude that

$$f(B_r(0)) \supset B_{r(1-\alpha)}(0).$$

(ii) Let $f(x) = f(y)$. Then $x + g(x) = y + g(y)$, or

$$|x - y| = |g(x) - g(y)| \le \alpha |x - y|,\ 0 \le \alpha < 1.$$

Therefore, $x = y$. ∎

Lemma 9.4 *Let $f \in C^1(\bar{\Omega}, \mathbb{R}^N)$, and $b \notin f(\partial\Omega) \cup f(S)$. Then there is a neighborhood U of f with respect to the topology of $C^1(\bar{\Omega}, \mathbb{R}^N)$, such that for $g \in U$, we have:*

(i) $b \notin g(\partial\Omega)$;

(ii) $x \in g^{-1}(b) \Rightarrow J_g(x) \ne 0$;

(iii) $\mathrm{d}(f, \Omega, b) = \mathrm{d}(g, \Omega, b)$.

Proof. Set $\epsilon_1 = \mathrm{dist}(b, f(\partial\Omega)) > 0$. Note that if $\|f - g\|_{C^1(\bar{\Omega})} < \epsilon_1$ then $b \notin g(\partial\Omega)$. Let $f^{-1}(b) = \{x_1, \cdots, x_k\}$ with $J_f(x_i) \ne 0$ for $i = 1, \cdots, k$. Take $r_1 > 0$ such that $\{B(x_i, r_1)\}$ are disjoint to each other and $f : B(x_i, r_1)) \to f(B(x_i, r_1))$ is a diffeomorphism such that

$$|J_f(x)| > \eta \quad \text{for } x \in \bigcup_{i=1}^{k} B(x_i, r_1), \tag{9.1}$$

where $\eta = \frac{1}{2} \min_{1 \le i \le k} |J_f(x_i)| > 0$. Note that $|J_f(x_i)| \ne 0$, so $f'(x_i)$ is invertible and bounded: $\|f'(x_i)^{-1}\| < a$, for $i = 1, \cdots, k$, for some $a > 0$. Because $f \in C^1(\Omega, \mathbb{R}^N)$, take $r_2 > 0$ such that for $x \in B(x_i, r_2)$ we have

$$\|f'(x_i)^{-1}[f'(x) - f'(x_i)]\| \le a \|f'(x) - f'(x_i)\| < \frac{1}{2}.$$

Let $r = \min(r_1, r_2)$, $B_i = B(x_i, r)$, and $D = \bar{\Omega} \backslash \bigcup\limits_{i=1}^{k} B_i$. Then D is compact and $b \notin f(D)$. Let $\epsilon_2 = \operatorname{dist}(b, f(D)) > 0$. If $g \in C^1(\Omega, \mathbb{R}^N)$, with $\|f - g\|_{C^1(\bar{\Omega})} < \epsilon_2$. Then $b \notin g(D)$. We claim that the equation $g(x) = b$ has one and only one solution $y_i \in B_i$ for each i. In fact, because $f' : \bar{\Omega} \to L(\mathbb{R}^N, \mathbb{R}^N)$, and the determinant map det: $L(\mathbb{R}^N, \mathbb{R}^N) \to \mathbb{R}$ is continuous, then det is uniformly continuous on a compact neighborhood K of $f'(\bar{\Omega})$ in $L(\mathbb{R}^N, \mathbb{R}^N)$. Take $\epsilon_3 > 0$, such that if $X, Y \in K$, $|X - Y| < \epsilon_3$ then $|\det X - \det Y| < \eta$, and if $X \in f'(\bar{\Omega})$ and $|Y - X| < \epsilon_3$, then $Y \in K$. For $g \in C^1(\Omega, \mathbb{R}^N)$, $\|f - g\|_{C^1(\bar{\Omega})} < \epsilon_3$. Then $x \in \bigcup\limits_{i=1}^{k} B_i$ implies that $f'(x) \in f'(\bar{\Omega})$, $g'(x) \in K$. Because $J_h(x) = \det h'(x)$,

$$|J_f(x) - J_g(x)| < \eta.$$

Then

$$0 < |J_f(x)| - \eta < |J_g(x)|.$$

Let $\epsilon = \min\{\epsilon_1, \epsilon_2, \epsilon_3\}$, and

$$M = \left\{ g \in C^1(\Omega, \mathbb{R}^N) \,\middle|\, \|f - g\|_{C^1(\bar{\Omega})} < \epsilon \right\}.$$

For $g \in M$, we have $J_g(x) \neq 0$ for $x \in \bigcup\limits_{i=1}^{k} B_i$. Note that both $f^{-1}(b) = \{x_1, x_2, \cdots, x_k\} \subset \bigcup\limits_{i=1}^{k} B_i$. If $|J_f(x) - J_g(x)| < \eta$, then by (9.1), for $x \in B_i$,

$$\operatorname{sgn}(J_g(x)) = \operatorname{sgn}(J_f(x)) = \operatorname{sgn}(J_f(x_i)).$$

Because $\|f'(x_i)^{-1}\| < a$ for $i = 1, 2, \cdots, k$, choose a C^1-neighborhood U of f such that $U \subset M$, and each $g \in U$ satisfies:

(a) $|g'(x_i)^{-1}| < a$ for $i = 1, 2, \cdots, k$;

(b) $|g'(x_i)^{-1}(g'(x) - g'(x_i))| < \frac{3}{4}$ for $x \in B_i$, $i = 1, 2, \cdots, k$;

(c) $\|f - g\|_{L^\infty} < \frac{r}{4a}$.

(i) If we set $\theta_i(x) = [g'(x_i)]^{-1} g(x)$, by (b), $|\theta_i'(x) - x| < \frac{3}{4}$, for $x \in B_i$, $i = 1, \cdots, k$. Thus, $\theta_i(x) - x$ is a $\frac{3}{4}$-contraction. By Lemma 9.3, θ_i is injective: g is injective on each B_i. Hence, we assert that the equation $g(x) = b$ has at most one solution in B_i.

(ii) The equation $g(x) = b$ has at least one solution in B_i : by Lemma 9.3,

$$\theta_i(B(x_i, r)) \supset B(\theta_i(x_i), \frac{r}{4}).$$

Then
$$g(B(x_i, r)) \supset g'(x_i)(B(\theta_i(x_i), \frac{r}{4})). \tag{9.2}$$

We claim that
$$g'(x_i)(B(\theta_i(x_i), \frac{r}{4})) \supset B(g(x_i), \frac{r}{4a}), \tag{9.3}$$

In fact, $y \in B(g(x_i), \frac{r}{4a})$ implies
$$\left| g'(x_i)^{-1} y - g'(x_i)^{-1} g(x_i) \right| \le |g'(x_i)|^{-1} |y - g(x_i)| < a \cdot \frac{r}{4a} = \frac{r}{4}.$$

Then
$$\left| g'(x_i)^{-1} y - \theta_i(x_i) \right| < \frac{r}{4}, \text{ and the claim follows.}$$

We conclude that by (9.2) and (9.3),
$$g(B(x_i, r)) \supset B(g(x_i), \frac{r}{4a}). \tag{9.4}$$

Now $|b - g(x_i)| = |f(x_i) - g(x_i)| < \frac{r}{4a}$, i.e., $b \in B(g(x_i), \frac{r}{4a})$. By (9.4), $b \in g(B(x_i, r)) = g(B_i)$.

(iii) We have proved that by (i) and (ii), $g(x) = b$ has one and only one solution in B_i. Let $g^{-1}(b) = \{y_1, y_2, \cdots, y_k\}$, where $y_i \in B_i$. Recall that, for $x \in B_i$, $\text{sgn } J_g(x) = \text{sgn } J_f(x_i)$. We have

$$d(g, \Omega, b) = \sum_{i=1}^{k} \text{sgn } J_g(y_i) = \sum_{i=1}^{k} \text{sgn } J_f(x_i) = d(f, \Omega, b). \qquad \blacksquare$$

Lemma 9.5 Let $O \subset \mathbb{R}^{N+1}$ be an open set, and $f \in C^2(\overline{O}, \mathbb{R}^N)$.

Set $D_i = \det(f_{x_0}, \cdots, \hat{f}_{x_i}, \cdots, f_{x_N})$, and let the ith column $f_{x_i} = \begin{pmatrix} \frac{\partial f_1}{\partial x_i} \\ \vdots \\ \frac{\partial f_N}{\partial x_i} \end{pmatrix}$ be

deleted. Then $\sum_{i=0}^{N} (-1)^i \frac{\partial D_i}{\partial x_i} = 0$.

Proof. Denote, for $i \neq j$,

$$D_{ij} = \det(f_{x_i x_j}, f_{x_0}, \cdots, \hat{f}_{x_i}, \cdots, \hat{f}_{x_j}, \cdots, f_{x_N}).$$

Then
$$D_{ij} = D_{ji},$$

$$\frac{\partial D_i}{\partial x_i} = \sum_{j=0}^{i-1} (-1)^j D_{ij} + \sum_{j=i+1}^{N} (-1)^{j-1} D_{ij} = \sum_{j=0}^{N} (-1)^j \alpha_{ij} D_{ij},$$

where
$$\alpha_{ij} = \begin{cases} 1 & \text{for} \quad i > j, \\ 0 & \text{for} \quad i = j, \\ -1 & \text{for} \quad i < j. \end{cases}$$

Therefore,
$$\sum_{i=0}^{N} (-1)^i \frac{\partial D_i}{\partial x_i} = \sum_{i=0}^{N} \sum_{j=0}^{N} (-1)^{i+j} \alpha_{ij} D_{ij} = 0. \qquad \blacksquare$$

Lemma 9.6 *Let $\Omega \subset \mathbb{R}^N$ be an open bounded set, let $f \in C^1(\mathbb{R}^N, \mathbb{R}^N)$ have compact support, and assume $g \in C^2(\bar{\Omega}, \mathbb{R}^N)$ with $(\operatorname{supp} f) \cap g(\partial \Omega) = \emptyset$. Then there is $u \in C^1(\mathbb{R}^N, \mathbb{R}^N)$ such that $\operatorname{supp} u \subset \Omega$, and $\operatorname{div} u(x) = [\operatorname{div} f(g(x))] J_g(x)$.*

Proof. Let $A_{ij}(x)$ and $M_{ij}(x)$ be the cofactor and the minor of $\frac{\partial g_i}{\partial x_j}$ with respect to $g'(x) = (\frac{\partial g_i}{\partial x_j})$, respectively. We have
$$A_{ij} = (-1)^{i+j} M_{ij},$$
and
$$\sum_{j=1}^{N} \frac{\partial g_i}{\partial x_j} A_{kj} = \delta_{ik} J_g(x).$$

Set $f = (f_1, \cdots, f_N)$, and
$$u_i(x) = \sum_{j=1}^{N} f_j(g(x)) A_{ji}(x), \quad 1 \leq i \leq N,$$

and $u = (u_1, \cdots, u_N)$. Then u vanishes near $\partial \Omega$. Extend u by 0 outside Ω. Then $u \in C^1(\mathbb{R}^N, \mathbb{R}^N)$, $\operatorname{supp} u \subset \Omega$, and

$$\operatorname{div} u(x) = \sum_i \frac{\partial u_i}{\partial x_i}(x) = \sum_{i,j} \frac{\partial}{\partial x_i}(f_j(g(x))) A_{ji}(x) + \sum_{i,j} f_j(g(x)) \frac{\partial}{\partial x_i} A_{ji}(x)$$

$$= \sum_{i,j,k} \frac{\partial f_j}{\partial g_k} \frac{\partial g_k}{\partial x_i} A_{ji}(x) + I = \sum_{j,k} \frac{\partial f_j}{\partial g_k} \delta_{jk} J_g(x) + I$$

$$= (\sum_j \frac{\partial f_j}{\partial g_j}) J_g(x) + I = [\operatorname{div} f(g(x))] J_g(x) + I,$$

where

$$I = \sum_{i,j} f_j(g(x)) \frac{\partial A_{ji}(x)}{\partial x_i} = \sum_j f_j(g(x)) \sum_i \frac{\partial A_{ji}(x)}{\partial x_i}$$

$$= \sum_j f_j(g(x)) (-1)^j \left(\sum_i (-1)^i \frac{\partial M_{ji}}{\partial x_i} \right).$$

We claim that

$$\sum_i (-1)^i \frac{\partial M_{ji}}{\partial x_i} = 0 \quad \text{for each } j.$$

In fact, if we let

$$h = \begin{pmatrix} g_1 \\ \vdots \\ \widehat{g}_j \\ \vdots \\ g_N, \end{pmatrix}$$

then

$$\det(h_{x_1}, \cdots, \hat{h}_{x_i}, \cdots, h_{x_N})$$

$$= \det \begin{pmatrix} \dfrac{\partial g_1}{\partial x_1} & \cdots & \dfrac{\partial \widehat{g}_1}{\partial x_i} & \cdots & \dfrac{\partial g_1}{\partial x_N} \\ \vdots & & \vdots & & \vdots \\ \dfrac{\partial \widehat{g}_j}{\partial x_1} & \cdots & \dfrac{\partial \widehat{g}_j}{\partial x_i} & \cdots & \dfrac{\partial \widehat{g}_j}{\partial x_1} \\ \vdots & & \vdots & & \vdots \\ \dfrac{\partial g_N}{\partial x_1} & \cdots & \dfrac{\partial \widehat{g}_N}{\partial x_i} & \cdots & \dfrac{\partial g_N}{\partial x_N} \end{pmatrix}$$

$$= M_{ji}(x).$$

Set $D_i = \det(h_{x_1}, \cdots, \hat{h}_{x_i}, \cdots, h_{x_N})$ as in Lemma 9.5 to obtain

$$0 = \sum_i (-1)^i \frac{\partial D_i}{\partial x_i} = \sum_i (-1)^i \frac{\partial M_{ji}}{\partial x_i},$$

so $I = 0$, i.e.,

$$\operatorname{div} u(x) = [\operatorname{div} f(g(x))] J_g(x). \qquad \blacksquare$$

Lemma 9.7 *Suppose $f \in C^1(\mathbb{R}^N, \mathbb{R})$ with compact support K, and $x_0 \neq 0$ in $\mathbb{R}^N$. Let $g(x) = f(x) - f(x - x_0)$ for $x \in \mathbb{R}^N$. Then there is $h \in C^1(\mathbb{R}^N, \mathbb{R}^N)$ such that $\operatorname{supp} h \subset \overline{\operatorname{conv}}(K \cup (K - x_0))$, and $g(x) = \operatorname{div} h(x)$.*

Proof.   Set

$$\varphi(x) = \int_{-\infty}^{0} f(x + tx_0)dt - \int_{-\infty}^{0} f(x + (1+t)x_0)dt$$

$$= \int_{-\infty}^{0} f(x + tx_0)dt - \int_{-\infty}^{1} f(x + tx_0)dt$$

$$= -\int_{0}^{1} f(x + tx_0)dt,$$

and

$$h(x) = \varphi(x)x_0 = (\varphi(x)x_0^1, \cdots, \varphi(x)x_0^N).$$

Let $x \in \mathbb{R}^N$ with $h(x) \neq 0$. Then $\varphi(x) \neq 0$. There is $t_0 \in [0,1]$ such that $x + t_0\,x_0 \in K$, i.e., $x = t_0(k - x_0) + (1 - t_0)k$ for some $k \in K$. We have

$$\mathrm{supp}\, h \subset \overline{\mathrm{conv}}(K \cup (K - x_0)),$$

and

$$\mathrm{div}\, h(x) = \sum_i \frac{\partial \varphi(x)}{\partial x_i} x_0^i = (\varphi'(x), x_0) = \frac{d}{ds}(\varphi(x + sx_0))|_{s=0}$$

$$= \frac{d}{ds}\left(\int_{-\infty}^{0} [f(x + (t+s)x_0) - f(x + (1+t+s)x_0]\, dt \right)|_{s=0}$$

$$= \int_{-\infty}^{0} \frac{d}{ds}[f(x + (t+s)x_0) - f(x + (1+t+s)x_0)]\, dt\,|_{s=0}$$

$$= \int_{-\infty}^{0} \frac{d}{dt}[f(x + tx_0) - f(x + (1+t)x_0)]\, dt$$

$$= f(x) - f(x + x_0) = g(x) \quad \text{for } x \in \mathbb{R}^N. \qquad \blacksquare$$

Lemma 9.8 *Suppose $f \in C^1(\mathbb{R}^N, \mathbb{R})$ has compact support K, $O \subset \mathbb{R}^N$ is a bounded open set, and $p \in C([0,1], \mathbb{R}^N)$ with $(\overline{\mathrm{conv}}K) - p(t) \subset O$ for $t \in [0,1]$. Then there is $h \in C^1(\mathbb{R}^N, \mathbb{R}^N)$, with $\mathrm{supp}\, h \subset O$, and $f(x + p(0)) - f(x + p(1)) = \mathrm{div}\, h(x)$.*

Proof.   For $s \in [0,1]$, $f(x + p(s))$ has support $K - p(s)$. By Lemma 9.7, there is $h_{st}(x) \in C^1(\mathbb{R}^N, \mathbb{R}^N)$ such that

$$f(x + p(s)) - f(x + p(t))$$
$$= f(x + p(s)) - f(x + p(s) + (p(t) - p(s)))$$
$$= \mathrm{div}\, h_{st}(x),$$

where

$$\mathrm{supp}\, h_{st} \subset \overline{\mathrm{conv}}[(K - p(s)) \cup (K - p(s) - (p(t) - p(s)))]$$
$$= \overline{\mathrm{conv}}[(K - p(s)) \cup (K - p(t))].$$

If we set

$$\epsilon_s = \text{dist}((\overline{\text{conv}}K) - p(s), \partial O),$$

then $\epsilon_s > 0$. Take $\alpha_s > 0$ such that $|t - s| < \alpha_s$ implies

$$|p(t) - p(s)| < \frac{\epsilon_s}{2}.$$

Choose a compact neighborhood $N_{\frac{\epsilon_s}{2}}((\overline{\text{conv}}K) - p(s))$ of $(\overline{\text{conv}}K) - p(s)$, such that $|t - s| < \alpha_s$ implies

$$\text{supp}\, h_{st} \subset N_{\frac{\epsilon_s}{2}}((\overline{\text{conv}}K) - p(s)) \subset N_{\epsilon_s}((\overline{\text{conv}}K) - p(s)) \subset O.$$

Similarly,

$$\text{supp}\, h_{ts} \subset \overline{\text{conv}}[(K - p(t)) \cup (K - p(s))],$$

and

$$|t - s| < \alpha_s \quad \text{implies supp}\, h_{ts} \subset O.$$

Because $\{(s - \alpha_s, s + \alpha_s)\}_{s \in [0,1]}$ forms an open covering of $[0, 1]$, there is a finite subcovering $\{(s_i - \alpha_{s_i}, s_i + \alpha_{s_i})\}_{0 \leq i \leq k}$ such that

$$s_0 = 0 < s_1 < \cdots < s_k = 1.$$

Set

$$\sigma_i \in (s_i, s_i + \alpha_{s_i}) \cap (s_{i+1} - \alpha_{s_{i+1}}, s_{i+1}), \quad \text{for } 0 \leq i \leq k - 1.$$

Then

$$\text{supp}\, k_{s_i \sigma_i} \subset O, \quad \text{supp}\, h_{\sigma_i s_{i+1}} \subset O.$$

Set

$$h = \sum_{i=0}^{k-1} (h_{s_i \sigma_i} + h_{\sigma_i s_{i+1}}).$$

Then $\text{supp}\, h \subset O$, and consequently,

$$\text{div}\, h = \sum_{i=0}^{k-1} (\text{div}\, h_{s_i \sigma_i} + \text{div}\, h_{\sigma_i s_{i+1}})$$

$$= \sum_{i=0}^{k-1} [f(x + p(s_i)) - f(x + p(\sigma_i)) + f(x + p(\sigma_i)) - f(x + p(s_{i+1}))]$$

$$= f(x + p(0)) - f(x - p(1)). \qquad \blacksquare$$

Lemma 9.9 *Assume $f \in C^2(\bar{\Omega}, \mathbb{R}^N)$, $b_i \in \mathbb{R}^N$, and $b_i \notin f(\partial \Omega) \cup f(S)$, $i = 1, 2$. If b_1 and b_2 belong to the connected component of $\mathbb{R}^N \backslash f(\partial \Omega)$, then $d(f, \Omega, b_1) = d(f, \Omega, b_2)$.*

Proof.   Because $f(\partial\Omega)$ is compact, $\mathbb{R}^N\backslash f(\partial\Omega)$ is open. The connected components of $\mathbb{R}^N\backslash f(\partial\Omega)$ are arcwise connected. Let O be a component of $\mathbb{R}^N\backslash f(\partial\Omega)$, $b_1, b_2 \in O$. Choose $q \in C([0,1], \mathbb{R}^N)$ such that $q(0) = b_1$, $q(1) = b_2$, and $q(t) \in O$ for $t \in [0,1]$. Take $\epsilon > 0$ such that $N_\epsilon(q[0,1]) \subset O$: i.e., $\overline{N_\epsilon(q(t))} \subset O$ for $t \in [0,1]$, but

$$B_\epsilon(q(t)) = q(t) + B_\epsilon(0)$$
$$= B_\epsilon(b_1) - (b_1 - q(t)).$$

Choose $j_\epsilon \in C^1(\mathbb{R}^N, \mathbb{R})$ such that $\operatorname{supp} j_\epsilon \subset \overline{B_\epsilon(b_1)}$, $\int_{\mathbb{R}^N} j_\epsilon(x)dx = 1$. If we apply Lemma 9.8 by considering the function j_ϵ,

$$\operatorname{supp} j_\epsilon = K, \quad p(t) = b_1 - q(t), \quad p(0) = 0, \quad p(1) = b_1 - b_2.$$

Then there is $h \in C^1(\mathbb{R}^N, \mathbb{R}^N)$ such that $\operatorname{supp} h \subset O$, and

$$q(x) = j_\epsilon(x + p(0)) - j_\epsilon(x + p(1))$$
$$= j_\epsilon(x) - j_\epsilon(x + b_1 - b_2) = \operatorname{div} h(x).$$

Now

$$(\operatorname{supp} h) \cap f(\partial\Omega) = \emptyset.$$

By Lemma 9.6, there is $u \in C^1(\mathbb{R}^N, \mathbb{R}^N)$ such that $\operatorname{supp} u \subset \Omega$, and

$$[\operatorname{div} h(f(x))]J_f(x) = [j_\epsilon(f(x)) - j_\epsilon(f(x) + b_1 - b_2)]J_f(x) = \operatorname{div} u.$$

Thus, u vanishes on $\partial\Omega$, and

$$\int_\Omega \operatorname{div} u\ dx = \int_{\partial\Omega} u \cdot n = 0.$$

Therefore,

$$\int_\Omega j_\epsilon(f(x))J_f(x)dx = \int_\Omega j_\epsilon(f(x) + b_1 - b_2)J_f(x)dx$$
$$= \int_\Omega (j_\epsilon)_{(b_2 - b_1)}(f(x))J_f(x)dx,$$

where

$$\operatorname{supp} j_{\epsilon(b_2 - b_1)} \subset \overline{B_\epsilon(b_2)}, \quad \int_{\mathbb{R}^N} j_{\epsilon(b_2 - b_1)}(x)dx = 1.$$

Let $\epsilon > 0$ be small enough and apply the Heinz integral of degree f: we obtain

$$\mathrm{d}(f, \Omega, b_1) = \mathrm{d}(f + b_1 - b_2, \Omega, b_1) = \mathrm{d}(f, \Omega, b_2).$$

Note that

$$b_1 \notin (f + b_1 - b_2)(\partial\Omega) = f(\partial\Omega) + b_1 - b_2 \quad \text{if and only if} \quad b_2 \notin f(\partial\Omega). \qquad \blacksquare$$

Now we are in a position to extend the definition of degree to $b \in f(\partial\Omega)$, but only as follows: let $f \in C^2(\bar{\Omega}, \mathbb{R}^N)$, and $b \in f(S) \setminus f(\partial\Omega)$. Let C_b be the connected component containing b in $\mathbb{R}^N \setminus f(\partial\Omega)$. Because C_b is a nonempty open set, by the Sard Theorem 3.18, $f(S)$ is a null set and C_b contains points not in $f(S)$. By Lemma 9.9, $\mathrm{d}(f, \Omega, x)$ is constant for each $x \in C_b \setminus f(S)$.

Definition 9.10 *For $b \in f(S) \setminus f(\partial\Omega)$, define $\mathrm{d}(f, \Omega, b) = \mathrm{d}(f, \Omega, a)$, for some $a \in C_b \setminus f(S)$.*

Hence, we obtain the following result.

Corollary 9.11 *Assume $f \in C^2(\bar{\Omega}, \mathbb{R}^N)$, $b \notin f(\partial\Omega)$. Then $\mathrm{d}(f, \Omega, x)$ is constant on each connected component of $\mathbb{R}^N \setminus f(\partial\Omega)$.*

Moreover, we have:

Corollary 9.12 *Assume $f \in C^2(\bar{\Omega}, \mathbb{R}^N)$, and $b \notin f(\partial\Omega)$. There is a C^1-neighborhood U of f in $C^2(\bar{\Omega}, \mathbb{R}^N)$ such that if $g \in U$ then:*

(i) $b \notin g(\partial\Omega)$;

(ii) $\mathrm{d}(g, \Omega, b) = \mathrm{d}(f, \Omega, b)$.

Proof. If $b \notin f(S)$, then we conclude the proof by Lemma 9.4. Suppose $b \in f(S)$. Set $r = \mathrm{dist}(b, f(\partial\Omega))$, with $r > 0$. By the Sard Theorem, there is $c \notin f(S)$ such that $|b - c| < \frac{r}{2}$, and consequently b and c both lie in a connected component of $\mathbb{R}^N \setminus f(\partial\Omega)$. By Corollary 9.11,

$$\mathrm{d}(f, \Omega, b) = \mathrm{d}(f, \Omega, c).$$

By Lemma 9.4, there is a C^1-neighborhood V of f satisfying, for $g \in V$,

(a) $c \notin g(\partial\Omega)$,

(b) $J_g(x) \neq 0$, for $x \in g^{-1}(c)$,

(c) $\mathrm{d}(g, \Omega, c) = \mathrm{d}(f, \Omega, c)$.

Take

$$U = V \cap C^2(\bar{\Omega}, \mathbb{R}^N) \cap \left\{ g \in C^2(\bar{\Omega}, \mathbb{R}^N) \,\Big|\, \|g - f\|_{L^\infty} < \frac{r}{2} \right\}.$$

For $g \in U$, then $\mathrm{dist}(b, g(\partial\Omega)) \geq \frac{r}{2}$. Because $|b - c| < \frac{r}{2}$, we have $c \notin g(\partial\Omega)$, and both b and c belong to a connected component of $\mathbb{R}^N \setminus g(\partial\Omega)$, so

$$\mathrm{d}(g, \Omega, b) = \mathrm{d}(g, \Omega, c) = \mathrm{d}(f, \Omega, c) = \mathrm{d}(f, \Omega, b). \qquad \blacksquare$$

Corollary 9.13 (*Homotopy Invariance in C^2*) *Assume $H(x,t) \in C^2(\bar{\Omega} \times [0,1], \mathbb{R}^N)$, and $b \notin H(\partial\Omega \times [0,1])$. Then $d(H(\cdot,t), \Omega, b)$ is constant on $[0,1]$.*

Proof. Note that H and $\frac{\partial H}{\partial x}$ are uniformly continuous on $\bar{\Omega} \times [0,1]$. Fix $t_0 \in [0,1]$. For $\epsilon > 0$ there is $\delta > 0$, such that $|t - t_0| < \delta$ implies $\|H(\cdot,t) - H(\cdot,t_0)\|_{C^1} < \epsilon$. If ϵ is small enough, by Corollary 9.11 we obtain

$$d(H(\cdot,t), \Omega, b) = d(H(\cdot,t_0), \Omega, b).$$

Therefore $t \to d(H(\cdot,t), \Omega, b)$ is locally constant and continuous on $[0,1]$. Because $[0,1]$ is connected, the degree $d(H(\cdot,t), \Omega, b)$ is constant on $[0,1]$. This completes the proof. ∎

Finally we can extend the definition of degree to $f \in C(\bar{\Omega}, \mathbb{R}^N)$ as follows. Suppose $f \in C(\bar{\Omega}, \mathbb{R}^N)$, and $b \notin f(\partial\Omega)$. Let $r = \text{dist}(b, f(\partial\Omega))$, $r > 0$. Take $g \in C^2(\bar{\Omega}, \mathbb{R}^N)$ with $\|g - f\|_\infty < \frac{r}{2}$. We have $b \notin g(\partial\Omega)$, and therefore $d(g, \Omega, b)$ is well-defined. We claim that $d(g, \Omega, b)$ is independent of the choice of g. In fact, if $g_1, g_2 \in C^2(\bar{\Omega}, \mathbb{R}^N)$, $\|g_1 - f\|_\infty < \frac{r}{2}$, $\|g_2 - f\|_\infty < \frac{r}{2}$. Set $H(x,t) = tg_1(x) + (1-t)g_2(x)$, $x \in \bar{\Omega}$, $t \in [0,1]$. Then $H \in C^2(\bar{\Omega} \times [0,1], \mathbb{R}^N)$, because

$$\|H(\cdot,t) - f\|_{L^\infty} \le t\|g_1 - f\|_{L^\infty} + (1-t)\|g_2 - f\|_{L^\infty} < \frac{r}{2} \quad \text{for } t \in [0,1].$$

Hence, $b \notin H(\partial\Omega \times [0,1])$. By Corollary 9.13, $d(H(\cdot,t), \Omega, b)$ is defined and constant for $t \in [0,1]$. Taking $t = 0$ or $t = 1$, we obtain

$$d(g_1, \Omega, b) = d(g_2, \Omega, b).$$

Definition 9.14 *Let $f \in C(\Omega, \mathbb{R}^N)$, $b \notin f(\partial\Omega)$, and $r = \text{dist}(b, f(\partial\Omega))$. Take $g \in C^2(\bar{\Omega}, \mathbb{R}^N)$ with $\|g - f\|_\infty < \frac{r}{2}$ and define*

$$d(f, \Omega, b) = d(g, \Omega, b).$$

Lemma 9.15 *Suppose $f \in C(\Omega, \mathbb{R}^N)$, and $b \notin f(\partial\Omega)$. Then*

$$d(f, \Omega, b) = d(f - b, \Omega, 0).$$

Proof. Set $r = \text{dist}(b, f(\partial\Omega)) = \text{dist}(0, (f - b)(\partial\Omega))$. Take $g \in C^2(\bar{\Omega}, \mathbb{R}^N)$ with $\|g - f\|_\infty < \frac{r}{2}$. Then

$$d(f, \Omega, b) = d(g, \Omega, b). \tag{9.5}$$

Note that

$$\|(g - b) - (f - b)\| < \frac{1}{2} \, \text{dist}(0, (f - b)(\partial\Omega)),$$

and it follows at once that

$$d(f - b, \Omega, 0) = d(g - b, \Omega, 0). \tag{9.6}$$

By the Sard Theorem, we can find c such that c and b belong to a connected component in $\mathbb{R}^N \backslash g(\partial\Omega)$, $|b - c| < \frac{r}{2}$, and $J_g(x) \neq 0$ for $x \in g^{-1}(c)$. We have

$$\mathrm{d}(g, \Omega, b) = \mathrm{d}(g, \Omega, c), \tag{9.7}$$

$$\mathrm{d}(g - b, \Omega, 0) = \mathrm{d}(g - c, \Omega, 0). \tag{9.8}$$

We claim that 0 is a regular value of $g - c$, and

$$\mathrm{d}(g, \Omega, c) = \mathrm{d}(g - c, \Omega, 0).$$

In fact, if $x \in (g - c)^{-1}(0)$, then $g(x) - c = 0$, i.e., $x \in g^{-1}(c)$ and consequently $J_{g-c}(x) \neq 0$. Moreover, let $j_\epsilon \in C(\mathbb{R}^N, \mathbb{R})$, $\mathrm{supp}\, j_\epsilon \subset B_\epsilon(c)$, $\int_{\mathbb{R}^N} j_\epsilon(x)\, dx = 1$. Then

$$\mathrm{d}(g, \Omega, c) = \int_{\bar{\Omega}} j_\epsilon(g(x)) J_g(x)\, dx$$

$$= \int_{\bar{\Omega}} (j_\epsilon(g(x) - c + c)) J_{g(x)-c}(x)\, dx$$

$$= \int_{\bar{\Omega}} (j_\epsilon)_{-c}(g(x) - c) J_{g(x)-c}(x)\, dx$$

$$= \mathrm{d}(g - c, \Omega, 0),$$

where $\mathrm{supp}(j_\epsilon)_{-c} \subset B_\epsilon(0)$, $\int_{\mathbb{R}^N} (j_\epsilon)_{-c}(x)\, dx = 1$. We conclude that by (9.5)–(9.8),

$$\mathrm{d}(f, \Omega, b) = \mathrm{d}(f - b, \Omega, 0).$$

This completes the proof. ∎

We are in a position to see the principal properties of degrees.

Theorem 9.16 *Let $f \in C(\Omega, \mathbb{R}^N)$, $b \notin f(\partial\Omega)$. Then:*

(i) *(Continuity of Function) Because there is a neighborhood U of f in $C(\Omega, \mathbb{R}^N)$ such that if $g \in U$, then $b \notin g(\partial\Omega)$, and $\mathrm{d}(g, \Omega, b) = \mathrm{d}(f, \Omega, b)$;*

(ii) *(Degree is Constant in Each Connected Component) If b and c belong to the same connected component of $\mathbb{R}^N \backslash f(\partial\Omega)$, then $\mathrm{d}(f, \Omega, b) = \mathrm{d}(f, \Omega, c)$;*

(iii) *(Additive Property) If $\Omega = \Omega_1 \cup \Omega_2$, $\Omega_1 \cap \Omega_2 = \varnothing$, Ω_1, Ω_2 are open, and $b \notin f(\partial\Omega_1) \cup f(\partial\Omega_2)$, then $\mathrm{d}(f, \Omega, b) = \mathrm{d}(f, \Omega_1, b) + \mathrm{d}(f, \Omega_2\, b)$;*

(iv) *(Homotopy Invariance) Let $H \in C(\bar{\Omega} \times [0, 1], \mathbb{R}^N)$, $b \notin H(\partial\Omega \times [0, 1])$. Then $\mathrm{d}(H(\cdot, t), \Omega, b)$ is constant on $[0, 1]$.*

Proof.

(*i*) Let $r = \text{dist}(b, f(\partial\Omega))$, $r > 0$. Consider the neighborhood U of f :

$$U = \left\{ g \in C(\Omega, \mathbb{R}^N) \,\Big|\, \|g - f\|_\infty < \frac{r}{4} \right\}.$$

For $g \in U$, we have $b \notin g(\partial\Omega)$, and

$$s = \text{dist}(b, g(\partial\Omega)) \geq \frac{3r}{4}.$$

If $h \in C^2(\Omega, \mathbb{R}^N)$ with $\|h - f\|_\infty < \frac{r}{8}$, then

$$\text{d}(f, \Omega, b) = \text{d}(h, \Omega, b). \tag{9.9}$$

In addition,

$$\|g - h\|_\infty \leq \|g - f\|_\infty + \|f - h\|_\infty < \frac{r}{4} + \frac{r}{8} = \frac{3r}{8} \leq \frac{s}{2},$$

so

$$\text{d}(g, \Omega, b) = \text{d}(h, \Omega, b). \tag{9.10}$$

We conclude that, by (9.9) and (9.10), if $\|g - f\|_{L^\infty} < \frac{r}{4}$, then

$$\text{d}(f, \Omega, b) = \text{d}(g, \Omega, b) \quad \text{for } g \in U.$$

(*ii*) Let C_b be a connected component of $\mathbb{R}^N \backslash f(\partial\Omega)$ containing b, and $r = \text{dist}(b, f(\partial\Omega))$. Let $c \in C_b$ with $|b - c| < \frac{r}{4}$, then

$$\|(f - b) - (f - c)\|_{L^\infty} = |b - c|,$$

and by (*i*),

$$\text{d}(f - b, \Omega, 0) = \text{d}(f - c, \Omega, 0). \tag{9.11}$$

By Lemma 9.15,

$$\text{d}(f, \Omega, b) = \text{d}(f - b, \Omega, 0), \tag{9.12}$$
$$\text{d}(f, \Omega, b) = \text{d}(f - c, \Omega, 0). \tag{9.13}$$

By (9.11)–(9.13) $\text{d}(f, \Omega, b) = \text{d}(f, \Omega, c)$, i.e., $b \in \mathbb{R}^N \backslash f(\partial\Omega) \rightarrow \text{d}(f, \Omega, b)$ is continuous and locally constant, so it is a global constant on each connected component of $\mathbb{R}^N \backslash f(\partial\Omega)$.

(*iii*) Because $b \notin f(\partial\Omega) = f(\partial\Omega_1) \cup f(\partial\Omega_2)$, $r = \text{dist}(b, f(\partial\Omega)) > 0$, also $r \leq \text{dist}(b, f(\partial\Omega_i))$, $i = 1, 2$. Take $g \in C^2(\Omega, \mathbb{R}^N)$ with $\|g - f\|_\infty < \frac{r}{2}$. We have

$$\text{d}(g, \Omega, b) = \text{d}(f, \Omega, b), \tag{9.14}$$
$$\text{d}(g, \Omega_i, b) = \text{d}(f, \Omega_i, b), \quad i = 1, 2. \tag{9.15}$$

Let $s = \text{dist}(b, g(\partial\Omega))$, $s \geq \frac{r}{2}$. By the Sard Theorem, there is c such that $|b - c| < \frac{s}{2}$, and $J_g(x) \neq 0$ for $x \in g^{-1}(b)$. Now b and c belong to a connected component with respect to $\mathbb{R}^N \backslash g(\partial\Omega)$, and $\mathbb{R}^N \backslash g(\partial\Omega_i)$, $i = 1, 2$. By (ii),

$$d(g, \Omega, b) = d(g, \Omega, c), \tag{9.16}$$

$$d(g, \Omega_i, b) = d(f, \Omega_i, c), i = 1, 2. \tag{9.17}$$

Let $B = B(b, \frac{s}{2})$, $j_\epsilon \in C^1(\mathbb{R}^N, \mathbb{R})$, $\text{supp}\, j_\epsilon \subset B$, $\int_{\mathbb{R}^N} j_\epsilon(x) dx = 1$,

$$\begin{aligned}
d(g, \Omega, c) &= \int_{\overline{\Omega}} j_\epsilon(g(x)) J_g(x)\, dx \\
&= \int_{\overline{\Omega_1}} j_\epsilon(g(x)) J_g(x)\, dx + \int_{\overline{\Omega_2}} j_\epsilon(g(x)) J_g(x)\, dx \\
&= \sum_{x \in (g^{-1}(c) \cap \Omega_1)} \text{sgn}\, J_g(x) + \sum_{x \in (g^{-1}(c) \cap \Omega_2)} \text{sgn}\, J_g(x) \\
&= d(g, \Omega_1, c) + d(g, \Omega_2, c).
\end{aligned}$$

By (9.14)–(9.17),

$$d(f, \Omega, b) = d(f, \Omega_1, b) + d(f, \Omega_2, b).$$

(iv) For $\epsilon = \frac{1}{4} \text{dist}(b, H(\partial\Omega \times [0,1])) > 0$ and some $\delta > 0$, $|t - s| < \delta$ implies $\|H(\cdot, t) - H(\cdot, s)\|_\infty < \epsilon$. By (i),

$$d(H(\cdot, t), \Omega, b) = d(H(\cdot, s), \Omega, b).$$

Thus, $t \to d(H(\cdot, t), \Omega, b)$ is continuous and locally constant on $[0, 1]$, so it is globally constant on $[0, 1]$. ∎

Corollary 9.17 *If we let $I : \bar{\Omega} \to \mathbb{R}^N$ be the identity map with $I(x) = x$, then*

$$d(I, \Omega, b) = \begin{cases} 1 & \text{if } b \in \Omega, \\ 0 & \text{if } b \notin \Omega. \end{cases}$$

Proof.   Note that

$$J_I(x) = \begin{vmatrix} 1 & & & 0 \\ & 1 & & \\ & & \ddots & \\ 0 & & & 1 \end{vmatrix} = 1.$$

For small $\epsilon > 0$, take $j_\epsilon \in C(\mathbb{R}^N, \mathbb{R})$, $\text{supp} j_\epsilon \subset B(b, \epsilon)$, $\int_{\mathbb{R}^N} j_\epsilon(x)dx = 1$ to obtain

$$
\begin{aligned}
\mathrm{d}(I, \Omega, b) &= \int_{\bar\Omega} j_\epsilon(x) J_I(x) dx \\
&= \int_\Omega j_\epsilon(x) dx \\
&= \begin{cases} 1 & \text{if } b \in \Omega, \\ 0 & \text{if } b \notin \bar\Omega. \end{cases}
\end{aligned}
$$

∎

Corollary 9.18 *Let* $f \in C(\bar\Omega, \mathbb{R}^N)$, $b \notin f(\partial\Omega)$. *Then:*

(i) *If* $b \notin f(\bar\Omega)$, *then* $\mathrm{d}(f, \Omega, b) = 0$;

(ii) *If* $\mathrm{d}(f, \Omega, b) \neq 0$, *then there is* $x_0 \in \Omega$ *such that* $f(x_0) = b$;

(iii) *If* $\mathrm{d}(f, \Omega, b) \neq 0$, *then* $f(\Omega)$ *is a neighborhood of* b;

(iv) *If* $f(\Omega)$ *is included in a proper subspace of* $\mathbb{R}^N$, *then* $\mathrm{d}(f, \Omega, b) = 0$;

(v) *(Excision Property) Letting* $K \subset \bar\Omega$ *be closed,* $b \notin f(K)$, *then* $\mathrm{d}(f, \Omega, b) = \mathrm{d}(f, \Omega \backslash K, b)$;

(vi) *Suppose* $\{\Omega_i\}_{i \in I}$ *is a family of open sets in* Ω, *disjoint to each other, and* $f^{-1}(b) \subset \bigcup_{i \in I} \Omega_i$. *Then* $\mathrm{d}(f, \Omega_i, b) = 0$ *except for finite* $i \in I$, *and* $\mathrm{d}(f, \Omega, b) = \sum_{i \in I} \mathrm{d}(f, \Omega_i, b)$.

Proof.

(i) Take $g \in C^2(\Omega, \mathbb{R}^N)$, $\|g - f\|_\infty < \frac{1}{2} \text{dist}(b, f(\partial\Omega)) = \epsilon$. Then $b \notin g(\bar\Omega)$. Let $j_\epsilon \in C(\mathbb{R}^N, \mathbb{R})$, $\text{supp } j_\epsilon \subset B(b, \epsilon)$, $\int_{\mathbb{R}^N} j_\epsilon(x)dx = 1$,

$$
\begin{aligned}
\mathrm{d}(f, \Omega, b) &= \mathrm{d}(g, \Omega, b) \\
&= \int_{\bar\Omega} j_\epsilon(g(x)) J_g(x) dx \\
&= 0.
\end{aligned}
$$

(ii) By (i).

(iii) By (ii), there is $x_0 \in \Omega$ such that $f(x_0) = b$. Let C_b be the connected component containing b in $\mathbb{R}^N \backslash f(\partial\Omega)$; we see, for $z \in C_b$,

$$
\mathrm{d}(f, \Omega, z) = \mathrm{d}(f, \Omega, b) \neq 0.
$$

By (ii), $C_b \subset f(\Omega)$. Because C_b is open, $f(\Omega)$ is a neighborhood of b.

(iv) Suppose $d(f, \Omega, b) \neq 0$, then $f(\Omega)$ is a neighborhood of b, or $f(\Omega) \supset B(b, r)$, for some $r > 0$, which contradicts the assumption that $f(\Omega)$ is included in a proper subspace of $\mathbb{R}^N$.

(v) Take $g \in C^2(\bar{\Omega}, \mathbb{R}^N)$ such that

$$d(f, \Omega, b) = d(g, \Omega, b). \tag{9.18}$$

$$d(f, \Omega \backslash K, b) = d(g, \Omega \backslash K, b),$$
$$\|f - g\|_\infty < \frac{1}{2} \min\{\operatorname{dist}(b, f(K)), \operatorname{dist}(b, f(\partial \Omega))\}. \tag{9.19}$$

Let $r = \frac{1}{2} \min\{\operatorname{dist}(b, g(K)), \operatorname{dist}(b, g(\partial \Omega)), \operatorname{dist}(b, g(\partial(\Omega \backslash K)))\}$, whence $r > 0$. By the Sard Theorem, there is $c \in B(b, r)$ such that for $x \in g^{-1}(c)$, $J_g(x) \neq 0$, and the result follows at once. $c \notin g(K)$, and b, c belong to a connected component of $\mathbb{R}^N \backslash g(\partial \Omega)$, and of $\mathbb{R}^N \backslash g(\partial(\Omega \backslash K))$. Hence,

$$d(g, \Omega, b) = d(g, \Omega, c), \tag{9.20}$$
$$d(g, \Omega \backslash K, b) = d(g, \Omega \backslash K, c). \tag{9.21}$$

However,

$$\begin{aligned}
d(g, \Omega, c) &= \sum_{x \in g^{-1}(c)} \operatorname{sgn} J_g(x) \\
&= \sum_{x \in (g^{-1}(c) \cap (\Omega \backslash K))} \operatorname{sgn}(J_g(x)) \\
&= d(g, \Omega \backslash K, c).
\end{aligned}$$

By (9.18)–(9.21),

$$d(f, \Omega, b) = d(f, \Omega \backslash K, b).$$

(vi) Because $f^{-1}(b) \subset \Omega$ and $f^{-1}(b) \subset \bigcup_{i \in I} \Omega_i$, the degrees $d(f, \Omega, b)$ and $d(f, \Omega_i, b)$ are defined. $f^{-1}(b)$ is compact, so there is a finite subset $I_0 \subset I$ with $f^{-1}(b) \subset \bigcup_{i \in I_0} \Omega_i$. Hence, by (i), $d(f, \Omega_i, b) = 0$ for $i \notin I_0$. Now $K = \bar{\Omega} \backslash (\bigcup_{i \in I_0} \Omega_i)$ is compact in $\bar{\Omega}$, $b \notin f(K)$, and by the Excision Property (v),

$$\begin{aligned}
d(f, \Omega, b) &= d(f, \bigcup_{i \in I_0} \Omega_i, b) \\
&= \sum_{i \in I_0} d(f, \Omega_i, b) \\
&= \sum_{i \in I} d(f, \Omega_i, b). \qquad \blacksquare
\end{aligned}$$

Corollary 9.19 *Assume $f, g \in C(\bar{\Omega}, \mathbb{R}^N)$. Then:*

(i) *If $f = g$ on $\partial\Omega$, $b \notin f(\partial\Omega) = g(\partial\Omega)$, then*

$$d(f, \Omega, b) = d(g, \Omega, b).$$

(ii) *If $\bar{H} \in C(\partial\Omega \times [0,1])$, there is $\mathbb{R}^N$) such that $b \notin \bar{H}(\partial\Omega \times [0,1])$, $\bar{H}(\cdot, 0) = f\vert_{\partial\Omega}$ and $\bar{H}(\cdot, 1) = g\vert_{\partial\Omega}$, then*

$$d(f, \Omega, b) = d(g, \Omega, b).$$

Proof.

(i) Set $H(x,t) = t f(x) + (1-t)g(x)$, for $x \in \bar{\Omega}$ and $t \in [0,1]$. Then $H \in C(\Omega \times [0,1], \mathbb{R}^N)$ and $b \notin H(\partial\Omega \times [0,1])$, because if $x_0 \in \partial\Omega$, then

$$H(x_0, t) = t f(x_0) + (1-t)g(x_0) = g(x_0) \neq b.$$

By the homotopy invariance,

$$d(f, \Omega, b) = d(g, \Omega, b).$$

(ii) By the Tietze Extension Theorem, there is an extension $H \in C(\bar{\Omega} \times [0,1], \mathbb{R}^N)$ of $\bar{H}$. Set

$$H(\cdot, 0) = F, \quad H(\cdot, 1) = G.$$

Then

$$d(F, \Omega, b) = d(G, \Omega, b).$$

By (i) $F = f$; $G = g$ on $\partial\Omega$, so

$$d(F, \Omega, b) = d(f, \Omega, b),$$
$$d(G, \Omega, b) = d(g, \Omega, b).$$

Thus,

$$d(f, \Omega, b) = d(g, \Omega, b). \qquad \blacksquare$$

Theorem 9.20 *Let $f \in C(\overline{B_1(0)}, \mathbb{R}^N)$ be such that $f(x)$ never points opposite to $x \in \partial B_1(0): f(x) + \lambda x \neq 0$ for all $\lambda \geq 0$, $x \in \partial B_1(0)$. Then $f(x) = 0$ has a solution in $B_1(0)$.*

Proof. By the hypothesis $f(x) \neq 0$ for $x \in \partial B_1(0)$, so the degree $d(f, B_1(0), 0)$ is defined. Deform f on $\partial B_1(0)$ by using the deformation

$$H(x,t) = t f(x) + (1-t)x \quad \text{for } x \in \overline{B_1(0)}, t \in [0,1].$$

By the hypothesis $H(x,t) \neq 0$ for $x \in \partial B_1(0)$, $t \in [0,1]$. By homotopic invariance 9.16 (iv),

$$d(f, B_1(0), 0) = d(I, B_1(0), 0) = 1.$$

Therefore $f(x) = 0$ has a solution in $B_1(0)$. $\qquad \blacksquare$

Theorem 9.21 *Assume $f \in C(\mathbb{R}^N, \mathbb{R}^N)$ such that*

$$\lim_{|x| \to \infty} \frac{(f(x), x)}{|x|} = \infty.$$

Then f is onto: for every $y \in \mathbb{R}^N$, the equation $f(x) = y$ has a solution.

Proof. If necessary, we may replace f by $f(x) - y$, and only consider whether the equation $f(x) = 0$ has a solution. Take $R > 0$ such that

$$(f(x), x) \geq 0 \quad \text{if } |x| = R.$$

Suppose $f(x) \neq 0$ on $|x| = R$, otherwise we are through. Then $(f(x), x) \geq 0$ on $|x| = R$ implies that $f(x)$ never points opposite to x for $|x| = R : f(x) + \lambda x \neq 0$ for $\lambda \geq 0$, $|x| = R$. Otherwise, let $f(x) + \lambda x = 0$.

(i) If $\lambda > 0$, then $(f(x), x) = (-\lambda x, x) = -\lambda |x|^2 < 0$.

(ii) If $\lambda = 0$, then $f(x) = 0$. We obtain a contradiction. Now the result follows from Theorem 9.20. ∎

Definition 9.22 *Let X and Y be topological spaces, $A \subset X$ a subspace.*

(i) *If $f : X \to A$ is continuous with $f \mid_A = I$, the identity map, then f is called a retraction of X on A.*

(ii) *If $f : A \to Y$ is continuous with no continuous extension $F : X \to Y$, then f is called essential.*

Notation Let $B_1(0)$ be the unit ball with center 0 in $\mathbb{R}^N$, and $S^{N-1} = \partial B_1(0)$.

Theorem 9.23 *(Non-retraction of unit ball on its boundary) There are no retractions $f : \overline{B_1(0)} \to S^{N-1}$.*

Proof. Suppose there is a retraction $f : \overline{B_1(0)} \to S^{N-1}$. $0 \notin f(\overline{B_1(0)}) = S^{N-1}$, so

$$d(f, B_1(0), 0) = 0,$$

but $f = I$, the identity map on S^{N-1}, so

$$d(f, B_1(0), 0) = d(I, B_1(0), 0) = 1,$$

, a contradiction. ∎

Theorem 9.24 *The following are equivalent:*

(i) *(Non-retraction of unit ball on its boundary) There are no retractions $f : \overline{B_1(0)} \to S^{N-1}$;*

(ii) (*Unit sphere essential on the unit ball*) *The inclusion* $i : S^{N-1} \to \mathbb{R}^N \backslash \{0\}$ *is essential;*

(iii) (*Brouwer Fixed Point Theorem*) *Let* $f : \overline{B_1(0)} \to \overline{B_1(0)}$ *be continuous. Then* f *admits at least one fixed point.*

Proof.   The equivalence of (i) and (ii) is clear.

(i) $\Rightarrow$ (iii). We prove by contradiction. Suppose there is a continuous function $f : \overline{B_1(0)} \to \overline{B_1(0)}$ with $f(x) \neq x$ for every $x \in \overline{B_1(0)}$. Denote by $g(x)$ the intersection between the projective line $\overrightarrow{f(x)x}$ and $\partial B_1(0)$:

$$\begin{aligned} g(x) &= \lambda(x)x + (1 - \lambda(x))f(x) \\ &= \lambda(x)(x - f(x)) + f(x), \quad \lambda(x) \geq 1, \\ &\quad |g(x)| = 1. \end{aligned}$$

Then

$$1 = g(x) \cdot g(x),$$

or

$$\begin{aligned} 0 &= g(x) \cdot g(x) - 1 \\ &= (\lambda(x)(x - f(x)) + f(x)) \cdot (\lambda(x)(x - f(x)) + f(x)) - 1 \\ &= \lambda^2(x)|x - f(x)|^2 + 2\lambda(x)[(x - f(x)) \cdot f(x)] + (|f(x)|^2 - 1). \end{aligned}$$

Let $\lambda_1(x)$, $\lambda_2(x)$ be its solutions:

$$\lambda_1(x)\lambda_2(x) = \frac{|f(x)|^2 - 1}{|x - f(x)|^2} \leq 0.$$

Because only one of $\lambda_1(x)$ and $\lambda_2(x)$ can be positive, say $\lambda_1(x)$, then

$$\lambda_1(x) = \frac{-2(x - f) \cdot f + 2[(x - f \cdot f)^2 - |x - f|^2 (|f|^2 - 1)]^{1/2}}{2|x - f|^2}.$$

Now if we consider $g(x) = \lambda_1(x)(x - f(x)) + f(x)$ for $x \in \overline{B_1(0)}$, then $g : \overline{B_1(0)} \to S^{N-1}$ is a retraction, a contradiction to (i).

(iii) $\Rightarrow$ (i). We prove by contradiction. Let $f : \overline{B_1(0)} \to S^{N-1}$ be a retraction. Consider the continuous function $-f : \overline{B_1(0)} \to \overline{B_1(0)}$. Then $x \in B_1(0)$ implies that $-f(x) \in \partial B_1(0)$, or $x \neq (-f)(x)$. If $x \in \partial B_1(0)$, then $x \neq -x = (-f)(x)$. Therefore $-f$ does not have any fixed point, a contradiction to (iii).   ∎

Theorem 9.25

(i) *If N is odd, then there is no continuous function* $H : S^{N-1} \times [0, 1] \to S^{N-1}$ *such that $H(x, 0) = x$, $H(x, 1) = -x$ for $x \in S^{N-1}$.*

(ii) *If N is odd, then there is no continuous $f : S^{N-1} \to \mathbb{R}^N$ such that $f(x) \neq 0$, $f(x) \cdot x = 0$ for $x \in S^{N-1}$.*

Proof.

(i) Suppose there is a continuous function $H : S^{N-1} \times [0,1] \to S^{N-1}$ such that $H(x,0) = x = I(x)$, $H(x,1) = -x = -I(x)$ for $x \in S^{N-1}$. Then by Corollary 9.19 (ii),

$$1 = \mathrm{d}(I, B_1(0), 0) = \mathrm{d}(-I, B_1(0), 0) = -1,$$

where

$$J_{-I}(0) = \begin{vmatrix} -1 & \cdots & & \cdots & 0 \\ & -1 & & & \\ & & \ddots & & \\ 0 & \cdots & & \cdots & -1 \end{vmatrix} = (-1)^N = -1, \quad \text{because } N \text{ is odd.}$$

We have a contradiction.

(ii) Suppose there is a continuous function $f : S^{N-1} \to \mathbb{R}^N$ such that $f(x) \neq 0$, $f(x) \cdot x = 0$ for $x \in S^{N-1}$. Set

$$h(x) = \frac{f(x)}{|f(x)|}$$

and $H(x,t) = (\cos \pi t)x + (\sin \pi t)h(x)$ for $x \in S^{N-1}$, $t \in [0,1]$. Then $H \in C(S^{N-1} \times [0,1], S^{N-1})$. In fact, because $x \in S^{N-1}$, we have $|x| = 1$, and it follows at once that $|h(x)| = 1$, $h(x) \cdot x = 0$, and for $x \in S^{N-1}$, $t \in [0,1]$,

$$\begin{aligned} |H(x,t)|^2 &= |(\cos \pi t)x + (\sin \pi t)h(x)|^2 \\ &= (\cos^2 \pi t)|x|^2 + 2(\cos \pi t \sin \pi t)(x \cdot h(x)) + (\sin^2 \pi t)|h(x)|^2 \\ &= 1. \end{aligned}$$

Now $H(x,0) = x$, $H(x,1) = -x$, which contradicts (i). ∎

Suppose $\Lambda = [\underline{\lambda}, \bar{\lambda}] \subset \mathbb{R}$, $\Lambda \times \mathbb{R}^N$ possesses the usual topology. Consider a bounded open set Ω in $\Lambda \times \mathbb{R}^N$ with the boundary $\partial \Omega$. Denote, for $\lambda \in [\underline{\lambda}, \bar{\lambda}]$,

$$\Omega_\lambda = \left\{ x \in \mathbb{R}^N \,\middle|\, (\lambda, x) \in \Omega \right\}.$$

Then Ω_λ is bounded open in $\mathbb{R}^N$. Moreover,

$$\partial(\Omega_\lambda) \subset (\partial \Omega)_\lambda.$$

In fact, if $x_0 \in \partial(\Omega_\lambda)$ and $\epsilon > 0$ takes $(x, \lambda) \in \Omega$ and $(y, \lambda) \notin \Omega$ such that $|x - x_0| < \epsilon$, then $|y - x_0| < \epsilon$, or $|(x, \lambda) - (x_0, \lambda)| < \epsilon$, $|(y, \lambda) - (x_0, \lambda)| < \epsilon$. Therefore, $(x_0, \lambda) \in \partial\Omega$, or $x_0 \in (\partial\Omega)_\lambda$. We see the following may occur:

$$\partial(\Omega_\lambda) \not\subset (\partial\Omega)_\lambda.$$
$$x_0 \in (\partial\Omega)_\lambda, \quad x_0 \notin \partial(\Omega_\lambda).$$

Lemma 9.26 *Let K be a metric compact space, and K_1 and K_2 two closed disjoint sets in K. Then either:*

(i) *there is a (maximal connected closed) component of K that meets K_1 and K_2, or;*

(ii) *there are two disjoint compact sets $\overline{K}_1$ and $\overline{K}_2$ such that $K = \overline{K}_1 \cup \overline{K}_2$, $K_i \subset \overline{K}_i$, $i = 1, 2$.*

K_1	K_2		$\widehat{K}_1$	$\widehat{K}_2$
	K		K_1	K_2
	(i)			(ii)

Proof. Suppose (i) is not true. We claim that there is $\epsilon > 0$ such that no ϵ-chain can meet K_1, K_2. In fact, otherwise, for each $\epsilon > 0$, there are $a \in K_1$, $b \in K_2$ and an ϵ-chain C_ϵ that connects a and b. Thus, there are $a_n \in K_1$, $b_n \in K_2$ and a $\frac{1}{n}$-chain $C_{\frac{1}{n}}$ that connects a_n and b_n. Because K_1, K_2 are compact, there are subsequences, still denoted by $\{a_n\}$, $\{b_n\}$, and $a \in K_1$, $b \in K_2$ such that

$$a_n \to a, \quad b_n \to b.$$

Consider $C_a = \{x \in K|$ for $\epsilon > 0$. a and x can be connected by an ϵ-chain$\}$. C_a is closed and connected. In fact, for x, $y \in C_a$ and $\epsilon > 0$ there are two ϵ-chains C_1 and C_2 : C_1 connects x and a, C_2 connects y and a, so $C_1 \cup C_2$ is an ϵ-chain that connects x and y. Note that $b \in C_a$, and C_a is contained in a connected component; this contradicts our hypothesis. Therefore, there is an $\epsilon > 0$ such that no ϵ-chain can meet K_1, K_2 simultaneously. Consider

$$\hat{K}_1 = \{y \in K \mid \text{ there is } x \in K_1, \text{ and a } \epsilon - \text{chain that connects } x \text{ and } y\}.$$

We see that $\hat{K}_1 \cap K_2 = \emptyset$. $\hat{K}_1$ is closed, because if $x \in \overline{\hat{K}_1}$, then there is $\epsilon_1 > 0$ such that $B(x, \epsilon_1) \cap \hat{K}_1 \neq \emptyset$, so $x \in \hat{K}_1$. $\hat{K}_1$ is open because for $x \in \hat{K}_1$, $B(x, \epsilon_2) \subset \hat{K}_1$ for some ϵ_2. Set $\hat{K}_2 = K \backslash \hat{K}_1$, $\hat{K}_1 \cap \hat{K}_2 = \emptyset$, $\hat{K}_1 \cup \hat{K}_2 = K$, $K_i \subset \hat{K}_i$ and $\hat{K}_i$ are closed, so are compact, and (ii) follows. ∎

Corollary 9.27 *Let Ω be a bounded open set in $\Lambda \times \mathbb{R}^N$, $H(\lambda, x) \in C(\Omega, \mathbb{R}^N)$, $b \notin H(\partial\Omega)$. Then:*

(i) *The degree* $\mathrm{d}(H(\lambda, \cdot), \Omega_\lambda, b)$ *is defined and constant on* $[\underline{\lambda}, \bar{\lambda}]$;

(ii) *Let* $A = \{(\lambda, x) \in \Omega \mid H(\lambda, x) = b\}$. *If* $\mathrm{d}(H(\lambda, \cdot), \Omega_\lambda, b) = \alpha$. *If* $\alpha \neq 0$, *then there is a connected component* C *containing* A, *which meets* $\{\underline{\lambda}\} \times \Omega_{\underline{\lambda}}$, *and* $\{\bar{\lambda}\} \times \Omega_{\bar{\lambda}}$.

Proof.

(i) Fix $\lambda \in [\underline{\lambda}, \bar{\lambda}]$, and let

$$A_\lambda = \{x \in \Omega_\lambda \mid H(\lambda, x) = b\}.$$

Then $A_\lambda \subset \Omega_\lambda$, and by the hypothesis $A_\lambda \cap (\partial\Omega)_\lambda = \varnothing$, and it follows at once that $A_\lambda \cap \partial(\Omega_\lambda) = \varnothing$. Because A_λ is compact in Ω_λ, take an open neighborhood O_λ of A_λ such that

$$A_\lambda \subset O_\lambda \subset \bar{O}_\lambda \subset \Omega_\lambda,$$

and choose $\epsilon > 0$ with

$$[\lambda - \epsilon, \lambda + \epsilon] \times O_\lambda \subset \Omega,$$

where $[\lambda - \epsilon, \lambda + \epsilon]$ is modulo Λ : that is, $[\lambda - \epsilon, \lambda + \epsilon] \cap \Lambda$, in particular for $\lambda = \underline{\lambda}$ or $\lambda = \bar{\lambda}$. If $\epsilon > 0$ is small enough, then all solutions (μ, x) of $H(\mu, x) = b$, $|\lambda - \mu| < \epsilon$, are contained in $[\lambda - \epsilon, \lambda + \epsilon] \times O_\lambda$. In fact, otherwise, there is a sequence (ϵ_n) such that $\epsilon_n \searrow 0$, and a sequence (μ_j, x_j) such that

$$|\mu_j - \lambda_j| \leq \epsilon_j,$$
$$H(\mu_j, x_j) = b,$$
$$(\mu_j, x_j) \notin [\lambda - \epsilon_j, \lambda + \epsilon_j] \times O_\lambda.$$

Because $A = \{(\lambda, x) \in \Omega \mid H(\lambda, x) = b\}$ is compact, there is a subsequence, still denoted by $\{(\mu_j, x_j)\}$ such that

$$b = H(\mu_j, x_j) \to H(\lambda, x)$$

or $H(\lambda, x) = b$. However, $x_j \in O_\lambda^c$, and O_λ^c is closed, so $x \in O_\lambda^c$, or $x \notin A_\lambda$, a contradiction. For $\mu \in [\lambda - \epsilon, \lambda + \epsilon]$, because $b \notin H(\mu, \partial O_\lambda)$, the degree $\mathrm{d}(H(\mu, \cdot), O_\lambda, b)$ is defined. By homotopy invariance, $\mathrm{d}(H(\mu, \cdot), O_\lambda b)$ is constant on $\mu \in [\lambda - \epsilon, \lambda + \epsilon]$. However, by the Excision Property, $x \in \Omega_\mu$, $H(\mu, x) = b$ implies $|\mu - \lambda| < \epsilon$, $x \in O_\lambda$, and consequently, $H(\mu, x) \neq b$ for $x \in \Omega_\mu \backslash O_\lambda$, and we have

$$\mathrm{d}(H(\mu, \cdot), O_\lambda, b) = \mathrm{d}(H(\mu, \cdot), \Omega_\mu, b).$$

Therefore $\mathrm{d}(H(\mu, \cdot), \Omega_\mu, b)$ is locally constant on $[\underline{\lambda}, \bar{\lambda}]$, and consequently it is a global constant on $[\underline{\lambda}, \bar{\lambda}]$.

(ii) By part (i), $\mathrm{d}(H(\lambda,\cdot),\Omega_\lambda,b) = d \neq 0$ for $\lambda \in \Lambda$. Because A is compact, and $b \notin H(\partial\Omega)$, we have $(\{\lambda\} \times \Omega_\lambda) \cap A = (\{\lambda\} \times \bar{\Omega}_\lambda) \cap A$, for each λ. Thus, $(\{\lambda\} \times \Omega_\lambda) \cap A$ is a closed subset of A. If we set

$$K_1 = A \cap (\{\underline{\lambda}\} \times \Omega_{\underline{\lambda}}), \quad K_2 = A \cap (\{\bar{\lambda}\} \times \Omega_{\bar{\lambda}}),$$

then K_1 and K_2 are two disjoint compact sets in A. Because $\mathrm{d}(H(\lambda,\cdot),\Omega_\lambda,b) = \alpha \neq 0$, for each $\lambda \in [\underline{\lambda},\bar{\lambda}]$, then K_1 and K_2 are non-empty. Suppose there is no connected component, C containing A, that meets $\{\underline{\lambda}\} \times \Omega_{\underline{\lambda}}$ and $\{\bar{\lambda}\} \times \Omega_{\bar{\lambda}}$. Apply Lemma 9.26 to obtain two compact subsets $\hat{K}_1$, $\hat{K}_2$ such that $K_i \subset \hat{K}_i$, $i = 1, 2$, $\hat{K}_1 \cap \hat{K}_2 = \emptyset$, $A = \hat{K}_1 \cup \hat{K}_2$. Set $\epsilon = \frac{1}{2}\mathrm{dist}(\hat{K}_1,\hat{K}_2)$, $O = \left\{(\lambda,x) \in \Omega \,\middle|\, \mathrm{dist}((\lambda,x),\hat{K}_1) < \epsilon \right\}$. Then $O \subset \Omega$ is a bounded open set, $O \cap K_2 = \emptyset$, $A \cap \partial O = \emptyset$. For $\lambda \in [\underline{\lambda},\bar{\lambda}]$, set $O_\lambda = \{x \in \mathbb{R}^N \mid (\lambda,x) \in O\}$. Consider the restriction of $H(x,\lambda)$ on O. The hypothesis in part (i) is still true, and $\mathrm{d}(H(\lambda,0),O_\lambda,b)$ is constant on $[\underline{\lambda},\bar{\lambda}]$. Because $O_{\underline{\lambda}}$ contains all solutions of $H(\underline{\lambda},x) = b$ in $\Omega_{\underline{\lambda}}$, it has no solutions on $\Omega_{\underline{\lambda}} \backslash O_{\underline{\lambda}}$, by the Excision Property, for $\lambda = \underline{\lambda}$,

$$\mathrm{d}(H(\underline{\lambda},\cdot),O_{\underline{\lambda}},b) = \mathrm{d}(H(\underline{\lambda},\cdot),\Omega_{\underline{\lambda}},b) = \alpha.$$

On the other hand, however, $b \notin H(\bar{\lambda},\cdot)(\overline{O_{\bar{\lambda}}})$, so

$$\mathrm{d}(H(\bar{\lambda},\cdot),O_{\bar{\lambda}},b) = 0,$$

, a contradiction.   ∎

There is a multiplicity result:

Lemma 9.28 *Suppose* $\Omega_i \subset \mathbb{R}^{N_i}$ *are bounded open sets,* $f_i \in C(\Omega_i,\mathbb{R}^{N_i})$, *and* $b_i \in \mathbb{R}^{N_i} \backslash f_i(\partial\Omega_i)$, $i = 1, 2$. *Then*

$$\mathrm{d}((f_1,f_2),\Omega_1 \times \Omega_2,(b_1,b_2)) = \mathrm{d}(f_1,\Omega_1,b_1)\,\mathrm{d}(f_2,\Omega_2,b_2).$$

Proof. It suffices to prove the regular case: $f_i \in C^2(\Omega_i,\mathbb{R}^{N_i})$, $b_i \notin f(S_i)$, where $S_i = \{x \in \Omega_i \mid J_{f_i}(x) = 0\}$, $i = 1, 2$. Now

$$\mathrm{d}((f_1,f_2),\Omega_1 \times \Omega_2,(b_1,b_2))$$

$$= \sum_{(x_1,x_2) \in f_1^{-1}(b_1) \times f_2^{-1}(b_2)} \mathrm{sgn}\, J_{(f_1,f_2)}((x_1,x_2)).$$

However, if $x_1 = (x_1^1,\cdots,x_1^{N_1})$, $x_2 = (x_2^1,\cdots,x_2^{N_2})$, $f_1(x_1) = (f_1^1(x_1),\cdots,f_1^{N_1}(x_1))$, $f_2(x_2) = (f_2^1(x_2),\cdots,f_2^{N_2}(x_2))$, $(f_1,f_2)(x_1,x_2) = (f_1^1(x_1),\cdots,f_1^{N_1}(x_1),f_2^1(x_2),\cdots$

$\cdot, f_2^{N_2}(x_2))$, then

$$
J_{(f_1, f_2)}(x_1, x_2) =
\begin{vmatrix}
\dfrac{\partial f_1^1}{\partial x_1^1} & \dfrac{\partial f_1^2}{\partial x_1^1} & \cdots & & & \\
\dfrac{\partial f_1^1}{\partial x_1^2} & \dfrac{\partial f_1^2}{\partial x_1^2} & \cdots & & 0 & \\
\cdots & \cdots & \cdots & & & \\
& & & \dfrac{\partial f_2^1}{\partial x_2^1} & \dfrac{\partial f_2^2}{\partial x_2^1} & \cdots \\
& 0 & & \dfrac{\partial f_2^1}{\partial x_2^2} & \dfrac{\partial f_2^2}{\partial x_2^2} & \cdots \\
& & & \cdots & \cdots & \cdots
\end{vmatrix}
$$

$$
=
\begin{vmatrix}
\dfrac{\partial f_1^1}{\partial x_1^1} & \dfrac{\partial f_1^2}{\partial x_1^1} & \cdots \\
\dfrac{\partial f_1^1}{\partial x_1^2} & \dfrac{\partial f_1^2}{\partial x_1^2} & \cdots \\
\cdots & \cdots & \cdots
\end{vmatrix}
\cdot
\begin{vmatrix}
\dfrac{\partial f_2^1}{\partial x_2^1} & \dfrac{\partial f_2^2}{\partial x_2^1} & \cdots \\
\dfrac{\partial f_2^1}{\partial x_2^2} & \dfrac{\partial f_2^2}{\partial x_2^2} & \cdots \\
\cdots & \cdots & \cdots
\end{vmatrix}
$$

$$
= J_{f_1}(x_1) J_{f_2}(x_2).
$$

Therefore

$$
\mathrm{d}((f_1, f_2), \Omega_1 \times \Omega_2, (b_1, b_2)) = \left[\sum_{x \in f_1^{-1}(b_1)} \operatorname{sgn} J_{f_1}(x_1)\right]\left[\sum_{x \in f_2^{-1}(b_2)} \operatorname{sgn} J_{f_2}(x_2)\right]
$$

$$
= \mathrm{d}(f_1, \Omega_1, b_1)\, \mathrm{d}(f_2, \Omega_2, b_2).
$$

Suppose $f \in C(\Omega, \mathbb{R}^N)$, $b \notin f(\partial\Omega)$. Let x_0 be an isolated solution of $f(x) = b$: take $\epsilon_0 > 0$ such that, in the ball $B_{\epsilon_0}(x_0)$, x_0 is the only solution of $f(x) = b$. For ϵ, $0 < \epsilon < \epsilon_0$, because $b \notin f((\overline{B_r(x_0)}\backslash B_\epsilon(x_0))$, by the Excision Property,

$$
\mathrm{d}(f, B_{\epsilon_0}(x_0), b) = \mathrm{d}(f, B_\epsilon(x_0), b).
$$

Therefore $\mathrm{d}(f, B_{\epsilon_0}(x_0), b)$ is constant on $\epsilon \in (0, \epsilon_0)$. If we denote

$$
\mathrm{i}(f, x_0, b) = \lim_{\epsilon \to 0} \mathrm{d}(f, B_\epsilon(x_0), b),
$$

then $\mathrm{i}(f, x_0, b)$ is called the index of f at x_0 relative to b. ∎

Recall that if A is an $n \times n$ matrix, $\mu = \frac{1}{\lambda}$ is called a characteristic value of A, and λ an eigenvalue of A, if there is $x \neq 0$ in $\mathbb{R}^N$ such that $x = \mu A x$ or $A x = \lambda x$.

Lemma 9.29 *Let $\Omega \subset \mathbb{R}^N$ be a bounded open set, $0 \in \Omega$. Assume $g \in C^1(\Omega, \mathbb{R}^N)$, $g(0) = 0$. Define $f(x) = x - g(x)$ for $x \in \Omega$. If $J_f(0) \neq 0$, then*

$$\mathrm{i}(f, 0, 0) = (-1)^\beta,$$

where β is the sum of the multiplicities of characteristic values of $g'(0)$ in $(0, 1)$.

Proof.   Because $J_f(0) \neq 0$, 0 is an isolated solution of $f(x) = 0$, the index $\mathrm{i}(f, 0, 0)$ is defined, and

$$\mathrm{i}(f, 0, 0) = \mathrm{sgn}\, J_f(0) = \mathrm{sgn}\det(I - g'(0)).$$

However, we may triangulate $g'(0)$ into the form

$$A = \begin{bmatrix} \dfrac{1}{\mu_1} & & * \\ & \ddots & \\ 0 & & \dfrac{1}{\mu_N} \end{bmatrix},$$

where the μ_i are characteristic values in the complex numbers C. $\det(I - g'(0)) = J_f(0) \neq 0$, so 0 and 1 are not characteristic values of $g'(0)$. Now,

$$\mathrm{i}(f, 0, 0) = \mathrm{sgn}(\det(I - A)) = \mathrm{sgn}\left(\prod_{i=1}^{N}(1 - \frac{1}{\mu_i})\right).$$

Note that if $\mu_i \notin \mathbb{R}$ is a characteristic value of A, then $\bar{\mu}_i$ is also a characteristic value of A. In this case, among the products $\prod_{i=1}^{N}(1 - \frac{1}{\mu_i})$ we obtain terms $\left|1 - \frac{1}{\mu_i}\right|^2 > 0$. On the other hand, if $\bar{\mu}_i \in \mathbb{R}$, $\mu_i \notin [0, 1]$, then $1 - \frac{1}{\mu_i} > 0$. We see that the terms occurring in both cases do not affect the sign of the product, and hence,

$$\mathrm{i}(f, 0, 0) = \mathop{\mathrm{sgn}}_{\substack{\mu_i \in \mathbb{R} \\ \mu_i \in (0,1)}} \prod_i (1 - \frac{1}{\mu_i}) = (-1)^\beta. \qquad \blacksquare$$

Lemma 9.30 *Let $\Omega \subset \mathbb{R}^N$ be a bounded open set, $0 \in \Omega$. Let $g \in C^1(\Omega, \mathbb{R}^N)$, $g(0) = 0$. Define $f_\lambda(x) = x - \lambda g(x)$ for $x \in \Omega$. Then $\mathrm{i}(f_\lambda, 0, 0)$ is defined for $\lambda \neq \mu_i$, where μ_i are characteristic values of $g'(0)$, multiplied by $(-1)^{m_i}$ when λ passes through the characteristic value μ_i of multiplicity m_i.*

Proof.   As in Lemma 9.29, we obtain, for $\lambda \neq \mu_i$, $i = 1, 2, \cdots\cdots, N$, $\mathrm{i}(f_\lambda, 0, 0) =$ $\mathrm{sgn} \prod_{i=1}^{N}(1 - \frac{\lambda}{\mu_i})$. Set $\mu_i - \epsilon < \lambda_1 < \mu_i < \lambda_2 < \mu_i + \epsilon$ for small $\epsilon > 0$. We have

$$\frac{\lambda_1}{\mu_i} < 1 < \frac{\lambda_2}{\mu_i}.$$

Therefore

$$\mathrm{sgn}\left(1 - \frac{\lambda_1}{\mu_j}\right) = \mathrm{sgn}\left(1 - \frac{\lambda_2}{\mu_j}\right), \quad i \neq j,$$

$$\mathrm{sgn}\left(1 - \frac{\lambda_1}{\mu_i}\right) = 1, \quad \mathrm{sgn}\left(1 - \frac{\lambda_2}{\mu_i}\right) = -1.$$

There are m_i occurrences of μ_i. Thus,

$$\mathrm{i}(f_{\lambda_2}, 0, 0) = (-1)^{m_i}\, \mathrm{i}(f_{\lambda_1}, 0, 0). \qquad \blacksquare$$

Remark 9.31 *In Lemmas 9.29 and 9.30, $g \in C^1(\Omega, \mathbb{R}^N)$ can be replaced because g is differentiable at 0. In fact,*

> g is differentiable at 0
>
> if and only if $\|g(x) - g(0) - g'(0)x\| = o(\|x\|)$
>
> if and only if $\|g(x) - g'(0)x\| = o(\|x\|)$ because $g(0) = 0$
>
> if and only if $\|(I - \lambda g)(x) - (I - \lambda g'(0))x\| = o(\lambda\,\|x\|)$.

Because $f \to \mathrm{d}(f, \Omega, 0)$ is continuous, we have for small $\epsilon > 0$,

$$\mathrm{d}(I - \lambda g, B_\epsilon(0), 0) = \mathrm{d}(I - \lambda g'(0), B_\epsilon(0), 0)$$

or

$$\mathrm{i}(f_\lambda, 0, 0) = \mathrm{i}(I - \lambda g'(0), 0, 0).$$

9.2   Brouwer Fixed Point Theorem

The Brouwer Fixed Point Theorem has a generalization.

Theorem 9.32 *If we let $K \subset \mathbb{R}^N$ be a compact convex set, $f : K \to K$ a continuous function, then f admits a fixed point.*

Proof.   If necessary we may replace $\mathbb{R}^N$ by $\mathbb{R}^M$, $M \leq N$, and by translation obtain $\overset{\circ}{K} \neq \varnothing$, with $0 \in \overset{\circ}{K}$. Consider the gauge $j_K : \mathbb{R}^N \to \mathbb{R}^+$ defined by

$$j_K(x) = \inf\{\lambda \geq 0 \,|\, x \in \lambda K\}.$$

Define $h(0) = 0$, $h(x) = j_K(x)\dfrac{x}{|x|}$ if $x \in K \backslash \{0\}$. Then $h : K \to \overline{B_1(0)}$ is a homeomorphism onto with $h(\partial K) \subset \partial B_1(0)$. Let $g = h \circ f \circ h^{-1} : \overline{B_1(0)} \to \overline{B_1(0)}$. Then g is continuous, and by the Brouwer Fixed Point Theorem, there is $x_0 \in \overline{B_1(0)}$ such that

$$x_0 = g(x_0) = h \circ f \circ h^{-1}(x_0)$$

or

$$h^{-1}(x_0) = f(h^{-1}(x_0)),$$

so $h^{-1}(x_0) \in K$ is a fixed point of f. $\qquad \blacksquare$

Remark 9.33 *It is easy to see that Theorem 9.32 is equivalent to the Brouwer Fixed Point Theorem.*

Theorem 9.34 *The following are equivalent.*

(i) *(Brouwer Fixed Point Theorem) Letting $f : \overline{B_1(0)} \to \overline{B_1(0)}$ be continuous, then f admits a fixed point.*

(ii) *(KKM Theorem, Knaster-Kuratowski-Mazurkiewicz Theorem) Let E be a Hausdorff Topological Vector Space, and let $x_1, \cdots, x_m \in E, X_1, \cdots, X_m$ be closed sets in E such that*

$$\text{conv}\{x_{i_1}, \cdots, x_{i_k}\} \subset X_{i_1} \cup \cdots \cup X_{i_k}$$

for $\{i_1, \cdots, i_k\} \subset \{1, 2, \cdots, m\}$. Then $\bigcap\limits_{i=1}^{m} X_i \neq \emptyset$.

(iii) *Let E be a Hausdorff Topological Vector Space and $X \subset E$ a subset such that for every $x \in X$, a closed set $F(x)$ of E is associated and there is at least one $x_0 \in X$ such that $F(x_0)$ is compact. Suppose for each finite family $\{x_1, \cdots, x_m\}$ in X, $\text{conv}\{x_1, \cdots, x_m\} \subset \bigcup\limits_{i=1}^{m} F(x_i)$. Then $\bigcap\limits_{x \in X} F(x) \neq \emptyset$.*

(iv) *(Ky Fan Minimax Inequality) Let E be a Hausdorff Topological Vector Space, $K \subset E$ a compact convex set, and let $f : K \times K \to \mathbb{R}$ satisfy:*
(a) for each $x \in K$ fixed, $y \to f(x, y)$ is a lower semicontinuous function;
(b) for each $y \in K$ fixed, $x \to f(x, y)$ is a quasiconcave function. Then

$$\min_{y \in K} \max_{x \in K} f(x, y) \leq \sup_{x \in K} f(x, x).$$

(v) *(Hartman-Stampacchia Theorem) Let $K \subset \mathbb{R}^N$ be a compact convex set, and $A : K \to \mathbb{R}^N$ continuous. Then there is $x \in K$ such that $(Ax, y - x) \geq 0$ for $y \in K$.*

Proof. We prove using the order $(i) \Rightarrow (ii) \Rightarrow (iii) \Rightarrow (iv) \Rightarrow (v) \Rightarrow (i)$.
$(i) \Rightarrow (ii)$. Because $\langle x_1, \cdots, x_m \rangle = \mathbb{R}^N$, we may consider $X_i' = X_i \cap \mathbb{R}^N$ and require $\bigcap\limits_{i=1}^{m} X_i' \neq \emptyset$. Hence, we may assume $E = \mathbb{R}^N$. Suppose

$$K = \text{conv}\{x_1, \cdots, x_m\} \subset X_1 \cup \cdots \cup X_m, \text{ and } \bigcap\limits_{i=1}^{m} X_i = \emptyset.$$

Set $\varphi_i(x) = \mathrm{d}(x, X_i)$. Because for each $x \in K$ there is $i \in \{1, \cdots, m\}$ such that $x \notin X_i$, we have $\sum_{i=1}^{m} \varphi_i(x) > 0$. Set

$$f(x) = \frac{\sum\limits_{i=1}^{m} \varphi_i(x)\, x_i}{\sum\limits_{i=1}^{m} \varphi_i(x)}.$$

Then $f : K \to K$ is continuous. Because K is a compact convex set, by the Brouwer Fixed Point Theorem, there is $x_0 \in K$ such that $f(x_0) = x_0$. Then there is $i_1 \in \{1, \cdots, m\}$ such that $x_0 \in X_{i_1}$, or $\varphi_{i_1}(x_0) = 0$,

$$x_0 = f(x_0) = \frac{\sum\limits_{i \neq i_1} \varphi_i(x)x_i}{\sum\limits_{i \neq i_1} \varphi_i(x)} \in \operatorname*{conv}_{i \neq i_1} \{x_i\} \subset \bigcup_{i \neq i_1} X_i.$$

Then there is i_2 such that $x_0 \in X_{i_2}$, or $\varphi_{i_2}(x_0) = 0$, so

$$x_0 \in \operatorname*{conv}_{i \neq i_1, i_2} \{x_i\} \subset \bigcup_{i \neq i_1, i_2} X_i.$$

Continuing this process we obtain $x_0 \in \bigcap_{i=1}^{m} X_i$, a contradiction.

$(ii) \Rightarrow (iii)$. Suppose $\bigcap_{x \in X} F(x) = \varnothing$. Because there is a compact set $F(x_0)$, by the Finite Intersection Property, there is a finite set $\{x_1, \cdots, x_m\}$ in X such that $\bigcap_{i=1}^{m} F(x_i) = \varnothing$, a contradiction to (ii).

$(iii) \Rightarrow (iv)$. Recall that $\varphi : K \to \mathbb{R}$ is quasiconcave if for each $t \in \mathbb{R}$, $\{x \in K \mid \varphi(x) > t\}$ is convex. Set $M = \sup\limits_{x \in K} f(x, x)$. For each $x \in K$, define $F(x) = \{y \in K \mid f(x, y) \leq M\} \subset K$. Clearly $F(x)$ is closed. For any set $\{x_1, \cdots, x_m\}$ in K, $\operatorname{conv}\{x_1, \cdots, x_m\} \subset \bigcup_{i=1}^{m} F(x_i)$. Otherwise, take $\sum_{j=1}^{m} \alpha_j x_j \notin \bigcup_{i=1}^{m} F(x_i)$, $\alpha_j \geq 0$, $\sum_{j=1}^{m} \alpha_j = 1$, so $f(x_i, \sum_{j=1}^{m} \alpha_j x_j) > M$ for $i = 1, 2, \cdots, m$. Because $x \to f(x, \sum_{j=1}^{m} \alpha_j x_j)$ is quasiconcave,

$$f\left(\sum_{j=1}^{m} \alpha_j x_j, \sum_{j=1}^{m} \alpha_j x_j\right) > M,$$

this contradicts $M = \sup\limits_{x \in K} f(x, x)$. Because each $F(x)$ is compact, by (iii) $\bigcap\limits_{x \in K} F(x) \neq \varnothing$: $y_0 \in \bigcap\limits_{x \in K} F(x)$, and consequently

$$\max_{x \in K} f(x, y_0) \leq M.$$

That is,

$$\min_{y \in K} \max_{x \in K} f(x, y) \leq \sup_{x \in K} f(x, x).$$

$(iv) \Rightarrow (v)$. Set $f(x, y) = (-Ay, x - y)$. Then $f(x, y)$ is continuous on y, and affine on x :

$$f((1 - \lambda)x_1 + \lambda x_2, y) = (1 - \lambda)f(x_1, y) + f(x_2, y).$$

By (iv)

$$\min_{y \in K} \max_{x \in K} f(x, y) \leq \sup_{x \in K} f(x, x) = 0.$$

There is $u \in K$ such that

$$f(v, u) = (-Au, v - u) \leq 0 \quad \text{for } v \in K,$$

or

$$(Au, v - u) \geq 0 \quad \text{for } v \in K.$$

$(v) \Rightarrow (i)$. Let $f : \overline{B_1(0)} \to \overline{B_1(0)}$ be continuous, and $A = I - f$, then $A : \overline{B_1(0)} \to \mathbb{R}$ is continuous. By (v) there is $u \in \overline{B_1(0)}$ such that

$$(u - f(u), v - u) \geq 0 \quad \text{for } v \in \overline{B_1(0)}.$$

(a) If $u \in B_1(0)$, then $\left\{ v - u \,\middle|\, v \in \overline{B_1(0)} \right\} \supset B_r(0)$ for some $r > 0$, and consequently, $(u - f(u), w) = 0$ for $w \in B_r(0)$. For $x \in \mathbb{R}^N$, $x \neq 0$, take $\rho > 0$ such that $\rho x \in B_r(0)$, so

$$(u - f(u), x) = \frac{1}{\rho}(u - f(u), \rho x) = 0.$$

Therefore $u - f(u) = 0$, or $f(u) = u$.

(b) If $u \in S^{N-1}$. By the inequality $(u - f(u), v - u) \geq 0$ for $v \in \overline{B_1(0)}$, let $u - fu = a$, $B_1(0) - u = W$. For $w \in W$

$$0 \leq (a, w) = |a| \, |w| \cos\theta, \ 0 \leq \theta \leq \frac{\pi}{2},$$

so $u - fu = a = \lambda u$, $\lambda \leq 0$. However, $f(u) = (1 - \lambda)u \in \overline{B_1(0)}$, so $-1 \leq 1 - \lambda \leq 1$, or $\lambda \geq 0$. We conclude that $\lambda = 0$, or $fu = u$. ∎

Remark 9.35 *We may prove* $(i) \Rightarrow (v)$, $(iv) \Rightarrow (i)$ *as follows.*

Proof. $(i) \Rightarrow (v)$ (by H. Brezis) Let $f(u) = \text{Proj}_K(-Au + u)$, for $u \in K$, then $f : K \to K$ is continuous. By the Brouwer Fixed Point Theorem, there is $u \in K$ such that

$$u = \text{Proj}_K(-Au + u).$$

Therefore

$$((-Au + u) - u, v - u) \leq 0 \quad \text{for } v \in K,$$

or

$$(Au, v - u) \geq 0 \quad \text{for } v \in K.$$

$(iv) \Rightarrow (i)$ Let $f \in C(K, K)$. Define $g : K \times K \to \mathbb{R}$ by

$$g(x, y) = \|f(y) - y\| - \|x - f(y)\|, \quad \text{for } x, y \in K.$$

Then

$$g(\lambda x_1 + (1 - \lambda)x_2, y) \geq \lambda g(x_1, y) + (1 - \lambda)g(x_2, y).$$

Now, g is continuous on y, and concave on x (in particular quasiconcave on x). By (iv)

$$\min_{y \in K} \max_{x \in K} g(x, y) \leq \max_{x \in K} g(x, x) = 0.$$

Take $y_0 \in K$ such that $g(x, y_0) \leq 0$ for $x \in K$ or

$$\|y_0 - f(y_0)\| \leq \|x - f(y_0)\| \quad \text{for } x \in K.$$

Let $x = f(y_0)$. Then $f(y_0) = y_0$. ∎

Theorem 9.36 (*Von Neumann Minimax Theorem*) *Let $E \subset \mathbb{R}^m$ and $F \subset \mathbb{R}^n$ be convex compact sets, and let $f : E \times F \to \mathbb{R}$ satisfy:*

(i) *For $y \in F$, $x \to f(x, y)$ is a continuous convex function;*

(ii) *For $x \in E$, $y \to f(x, y)$ is a continuous concave function.*
   Then $(x_0, y_0) \in E \times F$ is a saddle point of f, satisfying one of the following equivalent conditions,
   (a) $f(x_0, y) \leq f(x_0, y_0) \leq f(x, y_0)$ *for $x \in E$, $y \in F$;*
   (b) $\min_x(\max_y f(x, y)) = \max_y(\min_x f(x, y))$.

Proof. We claim that (a), (b) are equivalent: in fact,
$(a) \Rightarrow (b)$. Suppose

$$f(x_0, y) \leq f(x_0, y_0) \leq f(x, y_0) \quad \text{for } x \in E, \ y \in F.$$

Because

$$\min_x \max_y f(x, y) \leq \max_y f(x_0, y), \text{ and } \min_x f(x, y_0) \leq \max_y \min_x f(x, y).$$

We have

$$\min_x \max_y f(x, y) \leq \max_y f(x_0, y)$$
$$\leq f(x_0, y_0)$$
$$\leq \min_x f(x, y_0)$$
$$\leq \max_y \min_x f(x, y),$$

so
$$\min_x \max_y f(x,y) \leq \max_y \min_x f(x,y). \tag{9.22}$$

Fix $z \in E$
$$\min_x \ f(x,y) \leq f(z,y) \quad \text{for } y \in F,$$

so
$$\max_y \min_x f(x,y) \leq \max_y f(z,y) \leq \min_z \max_y f(z,y) = \min_x \max_y f(x,y),$$
$$\max_y \min_x f(x,y) \leq \min_x \max_y f(x,y). \tag{9.23}$$

By (9.22) and (9.23)
$$\min_x \max_y f(x,y) = \max_y \min_x f(x,y),$$

also by (9.22)
$$f(x_0, y_0) = \max_y f(x_0, y)$$
$$= \min_x \max_y f(x,y)$$
$$= \min_x f(x, y_0)$$
$$= \max_y \min_x f(x,y).$$

$(b) \Rightarrow (a)$. Let $\min_x \max_y f(x,y) = \max_y \min_x f(x,y) = \alpha$. Take $x_0 \in E$, $y_0 \in F$ such that
$$\max_y f(x_0, y) = \min_x \max_y f(x,y)$$
$$= \max_y \min_x f(x,y)$$
$$= \min_x f(x, y_0).$$

Then
$$f(x_0, y) \leq \alpha \leq f(x, y_0) \quad \text{for } x \in E, y \in F.$$

In particular,
$$f(x_0, y_0) \leq \alpha \leq f(x_0, y_0),$$

or
$$\alpha = f(x_0, y_0).$$

Thus,
$$f(x_0, y) \leq f(x_0, y_0) \leq f(x, y_0) \quad \text{for } x \in E, y \in F. \qquad \blacksquare$$

Theorem 9.37 *Let E be a Hausdorff Topological Vector Space, with $X \subset E$ a subset. For each $x \in X$, associate a set $F(x)$ in E satisfying:*

(i) *$\overline{F(x_0)} = L$ is compact for some $x_0 \in X$;*

(ii) *For $x_1, \cdots, x_n$ in X, $\operatorname{conv}\{x_1, \cdots, x_n\} \subset \bigcup_{i=1}^{n} F(x_i)$;*

(iii) *For each $x \in X$, $F(x) \cap M$ is closed for every finite-dimensional subspace M;*

(iv) *For each convex set D in E, we have*

$$\left(\overline{\bigcap_{x \in X \cap D} F(x)} \right) \bigcap D = \left(\bigcap_{x \in X \cap D} F(x) \right) \bigcap D.$$

Then $\bigcap_{x \in X} F(x) \neq \varnothing$.

Proof. If necessary we may replace E by $E - \{x_0\}$; we assume $x_0 = 0$. Let $\{M_i\}_{i \in I}$ be the family of all finite-dimensional subspaces of E, ordered by inclusion $M_i \subset M_j$ if $i \leq j$. Fix $i \in I$, $X \cap M_i \subset M_i$. By (iii), $x \in X \cap M_i \to F(x) \cap M_i$ is closed in M_i, so $L \cap M_i = F(0) \cap M_i$ is compact in M_i. By (ii), for $x_1, \cdots, x_k \in X \cap M_i$, $\operatorname{conv}\{x_1, \cdots, x_k\} \subset \bigcup_{j=1}^{k} (F(x_i) \cap M_i)$. By Theorem 9.34 (iii) :

$$\bigcap_{x \in X \cap M_i} (F(x) \cap M_i) \neq \varnothing.$$

Assume $u_i \in L \cap M_i$ and $u_i \in F(x)$ for each $x \in X \cap M_i$, so

$$u_j \in F(x) \quad \text{for } x \in X \cap M_j \supset X \cap M_i, j \geq i.$$

In particular,

$$u_j \in F(x) \quad \text{for } x \in X \cap M_i, \text{ for } j \geq i.$$

If we let $A_i = \bigcup_{j \geq i} \{u_j\}$, then $A_i \subset F(z)$ for $z \in X \cap M_i$, $A_i \subset L$. Then $A_i \subset \bigcap_{z \in X \cap M_i} F(z)$, $A_i \subset L$. For $\bar{A}_{i_1}, \cdots, \bar{A}_{i_n}$, there is $N \in N$ such that $i_j \leq N$, $j = 1, \cdots, n$, $u_N \in \bar{A}_{j_1}, \cdots, \bar{A}_{j_n}$. Thus, $\{\bar{A}_i\}_{i \in I}$ has the Finite Intersection Property, $\bar{A}_i \subset L$ for $i \in I$, and L is compact, so $\bigcap_{i \subset I} \bar{A}_i \neq \varnothing$. Let $\bar{x} \in \bigcap_{i \subset I} \bar{A}_i$. Let i_0 be with $\bar{x} \in M_{i_0}$. For $x \in X$, there is $i \geq i_0$ with $x \in M_i$. Then

$$\bar{x} \in \bar{A}_i \cap M_i \subset \left(\overline{\bigcap_{z \in X \cap M_i} F(z)} \right) \cap M_i = \left(\bigcap_{z \in X \cap M_i} F(z) \right) \cap M_i$$

or $\bar{x} \in F(z)$ for $z \in X \cap M_i$. In particular, $x \in X \cap M_{i_0}$, $\bar{x} \in F(x)$, so $\bar{x} \in \bigcap_{x \in X} f(x)$. ∎

Theorem 9.38 (*Brezis-Nirenberg-Stampacchia Theorem*) *Let E be a Hausdorff Topological Vector Space, $C \subset E$ a closed convex set. Let $f : C \times C \to \mathbb{R}$ satisfy:*

(i) $f(x, x) \leq 0$ *for $x \in C$;*

(ii) *for $x \in C$, $\{y \in C \mid f(x, y) > 0\}$ is convex;*

(iii) *for $y \in C$, $x \to f(x, y)$ is lower semicontinuous on $C \cap M$, for every finite-dimensional subspace M of E;*

(iv) *$x, y \in C$, $x \in \bar{G}$ for every set G in E such that*

$$f(z, (1-t)x + ty) \leq 0 \quad \text{for } 0 \leq t \leq 1, \ z \in G.$$

Then $f(x, y) \leq 0$;

(v) *there is a compact set L in E, $y_0 \in L \cap C$ such that*

$$f(x, y_0) > 0 \quad \text{for } x \in C, \ x \notin L.$$

Then there is $x_0 \in L \cap C$ such that

$$f(x_0, y) \leq 0 \quad \text{for } y \in C.$$

Proof. For $y \in C$, let $F(y) = \{x \in C \mid f(x, y) \leq 0\}$

(a) By (v), there is a compact set L in E, $y_0 \in L \cap C$ such that

$$f(x, y_0) > 0 \quad \text{for } x \in C, \ x \notin L,$$

and we have $F(y_0) \subset L$, or $\overline{F(y_0)} \subset L$. Thus, $\overline{F(y_0)}$ is compact.

(b) We claim that for $x_1, \cdots, x_n$ in X, $\text{conv}\{x_1, \cdots, x_n\} \subset \bigcup_{i=1}^{n} F(x_i)$. Suppose not, and $y_1, \cdots, y_n \in X$ and $\alpha_1, \cdots, \alpha_n \geq 0$, $\sum_{i=1}^{n} \alpha_i = 1$ exist such that

$$\sum_{i=1}^{n} \alpha_i y_i \notin \bigcup_{i=1}^{n} F(y_i), \text{ or } \sum_{i=1}^{n} \alpha_i y_i \in \bigcap_{i=1}^{n} F(y_i)^c,$$

so

$$f(\sum_{i=1}^{n} \alpha_i y_i, y_j) > 0 \quad \text{for } j = 1, \cdots, n.$$

By (ii)

$$f(\sum_{i=1}^{n} \alpha_i y_i, \sum_{i=1}^{n} \alpha_i y_i) > 0, \text{ contradicts (i).}$$

(c) By (iii), $x \to f(x,y)$ is lower semicontinuous on $C \cap M$, and $F(y) \cap M = \{x \in C \cap M \,|\, f(x,y) \le 0\}$ is closed.

(d) For the convex set D in E, we require $\left(\overline{\bigcap_{x \in X \cap D} F(x)}\right) \cap D = \left(\bigcap_{x \in X \cap D} F(x)\right)$

$\cap\, D$. It suffices to show that

$$\left(\overline{\bigcap_{x \in X \cap D} F(x)}\right) \cap D \subset \left(\bigcap_{x \in X \cap D} F(x)\right) \cap D.$$

For $u \in \left(\overline{\bigcap_{x \in X \cap D} F(x)}\right) \cap D$, $G = \bigcap_{x \in X \cap D} F(x)$, $u \in \bar{G}$, $u \in D$. For $0 \le t \le 1$, by definition,

$$f(z, (1-t)u + t\,x) \le 0 \quad \text{for } z \in G.$$

By (iv) $f(u,x) \le 0$ for $x \in X \cap D$, or $u \in F(x)$ for $x \in X \cap D$. Thus,

$$u \in \left(\bigcap_{x \in X \cap D} F(x)\right) \cap D.$$

By Theorem 9.37, let $x_0 \in \bigcap_{y \in C} F(y)$, $\bigcap_{y \in C} F(y) \ne O$ or $x_0 \in F(y)$ for $y \in C$. Thus,

$$f(x_0, y) \le 0 \quad \text{for } y \in C.$$

Note that $x_0 \in C \cap L$ because $F(y_0) \subset L$. ∎

Theorem 9.39 *Let E be a Hausdorff Topological Vector Space, C a convex set in E. Let $f : C \times C \to \mathbb{R}$ satisfy:*

(i) *$f(x,x) \le 0$ for $x \in C$;*

(iii) *for $y \in C$, $x \to f(x,y)$ is lower semicontinuous on $C \cap M$, for every finite-dimensional subspace M of E;*

(v) *there is a compact set L in E, $y_0 \in L \cap C$ such that*

$$f(x, y_0) > 0 \quad \text{for } x \in C, x \notin L;$$

(vi) *For $x \in C$, $k \ge 0$, the set $\{y \in C \,|\, f(x,y) \ge k\}$ is closed and convex;*

(vii) *For $x,y \in C$, $f(x,y) \le 0$, then $f(y,x) \ge 0$;*

$(viii)$ *If $f(x, y_1) > f(x, y_2) \ge 0$, then $f(x, t\,y_1 + (1-t)y_2) > f(x, y_2)$ for $0 < t \le 1$.*

Then there is $x_0 \in L \cap C$ such that $f(x_0, y) \le 0$ for $y \in C$.

Proof.

(ii) Consider

$$F = \{y \in C \,|\, f(x,y) > 0\} = \bigcup_{n=1}^{\infty} \left\{ y \in C \,\Big|\, f(x,y) \geq \frac{1}{n} \right\}.$$

Assume $y_1, y_2 \in F, t \in [0,1]$, $f(x, y_2) = \epsilon_2 > 0$. Take $n_0 \in N$ such that $\epsilon_1, \epsilon_2 \geq \frac{1}{n_0}$. Therefore, $y_1, y_2 \in \{y \in C|\, f(x,y) = \frac{1}{n_0}\}$ is a convex set by (vi). Thus, $(1-t)y_1 + ty_2 \in \{y \in C|\, f(x,y) = \frac{1}{n_0}\} \subset F$. Thus, F is convex.

(iv) We claim that $x, y \in C$, $x \in \overline{G}$ for every set G in E with $\overline{G} \subset C$ such that $f(z, (1-t)x + ty) \leq 0$ for $0 \leq t \leq 1$, $z \in G$, then $f(x,y) \leq 0$. Suppose not, and there is x, $y \in C$, $G \subset E$ such that $x \in \overline{G}$, $f(z, (1-t)x + ty) \leq 0$ for $0 \leq t \leq 1$, $z \in G$. However $f(x,y) > 0$.
By (vii), $f((1-t)x + ty, z) \geq 0$.
By (vi), $\{z \in C | f(x, z) \geq 0\}$ is closed and convex. Because $x \in \overline{G}$, $\overline{G} \subset C$, we have

$$f((1-t)x + ty, x) \geq 0 \quad \text{for } 0 \leq t \leq 1. \tag{9.24}$$

$v \to f(v, y)$ is lower semicontinuous on C modulus a finite-dimensional subspace, $v : (1-t)x + ty \to x$ as $t \to 0$, and $f(x,y) > 0$. Therefore $0 < f(x,y) \leq \lim_{t \to 0} f(v,y)$. Thus, $f(v,y) > 0$ for $0 < t < \delta$, $v = (1-t)x + ty$. For such v, we have $f(v,x) \geq 0$ by (9.24).

(a) If $f(v,x) > 0$, then $f(v,y) > 0$, $f(v,x) > 0$. By (ii), $\{y \in C \mid f(x,y) > 0\}$ is convex, so
$$f(v, (1-t)x + ty) = f(v,v) > 0.$$

(b) If $f(v,x) = 0$, then $f(v,y) > f(v,x) \geq 0$. By ($viii$), $f(v,v) = f(v, ty + (1-t)x) > f(v,x) \geq 0$. In both cases, $f(v,v) > 0$ contradicts (i). By Theorem 9.38, there is $x_0 \in L \cap C$ such that

$$f(x_0, y) \leq 0 \quad \text{for } y \in C. \qquad \blacksquare$$

Theorem 9.40 *Let E be a Hausdorff topological vector space, C a compact convex set in E, E^* the dual of E, and let $A : C \to E^*$ satisfy:*
(a) For $x, y \in C$, $(Ax, x - y) \leq 0$ implies $(Ay, y - x) \geq 0$;
(b) A is continuous on every finite-dimensional subspace.
Then there is $x_0 \in C$ such that $(Ax_0, x_0 - y) \leq 0$ for $y \in C$.

Proof. Let $f(x,y) = (Ax, x - y)$. Then f is convex in y.

(i) $f(x,x) = (Ax, x - x) = 0$.

(iii) For $y \in C$, $\{x_\alpha\} \subset C \cap M$ and $x \in C \cap M$, $x_\alpha \to x$. Then

$$
\begin{aligned}
|f(x_\alpha, y) - f(x, y)| &= |(Ax_\alpha, x_\alpha - y) - (Ax, x - y)| \\
&\leq |(Ax_\alpha, x_\alpha - y) - (Ax_\alpha, x - y)| \\
&\quad + |(Ax_\alpha, x - y) - (Ax, x - y)| \\
&\to 0 \text{ as } \alpha \to \infty \text{ because } A \text{ is continuous on } C \cap M.
\end{aligned}
$$

Therefore $x \to f(x, y)$ is continuous on $C \cap M$.

(v) Assume $y_0 \in C$,

$$
\begin{aligned}
L &= \{x \in C \,|\, f(x, y_0) = (Ax, x - y_0) \leq 0\} \\
&= \{x \in C \,|\, (Ay_0, y_0 - x) \geq 0\}.
\end{aligned}
$$

L is closed in C, so it is compact, and

$$
f(x, y_0) > 0 \quad \text{for } x \in C, x \notin L.
$$

(vi) For $x \in C$ and $k \geq 0$, $B = \{y \in C \,|\, f(x, y) \geq k\}$, $y_1, y_2 \in B$, $t \in [0, 1]$,

$$
\begin{aligned}
f(x, (1-t)y_1 + t\, y_2) &= (Ax, x - [(1-t)y_1 + t\, y_2]) \\
&= (Ax, [(1-t)(x - y_1) + t(x - y_2)]) \\
&= (1-t)\,(Ax, x - y_1) + t\,(Ax, x - y_2) \geq k.
\end{aligned}
$$

so $(1-t)y_1 + t\, y_2 \in B$, or B is convex. That B is closed can be proved as in (v).

(vii) For x and $y \in C$, $f(x, y) = (Ax, x - y) \leq 0$. By (a) $f(y, x) = (Ay, y - x) \geq 0$.

$(viii)$ If $f(x, y_1) = (Ax, x - y_1) > f(x, y_2) = (Ax, x - y_2) \geq 0$, then

$$
\begin{aligned}
f(x, t\, y_1 + (1-t)y_2) &= (Ax, x - (t\, y_1 + (1-t)y_2)) \\
&= (Ax, t(x - y_1) + (1-t)(x - y_2)) \\
&= t\,(Ax, x - y_1) + (1-t)\,(Ax, x - y_2) \\
&> t\,(Ax, x - y_2) + (1-t)\,(Ax, x - y_2) \\
&= (Ax, x - y_2) \\
&= f(x, y_2) \quad \text{for } 0 < t \leq 1.
\end{aligned}
$$

By Theorem 9.38, there is $x_0 \in L \cap C$ such that

$$
(Ax_0, x_0 - y) = f(x_0, y) \leq 0 \quad \text{for } y \in C.
$$
∎

9.3   The Borsuk Theorem

Lemma 9.41 *Let $K \subset \mathbb{R}^N$ be a compact set, and $f \in C(K, \mathbb{R}^M)$, $M > N$, with $0 \notin f(K)$. Then for each closed cube Q, $K \subset Q \subset \mathbb{R}^N$, there is $F \in C(Q, \mathbb{R}^M)$ such that $F|_K = f$, and $0 \notin F(Q)$.*

Proof.   Set $\alpha = \min\limits_{x \in K} |f(x)|$. Because $0 \notin f(K)$, $\alpha > 0$. Take ϵ, $0 < \epsilon < \frac{\alpha}{2}$. By the Tietze Extension Theorem, f can be extended continuously to Q, denoted by $\bar{f}$. Take $g \in C^1(Q, \mathbb{R}^M)$ such that

$$\left\| \bar{f} - g \right\|_{L^\infty(Q)} < \frac{\epsilon}{2}.$$

(a) If $0 \notin g(Q)$, take $h = g$,

(b) If $0 \in g(Q)$, then set $u(x, y) = g(x)$ for $x \in Q$, $y \in \mathbb{R}^{M-N}$. Hence, $u \in C^1(Q \times \mathbb{R}^{M-N}, \mathbb{R}^M)$, and $J_u(x, y) = 0$, for $(x, y) \in Q \times \mathbb{R}^{M-N}$.

By the Sard Theorem, $u(Q \times \mathbb{R}^{M-N}) = g(Q)$ is a null set in $\mathbb{R}^M$. Take $a \notin g(Q)$ with $|a| < \frac{\epsilon}{2}$, and $h(x) = g(x) - a$. Then $0 \notin h(Q)$, $\|h - f\|_{L^\infty(Q)} = \|g - f\|_{L^\infty} + |a| < \frac{\epsilon}{2} + \frac{\epsilon}{2} = \epsilon$. Define $\eta : \mathbb{R}^+ \to \mathbb{R}^+$ by

$$\eta(t) = \begin{cases} 1 & \text{if } t \geq \dfrac{\alpha}{2}, \\[2mm] \dfrac{2}{\alpha} t & \text{if } 0 < t < \dfrac{\alpha}{2}. \end{cases}$$

Set $w(x) = \frac{h(x)}{\eta(|h(x)|)}$, for $x \in Q$. Then

$$|w(x)| = \begin{cases} |h(x)| \geq \dfrac{\alpha}{2} & \text{if } |h(x)| \geq \dfrac{\alpha}{2}, \\[2mm] \dfrac{\alpha\,|h(x)|}{2\,|h(x)|} = \dfrac{\alpha}{2} & \text{if } |h(x)| \leq \dfrac{\alpha}{2}. \end{cases}$$

Thus, $0 \notin w(Q)$, $|w(x)| \geq \frac{\alpha}{2}$. For $x \in K$, we have

$$|h(x)| \geq |f(x)| - |f(x) - h(x)|$$
$$\geq \alpha - \epsilon$$
$$\geq \frac{\alpha}{2},$$

so on K, $\eta(|h(x)|) = 1 : w(x) = h(x)$. Thus,

$$\|w - f\|_{L^\infty(K)} < \epsilon.$$

Set $v = w - f$. By the Tietze Extension Theorem, there are $\bar{v} \in C(Q, \mathbb{R}^M)$ and $|\bar{v}| \le \epsilon$ on Q such that $\bar{v}|_K = v$. Set $F = w - \bar{v} \in C(Q, \mathbb{R}^M)$, $F|_K = w - h = f$, and

$$|F(x)| \ge |w(x)| - |\bar{v}(x)| \ge \frac{\alpha}{2} - \epsilon > 0.$$

Thus, $0 \notin \bar{f}(Q)$.   ∎

Lemma 9.42 *Let $\Omega \subset \mathbb{R}^N$ be a bounded open set, symmetric with respect to 0, $0 \notin \bar{\Omega}$. Suppose $f \in C(\partial\Omega, \mathbb{R}^M)$, where $M > N$, f is odd, and $0 \notin f(\partial\Omega)$. Then there is $F \in C(\bar{\Omega}, \mathbb{R}^M)$ such that F is odd, $F|_{\partial\Omega} = f$, and $0 \notin F(\bar{\Omega})$.*

Proof.

(a) $N = 1$: choose $0 < \epsilon < \delta$ such that $\Omega \subset [\epsilon, \delta] \cup [-\delta, -\epsilon]$. $\partial\Omega = K$ is compact, so by Lemma 9.41, there is $g \in C([\epsilon, \delta], \mathbb{R}^M)$ such that $0 \notin g([\epsilon, \delta])$, $g|_{\partial\Omega \cap [\epsilon, \delta]} = f$. Define

$$h(x) = \begin{cases} g(x) & \text{if } x \in [\epsilon, \delta], \\ -g(-x) & \text{if } x \in [-\delta, -\epsilon]. \end{cases}$$

If we set $F = h|_{\bar{\Omega}}$, then $F \in C(\bar{\Omega}, \mathbb{R}^M)$, F is odd, $F|_{\partial\Omega} = f$, $0 \notin F(\bar{\Omega})$.

(b) Suppose it is true for the dimensions $1, 2, \cdots, N - 1$.

(c) Let $\Omega \subset \mathbb{R}^N$, and identify $\mathbb{R}^{N-1}$ with the hyperplane $\{x \in \mathbb{R}^N \mid x_1 = 0\}$. Note that f is defined on $\partial(\Omega \cap \mathbb{R}^{N-1})$, and $\partial(\Omega \cap \mathbb{R}^{N-1}) \subset (\partial\Omega) \cap \mathbb{R}^{N-1}$. By part (b) there is $g \in C(\overline{\Omega \cap \mathbb{R}^M}, \mathbb{R}^M)$, and g is odd, $g|_{\partial(\Omega \cap \mathbb{R}^{N-1})} = f$, $0 \notin g(\overline{\Omega \cap \mathbb{R}^{N-1}})$. f is continuous on $\partial\Omega$, so we may have $g \in C(\partial\Omega \cup (\Omega \cap \mathbb{R}^M))$, and g is odd, $g|_{\partial\Omega} = f$, and $0 \notin g(\partial\Omega \cup (\Omega \cap \mathbb{R}^{N-1}))$. Set $\Omega^+ = \{x \in \Omega \mid x_1 > 0\}$, $\Omega^- = \{x \in \Omega \mid x_1 < 0\}$. Choose a cube Q, $\Omega^+ \subset Q \subset \mathbb{R}^+ \times \mathbb{R}^{N-1}$. By Lemma 9.41, there is $h \in C(Q, \mathbb{R}^M)$, $h|_{(\partial\Omega \cap \overline{\Omega^+}) \cup (\Omega \cap \mathbb{R}^{N-1})} = g$, and $0 \notin h(Q)$. Therefore $h \in C(\overline{\Omega^+}, \mathbb{R}^M)$. Set

$$F(x) = \begin{cases} h(x) & \text{if } x \in \Omega^+, \\ -h(-x) & \text{if } x \in \Omega^-. \end{cases}$$

We obtain $F \in C(\bar{\Omega}, \mathbb{R}^M)$; F is odd, $F|_{\partial\Omega} = f$, $0 \notin F(\bar{\Omega})$.   ∎

Lemma 9.43 *Let $\Omega \subset \mathbb{R}^N$ be a bounded open set, symmetric with respect to 0, $0 \notin \bar{\Omega}$. Suppose $f \in C(\partial\Omega, \mathbb{R}^N)$, f is odd, and $0 \notin f(\partial\Omega)$. Then for some $F \in C(\bar{\Omega}, \mathbb{R}^N)$, F is odd, $F|_{\partial\Omega} = f$, and $0 \notin F(\bar{\Omega} \cap \mathbb{R}^{N-1})$.*

Proof. By Lemma 9.42, $g \in C(\overline{\Omega \cap \mathbb{R}^{N-1}}, \mathbb{R}^N)$ exists, g is odd, and $g|_{\partial\Omega \cap \mathbb{R}^{N-1}} = f$, $0 \notin g(\overline{\Omega \cap \mathbb{R}^{N-1}})$. We can combine f and g to obtain $h \in C(\partial\Omega \cup (\bar{\Omega} \cap \mathbb{R}^{N-1}), \mathbb{R}^N)$, and h is odd, $0 \notin h(\partial\Omega \cup (\overline{\Omega \cap \mathbb{R}^{N-1}}))$. By the Tietze Extension Theorem, there is $u \in C(\overline{\Omega^+}, \mathbb{R}^N)$, and $u\big|_{\partial\Omega \cup (\overline{\Omega \cap \mathbb{R}^{N-1}})} = h$, set

$$F(x) = \begin{cases} u(x) & \text{if } x \in \overline{\Omega^+}, \\ -u(-x) & \text{if } x \in \overline{\Omega^-}. \end{cases}$$

Then $F \in C(\bar{\Omega}, \mathbb{R}^N)$, F is odd, and $F|_{\partial\Omega} = f$, $0 \notin F(\bar{\Omega} \cap \mathbb{R}^{N-1})$. ∎

Lemma 9.44 *Let $\Omega \subset \mathbb{R}^N$ be a bounded open set, symmetric with respect to 0, $0 \notin \bar{\Omega}$. Suppose $f \in C(\bar{\Omega}, \mathbb{R}^N)$, f is odd on $\partial\Omega$, and $0 \notin f(\partial\Omega)$. Then the degree $\mathrm{d}(f, \Omega, 0)$ is an even integer.*

Proof.   By Lemma 9.43, there is $F \in C(\bar{\Omega}, \mathbb{R}^N)$, F is odd, $F|_{\partial\Omega} = f$, and $0 \notin F(\bar{\Omega} \cap \mathbb{R}^{N-1})$. By Corollary 9.19 (i),

$$\mathrm{d}(f, \Omega, 0) = \mathrm{d}(F, \Omega, 0).$$

By Theorem 9.16 (i), there is $\epsilon > 0$ such that if $g \in C(\bar{\Omega}, \mathbb{R}^N)$, $\|g - F\|_{L^\infty(\Omega)} < \epsilon$, and then

$$\mathrm{d}(g, \Omega, 0) = \mathrm{d}(f, \Omega, 0).$$

Take $G \in C^2(\bar{\Omega}, \mathbb{R}^N)$, $\|G - F\|_{L^\infty(\Omega)} < \epsilon$, and consequently,

$$\mathrm{d}(F, \Omega, 0) = \mathrm{d}(G, \Omega, 0).$$

If we set $\bar{F}(x) = \frac{G(x) - G(-x)}{2}$, then $\bar{F} \in C^2(\bar{\Omega}, \mathbb{R}^N)$, $\bar{F}$ is odd and satisfies

$$\left\| \bar{F} - F \right\|_{L^\infty(\Omega)} = \left\| \frac{G(x) - F(x)}{2} - \frac{G(-x) - F(-x)}{2} \right\|_{L^\infty(\Omega)} < \epsilon.$$

Thus,

$$\mathrm{d}(F, \Omega, 0) = \mathrm{d}(\bar{F}, \Omega, 0).$$

In addition to the above, let ϵ satisfy $0 < \epsilon < \frac{1}{2} \mathrm{dist}(0, F(\bar{\Omega} \cap \mathbb{R}^{N-1}))$, and it follows at once that $0 \notin \bar{F}(\bar{\Omega} \cap \mathbb{R}^{N-1})$. By the Excision Property

$$\mathrm{d}(\bar{F}, \Omega, 0) = \mathrm{d}(\bar{F}, \Omega \backslash \mathbb{R}^{N-1}, 0).$$

However,

$$\Omega \backslash \mathbb{R}^{N-1} = \Omega^+ \cup \Omega^-.$$

By Theorem 9.16 (i),

$$\mathrm{d}(\bar{F}, \Omega, 0) = \mathrm{d}(\bar{F}, \Omega^+, 0) + \mathrm{d}(\bar{F}, \Omega^-, 0),$$

and consequently,
$$d(f, \Omega, 0) = d(\bar{F}, \Omega^+, 0) + d(\bar{F}, \Omega^-, 0). \tag{9.25}$$

By the Sard Theorem, there is b such that $J|_{\overline{F}}(x) \neq 0$ for $x \in \bar{F}^{-1}(b)$, and $|-b - 0| = |b - 0| < \frac{1}{2} \operatorname{dist}(0, \bar{F}(\bar{\Omega} \cap \mathbb{R}^{N-1}))$. Thus, $b, -b$, and 0 belong to the same connected component in $\mathbb{R}^N \backslash \bar{F}(\bar{\Omega} \cap \mathbb{R}^{N-1})$, and it follows that

$$d(\bar{F}, \Omega^+, 0) = d(\bar{F}, \Omega^+, b)$$
$$= \sum_{x \in (\overline{F}^{-1}(b)) \cap \Omega^+} \operatorname{sgn} J_{\overline{F}}(x),$$

and

$$d(\bar{F}, \Omega^-, 0) = d(\bar{F}, \Omega^-, -b)$$
$$= \sum_{y \in (\overline{F}^{-1}(-b)) \cap \Omega^-} \operatorname{sgn} J_{\overline{F}}(y).$$

Note that

$$x \in \bar{F}^{-1}(b) \cap \Omega^+ \text{if and only if } x \in \Omega^+, \bar{F}(x) = b$$
$$\text{if and only if } -x \in \Omega^-, \bar{F}(-x) = -b$$
$$\text{if and only if } y = -x \in \bar{F}^{-1}(-b) \cap \Omega^-.$$

Hence,
$$d(\bar{F}, \Omega^+, 0) = d(\bar{F}, \Omega^-, 0).$$

By (9.25), $d(f, \Omega, 0)$ is an even integer.   ∎

Theorem 9.45 (*Borsuk Theorem, or Borsuk Antipole Theorem*) *Let $\Omega \subset \mathbb{R}^N$ be a bounded open set, symmetric with respect to 0, $0 \in \Omega$. Suppose $f \in C(\bar{\Omega}, \mathbb{R}^N)$, f is odd on $\partial\Omega$, and $0 \notin f(\partial\Omega)$. Then the degree $d(f, \Omega, 0)$ is an odd integer.*

Proof.   By the Tietze Extension Theorem, there is a ball $\overline{B_r(0)}$ in Ω, a function $F \in C(\bar{\Omega}, \mathbb{R}^N)$, $F|_{\partial\Omega} = f$, and $F(x) = x$ on $\overline{B_r(0)}$. By Corollary 9.19 (*i*),

$$d(f, \Omega, 0) = d(F, \Omega, 0).$$

Apply Theorem 9.16 (*iii*) to obtain

$$d(F, \Omega, 0) = d(F, \Omega \backslash \overline{B_r(0)}, 0) + d(F, B_r(0), 0).$$

By Lemma 9.44, the degree $d(F, \Omega \backslash \overline{B_r(0)}, 0)$ is an even integer, and

$$d(F, B_r(0), 0) = d(I, B_r(0), 0) = 1.$$

Therefore $d(F, \Omega, 0)$ is an odd integer: $d(f, \Omega, 0)$ is an odd integer.   ∎

Corollary 9.46 *Let* $\Omega \subset \mathbb{R}^N$ *be a bounded open set, symmetric with respect to* 0, $0 \in \Omega$. *If we suppose* $f \in C(\bar{\Omega}, \mathbb{R}^N)$ *and* f *is odd on* $\partial\Omega$, *then;*

(i) *there is* $x \in \bar{\Omega}$ *such that* $f(x) = 0$;

(ii) *there is* $y \in \bar{\Omega}$ *such that* $f(y) = y$.

Proof. Let $g(x) = x - f(x)$ for $x \in \bar{\Omega}$. Then $g \in C(\bar{\Omega}, \mathbb{R}^N)$ and g is odd on $\partial\Omega$.
Case 1. If $0 \in f(\partial\Omega)$, then there is $x \in \partial\Omega$ such that $f(x) = 0$. If $0 \notin f(\partial\Omega)$, then by the Borsuk Theorem, $\mathrm{d}(f, \Omega, 0)$ is an odd integer, so there is $x \in \Omega$ such that $f(x) = 0$.
Case 2. If $0 \in g(\partial\Omega)$, then there is $y \in \partial\Omega$ such that $g(y) = 0 : f(y) = y$. If $0 \notin g(\partial\Omega)$, by the Borsuk Theorem, $\mathrm{d}(g, \Omega, 0)$ is an odd integer, so there is $y \in \Omega$ such that $g(y) = 0 : f(y) = y$. ∎

Corollary 9.47 *The Borsuk Theorem* 9.45 *implies the Brouwer Fixed Point Theorem* 9.34

Proof. By Theorem 9.34, it suffices to show that there is no retraction of the unit ball on its boundary. Suppose on the contrary there is a retraction $f : \overline{B_1(0)} \to S^{N-1}$. By the Borsuk Theorem 9.45, $\mathrm{d}(f, \Omega, 0)$ is an odd integer: there is $x \in B_1(0)$ such that $f(x) = 0$. This contradicts $f(B_1(0)) \subset S^{N-1}$. ∎

Corollary 9.48 *Let* $\Omega \subset \mathbb{R}^N$ *be a bounded open set, symmetric with respect to* 0, $0 \in \Omega$. *Suppose* $f \in C(\bar{\Omega}, \mathbb{R}^N)$, $0 \notin f(\partial\Omega)$, *and* f *does not point opposite on* $\partial\Omega$:

$$\frac{f(x)}{|f(x)|} \neq \frac{f(-x)}{|f(-x)|} \quad \text{for } x \in \partial\Omega.$$

Then $\mathrm{d}(f, \Omega, 0)$ *is an odd integer, and consequently* $f(\Omega)$ *contains a neighborhood of* 0 *in* $\mathbb{R}^N$.

Proof. Set $H(x, t) = f(x) - t f(-x)$ for $x \in \bar{\Omega}$, $t \in [0, 1]$. Then $H \in C(\bar{\Omega} \times [0, 1], \mathbb{R}^N)$. We claim that $0 \notin H(\partial\Omega \times [0, 1])$. In fact, if $0 \in H(\partial\Omega \times [0, 1])$, then $x \in \partial\Omega$, there is $t \in [0, 1]$ such that $f(x) = tf(-x)$, and consequently, $|f(x)| = t|f(-x)|$. Therefore $t = \frac{|f(x)|}{|f(-x)|}$, and $\frac{f(x)}{|f(x)|} = \frac{f(-x)}{|f(-x)|}$, a contradiction. Now $\mathrm{d}(f(\cdot, t), \Omega, 0)$ is constant on $[0, 1]$, or

$$\begin{aligned}
\mathrm{d}(f, \Omega, 0) &= \mathrm{d}(H(\cdot, 0), \Omega, 0) \\
&= \mathrm{d}(H(\cdot, 1), \Omega, 0) \\
&= \mathrm{d}(f(x) - f(-x), \Omega, 0) \\
&= \text{an odd integer, \quad because } f(x) - f(-x) \text{ is odd.}
\end{aligned}$$

Because $\mathrm{d}(f, \Omega, 0) \neq 0$, by Corollary 9.18 (iii), $f(\Omega)$ contains a neighborhood of 0 in $\mathbb{R}^N$. ∎

Corollary 9.49

(i) Let $\Omega \subset \mathbb{R}^N$ be a bounded open set, symmetric with respect to 0, $0 \in \Omega$. Suppose $f \in C(\partial\Omega, \mathbb{R}^N)$, f is odd on $\partial\Omega$, and $0 \notin f(\partial\Omega)$. Then there is no homotopy $H \in C(\partial\Omega \times [0,1], \mathbb{R}^N)$ such that $0 \notin H(\partial\Omega \times [0,1])$ and for some $x_0 \in \mathbb{R}^N \backslash \{0\}$, and for all $x \in \partial\Omega$, $H(x,0) = f(x)$, and $H(x,1) = x_0$.

(ii) $\partial B_r(0)$ cannot be deformed continuously to a point: there is no homotopy $H \in C(\partial B_r(0) \times [0,1], \partial B_r(0))$ such that for some $x_0 \in \partial B_r(0)$ and for all $x \in \partial B_r(0)$, $H(x,0) = x$, $H(x,1) = x_0$.

Proof.

(i) Extend f to $F \in C(\bar{\Omega}, \mathbb{R}^N)$ oddly. Then F satisfies the hypothesis of the Borsuk Theorem, and it follows at once that $\mathrm{d}(F, \Omega, 0)$ is an odd integer. Assume $x_0 \neq 0$, and $g(x) = x_0$ for $x \in \bar{\Omega}$. Because $0 \notin g(\bar{\Omega}) = x_0$, $\mathrm{d}(g, \Omega, 0) = 0$. Suppose there is a homotopy $H \in C(\partial\Omega \times [0,1], \mathbb{R}^N)$ such that $0 \notin H(\partial\Omega \times [0,1])$, and for some $x_0 \in \mathbb{R}^N \backslash \{0\}$, and for all $x \in \partial\Omega$, $H(x,0) = f(x)$, and $H(x,1) = g(x)$, then

$$\begin{aligned} \mathrm{d}(f, \Omega, 0) &= \mathrm{d}(F, \Omega, 0) \\ &= \mathrm{d}(g, \Omega, 0), \end{aligned}$$

a contradiction.

(ii) Let $\Omega = B_1(0)$, and $f = I$ in (i). ∎

Remark 9.50 *Corollary 9.48 is false in the infinite-dimensional Banach space case, see Dugundji [19].*

Corollary 9.51

(i) Let $\Omega \subset \mathbb{R}^N$ be a bounded open set, symmetric with respect to 0, $0 \in \Omega$. Suppose $f \in C(\partial\Omega, \mathbb{R}^N)$, $f(\partial\Omega)$ is contained in a proper subspace of $\mathbb{R}^N$. Then:
  (a) If f is odd, then there is $x \in \partial\Omega$ such that $f(x) = 0$;
  (b) (Borsuk-Ulam Theorem) There is $x \in \partial\Omega$ such that $f(x) = f(-x)$.

(ii) If $N > M$, then there is no odd function f in $C(S^N, S^M)$.

Proof.

$(i-a)$ Assume $f(\partial\Omega) \subset \mathbb{R}^{N-1}$. Suppose $0 \notin f(\partial\Omega)$. By the Tietz Extension Theorem, there is an extension $F \in C(\bar{\Omega}, \mathbb{R}^{N-1})$ of f. Then $F \in C(\Omega, \mathbb{R}^N)$, F is odd on $\partial\Omega$, and $0 \notin F(\partial\Omega)$. By the Borsuk Theorem, $\mathrm{d}(F, \Omega, 0)$ is odd. However, by Corollary 9.18 (iv) $\mathrm{d}(F, \Omega, 0) = 0$ because $F(\bar{\Omega})$ is contained in a proper subspace of $\mathbb{R}^N$, a contradiction.

$(i - b)$ Let $g(x) = f(x) - f(-x)$ for $x \in \partial\Omega$. Then g is odd, and $g(\partial\Omega)$ is contained in a proper subspace of $\mathbb{R}^N$. This follows from $(i - a)$.

(ii) Suppose there is an odd function f in $C(S^N, S^M)$, $N > M$. Because $f(S^N) \subset \mathbb{R}^{M+1} \underset{\neq}{\subseteq} \mathbb{R}^{N+1}$, and f is odd, by (i-a) there is $x \in S^N$ such that $f(x_0) = 0$, a contradiction.   ∎

Corollary 9.52 *Let $A_1, A_2, \cdots, A_{N+1}$ be closed sets in S^N.*

(i) *If $\{A_1, \cdots, A_{N+1}, -A_1, \cdots, -A_{N+1}\}$ covers S^N, and $A_i \cap (-A_i) = \emptyset$ for $i = 1, \cdots, N+1$, then $\bigcap\limits_{i=1}^{N+1} A_i \neq \emptyset$.*

(ii) *If $\{A_1, \cdots, A_{N+1}\}$ covers S^N, then at least one A_j contains a couple of anti-points;*

(iii) *If $\{A_1, \cdots, A_N, -A_1, \cdots, -A_N\}$ covers S^N, then at least one A_j contains a couple of antipoints.*

Proof.

(i) Because $A_i \cap (-A_i) = \emptyset$, by the Tietze Extension Theorem, for some $f_i \in C(S^N, \mathbb{R})$ satisfying $-1 \leq f_i \leq 1$:

$$f_i(x) = \begin{cases} 1 & \text{on } A_i, \\ -1 & \text{on } -A_i. \end{cases}$$

Set $g_i(x) = f_i(x) - f_i(-x)$ for $x \in S^N$, $i = 1, \cdots, N+1$. Then $g_i \in C(S^N, \mathbb{R})$, g is odd, and

$$g_i(x) = \begin{cases} 2 & \text{on } A_i, \\ -2 & \text{on } -A_i. \end{cases}$$

Let

$$h_i(x) = \frac{1}{1 + \text{dist}(x, A_i \cup (-A_i))}, \quad x \in S^N.$$

Then $h_i \in C(S^N, \mathbb{R})$, h_i is even because $A_i \cup (-A_i)$ is symmetric with respect to 0, $0 < h_i \leq 1$, and $h_i(x) = 1$ if and only if $x \in A_i \cup (-A_i)$. Set

$$u_i(x) = h_i(x)g_i(x).$$

Then $u_i \in C(S^N, \mathbb{R})$, u_i is odd, $u_i(x) = \pm 2$ if and only if $x \in \pm A_i$. Set $u(x) = (u_1(x), \cdots, u_{N+1}(x))$. Then $u \in C(S^N, \mathbb{R}^{N+1})$, and u is odd. Because

$x \in S^N$ then for some j such that $1 \leq j \leq N+1$, $x \in A_j \cup (-A_j) : u_j(x) = \pm 2$. We have $0 \notin u(S^N)$. Extend u to the unit ball $B_1(0)$ in $\mathbb{R}^{N+1}$ by

$$w(x) = \begin{cases} |x| \, u(\frac{x}{|x|}) & \text{if } x \neq 0, \\ 0 & \text{if } x = 0. \end{cases}$$

Then $w \in C(\overline{B_1(0)}, \mathbb{R}^{N+1})$, $0 \notin w(S^N)$, and w is odd on S^N. By the Borsuk Theorem 9.45, and Corollary 9.18 (iii), $w(B_1(0))$ contains a neighborhood of 0. In particular, if $\alpha > 0$ is small, then $w(B_1(0))$ contains the point $b = (\alpha, \cdots, \alpha)$. Let $x \neq 0$ in $w^{-1}(b)$, then $w(x) = (\alpha, \cdots, \alpha)$, or $u(\frac{x}{|x|}) = \frac{1}{|x|} w(x) = (\frac{\alpha}{|x|}, \cdots, \frac{\alpha}{|x|})$. j, $1 \leq j \leq N+1$ exists such that $\frac{x}{|x|} \in A_j \cup (-A_j)$, or $u_j(\frac{x}{|x|}) = 2$ or -2. Because $\frac{\alpha}{|x|} > 0$, $u_j(\frac{x}{|x|}) = 2$. We conclude that $u_i(\frac{x}{|x|}) = 2$ for $1 \leq i \leq N+1 : \frac{x}{|x|} \in \bigcap_{i=1}^{N+1} A_i$.

(ii) Otherwise, let $A_i \cap (-A_i) = O$ for $1 \leq i \leq N+1$. By (i), $\bigcap_{i=1}^{N+1} A_i \neq \varnothing$. Let $x \in \bigcap_{i=1}^{N+1} A_i$. Because $\{-A_i\}_{1 \leq i \leq N+1}$ covers S^N, we have $x \in -A_j$ for some j. Thus, $x \in A_j \cap (-A_j)$, a contradiction.

(iii) Otherwise, let $A_i \cap (-A_i) = O$ for $1 \leq i \leq N$. Set $A_{N+1} = -A_N$. By (i), $\bigcap_{i=1}^{N+1} A_i \neq \varnothing$: so there is $x \in A_N \cap A_{N+1} = A_N \cap (-A_N)$, a contradiction. $\blacksquare$

Lemma 9.53 *There are $N+1$ antipodal closed sets $\{B_1, \cdots, B_{N+1}\}$ that cover S^N :* $B_i = C_i \cup (-C_i)$, $C_i \cap (-C_i) = \varnothing$, $i = 1, \cdots, N+1$, *and* $S^{N-1} = \bigcup_{i=1}^{N+1} B_i$.

Proof. See Lemma 8.16 for the proof. $\blacksquare$

9.4 Multiplicative Property of Degrees

Let $\Omega \subset \mathbb{R}^N$ be a bounded open set, $f \in C(\Omega, \mathbb{R}^N)$. Recall that the degree $\mathrm{d}(f, \Omega, b)$ is constant on each connected component of $\mathbb{R}^N \backslash f(\partial\Omega)$. If A is contained in a component of $\mathbb{R}^N \backslash f(\partial\Omega)$, then define

$$\mathrm{d}(f, \Omega, A) = \mathrm{d}(f, \Omega, b) \quad \text{for each } b \in A.$$

Note that $f(\overline{\Omega})$ is compact, so there is a unique unbounded connected component of $\mathbb{R}^N \backslash f(\partial\Omega)$, in symbol A_∞. If b is large enough, then $b \notin f(\overline{\Omega}) : \mathrm{d}(f, \Omega, b) = 0$, and consequently $\mathrm{d}(f, \Omega, A_\infty) = 0$.

Theorem 9.54 (*Multiplicative Property of Degree*) *Assume* $f \in C(\Omega, \mathbb{R}^N)$, $g \in C(\mathbb{R}^N, \mathbb{R}^N)$, $b \in \mathbb{R}^N$, *and* $b \notin g \circ f(\partial\Omega)$. *Let* $\{A_i\}_{i\neq\infty}$ *be the family of bounded connected components of* $\mathbb{R}^N \backslash f(\partial\Omega)$. *Then*

$$\mathrm{d}((g \circ f), \Omega, b) = \sum_{i\neq\infty} \mathrm{d}(g, A_i, b)\, \mathrm{d}(f, \Omega, A_i).$$

In fact, $i \neq \infty$ in the above formula is unnecessary because $\mathrm{d}(f, \Omega, A_\infty) = 0$.

Proof. Because $\partial A_i \subset f(\partial\Omega)$ for each i, $b \notin g(\partial A_i)$. Note that $g^{-1}(b)$ is compact, $g^{-1}(b) \subset \mathbb{R}^N \backslash f(\partial\Omega) = \bigcup_i A_i$, where A_i are open and disjoint to each other, so $g^{-1}(b) \cap A_i = \varnothing$ except for possible finite numbers of i; that is, $\mathrm{d}(g, A_i, b) = 0$ except for possible finite numbers of i. Hence the formula is meaningful. Without loss of generality we assume that g, $f \in C^1$, and $J_{g \circ f}(x) \neq 0$ for $x \in (g \circ f)^{-1}(b)$. Now

$$J_{g \circ f}(x) = \det \begin{pmatrix} \sum_i \dfrac{\partial g_1}{\partial y_i}\dfrac{\partial y_i}{\partial x_1} & \sum_i \dfrac{\partial g_1}{\partial y_i}\dfrac{\partial y_i}{\partial x_2} & \cdots & \sum_i \dfrac{\partial g_1}{\partial y_i}\dfrac{\partial y_i}{\partial x_N} \\ \vdots & & & \vdots \\ \sum_i \dfrac{\partial g_N}{\partial y_i}\dfrac{\partial y_i}{\partial x_1} & \sum_i \dfrac{\partial g_N}{\partial y_i}\dfrac{\partial y_i}{\partial x_2} & \cdots & \sum_i \dfrac{\partial g_N}{\partial y_i}\dfrac{\partial y_i}{\partial x_N} \end{pmatrix}$$

$$= \det \left[\begin{pmatrix} \dfrac{\partial g_1}{\partial y_1} & \cdots & \dfrac{\partial g_1}{\partial y_N} \\ \vdots & & \vdots \\ \dfrac{\partial g_N}{\partial y_1} & \cdots & \dfrac{\partial g_N}{\partial y_N} \end{pmatrix} \cdot \begin{pmatrix} \dfrac{\partial y_1}{\partial x_1} & \cdots & \dfrac{\partial y_1}{\partial x_N} \\ \vdots & & \vdots \\ \dfrac{\partial y_N}{\partial x_1} & \cdots & \dfrac{\partial y_N}{\partial x_N} \end{pmatrix} \right]$$

$$= J_g(f(x))J_f(x).$$

Therefore,

$$\mathrm{d}(g \circ f, \Omega, b) = \sum_{x \in (g \circ f)^{-1}(b)} \mathrm{sgn}(J_{g \circ f}(x))$$

$$= \sum_{x \in (g \circ f)^{-1}(b)} \mathrm{sgn}\, J_g(f(x))\, \mathrm{sgn}\, J_f(x)$$

$$= \sum_{\substack{y \in \mathbb{R}^N \backslash f(\partial\Omega) \\ g(y)=b}} \mathrm{sgn}\, J_g(y) \sum_{\substack{x \in \Omega \\ f(x)=y}} \mathrm{sgn}\, J_f(x)$$

$$= \sum_{i\neq\infty} \left(\sum_{y \in g^{-1}(b) \cap A_i} \mathrm{sgn}\, J_g(y) \right) \mathrm{d}(f, \Omega, A_i)$$

$$= \sum_{i\neq\infty} \mathrm{d}(g, A_i, b)\, \mathrm{d}(f, \Omega, A_i). \qquad \blacksquare$$

Lemma 9.55 $d(H \circ F, A_i, A_j) = \sum_k d(H, B_k, A_j) \, d(F, A_i, B_k).$

Proof. Because $\partial A_i \subset \Lambda$, $H \circ F|_\Lambda = I_\Lambda$, $H \circ F(\partial A_i) \subset \Lambda$. Thus, A_j is a connected set in $\mathbb{R}^N \backslash H \circ F(\partial A_i)$. Thus, $d(H \circ F, A_i, A_j)$ is well-defined. Similarly $H(\partial B_k) \subset H(\Gamma) = \Lambda$, A_j is connected in $\mathbb{R}^N \backslash H(\partial B_k)$, $d(H, B_k, A_j)$ is well-defined, and $d(F, A_i, B_k)$ is also well-defined. If $\{B_k\}$ is the family of connected components in $\mathbb{R}^N \backslash F(\partial A_i)$, then the lemma is true by the Multiplicity Property of Degree. Let $\{B_k\}$ be a subfamily of connected components in $\mathbb{R}^N \backslash F(\partial A_i)$. Now $F(\partial A_i) \subset F(\Lambda) = \Gamma$, so $\mathbb{R}^N \backslash \Gamma \subset \mathbb{R}^N \backslash F(\partial A_i)$. Therefore, some components B_k of $\mathbb{R}^N \backslash \Gamma$ are contained in a component of $\mathbb{R}^N \backslash F(\partial A_i)$. Let $\{K_m\}$ be the family of connected components of $\mathbb{R}^N \backslash F(\partial A_i)$. In general, the inclusion is strict. The connected components of $\mathbb{R}^N \backslash F(\partial A_3)$ consist of $K_1 = B_1 \cup B_2 \cup \partial B_1$, $K_2 = B_3$, $K_3 = B_4 \cup B_\infty \cup \partial B_\infty = K_\infty$. Now let V_j be the union of connected components B_k that is included in K_j. The family of connected components will be denoted by $\{B_{js}\}_{s \in S_j}$. Thus,

$$V_m = \bigcup_{s \in S_m} B_{ms} \subset K_m,$$

V_m is open. Set $W_m = \bar{K}_m \backslash V_m$. W_m is closed in Γ. Suppose $b \in A_j$, $b \notin H(W_m)$ because $H(w_m) \subset H(\Gamma) = \Lambda$. The degree $d(H, K_m, b)$ is defined because $\partial K_m \subset F(\partial A_i) \subset \Gamma$, and we have $b \notin H(\partial K_m) \subset \Lambda$. By the Excision Property, $b \notin H(W_m)$,

$$d(H, K_m, b) = d(H, K_m \backslash W_m, b) = d(H, V_m, b).$$

By Theorem 9.16 (iii)

$$d(H, V_m, b) = \sum_{s \in S_m} d(H, B_{ms}, b),$$

because $\{K_m\}$ is the family of all connected components of $\mathbb{R}^N \backslash F(\partial A_i)$. By the Multiplicity Property,

$$\begin{aligned}
d(H \circ F, A_i, b) &= \sum_{m \neq \infty} d(H, K_m, b) \, d(F, A_i, K_m) \\
&= \sum_{m \neq \infty} d(H, V_m, b) \, d(F, A_i, K_m) \\
&= \sum_{m \neq \infty} \big(\sum_{s \in S_m} d(H, B_{ms}, b) \, d(F, A_i, K_m) \big),
\end{aligned}$$

but $d(F, A_i, K_m) = d(F, A_i, B_{ms})$ for $s \in S_m$, because $B_{ms} \subset K_m$. Therefore,

$$d(H \circ F, A_i, b) = \sum_{m \neq \infty} \big(\sum_{s \in S_m} d(H, B_{ms}, b) \, d(F, A_i, B_{ms}) \big)$$

or

$$d(H \circ F, A_i, b) = \sum_{m \neq \infty} d(H, B_k, b)\, d(F, A_i, B_k).$$

Because $b \in A_j$,

$$d(H \circ F, A_i, A_j) = \sum_{m \neq \infty} d(H, B_k, b)\, d(F, A_i, B_k). \qquad \blacksquare$$

Theorem 9.56 (*Separation Theorem of Jordan*) *Let* Λ, Γ *be two homeomorphic compact sets in* $\mathbb{R}^N$. *Then* $\mathbb{R}^N \backslash \Lambda$, *and* $\mathbb{R}^N \backslash \Gamma$ *have the same number of components: both numbers are infinite, or both numbers are finite and equal.*

Proof.   (By Leray).   Let $f : \Lambda \to \Gamma$ be a homeomorphism onto. By the Tietze Extension Theorem there are extensions F, $H \in C(\mathbb{R}^N, \mathbb{R}^N)$ such that $F|_\Lambda = f$, $H|_\Gamma = f^{-1}$. Let $\{A_i\}_i \cup \{A_\infty\}$, and let $\{B_j\}_j \cup \{B_\infty\}$ be the connected components of $\mathbb{R}^N \backslash \Lambda$, and $\mathbb{R}^N \backslash \Gamma$, respectively. Because $H \circ F\big|_{\partial A_i} = I$, the identity map, and noting that $\partial A_i \subset \Lambda$, we have

$$\begin{aligned}
d(H \circ F, A_i, A_j) &= d(I, A_i, A_j) \\
&= \delta_{ij} \\
&= \sum_{K \neq \infty} d(H, B_k, A_j)\, d(F, A_i, B_k).
\end{aligned}$$

Similarly,

$$d(F \circ H, B_i, B_j) = \delta_{ij} = \sum_{m \neq \infty} d(F, A_m, B_j)\, d(H, B_i, A_m).$$

Case 1. Let $\mathbb{R}^N \backslash \Lambda$, and $\mathbb{R}^N \backslash \Gamma$ have a finite number of connected components, denoted by, respectively, $\{A_i\}_{i=1,\cdots,r,\infty}$ and $\{B_j\}_{j=1,\cdots,s,\infty}$. Set

$$p_{ik} = d(F, A_i, B_k),\ \ q_{jm} = d(H, B_j, A_m),$$
$$P = (p_{ik}),\ \text{a } r \times s \text{ matrix},\ \ Q = (q_{jm}),\ \text{a } s \times r \text{ matrix}.$$

We have

$$\sum_{k=1}^{s} p_{ik}\, q_{kj} = \delta_{ij}, \quad i,\, j \in \{1, \cdots, r\},$$
$$\sum_{m=1}^{r} q_{im}\, p_{mj} = \delta_{ij}, \quad i,\, j \in \{1, \cdots, s\}. \qquad (9.26)$$

That is,

$$PQ = I_r, \quad OP = I_s,$$

so
$$Q = P^{-1}, \text{ and } r = s.$$

Case 2. Let $\mathbb{R}^N \backslash \Lambda$ and $\mathbb{R}^N \backslash \Gamma$ have infinite numbers of connected components. The connected components are in $\mathbb{R}^N$, so they are countable.

Case 3. Suppose $\mathbb{R}^N \backslash \Lambda$ has a finite number of connected components $\{A_i\}_{i=1,\cdots,r,\infty}$, and $\mathbb{R}^N \backslash \Gamma$ has an infinite number of connected components $\{B_j\}_{j \in \mathbf{N} \cup \{\infty\}}$. Fix i, $1 \le i \le r$, then by Theorem 9.16 (iii),

$$d(H, \bigcup_{k \neq \infty} B_k, A_i) = \sum_{k \neq \infty} d(H, B_k, A_i) = \sum_{k \neq \infty} q_{ki}.$$

For $i = 1, 2, .., r$, there is a finite subset $K(i) \subset N$ such that for $k \notin K(i)$, $q_{ki} = 0$. Hence, there is $n_0 \in N$ such that

$$q_{ki} = 0 \quad 1 \le i \le r, k \ge n_0.$$

By (9.26),

$$\delta_{ij} = \sum_{m=1}^{r} q_{im} \, p_{mj}.$$

Set $i = j = n_0$, we have

$$1 = \sum_{m=1}^{r} q_{n_0 m} \, p_{m n_0} = 0 \quad \text{because } q_{n_0 m} = 0 \text{ for } m = 1, \cdots, r.$$

This is a contradiction. ∎

Corollary 9.57 (*Closed Curve Theorem of Jordan*) *Let Γ be a simple closed curve in the plane. Then Γ separates the plane into two components, one bounded and the other unbounded.*

Proof. Because $\Gamma \cong S^1$. ∎

Theorem 9.58 (*Brouwer Invariant Domain Theorem*) *Let $\Omega \subset \mathbb{R}^N$ be an open set, and let $f \in C(\Omega, \mathbb{R}^N)$ be injective. Then f is an open map.*

Proof. Take $b \in \Omega$, $\rho > 0$ such that $B_\rho(b) \subset \Omega$. It suffices to claim that $f(B_\rho(b))$ is open. In fact, because $\overline{B_\rho(b)}$ is compact, we have $f\big|_{\overline{B_\rho(b)}} : \overline{B_\rho(b)} \to f(\overline{B_\rho(b)})$ which is a homeomorphism. By the Jordan Theorem, $\mathbb{R}^N \backslash f(\overline{B_\rho(b)})$ has only one connected component $B_\infty = \mathbb{R}^N \backslash f(\overline{B_\rho(b)})$, and $\mathbb{R}^N \backslash f(\partial B_\rho(b))$ has two connected components A_1 and A_∞. Note that $B_\infty = \mathbb{R}^N \backslash f(\overline{B_\rho(b)}) \subset \mathbb{R}^N \backslash f(\partial B_\rho(b))$, and B_∞ is unbounded and connected, so $B_\infty \subset A_\infty$, and it follows that $A_1 \subset \mathbb{R}^N \backslash A_\infty \subset \mathbb{R}^N \backslash B_\infty = f(\partial B_\rho(b)) \cup f(B_\rho(b))$. We conclude that

$$A_1 \subset f(B_\rho(b)). \tag{9.27}$$

On the other hand, because f is injective, $f(B_\rho(b))$ is connected in $\mathbb{R}^N \backslash f(\partial B_\rho(b))$, and consequently, $f(B_\rho(b))$ is contained in a connected component in $\mathbb{R}^N \backslash f(\partial B_\rho(b))$, by (9.27), which should be A_1 :

$$f(B_\rho(b)) \subset A_1. \tag{9.28}$$

By (9.27) and (9.28), $A_1 = f(B_\rho(b)) : f(B_\rho(b))$ is open.   ∎

Suppose V is a real vector space of dimension N. Identity $V = \mathbb{R}^N$ under a basis in V. Letting $\Omega \subset \mathbb{R}^N$ be a bounded open set, $f \in C(\Omega, V)$, $b \notin f(\partial\Omega)$, we define the degree $\mathrm{d}(f, \Omega, b)$ as usual.

Lemma 9.59 *The degree* $\mathrm{d}(f, \Omega, b)$ *is independent of the choice of the basis.*

Proof.   Let $A = \{a_1, \cdots, a_N\}$, $B = \{b_1, \cdots, b_N\}$ be two bases in V. If we let $P = (c_{ij})$ with

$$b_j = \sum_{i=1}^{N} c_{ij}\, a_i, \quad 1 \le j \le N,$$

then P is invertible. Letting $x \in V$, then

$$x = \sum_i \alpha_i\, a_i = \sum_i \beta_i\, b_i.$$

Letting $x_1 = (\alpha_1, \cdots, \alpha_N)$, $x_2 = (\beta_1, \cdots, \beta_N)$, then

$$\sum_i \alpha_i\, a_i = \sum_j \beta_i\, b_i = \sum_j \beta_j \Big(\sum_i c_{ij}\, a_i\Big) = \sum_i \Big(\sum_j c_{ij}\, \beta_j\Big) a_i.$$

Thus,
$$x_1 = Px_2, \text{ and consequently } x_2 = P^{-1}x_1.$$

Letting

$$V_1 = (V,\, A),\, V_2 = (V,\, B),$$
$$\Omega \subset V,\; f \in C(\bar\Omega,\, V),\; b \in V,$$
$$\Omega_1 \subset V_1,\; f_1 \in C(\bar\Omega_1,\, V_1),\; b_1 \in V_1,$$
$$\Omega_2 \subset V_2,\; f_2 \in C(\bar\Omega_2,\, V_2),\; b_2 \in V_2,$$

we may assume that $f_1, f_2 \in C^1$, $x_1 \in f_1^{-1}(b_1)$ implies $J_{f_1}(x_1) \ne 0$, and $x_2 \in f_2^{-1}(b_2)$ implies $J_{f_2}(x_2) \ne 0$. Note that

$$f_2(x_2) = y_2 = P^{-1}y_1 = P^{-1}f_1(x_1) = P^{-1} \circ f_1 \circ P(x_2).$$

Therefore,

$$f_2'(x_2) = P^{-1} \circ f_1'(x_1) \circ P,$$

and

$$\det f_2'(x_2) = \det(P^{-1} \circ f_1'(x_1) \circ P) = \det f_1'(x_1).$$

We have $J_{f_2}(x_2) = J_{f_1}(x_1)$. Now,

$$d(f_1, \Omega_1, b_1) = \sum_{x_1 \in f_1^{-1}(b_1)} \operatorname{sgn} J_{f_1}(x_1)$$

$$= \sum_{x_2 \in f_2^{-1}(b_2)} \operatorname{sgn} J_{f_2}(x_2)$$

$$= d(f_2, \Omega_2, b_2). \qquad \blacksquare$$

9.5 The Infinite-Dimensional Degree Theory

By Lemma 9.59, in finite-dimensional space, the degree is independent of the basis. For $M < N$, consider $\mathbb{R}^M$ as a vector subspace of $\mathbb{R}^N$ by identifying $x = (x_1, \cdots, x_M)$ in $\mathbb{R}^M$ with the element $x = (x_1, \cdots, x_M, 0, \cdots, 0)$ in $\mathbb{R}^N$. Let $\Omega \subset \mathbb{R}^N$ be a bounded open set, $f \in C(\bar{\Omega}, \mathbb{R}^M)$. We may identify f as a function $f \in C(\bar{\Omega}, \mathbb{R}^N)$. Then we have:

Lemma 9.60 *For $f \in C(\bar{\Omega}, \mathbb{R}^M)$. Let $g(x) = x - f(x)$. If $b \in \mathbb{R}^M$, $b \notin g(\partial\Omega)$, then*

$$d(g, \Omega, b) = d(g\big|_{\bar{\Omega} \cap \mathbb{R}^M}, \Omega \cap \mathbb{R}^M, b).$$

Proof. Because $\partial_{\mathbb{R}^M}(\Omega \cap \mathbb{R}^M) \subset \partial\Omega \cap \mathbb{R}^M$, we have $b \notin g\big|_{\bar{\Omega} \cap \mathbb{R}^M}(\partial_{\mathbb{R}^M}(\Omega \cap \mathbb{R}^M))$. It suffices to consider $g \in C^1(\bar{\Omega}, \mathbb{R}^N)$, $b \notin g(S)$, where $S = \{x \in \Omega \mid J_g(x) = 0\}$, and $b \notin g(\partial\Omega)$, $g\big|_{\bar{\Omega} \cap \mathbb{R}^M} \in C^1(\bar{\Omega} \cap \mathbb{R}^M, \mathbb{R}^M)$. For $x \in \Omega$, then

$$g(x) = (x_1 - f_1(x), \cdots, x_M - f_M(x), x_{M+1}, \cdots, x_N),$$

and it follows that

$$g'(x) = \left(\begin{array}{c|c} I_M - (f\big|_{\bar{\Omega} \cap \mathbb{R}^M})'(x) & * \\ \hline 0 & I_{N-M} \end{array} \right),$$

so

$$J_g(x) = J_{g\big|_{\bar{\Omega} \cap \mathbb{R}^M}}(x). \tag{9.29}$$

Note that $x \in \Omega$, $x \in g^{-1}(b)$ if and only if $x \in \Omega \cap \mathbb{R}^M$, $x \in g\big|_{\bar{\Omega} \cap \mathbb{R}^M}^{-1}(b)$ because

$$x - f(x) = g(x) = b \quad \text{if and only if} \quad x = f(x) + b \in \mathbb{R}^M.$$

Thus, by (9.29), if b is a regular value of g, then b is a regular value of $g|_{\overline{\Omega} \cap \mathbb{R}^M}$. Because $g^{-1}(b) = g|_{\overline{\Omega} \cap \mathbb{R}^M}^{-1}(b)$ and (9.29), we have

$$
\begin{aligned}
\mathrm{d}(g, \Omega, b) &= \sum_{x \in g^{-1}(b)} \operatorname{sgn} J_g(x) \\
&= \sum_{x \in g|_{\overline{\Omega} \cap \mathbb{R}^M}^{-1}(b)} \operatorname{sgn} J_{g|_{\overline{\Omega} \cap \mathbb{R}^M}}(x) \\
&= \mathrm{d}(g|_{\overline{\Omega} \cap \mathbb{R}^M}, \Omega \cap \mathbb{R}^M, b).
\end{aligned}
$$

Throughout this section, let E be an infinite-dimensional real Banach space. ∎

Definition 9.61 *Let $\Omega \subset E$ be a bounded open set. Let $T \in C(\overline{\Omega}, E)$ be a finite rank operator: $T(\overline{\Omega})$ is contained in a finite-dimensional subspace F of E. Let $f = I - T$, then f is called a finite-dimensional perturbation of the identity. Let $b \in F$, $b \notin f(\partial\Omega)$, and set*

$$
\mathrm{d}(f, \Omega, b) = \mathrm{d}(f|_{\overline{\Omega} \cap F}, \overline{\Omega} \cap F, b).
$$

Remark 9.62 *Definition 9.61 is well-defined. In fact, let F_1, F_2 be two finite-dimensional subspaces of E, such that*

$$
T(\overline{\Omega}) \subset F_i, \text{ and } b \in F_i \quad \text{for } i = 1, 2.
$$

Then

$$
T(\overline{\Omega}) \subset F_1 \cap F_2, \quad \text{and } b \in F_1 \cap F_2.
$$

By Lemma 9.60,

$$
\mathrm{d}(f|_{\overline{\Omega} \cap F_i}, \Omega \cap F_i, b) = \mathrm{d}(f|_{\overline{\Omega} \cap F_1 \cap F_2}, \Omega \cap F_1 \cap F_2, b).
$$

Thus,

$$
\mathrm{d}(f|_{\overline{\Omega} \cap F_1}, \Omega \cap F_1, b) = \mathrm{d}(f|_{\overline{\Omega} \cap F_2}, \Omega \cap F_2, b).
$$

Lemma 9.63 *Let $K \subset E$ be a compact set. For $\epsilon > 0$, there are finite-dimensional subspaces F_ϵ of E, and $g_\epsilon \in C(K, F_\epsilon)$, such that $\|x - g_\epsilon(x)\| < \epsilon$ for $x \in K$.*

Proof. For $\epsilon > 0$, there are $y_1, \cdots, y_{p(\epsilon)} \in K$ such that

$$
K \subset \bigcup_{1 \le i \le p(\epsilon)} B_\epsilon(y_i), we
$$

let $F_\epsilon = <y_1, y_2, \cdots, y_{p(\epsilon)}>$, define for $1 \le i \le p(\epsilon)$, $b_i : K \to \mathbb{R}$ by

$$
b_i(x) = \begin{cases} \epsilon - \|x - y_i\| & \text{for } x \in B_\epsilon(y_i), \\ 0 & \text{for } x \notin B_\epsilon(y_i). \end{cases}
$$

Then b_i are continuous, $\sum\limits_{i=1}^{p(\epsilon)} b_i(x) > 0$ on K. We set

$$g_\epsilon(x) = \frac{\sum\limits_{i=1}^{p(\epsilon)} b_i(x) y_i}{\sum\limits_{i=1}^{p(\epsilon)} b_i(x)}.$$

Then $g_\epsilon : K \to F_\epsilon$ is continuous and satisfies

$$\|x - g_\epsilon(x)\| \leq \frac{\sum\limits_{i=1}^{p(\epsilon)} b_i(x) \|y_i - x\|}{\sum\limits_{i=1}^{p(\epsilon)} b_i(x)} < \epsilon \quad \text{for } x \in K.$$

∎

Lemma 9.64 *Let E be a Banach space, $\Omega \subset E$ a bounded open set, $T \in Q(\bar{\Omega}, E)$, and $f = I - T$. Then:*

(i) f is closed: f maps closed sets to closed sets;

(ii) f is proper: f^{-1} maps compact sets to compact sets.

Proof. See Lemma 7.5 for the proof. ∎

Letting $T \in Q(\bar{\Omega}, E)$, and $f = I - T$, then $f \in C(\bar{\Omega}, E)$. Assume $b \in E$, $b \notin f(\partial\Omega)$. By Lemma 9.64, $f(\partial\Omega)$ is closed in E, letting $r = \text{dist}(b, f(\partial\Omega))$, then $r > 0$. Because T is compact, $K = \overline{T(\bar{\Omega})}$ is compact. By Lemma 9.63 there is a finite-dimensional subspace $F_{r/2}$ of E, and $g_{r/2} \in C(K, F_{r/2})$ such that

$$\|y - g_{r/2}(y)\| < \frac{r}{2} \quad \text{for } y \in K.$$

Let $T_r = g_{r/2} \circ T$. Then $T_r(\bar{\Omega})$ is contained in a finite-dimensional subspace $F_{r/2}$, and $h_r = I - T_r$ is a finite-dimensional perturbation of the identity. Note that, for $x \in \partial\Omega$, we have

$$\begin{aligned}
\|b - h_r(x)\| &= \|b - x + g_{r/2} \circ T(x)\| \\
&= \|b - x + Tx - Tx + g_{r/2}(T(x))\| \\
&\geq \|b - x + Tx\| - \|Tx - g_{r/2}(T(x))\| \\
&\geq r - \frac{r}{2} \\
&= \frac{r}{2}.
\end{aligned}$$

Thus, $\operatorname{dist}(b, h_r(\partial\Omega)) \geq \frac{r}{2} > 0$, and consequently, the degree $\operatorname{d}(h_r, \Omega, b)$ is defined.

Definition 9.65 *Assume* $T \in Q(\bar{\Omega}, E)$, $f = I - T$, *and* $r = \operatorname{dist}(b, f(\partial\Omega)) > 0$,. *Define*

$$\operatorname{d}(f, \Omega, b) = \operatorname{d}(f_r, \Omega, b),$$

where f_r *is a finite-dimensional perturbation of the identity such that*

$$\|f(x) - f_r(x)\| < \frac{r}{2} \quad \text{for } x \in \Omega.$$

One such f_r *is* $h_r = I - T_r = I - g_{r/2} \circ T$. *We call* $\operatorname{d}(f, \Omega, b)$ *the Leray-Schauder degree of the function* f *with respect to the bounded open set* Ω *and* $b \in E \backslash f(\partial\Omega)$.

Remark 9.66 *The degree is independent of* f_r.

Proof.   Let $r = \operatorname{dist}(b, f(\partial\Omega))$, $f_i = I - T_i$, $i = 1, 2$, and let F_1, F_2 be two finite-dimensional subspaces of E such that, for $i = 1, 2$,

$$T_i(\bar{\Omega}) \subset F_i,$$

$$\|f - f_i\|_\infty = \sup_{x \in \bar{\Omega}} \|f(x) - f_i(x)\| < \frac{r}{2}.$$

Let F be a finite-dimensional subspace of E such that

$$F_1 + F_2 \subset F, \quad \text{and } b \in F.$$

We have by Definition 9.61, and Lemma 9.60,

$$\operatorname{d}(f_i, \Omega, b) = \operatorname{d}(f_i|_{\bar{\Omega} \cap F}, \Omega \cap F, b) \quad i = 1, 2.$$

Let

$$H(x, t) = t\, f_1|_{\bar{\Omega} \cap F}(x) + (1 - t) f_2|_{\bar{\Omega} \cap F}(x) \quad \text{for } x \in \bar{\Omega} \cap F, t \in [0, 1].$$

Then for $x \in \partial\Omega \cap F$, $t \in [0, 1]$,

$$\begin{aligned}
\|b - H(x, t)\| &= \|b - t f_1(x) - (1 - t) f_2(x)\| \\
&= \|b - f(x) + t(f(x) - f_1(x)) + (1 - t)(f(x) - f_2(x))\| \\
&\geq \|b - f(x)\| - t\|f(x) - f_1(x)\| - (1 - t)\|f(x) - f_2(x)\| \\
&\geq r - t\frac{r}{2} - (1 - t)\frac{r}{2} \\
&\geq \frac{r}{2}.
\end{aligned}$$

Note that $\partial_F(\Omega \cap F) \subset \partial\Omega \cap F$, therefore,

$$\operatorname{dist}(b, H(\partial_F(\Omega \cap F), t)) \geq \frac{r}{2} \quad \text{for } t \in [0, 1].$$

Because the Brouwer degree is homotopic invariant, we have

$$\mathrm{d}(f_1\,|_{\overline{\Omega}\cap F}\,,\Omega\cap F,b) = \mathrm{d}(f_2\,|_{\overline{\Omega}\cap F}\,,\Omega\cap F,b),$$

and consequently $\mathrm{d}(f_1,\Omega,b) = \mathrm{d}(f_2,\Omega,b)$. ∎

Proposition 9.67 *Let $t \to S(\cdot,t)$ be a continuous map of $[0,1]$ into $Q(\bar{\Omega},E)$. Then, $S \in Q(\bar{\Omega}\times[0,1],E)$.*

Proof. See Proposition 7.14. ∎

Lemma 9.68 *Let $T \in Q(\bar{\Omega},E)$, $f = I - T$, and $b \notin f(\partial\Omega)$, then*

$$\mathrm{d}(f,\Omega,b) = \mathrm{d}(f - b,\Omega,0).$$

Proof. Letting $r = \mathrm{dist}(b,f(\partial\Omega)) > 0$, and taking $T_r \in Q(\bar{\Omega},F)$ for some finite-dimensional subspace F of E, $f_r = I - T_r$, $b \in F$ such that $\|f - f_r\|_\infty < \frac{r}{2}$, then

$$\mathrm{d}(f,\Omega,b) = \mathrm{d}(f_r,\Omega,b) = \mathrm{d}(f_r\,|_{\overline{\Omega}\cap F}\,,\Omega\cap F,b)$$
$$= \mathrm{d}(f_r - b\,|_{\overline{\Omega}\cap F}\,,\Omega\cap F,0) = \mathrm{d}(f_r - b,\Omega,0) = \mathrm{d}(f - b,\Omega,0),$$

where $\|(f_r - b) - (f - b)\|_\infty = \|f_r - f\|_\infty < \frac{r}{2}$. ∎

We come to the fundamental properties of the Leray-Schauder degree.

Theorem 9.69 *Let $T \in Q(\bar{\Omega},E)$, $f = I - T$, and $b \notin f(\partial\Omega)$. Then:*

(i) *(Continuity of Function) There is a neighborhood A of T in $Q(\bar{\Omega},E)$, such that if $U \in A$ and $g = I - U$, then $b \notin g(\partial\Omega)$, and $\mathrm{d}(g,\Omega,b) = \mathrm{d}(f,\Omega,b)$;*

(ii) *(Compact Homotopy Invariance) Letting $S \in Q(\bar{\Omega}\times[0,1],E)$, and $H(x,t) = x - S(x,t)$, and $b \notin H(\partial\Omega\times[0,1])$, then the degree $\mathrm{d}(H(\cdot,t),\Omega,b)$ is constant on $[0,1]$;*

(iii) *(Constant in Component) The degree $\mathrm{d}(f,\Omega,b)$ is constant for b in each connected component of $E\backslash f(\partial\Omega)$;*

(iv) *(Additive Property) Let Ω_1, $\Omega_2 \subset E$ be bounded open sets, $\Omega = \Omega_1 \cup \Omega_2$, $\Omega_1 \cap \Omega_2 = \emptyset$, $b \notin f(\partial\Omega_1) \cup f(\partial\Omega_2)$. We have*

$$\mathrm{d}(f,\Omega,b) = \mathrm{d}(f,\Omega_1,b) + \mathrm{d}(f,\Omega_2,b).$$

Proof.

(*i*) Let $r = \text{dist}(b, f(\partial\Omega)) > 0$, and

$$A = \{U \in Q(\bar{\Omega}, E) \mid \|U - T\|_\infty < \frac{r}{4}\}.$$

For $U \in A$, and $g = I - U$, then $\|f - g\|_\infty < \frac{r}{4}$, and $\text{dist}(b, g(\partial\Omega)) > \frac{r}{2}$. Thus, $b \notin g(\partial\Omega)$. By the arguments preceding Definition 9.65, there are two finite-dimensional perturbations of the identity $f_1 = I - T_1$, $g_1 = I - U_1$ such that

$$\|f - f_1\|_\infty < \frac{r}{4},$$
$$\|g - g_1\|_\infty < \frac{r}{4}.$$

Choose a finite-dimensional subspace F of E, containing $T_1(\bar{\Omega})$, and $U_1(\bar{\Omega})$, $b \in F$. Because $\text{dist d}(b, g(\partial\Omega)) > \frac{r}{2}$, and $\|g - g_1\|_\infty < \frac{r}{4}$, we have

$$d(f, \Omega, b) = d(f_1\mid_{\overline{\Omega \cap F}}, \Omega \cap F, b),$$
$$d(g, \Omega, b) = d(g_1\mid_{\overline{\Omega \cap F}}, \Omega \cap F, b).$$

Set $H(x, t) = t\,f_1\mid_{\overline{\Omega \cap F}}(x) + (1 - t)g_1\mid_{\overline{\Omega \cap F}}(x)$ for $x \in \bar{\Omega} \cap F$, $t \in [0, 1]$. Because $\partial_F(\Omega \cap F) \subset \partial\Omega \cap F$, we have, for $x \in \partial_f(\Omega \cap F)$, $t \in [0, 1]$,

$$
\begin{aligned}
\|b - H&(x, t)\| \\
&= \|b - f(x) + tf(x) + (1 - t)f(x) + (1 - t)g(x) \\
&\quad - tf_1(x) - (1 - t)g_1(x)\| \\
&\geq \|b - f(x)\| - t\,\|f(x) - f_1(x)\| - (1 - t)\,\|g(x) - g_1(x)\| \\
&\quad - (1 - t)\,\|f(x) - g(x)\| \\
&\geq \frac{r}{4}.
\end{aligned}
$$

By the homotopic invariance of the Brouwer degree,

$$d(f_1\mid_{\overline{\Omega \cap F}}, \Omega \cap F, b) = d(g_1\mid_{\overline{\Omega \cap F}}, \Omega \cap F, b),$$

and consequently, $d(f, \Omega, b) = d(g, \Omega, b)$.

(*ii*) Let $r = \text{dist}(b, H(\partial\Omega \times [0, 1]))$. Then $r > 0$. In fact, consider $\bar{H}(x, t) : \bar{\Omega} \times [0, 1] \to E \times \mathbb{R}$ defined by $\bar{H}(x, t) = (x, t) - (S(x, t), t)$. $\bar{H}(x, t)$ is a compact perturbation of the identity. By Lemma 9.65, $\bar{H}(\partial\Omega \times [0, 1]) = H(\partial\Omega \times [0, 1]) \times \{0\}$ is closed in $E \times \mathbb{R} : H(\partial\Omega \times [0, 1])$ is closed in E. Let

$$K = \overline{S(\bar{\Omega} \times [0, 1])}.$$

Thus, K is compact. By Lemma 9.63, there is a finite-dimensional subspace $F_{r/2}$ in E, $g_{r/2} \in C(K, F_{r/2})$, such that

$$b \in F_{r/2},$$
$$\|x - g_{r/2}(x)\| < \frac{r}{2} \quad \text{for } x \in K.$$

Set

$$H_1(x, t) = x - g_{r/2} \circ S(x, t).$$

For $t \in [0, 1]$, $\|H - H_1\|_\infty < \frac{r}{2}$, so $b \notin H_1(\partial\Omega \times [0, 1])$, and consequently

$$d(H(\cdot, t), \Omega, b) = d(H_1(\cdot, t), \Omega, b).$$

However, H_1 is a finite-dimensional perturbation of the identity, so

$$d(H_1(\cdot, t), \Omega, b) = d(H_1(\cdot, t)\big|_{\overline{\Omega} \cap F_{r/2}}, \Omega \cap F_{r/2}, b).$$

Note that $H_1 : \overline{\Omega} \cap F_{r/2} \to F_{r/2}$ is continuous, $b \notin H_1(\partial\Omega \times [0, 1])$. The degree $d(H(\cdot, t), \Omega, b)$ is constant by the homotopic invariance of the Brouwer degree of $d(H_1(0, t)\big|_{\overline{\Omega} \cap F_{r/2}}, \Omega \cap F_{r/2}, b)$.

(iii) and (iv) By the approximation of the Brouwer degree:

$$d(I, \Omega, b) = \begin{cases} 1 & \text{if } b \in \Omega, \\ 0 & \text{if } b \notin \Omega. \end{cases}$$

In fact, let $Rb = \langle b \rangle$, $I = I - 0$, $0(\Omega) = \{0\} \subset Rb$ and note that

$$d(I, \Omega, b) = d(I\big|_{\overline{\Omega} \cap \mathbb{R}b}, \Omega \cap \mathbb{R}b, b). \qquad \blacksquare$$

Corollary 9.70 *Let* $T \in Q(\overline{\Omega}, E)$, *and* $f = I - T$.

(i) *If* $b \notin f(\overline{\Omega})$, *then* $d(f, \Omega, b) = 0$;

(ii) *If* $d(f, \Omega, b) \neq 0$, *then there is* $x_0 \in \Omega$ *such that* $f(x_0) = b$.

Proof. Note that (i) and (ii) are equivalent. It suffices to prove (i). By Lemma 9.64, f is a closed map. Let $\alpha = \text{dist}(b, f(\overline{\Omega}))$, and $r = \text{dist}(b, f(\partial\Omega))$, then $0 < \alpha \leq r$. There is $T_\alpha \in Q(\overline{\Omega}, F)$ for some finite-dimensional subspace F of E, and $f = I - T_\alpha$, such that

$$\|f(x) - f_\alpha(x)\| < \frac{\alpha}{2} \quad \text{for } x \in \overline{\Omega},$$

so $b \notin f_\alpha(\overline{\Omega})$. We have $d(f, \Omega, b) = d(f_\alpha, \Omega, b) = d(f_\alpha\big|_{\overline{\Omega} \cap F}, \Omega \cap F, b) = 0. \qquad \blacksquare$

Corollary 9.71 *Let $T \in Q(\bar{\Omega}, E)$, and $f = I - T$.*

(i) *If* $\mathrm{d}(f, \Omega, b) \neq 0$*, then* $f(\Omega)$ *is a neighborhood of* b*.*

(ii) *If* $f(\Omega)$ *is contained in a proper subspace of* E*, and* $b \notin f(\partial\Omega)$*, then* $\mathrm{d}(f, \Omega, b) = 0$*.*

(iii) *(Excision Property) Let* $K \subset \Omega$ *be a closed set, and* $b \notin f(K)$*, then* $\mathrm{d}(f, \Omega, b) = \mathrm{d}(f, \Omega \backslash K, b)$*.*

(iv) *Let* $\{\Omega_i\}_{i \in I}$ *be a family of open sets in* Ω*, disjoint to each other, and* $f^{-1}(b) \subset \bigcup\limits_{i \in I} \Omega_i$*. Then the degree* $\mathrm{d}(f, \Omega_i, b)$ *is 0 except possibly for finite indices* $i \in I$*, and* $\mathrm{d}(f, \Omega, b) = \sum\limits_{i \in I} \mathrm{d}(f, \Omega_i, b)$*.*

Proof.   By the approximation of the Brouwer degree.   ∎

Corollary 9.72 *Let* $T, U \in Q(\bar{\Omega}, E)$*,* $f = I - T$*,* $g = I - U$*,* $f(x) = g(x)$ *for* $x \in \partial\Omega$*, and* $b \notin f(\partial\Omega)$*. Then*
$$\mathrm{d}(f, \Omega, b) = \mathrm{d}(g, \Omega, b).$$

Proof.   Let $H(x, t) = t\,f(x) + (1-t)g(x)$ for $u \in \bar{\Omega}$, $t \in [0,1]$. For $t \in [0,1]$, $H(x,t)$ is a compact perturbation of the identity:
$$\begin{aligned} H(x,t) &= t\,f(x) + (1-t)g(x) \\ &= I(x) - [t\,T(x) + (1-t)U(x)] \\ &= x - S(x,t). \end{aligned}$$

Because $f(x) = g(x)$ for $x \in \partial\Omega$, $b \notin H(\partial\Omega \times [0,1])$. By the compact homotopic invariance of the degree
$$\mathrm{d}(f, \Omega, b) = \mathrm{d}(g, \Omega, b).$$
                                                                            ∎

Corollary 9.73 *There is no retraction* $f \in C(\overline{B_1(0)}, \partial B_1(0))$*,* $f\big|_{\partial B_1(0)} = I$*, where* $T \in Q(\overline{B_1(0)}, E)$*, and* $f = I - T$*.*

Proof.   Suppose such f does exist, then by Corollary 9.72,
$$\mathrm{d}(f, B_1(0), 0) = \mathrm{d}(I, B_1(0), 0) = 1.$$

By Corollary 9.71, there is $x_0 \in B_1(0)$ such that $f(x_0) = 0$, which contradicts the assumption that $f(\overline{B_1(0)}) \subset \partial B_1(0)$.   ∎

Remark 9.74 *Corollary 9.73 is false if we replace* $T \in Q(\overline{B_1(0)}, E)$ *by* $T \in C(\overline{B_1(0)}, E)$ *(see Dugundji) [19].*

Let $B = \mathbb{R} \times E$ be a Banach space under the sum norm

$$\|(\lambda,\, x)\|_B = |\lambda| + \|x\|_E \quad \text{for } \lambda \in \mathbb{R},\ x \in E.$$

Let $\Lambda = [\underline{\lambda}, \bar{\lambda}]$. Consider a bounded open set Ω in $\Lambda \times E$, and $T \in Q(\bar{\Omega}, E)$. Set $H(\lambda, x) = x - T(\lambda, x)$ for $(\lambda, x) \in \bar{\Omega}$. If $\lambda \in \Lambda$, let $\Omega_\lambda = \{x \in E \mid (\lambda, x) \in \Omega\}$. Then, Ω_λ is a bounded open set in E.

Corollary 9.75 *Suppose $b \in E \backslash H(\partial\Omega)$, then:*

(i) $\mathrm{d}(H(\lambda, \cdot), \Omega_\lambda, b) = \alpha$, *for each $\lambda \in \Lambda$.*

(ii) *Let $A = \{(\lambda, x) \in \Omega \mid H(\lambda, x) = b\} = H^{-1}(b)$. If $\alpha \neq 0$ in (1), then there is a connected component C containing A that meets $\{\underline{\lambda}\} \times \Omega_{\underline{\lambda}}$ and $\{\bar{\lambda}\} \times \Omega_{\bar{\lambda}}$.*

Proof. Note that $A = H^{-1}(b)$ is compact because H is a proper map, and the corollary follows from the approximation of the Brouwer degree. ∎

Corollary 9.76 *Let E_i be Banach spaces, and let $\Omega_i \subset E_i$ be open sets, $T_i \in Q(\bar{\Omega}_i, E_i)$, $f_i = I_{E_i} - T_i$, and $b_i \in E_i \backslash f_i(\partial\Omega_i)$, $i = 1, 2$. Then*

$$\mathrm{d}((f_1, f_2), \Omega_1 \times \Omega_2, (b_1, b_2)) = \mathrm{d}(f_1, \Omega_1, b_1)\,\mathrm{d}(f_2, \Omega_2, b_2).$$

Proof. By the approximation of the Brouwer degree. ∎

Corollary 9.77 *Let F be a closed subspace of E, $b \in F$, $T \in Q(\bar{\Omega}, F)$, and $f = I - T$. Then*

$$\mathrm{d}(f, \Omega, b) = \mathrm{d}(f\,|_{\bar{\Omega} \cap F}, \Omega \cap F, b).$$

Proof. Let $K = \overline{T(\bar{\Omega})}$. Then $K \subset F$ is compact. By Lemma 9.63, there is a finite-dimensional subspace $F_{r/2}$ of F, $g_{r/2} \in C(K, F_{r/2})$ such that

$$\|x - g_{r/2}(x)\| < \frac{r}{2} \quad \text{for } x \in K,$$

where $r = dist(b, f(\partial\Omega))$. Let $s = dist(b, f\,|_{\bar{\Omega} \cap F}\, (\partial_F(\Omega \cap F)))$. Then $s \geq r$. Set $f_r = I - g_{r/2} \circ T$. Because

$$\|f - f_r\|_{L^\infty(\Omega \cap F)} \leq \|f - f_r\|_{L^\infty(\Omega)} \leq \frac{r}{2} \leq \frac{s}{2},$$

we have

$$\mathrm{d}(f\,|_{\bar{\Omega} \cap F}, \Omega \cap F, b) = \mathrm{d}(f_r\,|_{\bar{\Omega} \cap F_{r/2}}, \Omega \cap F_{r/2}, b) = \mathrm{d}(f, \Omega, b).$$

Now we study the infinite-dimensional Borsuk Theorem. ∎

Theorem 9.78 (*Borsuk*) *Let* $\Omega \subset E$ *be a bounded open set, symmetric with respect to* 0, *and* $0 \in \Omega$. *If* $T \in Q(\bar{\Omega}, E)$ *is odd on* $\partial\Omega$, $f = I - T$, *and* $0 \notin f(\partial\Omega)$, *then the degree* $d(f, \Omega, 0)$ *is odd.*

Proof. Set

$$\bar{T}(x) = \frac{T(x) - T(-x)}{2} \quad \text{for } x \in \bar{\Omega},$$

then $\bar{T}$ is odd on $\bar{\Omega}$, $\bar{T}(x) = T(x)$ for $x \in \partial\Omega$, so

$$d(I - \bar{T}, \Omega, 0) = d(I - T, \Omega, 0).$$

Hence, we may assume that T is odd on $\bar{\Omega}$. Let $K = \overline{T(\bar{\Omega})}$. Then K is compact, and symmetric with respect to 0. By Lemma 9.63, there is a finite-dimensional subspace F of E, $g \in C(K, F)$ such that

$$\|x - g(x)\| \le \frac{r}{2},$$

where $r = dist(0, f(\partial\Omega))$. Define $g_1(x) = (g(x) - g(-x))/2$ for $x \in K$. Then

$$\|x - g_1(x)\| \le \frac{1}{2}\|x - g(x)\| + \frac{1}{2}\|(-x) - g(-x)\| \le \frac{r}{2}.$$

Let $f_1 = I - g_1 \circ T$. Then f_1 is odd, and

$$d(f, \Omega, 0) = d(f_1 \mid_{\overline{\Omega \cap F}}, \Omega \cap F, 0).$$

The right-hand side is odd by the finite-dimensional Borsuk Theorem. ∎

9.6   Leray-Schauder Index Theory

We study the index by linearization. Let E be a Banach space, $T \in Q(E, E)$, and $f = I - T$. Consider an isolated solution x_0 of the equation $f(x) = b$. Let x_0 be the only solution of the equation in the ball $\bar{B}_{r_0}(x_0)$. By the Excision Property, for $0 < r \le r_0$, we have

$$d(f, B_r(x_0), b) = d(f, B_{r_0}(x_0), b).$$

The index of the isolated solution x_0 of the equation $f(x) = b$ is defined by

$$i(f, x_0, b) = \lim_{r \to 0} d(f, B_r(x_0), b).$$

See Definition 7.2, and Proposition 7.3 for the properties of compact linear operators.

Lemma 9.79 *Let* $T : \mathbb{R}^N \to \mathbb{R}^N$ *be a linear operator, with* λ_0 *a real eigenvalue of* T *of multiplicity* n_0. *Then for small* $\epsilon > 0$,

$$\text{sgn} \det((\lambda_0 + \epsilon)I - T) = (-1)^{n_0} \text{sgn} \det((\lambda_0 - \epsilon)I - T).$$

Proof. Suppose λ_i are the eigenvalues of T of multiplicities n_i respectively. Then,

$$\det(\lambda I - T) = \det \begin{pmatrix} \lambda - a_{11} & -a_{12} & \cdots & -a_{1N} \\ -a_{21} & \lambda - a_{22} & \cdots & -a_{2N} \\ \vdots & \vdots & & \vdots \\ -a_{N1} & -a_{N2} & \cdots & \lambda - a_{NN} \end{pmatrix}$$
$$= \prod_i (\lambda - \lambda_i)^{n_i}.$$

(i) If λ_i is a complex eigenvalue of multiplicity n_i, then $\bar{\lambda}_i$ is also an eigenvalue of multiplicity n_i. Let $\lambda = \lambda_0 + \epsilon$, or $\lambda_0 - \epsilon$. Then λ is real; so, in the expression $\prod_i (\lambda - \lambda_i)^{n_i}$, we have

$$(\lambda - \lambda_i)^{n_i}(\lambda - \bar{\lambda}_i)^{n_i} = |\lambda - \lambda_i|^{2n_i} > 0,$$

which does not affect the sign of $\prod_i (\lambda - \lambda_i)^{n_i}$.

(ii) For the real eigenvalues λ_i, take $\epsilon > 0$ small such that

$$[\lambda_0 - \epsilon, \lambda_0 + \epsilon] \cap \{\lambda_i\} = \lambda_0.$$

Thus,

$$\mathrm{sgn}(\lambda_0 + \epsilon - \lambda_i) = \mathrm{sgn}(\lambda_0 - \epsilon - \lambda_i) \quad \text{for } i \neq 0,$$
$$\mathrm{sgn}(\lambda_0 + \epsilon - \lambda_0) = (-1)\mathrm{sgn}(\lambda_0 - \epsilon - \lambda_0).$$

Therefore,

$$\mathrm{sgn}\det((\lambda_0 + \epsilon)I - T) = \mathrm{sgn}\prod_i (\lambda_0 + \epsilon - \lambda_i)^{n_i}$$
$$= (-1)^{n_0}\mathrm{sgn}\prod_i (\lambda_0 - \epsilon - \lambda_i)^{n_i}$$
$$= (-1)^{n_0}\mathrm{sgn}\det((\lambda_0 - \epsilon)I - T). \quad \blacksquare$$

Lemma 9.80 *Let A be a $N \times N$ real nonsingular matrix, and let $A = I - B$. Then*

$$\mathrm{sgn}\det A = (-1)^\beta,$$

where $\beta = \sum_i m_i(B) = \sum_j n_j(B)$, $m_i(B)$ is the multiplicity of the characteristic value μ_i of B, $0 < \mu_i < 1$, and $n_j(B)$ is the multiplicity of the eigenvalues λ_j of B, $\lambda_j > 1$.

Proof. Because $A = I - B$ is nonsingular, 1 is not an eigenvalue of B. Now,

$$\text{sgn} \det A = \text{sgn} \det(I - B)$$

$$= \text{sgn} \prod_j (1 - \lambda_j)^{n_j}$$

$$= \text{sgn} \prod_{\lambda_j > 1} (1 - \lambda_j)^{n_j} \tag{9.30}$$

$$= (-1)^{n_1} \cdots (-1)^{n_\ell}$$

$$= (-1)^\beta.$$

where $\beta = \sum_j n_j(B)$, and in the expression (9.30), $\lambda_j < 1$ or complex λ_j are ruled out as in the proof of Lemma 9.79. ∎

Theorem 9.81 (*Leray-Schauder Index Theorem, Linear Case*) *Let $T : E \to E$ be a compact linear operator, and $f = I - T$. If 1 is not a characteristic value of T, then 0 is an isolated solution of f in E, and*

$$i(f, 0, 0) = (-1)^\beta,$$

where $\beta = \sum_i m_i(B) = \sum_j n_j(B)$, $m_i(B)$ is the multiplicity of the characteristic value μ_i of B, $0 < \mu_i < 1$, and $n_j(B)$ is the multiplicity of the eigenvalues λ_j of B, $\lambda_j > 1$.

Proof. Note that the sets

$$C = \{\lambda_1, \cdots, \lambda_s \mid \lambda_j \text{ eigenvalues of } T, \ \lambda_j > 1\},$$

and

$$D = \{\mu_1, \cdots, \mu_s \mid \mu_i \text{ characteristic values of } T, 0 < \mu_i < 1\}.$$

are equal. For $\mu_i \in D$, let N_i be the characteristic subspace associated with μ_i :, that is,

$$N_i = \bigcup_{k=1}^\infty Ker(\mu_i T - I)^k = Ker(\mu_i T - I)^{\alpha_i}, \dim N_i = m_i,$$

where $\alpha_1, \cdots, \alpha_s \in N$. Let $N = \bigoplus_{i=1}^s N_i$, then N is a subspace of E of dimension $m = m_1 + \cdots + m_s$. Moreover, N is invariant with respect to T :, when we let $x = \sum_{i=1}^s x_i \in N$. Then, $Tx = \sum_{i=1}^s Tx_i$. However, $x_i \in N_i$ implies $(\mu_i T - I)^{\alpha_i} x_i = 0$, and it follows that

$$(\mu_i T - I)^{\alpha_i}(Tx_i) = T(\mu_i T - I)^{\alpha_i} x_i = 0,$$

and consequently, $Tx_i \in N_i : Tx \in N$. Write $E = N \oplus F$ with the continuous projections $P : E \to N$, $Q : E \to F$. Assume $H : E \times [0, 1] \to E$ defined by

$H(x,t) = x - S(x,t) = x - T(Px) - tT(Qx)$, for $x \in E$, $t \in [0,1]$. Because TQ is bounded on $\overline{B_1(0)}$, we have, for fixed $t_0 \in [0,1]$, $t \in [0,1]$, and $x \in \overline{B_1(0)}$,

$$\|S(x,t) - S(x,t_0)\| = \|T(P(x)) + tT(Q(x)) - [T(P(x)) + t_0 T(Q(x))]\|$$
$$= |t - t_0| \, \|TQ\| \leq c \, |t - t_0| \, .$$

Thus, $\alpha : [0,1] \to S(\cdot, t) = TP + tTQ \in Q(\overline{B_1(0)}, E)$ is continuous. By Proposition 9.67, $S \in Q(B_1(0) \times [0,1], E)$ and consequently $H(x,t)$ is a compact perturbation of the identity. Let $A_i = (\mu_i T - I)^{\alpha_i}$, and $A = A_s \cdots A_1$, where $A_i A_j = A_j A_i$ for i, j.

(i) Ker $A = N$: Clearly $Ax = 0$ for $x \in N$. On the other hand, let $p_i(x) = \sum_{j \neq i} (\mu_j x - 1)^{\alpha_j}$, then $p_1, \cdots, p_s$ are relatively prime. Take polynomials $q_1(x), \cdots, q_s(x)$ such that

$$\sum_{i=1}^{s} p_i(x) q_i(x) = 1$$

and consequently,

$$\sum_{i=1}^{\ell} p_i(T) q_i(T) = I.$$

For $x \in$ Ker$A$, let $x_i = p_i(T) q_i(T) x$, then $x = \sum_{i=1}^{s} x_i$. Note that

$$(\mu_i T - I)^{\alpha_i} x_i = (\mu_i T - I)^{\alpha_i} p_i(T) q_i(T) x = A q_i(T)(x) = q_i(T) A x = 0,$$

so $x_i \in N_i : x \in N$.

(ii)

$$H(x,t) \neq 0 \, for x \in \partial B_1(0), t \in [0,1],$$
$$H(x,t) = 0 \quad \text{if and only if} \quad x - TP(x) - tTQ(x) = 0$$
$$\text{if and only if} \quad Px + Qx - TP(x) - tTQ(x) = 0$$
$$\text{if and only if} \quad (tT - I)Q(x) = P(x) - TP(x).$$

Because $Px - TP(x) \in N$, we have

$$A(tT - I)Q(x) = (tT - I)(A(Q(x))) = 0$$

(a) If t is not a characteristic value of T, then

$$AQ(x) = 0 : Q(x) \in N.$$

(b) If $t = \mu_k$ for some k, $1 \leq k \leq s$

$$Q(x) \in Ker(\mu_k T - I)A = Ker \, A = N.$$

In both cases $Q(x) \in N : x \in N$, so

$$H(x,t) = 0 \quad \text{if and only if} \quad x - TPx - tTQ(x) = 0, \ x \in N$$
$$\text{if and only if} \quad x - Tx = 0$$
$$\text{if and only if} \quad x = 0.$$

We conclude that

$$H(x,t) \neq 0 \quad \text{for } x \in \partial B_1(0), \ t \in [0,1].$$

Note that

$$H(x,0) = x - TP(x) = (I - TP)(x),$$
$$H(x,1) = x - TP(x) - TQ(x) = x - Tx = f(x).$$

By the compact homotopic invariance of the degree,

$$d(f, B_1(0), 0) = d(I - TP, B_1(0), 0),$$

but $I - TP$ is a finite-dimensional perturbation of the identity. Therefore, by Lemma 9.80,

$$
\begin{aligned}
i(f,0,0) &= d(I - TP, B_1(0), 0) \\
&= d\left((I - TP) \big|_{\overline{B_1(0)} \cap N}, B_1(0) \cap N, 0 \right) \\
&= \operatorname{sgn} \det \ J_{I-TP}(0) \\
&= \operatorname{sgn} \det (I - TP)'(0) \\
&= \operatorname{sgn} \det (I - TP) \\
&= (-1)^{\beta},
\end{aligned}
$$

where $\beta = \sum_i m_i(B) = \sum_j n_j(B)$, $m_i(B)$ is the multiplicity of the characteristic value μ_i of B, $0 < \mu_i < 1$, and $n_j(B)$ is the multiplicity of the eigenvalues λ_j of B, $\lambda_j > 1$. ∎

Theorem 9.82 *Let $T : B_r(0) \to E$ be a compact operator, Fréchet -differentiable at 0. Then $T'(0) : E \to E$ is a compact linear operator.*

Proof. See Theorem 7.4 for the proof. ∎

Theorem 9.83 (*Leray-Schauder Index Theorem, Nonlinear Case*) *Let $T \in Q(\overline{B_r(0)}, E)$ be Fréchet-differentiable near 0, $T(0) = 0$, and $f = I - T$. If 1 is not a characteristic value of $T'(0)$, then 0 is the isolated solution of the equation $f(x) = 0$, and*

$$i(f, 0, 0) = (-1)^{\beta},$$

where $\beta = \sum\limits_i m_i(B) = \sum\limits_j n_j(B)$, $m_i(B)$ is the multiplicity of the characteristic value μ_i of B, $0 < \mu_i < 1$, and $n_j(B)$ is the multiplicity of the eigenvalues λ_j of B, $\lambda_j > 1$.

Proof. Write, for small $r > 0$, $x \in \overline{B_r(0)}$,

$$f(x) = f(0) + f'(0)x + r(x) = x - T'(0)x + r(x),$$

where $r(x) = o(\|x\|)$. By Theorem 9.82, $T'(0)$ is a compact linear operator. Set

$$H(x,t) = x - T'(0)x + t\,r(x) \quad \text{for } x \in \overline{B_r(0)}, \ t \in [0,1].$$

Note that because 1 is not a characteristic value of $T'(0)$, we have $\left\|(I - T'(0))^{-1}\right\| \le c$, and

$$
\begin{aligned}
H(x,t) = 0 \ &\text{if and only if } x - T'(0)x + t\,r(x) = 0 \\
&\text{if and only if } (I - T'(0))x = (-t)r(x) \qquad\qquad (9.31) \\
&\text{if and only if } x = (I - T'(0))^{-1}((-t)r(x)).
\end{aligned}
$$

From $r(x) = o(\|x\|)$, choose $\epsilon > 0$ such that $\|r(x)\| < \frac{1}{c}\|x\|$ for $x \in \overline{B_\epsilon(0)}$. Now for $t \in (0,1]$, $x \neq 0$ in $\overline{B_\epsilon(0)}$,

$$\|x\| \le \left\|(I - T'(0))^{-1}(-t\,r(x))\right\| \le c\,\|-t\,r(x)\| < \|x\|. \qquad (9.32)$$

Letting $t = 0$, $x \neq 0$, we have

$$\|x\| \le 0. \qquad (9.33)$$

By (9.33), the Inverse Function Theorem, and the Fredholm Alternative Theorem, 0 is an isolated solution of the equation $f(u) = 0$:

$$i(f, 0, 0) = d(f, B_\epsilon(0), 0).$$

By (9.32) and (9.33), $H(x,t) \neq 0$ for $x \in \partial B_\epsilon(0)$, $0 \in [0,1]$. Now,

$$x - T(x) = f(x) = x - T'(0)x + r(x)$$

or

$$r(x) = (T'(0) - T)x.$$

Thus, r is a compact operator of $\overline{B_\epsilon(0)}$ into E, and consequently,

$$S(x,t) = T'(0)x - t\,r(x)$$

is a compact operator; $H(u,t)$ is a compact perturbation of the identity. By the compact homotopic invariance of the degree, the degree $d(H_t, B_\epsilon(0), 0)$ is constant for $t \in [0,1]$. Note that

$$
\begin{aligned}
H(x,0) &= x - T'(0)x, \\
H(x,1) &= x - T'(0)x + r(x) = f(x).
\end{aligned}
$$

Thus, by Theorem 9.82,

$$i(f, 0, 0) = i(I - T'(0), 0, 0) = (-1)^\beta,$$

where $\beta = \sum_i m_i(B) = \sum_j n_j(B)$, $m_i(B)$ is the multiplicity of the characteristic value μ_i of B, $0 < \mu_i < 1$, and $n_j(B)$ is the multiplicity of the eigenvalues λ_j of B, $\lambda_j > 1$.   ∎

Theorem 9.84 *Let $T : \overline{B_r(0)} \to E$ be a nonlinear compact operator, Fréchet-differentiable at 0, $T(0) = 0$, and $f_\lambda = I - \lambda T$ for $\lambda \in \mathbb{R}$. Then $i(f_\lambda, 0, 0)$ is defined when λ is not a characteristic value of $T'(0)$, and is a multiple of $(-1)^{m_j}$ when λ passes through the characteristic value μ_j of multiplicity m_j.*

Proof.   By the approximation of the Brouwer degree.   ∎

9.7   Leray-Schauder Fixed Point Theorems

We study the fixed point theorems in this section. Let E be a Banach space.

Theorem 9.85 (*Schauder Fixed Point Theorem*) *Let $K \subset E$ be a compact convex set, and $T \in C(K, K)$. Then T admits a fixed point in K.*

Proof.   By Lemma 9.63, for $\epsilon > 0$, there is a finite-dimensional subspace F_ϵ of E, and $g_\epsilon \in C(K, F_\epsilon)$ such that

$$\|x - g_\epsilon(x)\| < \epsilon \quad \text{for } x \in K. \tag{9.34}$$

In fact, there are $y_1, \cdots, y_{p(\epsilon)} \in K$ such that

$$K \subset \bigcup_{i=1}^{p(\epsilon)} B_\epsilon(y_i),$$

$$g_\epsilon(x) = \frac{\displaystyle\sum_{i=1}^{p(\epsilon)} b_i(x) y_i}{\displaystyle\sum_{i=1}^{p(\epsilon)} b_i(x)}$$

where $b_i \in C(K, \mathbb{R})$, $b_i \geq 0$, for $i = 1, 2, \cdots, p(\epsilon)$, and

$$\sum_{i=1}^{p(\epsilon)} b_i(x) > 0 \quad \text{for } x \in K.$$

Let $F_\epsilon = \langle y_1, \cdots, y_{p(\epsilon)} \rangle, K_\epsilon = \overline{conv}\{y_1, \cdots, y_{p(\epsilon)}\}$. Then K_ϵ is compact convex, $K_\epsilon \subset K \cap F_\epsilon$, and $g_\epsilon \in C(K, K_\epsilon)$ satisfy (9.34). Consider the continuous map $(g_\epsilon T)|_{K_\epsilon} : K_\epsilon \to K_\epsilon$. By the Brouwer Fixed Point Theorem, there is $y_\epsilon \in K_\epsilon \subset K$ such that

$$y_\epsilon = (g_\epsilon \circ T)(y_\epsilon).$$

Because K is compact, there is a sequence $\{\epsilon_n\}$ such that $\epsilon_n \to 0$ and $y_{\epsilon_n} \to y$ for some $y \in K$. Now by (9.34),

$$\begin{aligned}
\|y - T(y)\| &= \|y - y_{\epsilon_n} + y_{\epsilon_n} - T(y_{\epsilon_n}) + T(y_{\epsilon_n}) - T(y)\| \\
&\leq \|y - y_{\epsilon_n}\| + \|g_{\epsilon_n} \circ T(y_{\epsilon_n}) - T(y_{\epsilon_n})\| + \|T(y_{\epsilon_n}) - T(y)\| \\
&\to 0 \quad \text{as } n \to \infty.
\end{aligned}$$

Thus, $y = T(y)$.   ∎

Theorem 9.86 (*Schauder Fixed Point Theorem*) *Let $A \subset E$ be a bounded closed convex set and $T \in Q(A, A)$. Then T admits a fixed point in A.*

Proof. Method 1. Because $T(A)$ is relatively compact, let $K = \overline{conv}(T(A))$. Then $K \subset A$ is a compact convex set, and $T : K \to K$ is a continuous map. By Theorem 9.85, T admits a fixed point in $K \subset A$.

Method 2. (a) Let $A = \overline{B_r(0)}$, then either: (i) T admits a fixed point on $\partial B_1(0)$, or (ii) $d(I - T, B_r(0), 0)$ is defined. Suppose

$$(I - T)(x) \neq 0 \quad \text{for } x \in \partial B_1(0). \tag{9.35}$$

Consider $H(x, t) = x - tT(x)$ for $x \in \overline{B_r(0)}$, $t \in [0, 1]$. Note that $x \in \partial B_1(0)$,

$$H(x, t) = 0,\ 0 \leq t < 1 \quad \text{if and only if} \quad x = tT(x)$$

or $1 = \|x\| \leq t\,\|T(x)\| \leq t < 1$, a contradiction. Therefore,

$$H(x, t) \neq 0 \quad \text{for } x \in \partial B_1(0),\ 0 \leq t < 1. \tag{9.36}$$

By (9.35) and (9.36), $H(x, t) \neq 0$ for $x \in \partial B_1(0)$, $t \in [0, 1]$. By the compact homotopic invariance of the degree

$$d(I - T, B_r(0), 0) = d(I, B_r(0), 0) = 1.$$

Then for some $x \in B_r(0)$, $(I - T)(x) = 0 : x = T(x)$.

(b) In the general case, let $K = \overline{conv}(T(A))$, then K is a compact convex bounded set in A. Assume that there is $r > 0$ satisfying $K \subset \overline{B_r(0)}$. Because $T : K \to K$ is continuous, by the Dugundji Extension Theorem 1.20 there is a continuous extension $\bar{T} : \overline{B_r(0)} \to K \subset \overline{B_r(0)}$. By part (a), there is $x \in B_r(0)$ such that

$$x = \bar{T}(x).$$

However, $\bar{T}(x) \in K \subset \overline{B_r(0)}$, and $\bar{T}|_K = T$, so $x \in K$, and $x = T(x)$.   ∎

Remark 9.87 *In Theorem 9.85, the continuity of T is not sufficient for the existence of a fixed point.*

(i) *In an infinite-dimensional space, there is a continuous retraction f of $\overline{B_1(0)}$ to $\partial B_1(0)$. Then the continuous map $g = -f : \overline{B_1(0)} \to \overline{B_1(0)}$ has no fixed point.*

(ii) *Consider the map $f : \ell_2 \to \ell_2$ defined by $f(x) = y$, $x = (x_1, x_2, \cdots)$ and $y = (\sqrt{1 - \|x\|^2}, x_1, x_2, \cdots)$. For $x \in \overline{B_1(0)}$, we have*

$$\|f(x)\| = \sqrt{1 - \|x\|^2 + \|x\|^2} = 1.$$

Then, $f \in C(\overline{B_1(0)}, \overline{B_1(0)})$. If $x = f(x)$, then $(x_1, x_2, \cdots) = (\sqrt{1 - \|x\|^2}, x_1, x_2, \cdots)$. Because $x \in \ell_2$,

$$\sqrt{1 - \|x\|^2} = x_1 = x_2 = \cdots = 0 : 1 = 0.$$

We obtain a contradiction.

Theorem 9.88 (*Leray-Schauder Fixed Point Theorem*) *Assume $T \in Q(E, E)$. Suppose there is $r > 0$ such that if $x \in E$, $\sigma \in [0, 1]$ and $x = \sigma T(x)$. Then, $\|x\| < r$. Then for each $\sigma \in [0, 1]$, σT admits a fixed point in $B_r(0)$.*

Proof.   Let P be the radial projection of E into $\overline{B_r(0)}$:

$$P(x) = \begin{cases} x & \text{if } \|x\| \leq r, \\ r\dfrac{x}{\|x\|} & \text{if } \|x\| \geq r. \end{cases}$$

Consider the map $PT_\sigma : \overline{B_r(0)} \to \overline{B_r(0)}$, where $T_\sigma = \sigma T$. Then PT_σ is a compact operator. By Theorem 9.86, there is $x \in \overline{B_r(0)}$ such that

$$x = PT_\sigma(x).$$

Suppose $\|T_\sigma(x)\| \geq r$, then $\frac{r}{\|T_\sigma(x)\|} \leq 1$. Now,

$$x = P(T_\sigma(x)) = \frac{r}{\|T_\sigma(x)\|} T_\sigma(x),$$

where $\frac{r}{\|T_\sigma(x)\|} \in [0, 1]$. By the hypothesis $\|x\| < r$, but $\|x\| = r\frac{\|T_\sigma(x)\|}{\|T_\sigma(x)\|} = r$, a contradiction. Therefore, $\|T_\sigma(x)\| < r : x = PT_\sigma(x) = T_\sigma(x)$, and

$$\|x\| = \|T_\sigma(x)\| < r,$$

so $x \in B_r(0)$.   ∎

Theorem 9.89 *Assume that there is $T \in Q(\overline{B_r(0)}, E)$ satisfying $T(\partial B_r(0)) \subset B_r(0)$. Then T admits a fixed point in $B_r(0)$.*

Proof. Let P be the radial projection of E onto $\overline{B_r(0)}$. Then, $TP \in Q(E, E)$. Let $x \in E$, $\sigma \in [0,1]$, $x = \sigma T(P(x))$. We claim that $\|x\| < r$. In fact, if $\|x\| \geq r$ then $P(x) \in \partial B_r(0)$, and consequently $T(P(x)) \in B_r(0)$; so $r \leq \|x\| \leq \sigma \|T(P(x))\| < r$, a contradiction. Thus, $\|x\| < r$. By Theorem 9.88, for some $x \in B_r(0)$,

$$x = T(P(x)) = Tx. \qquad \blacksquare$$

Theorem 9.90 (*Leray-Schauder Fixed Point Theorem*) *Let $T \in Q(E \times [0,1], E)$, $T(x,0) = 0$ for $x \in E$. Suppose there is $r > 0$ such that if $x \in E$, $\sigma \in [0,1]$, $x = T(u,\sigma)$, then $\|x\| < r$. Then for $\sigma \in [0,1]$, $T(\cdot, \sigma)$ admits a fixed point in $B_r(0)$.*

Proof. Because $T \in Q(\overline{B_r(0)} \times [0,1], E)$., let $H(x,\sigma) = x - T(x,\sigma)$. Then H is a compact perturbation of the identity. Moreover, by the hypothesis

$$H(x,\sigma) = x - T(x,\sigma) \neq 0 \quad \text{for } x \in \partial B_r(0), \ \sigma \in [0,1],$$

so the degree $d(H(\cdot, t), B_r(0), 0)$ is defined and constant for $\sigma \in [0,1]$. We have

$$\mathrm{d}(I - T(\cdot, \sigma), B_r(0), 0) = \mathrm{d}(I, B_r(0), 0) = 1.$$

Thus, $T(\cdot, \sigma)$ admits a fixed point in $B_r(0)$. $\qquad \blacksquare$

Notes The materials in this chapter were adapted from Rabinowitz [37]. For related materials, see Berger [6], Browder [11], Deimling [17], Dugundji [19], Krasnoselskii [28], Nirenberg [36], Rabinowitz [37], Schwartz [44], and Temme [49].

Chapter 10

TOPOLOGICAL DEGREE APPLICATIONS

In this chapter we present various applications of degree theory.

10.1 Regularity Theorems and Maximum Principles

First we introduce the Hölder spaces $C^{k,\alpha}(\Omega)$. Let Ω be a bounded domain in $\mathbb{R}^N$, $k = 0, 1, 2, \cdots$, and $\alpha \in (0, 1]$. Define the seminorm,

$$[u]_{k,\alpha} = \sum_{\substack{x,\, y \in \Omega \\ |\sigma| = k \\ x \neq y}} \frac{|D^\sigma u(x) - D^\sigma u(y)|}{|x - y|^\alpha}.$$

The Hölder space $C^{k,\alpha}(\Omega)$ is the Banach space

$$C^{k,\alpha}(\Omega) = \left\{ u \in C^k(\Omega) \mid [u]_{k,\alpha} < \infty \right\},$$

under the norm

$$\|u\|_{k,\alpha} = \sum_{|\alpha| \le k} \|D^\alpha u\|_{L^\infty} + [u]_{k,\alpha}.$$

Denote by $C^{0,\alpha}(\Omega) = C^\alpha(\Omega)$. Consider the second-order linear operator L :

$$Lu = \sum_{i,j} a_{ij}(x) \frac{\partial^2 u}{\partial x_i \, \partial x_j} + \sum_i b_i(x) \frac{\partial u}{\partial x_i} + c(x)u,$$

where $a_{ij}(x) = a_{ji}(x)$ for $x \in \bar{\Omega}$. L is elliptic at $x \in \Omega$ if the coefficient matrix $(a_{ij}(x))$ is positive-definite; the minimum eigenvalue $\lambda(x)$, and the maximum eigenvalue $\Lambda(x)$ of $(a_{ij}(x))$ satisfy

$$0 < \lambda(x) |\xi|^2 \le \sum_{i,j} a_{ij}(x)\xi_i \, \xi_j \le \Lambda(x) |\xi|^2,$$

for each $\xi = (\xi_1, \cdots, \xi_N)$ in $\mathbb{R}^N \backslash \{0\}$. If $\lambda(x) > 0$ for $x \in \Omega$, then L is called elliptic in Ω. If $\lambda(x) \ge \alpha > 0$ for some constant α, then L is called strictly elliptic in Ω. If $\dfrac{\Lambda(x)}{\lambda(x)}$ is bounded in Ω, L is called uniformly elliptic in Ω.

For the proofs of theorems in this section, see Gilbarg-Trudinger [24].

Theorem 10.1 *(Schauder Estimation) Assume $\alpha \in (0,1)$, Ω is a bounded domain in $\mathbb{R}^N$ of $C^{2,\alpha}$, and L is strictly elliptic in Ω. Suppose $u \in C^{2,\alpha}(\overline{\Omega})$, a_{ij}, b_i, $c \in C^\alpha(\Omega)$, satisfies*

$$\begin{cases} Lu = f & \text{in} \quad \Omega \\ u = \varphi & \text{on} \quad \partial\Omega, \end{cases}$$

Then $f \in C^\alpha(\Omega)$, $\varphi \in C^{2,\alpha}(\partial\Omega)$, satisfies the Schauder estimation

$$\|u\|_{2,\alpha} \leq d(\|f\|_\alpha + \|\varphi\|'_{2,\alpha}),$$

where d is a positive constant.

Theorem 10.2 *(Schauder Estimation) Assume $\alpha \in (0,1)$, Ω is a bounded domain in $\mathbb{R}^N$ of $C^{2,\alpha}$, and L is strictly elliptic in Ω, with $c(x) \leq 0$ for $x \in \bar{\Omega}$. Suppose a_{ij}, b_i, $c \in C^\alpha(\Omega)$, $f \in C^\alpha(\Omega)$, and $\varphi \in C^{2,\alpha}(\partial\Omega)$. Then there is a unique $u \in C^{2,\alpha}(\bar{\Omega})$ satisfying*

$$\begin{cases} Lu = f & \text{in} \quad \Omega \\ u = \varphi & \text{on} \quad \partial\Omega, \end{cases}$$

and

$$\|u\|_{2,\alpha} \leq d(\|f\|_\alpha + \|\varphi\|'_{2,\alpha}),$$

where d is a positive constant.

Theorem 10.3 *(L^p-Estimation) Let Ω be a $C^{1,1}$ domain in $\mathbb{R}^N$, and suppose the operator L is elliptic satisfying $a_{ij} \in C(\bar{\Omega})$, b_i, $c \in L^\infty$, $f \in L^p(\Omega)$, $|a_{ij}|, |b_i|, |c| \leq \Lambda$, where $i, j = 1, ..., N$. Then if $u \in W^{2,p}(\Omega) \cap W_0^{1,p}(\Omega)$, $1 < p < \infty$, we have*

$$\|u\|_{W^{2,p}(\Omega)} \leq d \|Lu - \sigma u\|_{L^p(\Omega)},$$

for all $\sigma \geq \sigma_0$, where d and σ_0 are positive constants depending only on N, p, λ, Λ, Ω and the moduli of continuity of the coefficients a_{ij}.

Theorem 10.4 *(Agmon-Douglis-Nirenberg Theorem) Let Ω be a $C^{1,1}$ domain in $\mathbb{R}^N$, and let the operator L be strictly elliptic in Ω with coefficients $a_{ij} \in C(\bar{\Omega})$, $b_i, c \in L^\infty$, with $i, j = 1, \cdots, N$ and $c \leq 0$. Then , if $f \in L^p(\Omega)$ and $\varphi \in W^{2,p}(\Omega)$, with $1 < p < \infty$, the Dirichlet problem $Lu = f$ in Ω, $u - \varphi \in W_0^{1,p}(\Omega)$ has a unique solution $u \in W^{2,p}(\Omega)$.*

Definition 10.5 *An a priori estimate is an estimate, in terms of given data, valid for all possible solutions of a class of problems even if the hypotheses do not guarantee the existence of such solutions.*

Proposition 10.6 *If $A = (a_{ij})$ is positive-definite and $B = (b_{ij})$ is nonpositive-definite, then $(a_{ij}b_{ij})$ is nonpositive-definite.*

Proof. Because A is positive-definite, there is $C = (c_{ij})$, satisfying $A = C^*C$: $a_{ij} = \sum_k c_{ki}c_{kj}$. Let $\xi = (\xi_1, \cdots, \xi_N)$ in $\mathbb{R}^N \backslash \{0\}$. Then

$$
\begin{aligned}
\sum_{i,j} a_{ij}b_{ij}\xi_i\xi_j &= \sum_{i,j}\sum_k c_{ki}c_{kj}b_{ij}\xi_i\xi_j \\
&= \sum_k \sum_{i,j} b_{ij}(c_{ki}\xi_i)(c_{kj}\xi_j) \qquad (10.1) \\
&\leq 0 \quad \text{because } (b_{ij}) \text{ is nonpositive-definite.}
\end{aligned}
$$

We conclude that $(a_{ij}b_{ij})$ is nonpositive-definite. In particular, in (10.1), set $\xi_i = 1$ for $i = 1, 2, \cdots, N$, then

$$
\sum_{i,j} a_{ij}\, b_{ij} \leq 0. \qquad \blacksquare
$$

Let us recall the Maximum Principles.

Theorem 10.7 (*Maximum Principle*) *Let Ω be a bounded domain in $\mathbb{R}^N$, and L elliptic in Ω. Suppose $u \in C^2(\Omega) \cap C^0(\bar{\Omega})$,, then:*

(*i*) *If $Lu \geq 0$, $c \leq 0$, then $\sup_\Omega u \leq \sup_{\partial\Omega} u^+$;*

(*ii*) *If $Lu \geq 0$, $c = 0$, then $\sup_\Omega u = \sup_{\partial\Omega} u$;*

(*iii*) *If $Lu = 0$, $c \leq 0$, then $\sup_\Omega |u| = \sup_{\partial\Omega} |u|$;*

(*iv*) *If $Lu > 0$, $c = 0$, then we have the strong maximum principle: u cannot achieve an interior maximum in $\bar{\Omega}$.*

Proof. We prove (*iv*) only. Suppose $c = 0$, and

$$
Lu = \sum_{i,j} a_{ij}\, D_{ij}\, u(x) + \sum b_i\, D_i u(x) > 0 \quad \text{for } x \in \Omega.
$$

If u achieves its maximum at $x_0 \in \Omega$, then

$$
D_i\, u(x_0) = 0 \quad \text{for } i = 1, \cdots, N, \quad (D_{ij}u(x_0)) \text{ is nonpositive-definite.}
$$

By the hypothesis, L is elliptic, and (a_{ij}) is positive-definite. By Proposition 10.6, $\sum_{i,j} a_{ij}D_{ij}u(x_0) \leq 0$, and consequently

$$
Lu(x_0) = \sum_{i,j} a_{ij}D_{ij}u(x_0) \leq 0,
$$

which contradicts $Lu > 0$ in Ω. $\blacksquare$

Theorem 10.8 (*Maximum Principle*) *Let Ω be a bounded domain in $\mathbb{R}^N$, and L elliptic in Ω, and $c \leq 0$. Suppose $u \in C^2(\Omega) \cap C^0(\bar{\Omega})$, then:*

(*i*) *If $Lu \geq f$, then $\sup\limits_{\Omega} u \leq \sup\limits_{\partial\Omega} u^+ + \alpha \sup\limits_{\Omega} \frac{|f^-|}{\lambda}$;*

(*ii*) *If $Lu = f$, then $\sup\limits_{\Omega} |u| \leq \sup\limits_{\partial\Omega} |u| + \alpha \sup\limits_{\Omega} \frac{|f|}{\lambda}$, where α depends only on diam Ω, and $\beta = \sup\limits_{x \in \Omega} \frac{|(b_1(x), \cdots, b_N(x))|}{\lambda(x)}$.*

Theorem 10.9 (*Hopf Maximum Principle*) *Let Ω be a domain (not necessarily bounded) in $\mathbb{R}^N$, L uniformly elliptic in Ω, $c = 0$, and $Lu \geq 0$ in Ω. Let $x_0 \in \partial\Omega$ be such that:*

(*i*) *u is continuous at x_0;*

(*ii*) *$u(x_0) > u(x)$ for all $x \in \Omega$;*

(*iii*) *$\partial\Omega$ satisfies an interior sphere condition at x_0.*

Then the outer normal derivative of u at x_0, if it exists, satisfies the strict inequality

$$\frac{\partial u}{\partial n}(x_0) > 0.$$

If $c \leq 0$ and $\frac{c}{\lambda}$ is bounded, the same conclusion holds, provided $u(x_0) \geq 0$; and if $u(x_0) = 0$, the same conclusion holds irrespective of the sign of c.

Theorem 10.10 (*Strong Maximum Principle*) *Let Ω be a domain (not necessarily bounded) in $\mathbb{R}^N$, L uniformly elliptic in Ω, $c = 0$ and $Lu \geq 0$ in Ω. Then:*

(*i*) *If u achieves its maximum in the interior of Ω, then u is a constant;*

(*ii*) *If $c \leq 0$ and $\frac{c}{\lambda}$ is bounded, then u cannot achieve a nonnegative maximum in the interior of Ω unless it is a constant.*

10.2   Elliptic Quasilinear Equations

Consider the quasilinear elliptic equation (Q) :

$$\begin{cases} \sum\limits_{i,j} A_{ij}(x, u, \nabla u) \frac{\partial^2 u}{\partial x_i \partial x_j} + \sum\limits_{i} B_i(x, u, \nabla u) \frac{\partial u}{\partial x_i} + C(x, u, \nabla u)u \\ \quad = F(x, u, \nabla u) \text{ in } \Omega, \\ u = \varphi \text{ on } \partial\Omega. \end{cases} \quad (Q)$$

The problem (Q) is (Q_1) of the problem (Q_σ) : for $\sigma \in [0, 1]$,

$$
\begin{cases}
\displaystyle\sum_{i,j} A_{ij}(x, \sigma u, \sigma \nabla u)\frac{\partial^2 u}{\partial x_i \partial x_j} + \sum_i B_i(x, \sigma u, \sigma \nabla u)\frac{\partial u}{\partial x_i} + C(x, \sigma u, \sigma \nabla u)u \\
\quad = \sigma F(x, u \nabla u) \text{ in } \Omega, \\
u = \sigma \varphi \text{ on } \partial\Omega,
\end{cases} \tag{Q_σ}
$$

where Ω is a domain of $C^{2,\alpha}$, $\varphi \in C^{2,\alpha}(\partial\Omega)$, A_{ij}, B_i, C, $F \in C^1(\bar{\Omega} \times \mathbb{R} \times \mathbb{R}^N)$, $C(x, \eta, p) \leq 0$ for $x \in \bar{\Omega}$, $\eta \in \mathbb{R}$, $p \in \mathbb{R}^N$, and some $\alpha > 0$ satisfies

$$
\sum_{i,j} A_{ij}(x, \eta, p)\xi_i\xi_j \geq \alpha |\xi|^2 \quad \text{for } x \in \bar{\Omega}, \ \eta \in \mathbb{R}, \ p, \ \xi \in \mathbb{R}^N.
$$

Let $E = C^{1,\alpha}(\Omega)$ fix u in E, and consider the linear elliptic equation

$$
\begin{cases}
\displaystyle\sum_{i,j} A_{ij}(x, u, \nabla u)\frac{\partial^2 v}{\partial x_i \partial x_j} + \sum_i B_i(x, u, \nabla u)\frac{\partial v}{\partial x_i} + C(x, u, \nabla u)v \\
\quad = F(x, u, \nabla u) \text{ in } \Omega, \\
v = \varphi \text{ on } \partial\Omega.
\end{cases}
$$

Lemma 10.11 *Let Ω be a bounded convex domain in $\mathbb{R}^N$ of $C^{0,1}$: Ω a Lipschitz domain, and $f \in C^1(\bar{\Omega})$. Then $f \in C^{0,1}(\bar{\Omega})$: f is Lipschitz-continuous on $\bar{\Omega}$, and satisfies*

$$
\|f\|_{0,1} \leq \|f\|_{C^1}.
$$

Proof. By the Fundamental Theorem of Calculus,

$$
\begin{aligned}
f(y) - f(x) &= \int_0^1 \nabla f(x + t(y - x) \cdot (y - x)dt \\
&\leq \|y - x\| \, \|f\|_{C^1} \, ;
\end{aligned}
$$

so if $x \neq y$, then

$$
\|f\|_{0,1} = \sup_{\substack{x, y \in \bar{\Omega} \\ x \neq y}} \frac{|f(y) - f(x)|}{\|y - x\|} \leq \|f\|_{C^1} \, .
$$

Let Ω be a bounded domain in $\mathbb{R}^N$ of $C^{2,\alpha}$. Given $u \in E$, by the Schauder Theorem 10.2, there is an unique solution v in $C^{2,\alpha}(\Omega)$ of (L_u). Write $v = Tu$. We intend to find a fixed point of $T : Tu = u$. ∎

Lemma 10.12 *The injection $i : C^{2,\alpha}(\Omega) \to C^{1,\alpha}(\Omega)$ is compact.*

Proof. For $b > 0$, let $B = \{u \in C^{2,\alpha}(\Omega) \mid \|u\|_{2,\alpha} \leq b\}$. Assume $u \in B$, then

$$|D^\sigma u(y) - D^\sigma u(x)| \leq b\,|x - y|^\alpha \quad \text{for } x,\, y \in \Omega, \sigma = 2.$$

Thus, B is bounded and uniformly equicontinuous in C^2, and consequently it is relatively compact in $C^2(\Omega)$. Letting $\{u_n\}$ be a sequence in B, there is a subsequence of $\{u_n\}$, still denoted by $\{u_n\}$, and $u \in C^2(\Omega)$ such that $\|u_n - u\|_{C^2} \to 0$ as $n \to \infty$. We have $u \in E$, and $\|u_n - u\|_E \to 0$ as $n \to \infty$. ∎

Lemma 10.13 *The operator $T : E \to E$ is compact.*

Proof. For $b > 0$, let $B = \{u \in E \mid \|u\|_{1,\alpha} \leq b\}$. By the Schauder Theorem 10.2, for $u \in B$,

$$\|Tu\|_{2,\alpha} \leq d(\|F(\cdot, u(\cdot), \nabla u(\cdot)\|_\alpha + \|\varphi\|_{2,\alpha}').$$

Because F is Lipschitz on $\bar\Omega \times \mathbb{R} \times \mathbb{R}^N$, it follows by Lemma 10.12,

$$\|F(x, u(x), \nabla u(x))\|_\alpha = \sup_{\substack{x,y \in \Omega \\ x \neq y}} \frac{|F(x, u(x), \nabla u(x)) - F(y, u(y), \nabla u(y))|}{|x - y|^\alpha}$$

$$\leq c_1 \sup_{\substack{x,y \in \Omega \\ x \neq y}} \frac{|x - y| + |u(x) - u(y)| + |\nabla u(x) - \nabla u(y)|}{|x - y|^\alpha}$$

$$\leq c_2.$$

Therefore,

$$\|Tu\|_{2,\alpha} \leq c \quad \text{for } u \in B. \tag{10.2}$$

Apply (10.2) and Lemma 10.12 to the map $i \circ T$, where $E \xrightarrow{T} C^{2,\alpha}(\Omega) \xrightarrow{i} E$ to show that T is compact. ∎

Lemma 10.14 $T : E \to E$ *is continuous.*

Proof. Let $\{u_n\}$ be a sequence in E, $u \in E$ such that $u_n \to u$ in E. Let $v_n = Tu_n$:

$$\begin{cases} \displaystyle\sum_{i,j} A_{ij}(x, u_n, \nabla u_n)\frac{\partial^2 v_n}{\partial x_i \partial x_j} + \sum_i B_i(x, u_n, \nabla u_n)\frac{\partial v_n}{\partial x_i} \\ \quad + C(x, u_n, \nabla u_n)v_n = F(x, u_n, \nabla u_n) \text{ in } \Omega, \\ v_n = \varphi \text{ on } \partial\Omega. \end{cases} \tag{10.3}$$

Because $\{u_n\}$ is bounded in E, $\{v_n\}$ is bounded in $C^{2,\alpha}(\Omega)$, and consequently $\{v_n\}$ is relatively compact in $C^2(\Omega)$. There is a subsequence, still denoted by $\{v_n\}$, $v \in C^2(\Omega)$, such that $v_n \to v$ in $C^2(\Omega)$. Letting $n \to \infty$ in (10.3),

$$
\begin{cases}
\displaystyle\sum_{i,j} A_{ij}(x, u, \nabla u)\frac{\partial^2 v}{\partial x_i \partial x_j} + \sum_i B_i(x, u, \nabla u)\frac{\partial v}{\partial x_i} + C(x, u, \nabla u)v \\
\quad = F(x, u, \nabla u) \text{ in } \Omega, \\
v = \varphi \text{ on } \partial\Omega.
\end{cases}
$$

By uniqueness, $v = T(u)$. Note that $v_n \to v$ in $C^2(\Omega)$ implies that $v_n \to v$ in $C^{1,\alpha}(\Omega)$. ∎

Theorem 10.15 *Suppose there is $r > 0$ such that for each $\sigma \in [0,1]$, and for each solution u in $C^{1,\alpha}(\Omega)$ of the problem (Q_σ), we have $\|u\|_{1,\alpha} < r$. Then for each $\sigma \in [0,1]$, there is a solution $u_\sigma \in C^{2,\alpha}(\Omega)$ of (Q_σ). In particular, in this case, the problem (Q) admits a solution.*

Proof. For $\sigma \in [0,1]$, $u \in E$, there is a solution v of problem $(Q_{\sigma,u})$:

$$
\begin{cases}
\displaystyle\sum_{i,j} A_{ij}(x, \sigma u, \nabla u)\frac{\partial^2 v}{\partial x_i \partial x_j} + \sum_i B_i(x, \sigma u, \sigma\nabla u)\frac{\partial v}{\partial x_i} \\
\quad + C(x, \sigma u, \sigma\nabla u)v = \sigma F(x, u, \nabla u) \text{ in } \Omega, \\
v = \sigma\varphi \text{ on } \partial\Omega.
\end{cases}
$$

Let $v = T(\sigma, u)$. By Lemmas 10.13-10.14, $T : [0,1] \times E \to E$ is a continuous compact operator. The theorem follows from the Leray-Schauder Fixed Point Theorem. ∎

In practice, we can break the a priori estimates of Theorem 10.15 into four stages:

(i) Estimation of $\sup_{\Omega} |u|$;

(ii) Estimation of $\sup_{\partial\Omega} |\nabla u|$ in terms of $\sup_{\Omega} |u|$;

(iii) Estimation of $\sup_{\Omega} |\nabla u|$ in terms of $\sup_{\partial\Omega} |\nabla u|$, and $\sup_{\Omega} |u|$;

(iv) Estimation of $[\nabla u]_{\beta,\Omega}$ for some $\beta > 0$, in terms of $\sup_{\Omega} |\nabla u|$, and $\sup_{\Omega} |u|$. For example, assume $\alpha \in (0,1)$, Ω is a bounded domain in $\mathbb{R}^N$ of $C^{2,\alpha}$, and $\varphi \in C^{2,\alpha}(\bar{\Omega})$.

Definition 10.16 *The boundary manifold $(\partial\Omega, \varphi)$:*

$$
\Gamma = (\partial\Omega, \varphi) = \{(x, z) \in \partial\Omega \times \mathbb{R} \mid z = \varphi(x)\}
$$

satisfies a bounded slope condition for $p \in \Gamma$ if there are two planes in $\mathbb{R}^{N+1}$, $z = \pi_p^+(x)$ and $z = \pi_p^-(x)$, passing through p such that:

(i) $\pi_p^+(x) \leq \varphi(x) \leq \pi_p^-(x)$ for $x \in \partial\Omega$;

(ii) The slopes of the planes are uniformly bounded, independent of p, by a constant d :

$$\left|D\pi_p^\pm\right| \leq d \quad \text{for } p \in \Gamma.$$

Lemma 10.17 Let $\partial\Omega \in C^2$, $\varphi \in C^2(\bar{\Omega})$. If $\partial\Omega$ is uniformly convex such that its principal curvatures are bounded away from 0, then $\Gamma = (\partial\Omega, \varphi)$ satisfies a bounded slope condition.

Consider the quasilinear elliptic equation

$$\begin{cases} Q_u = \displaystyle\sum_{i,j} A_{ij}(x, u, \nabla u)\frac{\partial^2 u}{\partial x_i \partial x_j} = 0 & \text{in } \Omega \\ u = \varphi & \text{on } \partial\Omega. \end{cases} \tag{10.4}$$

Theorem 10.18 If $A_{ij} \in C^\alpha(\bar{\Omega} \times \mathbb{R} \times \mathbb{R}^N)$, $\varphi \in C^{2,\alpha}(\bar{\Omega})$, and $\Gamma = (\partial\Omega, \varphi)$ satisfies a bounded slope condition, then the Dirichlet problem of (10.4) is solvable in $C^{2,\alpha}(\bar{\Omega})$.

Next we consider the following semilinear elliptic problem.

Theorem 10.19 Let Ω be a bounded domain in $\mathbb{R}^N$ of $C^{2,\alpha}$, and $f \in C^1(\bar{\Omega} \times \mathbb{R})$ a bounded function on $\bar{\Omega} \times \mathbb{R}$. Then the equation

$$\begin{cases} \triangle u &= f(x, u) & \text{in } \Omega, \\ u &= 0 & \text{on } \partial\Omega, \end{cases}$$

admits a solution.

Proof. For $\sigma \in [0, 1]$, consider the equation (S_σ)

$$\begin{cases} \triangle u &= \sigma f(x, u) & \text{in } \Omega, \\ u &= 0 & \text{on } \partial\Omega. \end{cases} \tag{S_σ}$$

Suppose $u \in C^{2,\alpha}(\Omega)$ is a solution of (S_σ). By the a priori estimates,

$$\|u\|_{L^\infty} \leq c\|\sigma f(\cdot, u(\cdot))\|_{L^\infty} \leq M,$$

and it follows from the Schauder Theorem 10.2 that

$$\|u\|_{2,\alpha} \leq c\|f(x, u(x))\|_\alpha. \tag{10.5}$$

Because f is C^1 on $\bar{\Omega} \times [-M, M]$, f is Lipschitz continuous on $\bar{\Omega} \times [-M, M]$. Now, for x, $y \in \Omega$, $x \neq y$,

$$
\begin{aligned}
& \frac{|f(x, u(x)) - f(y, u(y))|}{|x - y|^\alpha} \\
& \leq \frac{|f(x, u(x)) - f(x, u(y))| + |f(x, u(y)) - f(y, u(y))|}{|x - y|^\alpha} \\
& \leq d \left(\frac{|u(x) - u(y)|}{|x - y|^\alpha} + \frac{|x - y|}{|x - y|^\alpha} \right) \qquad (10.6) \\
& \leq d \left(\frac{|u(x) - u(y)|^2}{|x - y|^\alpha} |u(x) - u(y)|^{1-\alpha} + |x - y|^{1-\alpha} \right) \\
& \leq d(2\|u\|_1^\alpha \|u\|_0^{1-\alpha} + \delta^{1-\alpha}) \quad \text{where } \delta = \operatorname{diam}\Omega.
\end{aligned}
$$

Let $a = \|u\|_1$, $b = \|u\|_0$, and apply the Young inequality to obtain, for $\epsilon > 0$,

$$
\begin{aligned}
a^\alpha b^{1-\alpha} & \leq \alpha \left[(\epsilon a)^\alpha \right]^{1/\alpha} + (1 - \alpha) \left[\epsilon^{-\alpha} b^{1-\alpha} \right]^{1/(1-\alpha)} \\
& \leq \alpha \epsilon a + (1 - \alpha) \epsilon^{-\alpha/(1-\alpha)} b.
\end{aligned}
$$

Therefore, we can write (10.6) as

$$
\|f(\cdot, u(\cdot))\|_\alpha \leq c\epsilon \|u\|_{2,\alpha} + c(\epsilon).
$$

From (10.5), choose ϵ small such that $c\epsilon < \frac{1}{2}$ to obtain

$$
\|u\|_{2,\alpha} \leq c.
$$

Then the theorem follows from the proof of Theorem 10.15, and the Leray-Schauder Fixed Point Theorem. ∎

Remark 10.20

(i) *We may replace $\triangle$ in Theorem 10.19 by*

$$
L = \sum a_{ij}(x) \frac{\partial^2}{\partial x_i \partial x_j} + \sum b_i(x) \frac{\partial}{\partial x_i} + c(x),
$$

where L is uniformly elliptic in Ω, $c \leq 0$ in $\bar{\Omega}$.

(ii) *The hypothesis that f is bounded on $\bar{\Omega} \times \mathbb{R}$ is essential. For example, let $\Omega = (0, 2\pi)$, $f(x, u) = -u + x$, then the equation*

$$
\begin{cases}
u'' & = -u + x, \\
u(0) & = u(2\pi) = 0,
\end{cases} \qquad (10.7)
$$

does not admit any solution.

Proof.

(ii) Suppose u is a solution of (10.7), then

$$(u - x)'' = -(u - x),$$

so

$$u - x = c_1 \cos x + c_2 \sin x \qquad x \in [0, 2\pi].$$

Thus,

$$u(0) = c_1 = 0,$$
$$u(2\pi) = 2\pi + c_1 = 2\pi \neq 0,$$

a contradiction.   ■

10.3   Connectedness of Solution Sets

Let E be a real Banach space, $\Omega \subset E$ a bounded open subspace, $T : \bar{\Omega} \to E$ a compact operator, and $f(u) = u - T(u)$, for $u \in \bar{\Omega}$. As before, let $Q(\bar{\Omega}, E)$ denote the family of all compact operators from $\bar{\Omega}$ into E.

Definition 10.21 T satisfies the condition (P) if for $\epsilon > 0$, there is $T_\epsilon \in Q(\Omega, E)$ such that $\|Tu - T_\epsilon u\| < \epsilon$ for $u \in \bar{\Omega}$, and $u = T_\epsilon u + b$ admits at most one solution with $\|b\| < \epsilon$.

Definition 10.22 Let E and F be Banach spaces, and $f \in C(E, F)$. f is called proper if, for each compact set K in F, we have that $f^{-1}(K)$ is compact in E.

Theorem 10.23 Let $f \in C(E, F)$. Then the following are equivalent:

(i) f is proper;

(ii) f is a closed map, and the solution set

$$S_p = \{x \in E \mid f(x) = p\},$$

   is compact for each $p \in F$;

(iii) If E, F are finite-dimensional, then f is coercive:

$$\|f(x)\| \to \infty \quad \text{as} \quad \|x\| \to \infty.$$

Proof. $(i) \Rightarrow (ii)$. Because $\{p\}$ is compact in F, and S_p is compact in E :, let K be closed in E, $\{y_n\}$ a sequence in $f(K) : y_n = f(x_n)$, for $x_n \in K$, and $y_n \to y$ in F. Because $\overline{\{y_n\}}$ is compact, $\Lambda = f^{-1}(\overline{\{y_n\}})$ is compact in E. Now, in $\{x_n\} \subset \Lambda$, there is a subsequence, still denoted by $\{x_n\}$, $x \in E$ such that $x_n \to x$. Because K is closed, we have $x \in K$. Because f is continuous, and $x_n \to x$, we have $f(x_n) \to f(x)$. Thus, $y = f(x) \in f(K)$.

$(ii) \Rightarrow (i)$. Let C be compact in F, and $K = f^{-1}(C)$. Suppose K is covered with closed sets $\{K_\lambda\}_{\lambda \in \Lambda}$ that have the Finite Intersection Property. Let $\beta = \{\lambda_1, \cdots, \lambda_k\}$ be a subset of Λ, and $G_\beta = \bigcap_{i=1}^{k} K_{\lambda_i}$. Then G_β is closed and nonempty. Then $f(G_\beta)$ have the Finite Intersection Property because for each finite subset γ of 2^Λ,

$$\bigcap_{\beta \in \gamma} f(G_\beta) \supset f(\bigcap_{\beta \in \gamma} G_\beta) \neq \emptyset.$$

By the compactness of C, $\delta = \bigcap_{\beta \in 2^\Lambda} f(G_\beta) \neq \emptyset$. For $y \in \delta$, by the hypothesis, $f^{-1}(y)$ is compact. For $\gamma = \{\lambda_1, \cdots, \lambda_j\}$ and $y \in \bigcap_{\beta \in 2^\alpha} f(G_\beta)$, then

$$\bigcap_{i=1}^{j} \{K_{\lambda_i} \cap f^{-1}(y)\} = G_\gamma \cap f^{-1}(y) \neq \emptyset,$$

and consequently,

$$\bigcap_{\lambda \in \Lambda} \{K_\lambda \cap f^{-1}(y)\} \neq \emptyset.$$

Now,

$$\bigcap_{\lambda \in \Lambda} K_\lambda \supset \bigcap_{\lambda \in \Lambda} \{K_\lambda \cap f^{-1}(y)\} \neq \emptyset.$$

Therefore, K is compact.

$(iii) \Rightarrow (i)$. Let E, F be finite-dimensional. Suppose f is coercive, C is compact in F, and $K = f^{-1}(C)$. Then $K = f^{-1}(C)$ is closed and bounded in E, so K is compact in E.

$(i) \Rightarrow (iii)$ Let E, F be finite-dimensional. Suppose C is bounded in F, and $K = f^{-1}(C)$. Note that $f^{-1}(\bar{C})$ is compact, and $K \subset f^{-1}(\bar{C})$, so K is bounded in E. ∎

Theorem 10.24 *Let E and F be real Banach spaces, $f \in C(E, F)$, $\|f(x)\| \to \infty$ as $\|x\| \to \infty$, and let f be a compact perturbation of a proper map: $f = g + h$ where $g \in C(E, F)$, a proper map and $h \in Q(E, F)$. Then f is a proper map.*

Proof. Let $\{x_n\}$ be a sequence in E, and $y_n = f(x_n) \to y$ in F. Then $\{x_n\}$ is bounded because f is coercive, and consequently $\{h(x_n)\}$ contains a convergent subsequence, still denoted by $\{h(x_n)\}$. Let $h(x_n) \to z$, then

$$g(x_n) = f(x_n) - h(x_n) \to y - z.$$

Because g is proper, $\{x_n\}$ contains a convergent subsequence, still denoted by $\{x_n\}$: $x_n \to x$. Because f is continuous, $f(x) = y$, and it follows at once that f is proper. ∎

Theorem 10.25 *Let E be a real Banach space, Ω a bounded open set in E, $T \in Q(\bar{\Omega}, E)$, and $f = I - T$. Then*

(i) f *is a proper map;*

(ii) f *is a closed map.*

Proof. See Lemma 7.13.   ∎

Theorem 10.26 *(Krasnosellski-Perov Theorem) Let E be a real Banach space, Ω a bounded open set in E, let $T \in Q(\bar{\Omega}, E)$ satisfy condition (P) in Definition 10.21, $f = I - T$, $0 \notin f(\partial\Omega)$, and $d(f, \Omega, 0) \neq 0$. Then the set $N = \{x \in \bar{\Omega} \mid f(x) = 0\}$ is connected.*

Proof. N is nonempty because $d(f, \Omega, 0) \neq 0$. Moreover, by Theorem 10.23, N is compact. Suppose N is disconnected: there are open sets V, W in Ω such that $N \cap V \neq \emptyset$, $N \cap W \neq \emptyset$, $N \subset V \cup W$, $V \cap W = \emptyset$. By Corollary 9.18 (v), and Theorem 9.16 (iii),

$$\mathrm{d}(f, \Omega, 0) = \mathrm{d}(f, V, 0) + \mathrm{d}(f, W, 0).$$

Suppose $v \in V \cap N$: $f(v) = 0$ and $v = Tv$. Apply the condition that T satisfies condition (P) to set

$$g_\epsilon(u) = u - T_\epsilon u - (v - T_\epsilon v),$$
$$H(t, u) = t\, g_\epsilon(u) + (1 - t)f(u) \quad \text{for } t \in [0, 1], \ u \in \bar{\Omega}.$$

Then,

$$\begin{aligned}
H(t, u) &= f(u) + t\, g_\epsilon(u) - t\, f(u) \\
&= f(u) + t[u - T_\epsilon u - (v - T_\epsilon v) - (u - Tu)] \\
&= f(u) + t(Tu - T_\epsilon u) - t(v - T_\epsilon v).
\end{aligned}$$

Thus,

$$\begin{aligned}
\|H(t, u)\| &\geq \|f(u)\| - t\, \|(T\, u - T_\epsilon u) - (v - T_\epsilon v)\| \\
&\geq \|f(u)\| - t\, \|Tu - T_\epsilon u\| - \|v - T_\epsilon v\|.
\end{aligned}$$

Because f is closed, and $0 \notin f(\partial W)$, we have

$$\alpha = \inf_{u \in \partial W} \|f(u)\| > 0.$$

Now,

$$\|T\,u - T_\epsilon u\| < \epsilon \quad \text{for } u \in \bar{\Omega},$$
$$\|v - T_\epsilon v\| = \|T\,v - T_\epsilon v\| < \epsilon.$$

If we set $\epsilon < \frac{\alpha}{4}$, we obtain, for $u \in \partial W$,,

$$\|H(t, u)\| \geq \alpha - 2\epsilon \geq \frac{\alpha}{2} > 0.$$

Therefore, $d(H(t, 0), W, 0)$ is well-defined and constant for $t \in [0, 1]$:

$$\mathrm{d}(f, W, 0) = \mathrm{d}(g_\epsilon, W, 0).$$

Now,

$$g_\epsilon(u) = (u - T_\epsilon u) - (v - T_\epsilon v) = u - T_\epsilon u - b,$$

where $b = v - T_\epsilon v$, and consequently

$$\|b\| = \|v - T_\epsilon v\| < \epsilon.$$

By condition (P), $g_\epsilon(u) = 0$ admits at most one solution. However, $g_\epsilon(v) = 0$, so $g_\epsilon(u) \neq 0$ for $u \in W$. That $0 \notin g_\epsilon(\bar{W})$ implies

$$\mathrm{d}(g_\epsilon, W, 0) = 0,$$

and it follows at once that

$$\mathrm{d}(f, W, 0) = 0.$$

Similarly, $d(f, V, 0) = 0$, and we conclude that $d(f, \Omega, 0) = 0$, a contradiction. ∎

Consider the initial value problem

$$\begin{cases} \dfrac{du}{dt} = f(t,\,u), \\[2mm] u(0) = 0, \end{cases} \tag{10.8}$$

where $B_c(0)$ is a ball in $\mathbb{R}^N$, and $f : [-a,\,a] \times \overline{B_c(0)} \to \mathbb{R}^N$ a continuous function. Set

$$M = \max_{\substack{|t| \leq a \\ |u| \leq c}} |f(t, u)|, \quad \text{and} \quad \alpha = \min(a, \frac{c}{M}).$$

It is known that

Theorem 10.27

(i) *If f is Lipschitz-continuous on u, then the equation (10.8) admits a unique local solution;*

(ii) *If f is only continuous on $J \times \Omega$, then the equation (10.8) admits a solution, but in general it is not unique.*

Example 10.28 *Note that $u \equiv 0$ and $u(t) = (\frac{t}{3})^3$ are the solution of the equation*

$$\begin{cases} u' = u^{2/3} \\ u(0) = 0. \end{cases}$$

Let N be the solution set of all solutions of equation (10.8), and for $|t| \leq \alpha$, let function $K_t = \{u(t) \mid u$ be a solution of equation (10.8)\}. Let E be the Banach space of all continuous functions $g : [-a, a] \times \mathbb{R}^N \to \mathbb{R}^N$ under the norm $\|g\| = \max\limits_{\substack{|t| \leq a \\ u \in \mathbb{R}^N}} |g(t, u)|$.

We intend to prove that N is connected in E, and K_t is a connected set in $\mathbb{R}^N$, for each t, $|t| \leq \alpha$.

Theorem 10.29 *(Kneser-Hukuhara Theorem) Let $f : [-a, a] \times \overline{B_c(0)} \to \mathbb{R}^N$ be a continuous function. Then the set N of all solutions u of equation (10.8) forms a connected set in E.*

Proof.   Let g be a continuous extension of f to $[-a, a] \times \mathbb{R}^N$, with $\|g\| = \|f\|$. Let $u \in E$, and define, for $|t| \leq \alpha$,

$$Tu(t) = \int_0^t g(s, u(s))ds.$$

Then,

$$\|Tu(t)\| = \| \int_0^t g(s, u(s))ds\| \leq Mt \leq M\alpha \leq c,$$

$$\|Tu(t_2) - Tu(t_1)\| = \| \int_{t_1}^{t_2} g(s, u(s))ds\| \leq M |t_2 - t_1|.$$

Thus, $T(E)$ is bounded and uniformly equicontinuous. Apply the Ascoli Theorem to show that $T : E \to E$ is compact. Let $f = I - T$. Then the solutions u of equation (10.8) are the fixed point u of T: $f(u) = 0$. Suppose $f(u) = 0$, then

$$\|u\| = \|Tu\| \leq M\alpha \leq c, \tag{10.9}$$

and it follows at once that $g(t, u) = f(t, u)$. Letting $\Omega = B_{c+1}(0)$, we require:

(i) T satisfies condition (P);

(ii) $0 \notin f(\partial \Omega)$;

(iii) $d(f, \Omega, 0) \neq 0$.

By (10.9), $f(u) \neq 0$ for $u \in \partial\Omega$, and then (ii) follows. Suppose $(I - \sigma T)u = 0$, $0 \leq \sigma \leq 1$, then $\|u\| = \|\sigma T u\| \leq \sigma M \alpha \leq c$. Therefore, $d(I - \sigma T, \Omega, 0)$ is defined and constant for $\sigma \in [0, 1]$:

$$d(f, \Omega, 0) = d(I, \Omega, 0) = 1, \text{ and } (iii) \text{ follows.}$$

Finally let us prove (i). For $\epsilon > 0$, take $f_\epsilon \in C^1([-\alpha, \alpha] \times \mathbb{R}^N, \mathbb{R}^N)$ such that

$$|f_\epsilon(t, z) - g(t, z)| \leq \frac{\epsilon}{\alpha} \quad \text{for } t \in [-\alpha, \alpha], \ |z| \leq c + 1.$$

Set

$$T_\epsilon u(t) = \int_0^t f_\epsilon(s, u(s)) ds.$$

Then

$$\|T_\epsilon u - T u\| < \epsilon \quad \text{for } \|u\| \leq c + 1.$$

Consider the equation $u = T_\epsilon u + b$:

$$u(t) = \int_0^t f_\epsilon(s, u(s)) ds + b. \tag{10.10}$$

Suppose u, v are solutions of (10.10). Then

$$|u(t) - v(t)| \leq K \int_0^t |u(s) - v(s)| \, ds,$$

where

$$K = \max_{\substack{|t| \leq \alpha \\ |z| c + 1}} |(f_\epsilon)_z(t, z)|.$$

Let $g(t) = \int_0^t |u(s) - v(s)| \, ds$, then

$$g'(t) \leq Kg(t), \quad \text{and } g(0) = 0.$$

By the Gronwall inequality,

$$(g(t)e^{-Kt})' = e^{-Kt}(g'(t) - Kg(t)) \leq 0,$$

so

$$0 \leq g(t)e^{-Kt} \leq g(0) = 0.$$

We obtain $u = v$, and conclude that T satisfies condition (P). Apply (i), (ii), (iii) and Theorem 10.26 to show that N is connected in E. ∎

Theorem 10.30 (*Kneser-Hukuhara Theorem*) *For each t, $|t| \leq \alpha$, the set $K_t = \{u(t) \mid u \text{ is a solution of equation } (10.8)\}$, and is connected in $\mathbb{R}^N$.*

Proof. Consider the continuous map

$$u \in E \to \delta_t(t) = u(t).$$

Then $K_t = \delta_t(N)$. By Theorem 10.29, K_t is connected in $\mathbb{R}^N$. ∎

10.4   Global Results for Eigenvalue Problems

Let E be a real Banach space, $T : \mathbb{R} \times E \to E$ a compact operator, and $T(0, u) = 0$ for $u \in E$. Consider the nonlinear eigenvalue problem:

$$u = T(\lambda, u). \tag{10.11}$$

A solution of (10.11) is a couple $(\lambda, u) \in \mathbb{R} \times E$ satisfying (10.11). Note that $(0, 0)$ is the unique solution of (10.11) in $\{0\} \times E$. There are two methods to prove the local existence of solutions of (10.11) near $\lambda = 0$.

(i) Suppose T is differentiable with respect to u at 0, and the Fréchet derivative $T_u(0, 0)$ of T is an isomorphism of E onto E. Apply the Implicit Function Theorem to obtain a unique solution curve of (10.11) $\{(\lambda, u(\lambda)) \mid |\lambda| \leq a\}$, for some $a > 0$.

(ii) Suppose $K : E \to E$ is a compact operator, and $T(\lambda, u) = \lambda K(u)$. Let $\|u\| \leq 1$, then

$$\|\lambda K(u)\| \leq |\lambda| \, \|K\| \, \|u\| \leq |\lambda| \, \|K\| \leq 1 \quad \text{if } |\lambda| \leq \frac{1}{\|K\|} = \lambda_0.$$

Therefore, if $|\lambda| \leq \lambda_0$, then as a compact operator, $\lambda K : \overline{B_1(0)} \to \overline{B_1(0)}$. By the Schauder Fixed Point Theorem, for each λ, $|\lambda| \leq \lambda_0$, there is $u_\lambda \in \overline{B_1(0)}$ such that

$$u_\lambda = \lambda K(u_\lambda) = T(\lambda, u_\lambda).$$

Theorem 10.31 (*Leray-Schauder Theorem*) *Let E be a real Banach space, $T : \mathbb{R} \times E \to E$ a compact operator with $T(0, u) = 0$ for $u \in E$. Let C be the connected component of the solution set of the equation $u = T(\lambda, u)$ with $(0, 0) \in C$. Then $C = C^+ \cup C^-$ where C^+ (respectively C^-) is contained in $\mathbb{R}^+ \times E(\mathbb{R}^- \times E)$, where $\mathbb{R}^+ = \{x \in \mathbb{R} \mid x \geq 0\}$, and $\mathbb{R}^- = \{x \in \mathbb{R} \mid x \leq 0\}$, and is unbounded, and $C^+ \cap C^- = \{(0, 0)\}$.*

Proof.

(i) Let $N = \{(\lambda, u) \in \mathbb{R} \times E \mid u = T(\lambda, u)\}$. If B is closed and bounded in N, then B is compact. In fact, let $u_n = T(\lambda_n, u_n)$, for $n = 1, 2, \cdots$, and $(\lambda_n, u_n) \in B$. Because $\{\lambda_n\}$ and $\{u_n\}$ are bounded, and T is compact, there are subsequences, still denoted by $\{\lambda_n\}$ and $\{u_n\}$, such that

$$\lambda_n \to \lambda,$$
$$u_n = T(\lambda_n, u_n) \to u.$$

That T is continuous implies $u_n = T(\lambda_n, u_n) \to T(\lambda, u)$, and consequently $u = T(\lambda, u)$. We have $(\lambda, u) \in B$, and $(\lambda_n, u_n) \to (\lambda, u)$.

(ii) Suppose C^+ is bounded in $\mathbb{R}^+ \times E$, then C^+ is also closed in $\mathbb{R}^+ \times E$. Because $T(0, u) = 0$ for $u \in E : N \cap (\{0\} \times E) = \{(0, 0)\}$, for $\epsilon > 0$ small, let

$$C_\epsilon = \{x \in \mathbb{R}^+ \times E \mid \text{dist}(x, y) \leq \epsilon \quad \text{for some } y \in C^+\}.$$

Let $K = \overline{C_\epsilon} \cap N$, then K is a metric compact set under the induced topology. Let $K_1 = C^+$, and $K_2 = N \cap \partial C_\epsilon$ be two disjoint compact sets in K. There is no connected component of K that meets K_1 and K_2. By Lemma 9.24, there are two disjoint compact sets $\bar{K}_1$, $\bar{K}_2$ such that $K = \bar{K}_1 \cup \bar{K}_2$, $K_i \subset \bar{K}_i$, $i = 1, 2$, and $\bar{K}_1 \cap \bar{K}_2 = \emptyset$. Let δ be the minimum of the distance of $\bar{K}_1$ and $\bar{K}_2$ in $\mathbb{R}^+ \times E$ and the distance of $\bar{K}_1$ and ∂C_ϵ. Then, $\delta > 0$. Set $G = N_{\delta/2}(\bar{K}_1)$ in $\mathbb{R}^+ \times E$, and $G_\lambda = \{u \in E \mid (\lambda, u) \in G\}$, and $f(\lambda, u) = u - T(\lambda, u)$. Because $0 \notin f(\lambda, \cdot)(\partial G_\lambda)$ for each λ, $d(f(\lambda, \cdot), G_\lambda, 0)$ is well-defined and constant on λ : If λ is large enough, then $G_\lambda = \emptyset$, and consequently $d(f(\lambda, \cdot), G_\lambda, 0) = 0$. If $\lambda = 0$, then $f(0, u) = u$, and $d(f(0, \cdot), G_\lambda, 0) = 1$, a contradiction. Therefore, the connected component C^+ passing through $(0, 0)$ is unbounded in $\mathbb{R}^+ \times E$.

(iii) Similarly, C^- is unbounded in $\mathbb{R}^- \times E$.   ∎

Theorem 10.32 (*Dugundji Extension Theorem*) *Let X be a metric space and A a closed subset of X. Let E be a locally convex topological vector space over real or complex numbers and C a convex subset of E. Then any continuous map $f : A \rightarrow C$ admits a continuous extension $F : X \rightarrow C$.*

Proof.   See Theorem 1.26.   ∎

Theorem 10.33 (*Mazur Theorem*) *Let E be a Banach space, and $A \subset E$ a relatively compact subset. Then its convex closure conv A is compact.*

Proof.   See Theorem 1.13.   ∎

Recall that the Hilbert cube I^∞ can be characterized by

$$I^\infty = I \times I \times \cdots = \{x = \{x_n\} \in \ell^2 \mid |x_n| \leq \frac{1}{n}\},$$

and I^∞ is a compact metric space.

Theorem 10.34 (*Urysohn Theorem*) *Any compact metric space admits an embedding in the Hilbert cube I^∞.*

Theorem 10.35 *Let X be a metric space, and $A \subset X$ a closed bounded subset. Let E be a Banach space, and $C \subset E$ a convex set. Then any compact operator $f : A \rightarrow C$ admits a compact extension $g : X \rightarrow C$.*

Proof. f is a compact operator, so $\overline{f(A)}$ is a compact subset of E. Let $C = \overline{\text{conv}}(\overline{f(A)})$, then C is a compact convex subset of E. By the Urysohn Theorem 10.34, there is an isometric embedding u of $\overline{f(A)}$ onto $Q \subset I^\infty$. By the Dugundji Extension Theorem, there are continuous extensions $V : X \to I^\infty$ of v, and $U : I^\infty \to C$ of u^{-1}. Because U is continuous, and I^∞ is compact, we have U is a compact operator, so $U \circ V$ is a compact operator, as an extension of f.   ∎

   We extend Theorem 10.35 to the following.

Theorem 10.36 *Let X be a metric space, and $A \subset X$ a closed subset. Let E be a Banach space, and $C \subset E$ a convex set. Then any compact operator $f : A \to C$ has a compact extension $g : X \to C$.*

Proof.   Let $B \subset X$ be a bounded set, $B_1 = B \cap A$, $B_2 = B \cap (X \backslash A)$. For $x \in B_2$, take a neighborhood V_x of x such that $\bigcup\limits_{x \in B_2} V_x$ is bounded, and

$$c_1 \, \text{diam} V_x \leq \text{dist}(V_x, A) \leq c_2 \, \text{diam} V_x.$$

Let $\{U\}$ be a neighborhood-finite refinement of $\{V_x\}$, and let $\{a_U\} \subset A$ be such that

$$g(x) = \sum\nolimits_U h_U(x) f(a_U) \quad \text{for } x \in X,$$

where $\{h_U\}$ is a partition of unity subordinated to $\{U\}$; see Dugundji [20, p. 188], and Stein [45, p. 172]. Let $B_3 = \{a_U\}$. Then B_3 is bound in A. Because $B' = B_1 \cup B_3$ is bounded in A, $\overline{f(B')}$ is compact in $C = \overline{\text{conv}}(\overline{f(A)})$. Thus,

$$g(B) \subset \overline{\text{conv}} \overline{f(B')} \subset \overline{\text{conv}} f(B') \subset C,$$

so g is a compact operator.   ∎

Corollary 10.37 *Let E be a Banach space, $K \subset E$ a closed convex cone in E with vertex 0, and $T : \mathbb{R}^+ \times K \to K$ a compact operator satisfying $T(0, u) = 0$ for $u \in K$. Suppose C^+ is the connected component in $\mathbb{R}^+ \times K$ of the set of solutions of (10.8) containing $(0, 0)$. Then C^+ is unbounded.*

Proof.   By Theorem 10.36, there is a compact extension $\bar{T} : \mathbb{R} \times E \to K$ satisfying $\bar{T}(0,) = 0$ for $u \in E$. Note that for $\lambda \geq 0$, the set of solutions of $u - T(\lambda, u) = 0$ and the set of solutions of $u - \bar{T}(\lambda, u) = 0$ are identical. By Theorem 10.31, C^+ is unbounded.   ∎

   We apply the above theory to second-order quasilinear elliptic equations. Let Ω be a bounded regular domain, $\lambda \in \mathbb{R}$, a_{ij}, b_i, $c \in C^1(\bar{\Omega} \times \mathbb{R} \times \mathbb{R}^N)$, $c(x, u, p) \geq 0$ for $x \in \bar{\Omega}$, $u \in \mathbb{R}$, $p \in \mathbb{R}^N$, $F \in C^1(\bar{\Omega} \times \mathbb{R} \times \mathbb{R}^N \times \mathbb{R})$, and let $\alpha, \Lambda > 0$ exist such that

$$\alpha \, |\xi|^2 \leq \sum_{i, j} a_{ij}(x, u, p) \xi_i \xi_j \leq \Lambda \, |\xi|^2 \quad \text{for } x \in \bar{\Omega}, u \in \mathbb{R}, \ p, \ \xi \in \mathbb{R}^N.$$

Consider the equation

$$
\begin{cases}
-\sum_{i,j} a_{ij}(x, u, \nabla u) \frac{\partial^2 u}{\partial x_i \partial x_j} + \sum_i b_i(x, u, \nabla u) \frac{\partial u}{\partial x_i} + c(x, u, \nabla u)u \\
\quad = F(x, u, \nabla u, \lambda) \ \text{ in } \Omega, \\
u = 0 \ \text{ on } \partial\Omega.
\end{cases}
\tag{10.12}
$$

Let $E = C^{1,\alpha}(\Omega)$ be a Hölder space. For $(\lambda, u) \in \mathbb{R} \times E$, there is a solution $w \in E$ of the equation

$$
\begin{cases}
-\sum_{i,j} a_{ij}(x, u, \nabla u) \frac{\partial^2 w}{\partial x_i \partial x_j} + \sum_i b_i(x, u, \nabla u) \frac{\partial w}{\partial x_i} + c(x, u, \nabla u)w \\
\quad = F(x, u, \nabla u, \lambda) \ \text{ in } \Omega, \\
w = 0 \ \text{ on } \partial\Omega.
\end{cases}
\tag{10.13}
$$

Let $w = T(\lambda, u)$. As before, $T : \mathbb{R} \times E \to E$ is a compact operator. Note that u is a solution of equation (10.12) if and only if $u = T(\lambda, u)$.

(i) If $F(x, u, p, 0) \equiv 0$, then apply the Schauder Estimation to show that $T(0, u) = 0$ for $u \in E$. By Theorem 10.31, there is a connected component C of the set of solutions of equation (10.12) in $\mathbb{R} \times E$ containing $(0, 0)$, such that $C = C^+ \cup C^-$ with $C^+ \subset \mathbb{R}^+ \times E$ unbounded, and $C^- \subset \mathbb{R}^- \times E$ unbounded.

(ii) If $F(x, u, \nabla u, \lambda) \geq 0$ for $\lambda \in \mathbb{R}$, $u \in E$, then by the Maximum Principle $w = T(\lambda, u) \geq 0$ in $\bar{\Omega}$. If $F(x, u, \nabla u, \lambda) > 0$, then by the Strong Maximum Principle, $T(\lambda, u) = w > 0$ in $\bar{\Omega}$, and $\frac{\partial w}{\partial n} < 0$ in $\partial\Omega$.
Let $P^+ = \{u \in E \mid u > 0 \text{ in } \Omega, \ \frac{\partial u}{\partial n} < 0 \text{ on } \partial\Omega\}$. If $F(x, u, p, \lambda) > 0$ for $x \in \bar{\Omega}$, $u \in \mathbb{R}$, $p \in \mathbb{R}^N$, $\lambda \geq 0$, then because $\overline{P^+}$ is a closed cone in E with vertex 0, we have by (ii), $C^+ \subset \mathbb{R}^+ \times \overline{P^+}$.

Theorem 10.38 *If $F(x, u, p, \lambda) = \lambda G(x, u, p)$ and $G(x, 0, 0) > 0$ for $x \in \bar{\Omega}$, there is a connected component C^+ of solutions of equation (10.12) containing $(0,0)$, which is unbounded and contained in $(\mathbb{R}^+ \times P^+) \cup \{0, 0\}$.*

Proof. By the continuity of $G(x, u, p)$ near $u = 0$, $p = 0$, and the Strong Maximum Principle, there is a neighborhood H of $(0, 0)$ in $\mathbb{R}^+ \times E$ such that

$$
C^+ \cap \mathcal{H} \subset (\mathbb{R}^+ \times P^+) \cup \{0, 0\}.
$$

We claim that $C^+ \subset (\mathbb{R}^+ \times P^+) \cup \{0, 0\}$. Otherwise, because C^+ is connected, there exist $\bar{\lambda} > 0$ and $\bar{u} \in E$ such that $(\bar{\lambda}, \bar{u}) \in C^+$, $\bar{u} \in \partial P^+$. Either: there is (i) $\xi \in \Omega$ such that $\bar{u}(\xi) = 0$, or there is (ii) $\eta \in \partial\Omega$ such that $\frac{\partial \bar{u}}{\partial n}(\eta) = 0$.

(i) Note that $\bar{u} \geq 0$ in Ω. If $\bar{u}$ attains its minimum at ξ, $\bar{u}(\xi) = 0$, then $\nabla\bar{u}(\xi) = 0$. Because G is continuous and $G(x, 0, 0) > 0$ for $x \in \bar{\Omega}$, there is an open ball $B(\xi, r) \subset \Omega$ such that $\bar{u}$, and $\nabla\bar{u}$ on B are small enough, and $\bar{u}$ and $\nabla\bar{u}$ are continuous at ξ, such that $G(x, \bar{u}, \nabla\bar{u}) > 0$ for $x \in B$. By the Strong Maximum Principle, $\bar{u} \equiv 0$ on B, so $\{x \in \Omega \mid \bar{u}(x) = 0\}$ is nonempty and open. Clearly, however, $\{x \in \Omega \mid \bar{u}(x) = 0\}$ is closed, and it follows at once that $\bar{u} = 0$ on Ω because Ω is connected. Now $\bar{\lambda} > 0$, $\bar{u} = 0$ satisfy

$$0 = \sum_{i,j} a_{ij}(x, \bar{u}, \nabla\bar{u})\frac{\partial^2\bar{u}}{\partial x_i \partial x_j} + \sum b_i(x, \bar{u}, \nabla\bar{u})\frac{\partial\bar{u}}{\partial x_i} + c(\bar{u})\bar{u}$$
$$= \bar{\lambda}G(x, \bar{u}, \nabla\bar{u}) > 0 \quad \text{in } \Omega,$$

a contradiction.

(ii) Suppose $\eta \in \partial\Omega$ with $\frac{\partial\bar{u}}{\partial n}(\eta) = 0$. By the Hopf Maximum Principle, there exists $x_0 \in \Omega$ such that $\bar{u}(x_0) \leq \bar{u}(\eta)$. However, $\bar{u} = 0$ on $\partial\Omega$, and $\bar{u} \geq 0$ on Ω, then $\bar{u}(x_0) = 0$. By part (i), we have a contradiction. ∎

Let E be a real Banach space, and K a closed convex cone in E such that $K \cap (-K) = \{0\}$. E is ordered by K in the way $x \geq y$ if and only if $x - y \in K$. Suppose $\overset{\circ}{K} \neq \emptyset$, a linear operator $L : E \to E$ is strictly positive if $L(K\backslash\{0\}) \subset \overset{\circ}{K}$.

Theorem 10.39 (*Krein-Rutman Theorem*) *If E is ordered by K, and $L : E \to E$ is a strictly positive linear compact operator, then L admits uniquely an eigenvector $x_0 \in \overset{\circ}{K}$, $\|x_0\| = 1$, which is associated with a simple characteristic value $\mu_0 > 0$, and $\mu_0 < |\mu|$ for any other real or complex characteristic values μ.*

Proof.

(i) Existence of x_0 : fix $u \in K\backslash\{0\}$, then for some $d > 0$, $dL(u) \geq u$. Otherwise $Lu - \frac{1}{d}u \notin K$ for each $d > 0$. Letting $d \to \infty$, we have $Lu \notin \overset{\circ}{K}$, a contradiction. For $\epsilon > 0$ define $T_\epsilon(\lambda, x) = \lambda L(x + \epsilon u)$, then $\mathbb{R} \times K \to K$ is compact, and $T_\epsilon(0, u) = 0$ for $u \in K$. By Corollary 10.37, there is an unbounded connected component $C_\epsilon \subset \mathbb{R}^+ \times K$ of the set of solutions of $x = T_\epsilon(\lambda, x)$. We claim that $C \subset [0, d] \times K$. In fact, let $(\lambda, x) \in C_\epsilon$,

$$x = T_\epsilon(\lambda, x) = \lambda L(x + \epsilon u)$$
$$= \lambda Lx + \lambda\epsilon Lu \geq \lambda\epsilon Lu \geq \frac{\lambda}{d}\epsilon u,$$

and consequently

$$Lx \geq (\frac{\lambda}{d}\epsilon)Lu \geq \frac{\lambda}{d^2}\epsilon u.$$

Hence, $x = \lambda Lx + \lambda \epsilon Lu \geq \lambda Lx \geq (\frac{\lambda}{d})^2 \epsilon u$, and the recurrence becomes

$$x \geq (\frac{\lambda}{d})^n \epsilon u \quad \text{for } n \in \mathbf{N}.$$

Letting $\lambda > d$ and $n \to \infty$,

$$0 \leftarrow (\frac{\lambda}{d})^n x \geq \epsilon u.$$

Thus, $-\epsilon u \in K$, a contradiction. We have $\lambda \leq d : C_\epsilon \subset [0, d] \times K$. Because C_ϵ is unbounded, there is at least $x_\epsilon \in K, \|x_\epsilon\| = 1, \lambda_\epsilon \in [0, M]$, such that

$$x_\epsilon = T_\epsilon(\lambda_\epsilon, x_\epsilon) = \lambda_\epsilon L(x_\epsilon + \epsilon u),$$

$\{\lambda_\epsilon\}, \{x_\epsilon + \epsilon u\}$ are bounded, so there is $\mu_0 \in [0, d]$, subsequences $\{\lambda_{\epsilon_n}\}, \{x_{\epsilon_n}\}$ and $y \in K$ such that $\lambda_{\epsilon_n} \to \mu_0$, and $L(x_{\epsilon_n} + \epsilon_n u) \to y$ as $\epsilon_n \to 0$, and consequently

$$x_{\epsilon_n} = \lambda_{\epsilon_n} L(x_{\epsilon_n} + \epsilon_n u) \to \mu_0 y = x_0 \in K \quad \text{as } n \to \infty.$$

Because $x_{\epsilon_n} + \epsilon_n u \to x_0$, we have

$$\lambda_{\epsilon_n} L(x_{\epsilon_n} + \epsilon_n u) \to \mu_0 L(x_0).$$

Thus, $x_0 = \mu_0 L(x_0)$, and

$$1 = \|x_{\epsilon_n}\| \to \|x_0\|,$$

so $\|x_0\| = 1$, and $x_0 \in K \backslash \{0\}$, so $\mu_0 \neq 0$ and $L(x_0) \in \overset{\circ}{K}$. Thus, $\mu_0 > 0$, $x_0 \in \overset{\circ}{K}$.

(ii) Suppose $y_0 \in \overset{\circ}{K}$. For $y \notin K$ there is uniquely a number $\delta(y) > 0$ such that $y_0 + \lambda y \in K$ for $0 \leq \lambda \leq \delta(y)$, and $y_0 + \lambda y \notin K$ for $\lambda > \delta(y) : y_0 + \lambda y \in \overset{\circ}{K}$ implies $\lambda < \delta(y)$. Moreover, the map $\delta : E \backslash K \to \mathbb{R}^+$ is continuous. Set

$$\delta(y) = \sup_{\lambda \geq 0} \{y_0 + \lambda y \in K\}.$$

Note that for $y \in E \backslash K$ and $\epsilon > 0$, there exists $\eta > 0$ such that for $|z - y| < \eta$ we have

$$y_0 + (\delta(y) - \epsilon)z \in \overset{\circ}{K}, \quad \text{and} \quad z_0 + (\delta(y) + \epsilon)z \notin K.$$

Thus, $|\delta(z) - \delta(y)| < \epsilon : \delta$ is continuous.

(iii) The only eigenvectors of L in K are λx_0, $\lambda > 0$. Let $x \in K$, $x \neq 0$, be another eigenvector of L in $K : x = \mu L x$. Then $\mu > 0$, and $x \in \overset{\circ}{K}$, because L is strictly positive. Set $r_1 = \delta_{x_0}(-x)$, $r_2 = \delta_x(-x_0)$. Then

$$L(x_0 - r_1 x) = \frac{1}{\mu_0}[x_0 - r_1 \frac{\mu_0}{\mu} x],$$

$$L(x - r_2 x_0) = \frac{1}{\mu}[x - r_2 \frac{\mu}{\mu_0} x_0].$$

Suppose $x - r_2 x_0 \neq 0$, by (ii), $x - r_2 x_0 \in K$, and consequently $L(x - r_2 x_0) \in \overset{\circ}{K}$, or $x - r_2 \frac{\mu}{\mu_0} x_0 \in \overset{\circ}{K}$. We have $r_2 \frac{\mu}{\mu_0} < r_2 : \frac{\mu}{\mu_0} < 1$, or $\mu < \mu_0$. Because $x_0 - r_1 x \in K$, we have $L(x_0 - r_1 x) \in K : x_0 - r_1 \frac{\mu_0}{\mu} x \in K$, and consequently $r_1 \frac{\mu_0}{\mu} \leq r_1 : \mu_0 \leq \mu$, a contradiction. Therefore, $x = r_2 x_0$. Suppose $c \neq 0$, then $\mu_0 L(c x_0) = c x_0 : c x_0$ is still the associated characteristic value μ_0.

(iv) Set $\delta(\cdot) = \delta_{x_0}(\cdot)$, and let $x \notin K \cup (-K)$ be an eigenvector of L associated with the real characteristic value μ. Then $\mu_0 \leq |\mu|$. Let $x \notin K \cup (-K)$, $\mu \in \mathbb{R}$ be such that $x = \mu L x$. Then $0 \neq x_0 \pm \delta(\pm x) x \in K$, we have

$$L[x_0 \pm \delta(\pm x) x] = \frac{1}{\mu_0}[x_0 \pm \frac{\mu_0}{\mu} \delta(\pm x) x] \in \overset{\circ}{K}.$$

If $\mu > 0$ then $\frac{\mu_0}{\mu} \delta < \delta : \mu_0 < \mu = |\mu|$. If $\mu < 0$, then $x_0 + (-\frac{\mu_0}{\mu} \delta(x))(-x) \in \overset{\circ}{K}$, $x_0 + (-\frac{\mu_0}{\mu} \delta(-x)) x \in \overset{\circ}{K}$, or $-\frac{\mu_0}{\mu} \delta(x) < \delta(-x)$, and $-\frac{\mu_0}{\mu} \delta(-x) < \delta(x)$. Thus, $(-\frac{\mu_0}{\mu})^2 \delta(x) < \delta(x) : \mu_0^2 < \mu^2$, or $\mu_0 < -\mu = |\mu|$. In any case $\mu_0 < |\mu|$.

(v) For any nonreal characteristic value μ of L, we have $\mu_0 < |\mu| :$ Let $\mu = |\mu| e^{i\theta}$ be a characteristic value of L, $\theta \not\equiv 0 \pmod{\pi}$, with eigenvector $z_1 + i z_2$, with z_1, $z_2 \in E$, and z_1 and z_2 independent:

$$z_1 + i z_2 = \mu(L z_1 + i L z_2).$$

Let P be the plane generated by $\{z_1, z_2\}$, then

$$L z_1 + i L z_2 = \frac{1}{\mu}(z_1 + i z_2)$$

$$= \frac{1}{|\mu|} e^{-i\theta}(z_1 + i z_2)$$

$$= \frac{1}{|\mu|}[(\cos\theta) z_1 + (\sin\theta) z_2 + i((\cos\theta) z_2 - (\sin\theta) z_1)],$$

$$L z_1 = \frac{1}{|\mu|}((\cos\theta) z_1 + (\sin\theta) z_2),$$

$$L z_2 = \frac{1}{|\mu|}((-\sin\theta) z_1 + (\cos\theta) z_2),$$

$$L \begin{bmatrix} z_1 \\ z_2 \end{bmatrix} = \frac{1}{|\mu|} \begin{bmatrix} \cos\theta & \sin\theta \\ -\sin\theta & \cos\theta \end{bmatrix} \begin{bmatrix} z_1 \\ z_2 \end{bmatrix}.$$

That is, $L|_P = \frac{1}{|\mu|} R_\theta$, where

$$R_\theta = \begin{bmatrix} \cos\theta & \sin\theta \\ -\sin\theta & \cos\theta \end{bmatrix}.$$

We claim that $K \cap P = \{0\}$. Otherwise, if $K \cap P \underset{\neq}{\supset} \{0\}$ is a closed convex cone in P, then $L|_P$ is strictly positive in $K \cap P$. By (iii), $L|_P$ possesses an eigenvector in $K \cap P$: there exists $x \in K \cap P$ such that $x = \mu L x$. By (iii), $x = \lambda x_0$, and $\mu = \mu_0$ is real, which contradicts $\theta \not\equiv 0 \ (mod \pi)$. Therefore, for each $z \in P \backslash \{0\}$, $\delta(z)$ is defined and

$$L(x_0 + \delta(z)z) = \frac{1}{\mu_0}[x_0 + \frac{\mu_0}{|\mu|}\delta(z)R_\theta z],$$

so $\frac{\mu_0}{|\mu|}\delta(z) < \delta(R_\theta z)$. Consider the ellipse

$$C = \left\{ T_\alpha \left[\begin{array}{c} z_1 \\ z_2 \end{array} \right] = (\cos\alpha)z_1 + (\sin\alpha)z_2, \alpha \equiv 0(\mathrm{mod}\ 2\pi) \right\},$$

then

$$R_\theta \circ T_\alpha = \left[\begin{array}{cc} \cos\theta & \sin\theta \\ -\sin\theta & \cos\theta \end{array} \right] \left[\begin{array}{c} \cos\alpha \\ \sin\alpha \end{array} \right] = \left[\begin{array}{c} \cos(\alpha-\theta) \\ \sin(\alpha-\theta) \end{array} \right] = T_{\alpha-\theta}.$$

We have shown that C is compact and invariant under R_θ. Then there exists $z_0 \in C$ such that
$$\delta(z_0) = \sup_{z\in C} \delta(z),$$

we obtain $\frac{\mu_0}{|\mu|}\delta(z_0) < \delta(R_\theta z_0) \le \delta(z_0) : \mu_0 < |\mu|$.

(vi) μ_0 is simple: by (iii), $\mathrm{Ker}(I - \mu_0 L) = \mathbb{R}x_0$. We require $\mathrm{Ker}(I - \mu_0 L)^2 = \mathrm{Ker}(I - \mu_0 L)$. Clearly $\mathrm{Ker}(I - \mu_0 L)^2 \supset \mathrm{Ker}(I - \mu_0 L)$. Suppose $x \in \ker(I - \mu_0 L)^2$: $(I - \mu_0 L)^2 x = 0$, or $x - \mu_0 L x \in \ker(I - \mu_0 L) = \mathbb{R}x_0$. Then there exists $\lambda \in \mathbb{R}$ such that $x - \mu_0 L x = \lambda x_0$. If $\lambda > 0$ (for $\lambda < 0$ replace x by $-x$),

$$x = \mu_0 L x + \lambda x_0 = \mu_0 L x + \mu_0 \lambda L x_0 = \mu_0 L(x + \lambda x_0).$$

Thus,

$$x + \lambda x_0 = \mu_0 L(x + \lambda x_0) + \mu_0 \lambda L(x_0) = \mu_0 L(x + 2\lambda x_0),$$

or for $n = 1, 2, ...,$

$$x = \mu_0 L(x + \lambda x_0) = \mu_0^2 L^2(x + 2\lambda x_0) = ... = \mu_0^n L^n(x + n\lambda x_0), \qquad (10.14)$$

$$x_0 = \mu_0 L(x_0) = \mu_0 L(\mu_0 L(x_0)) = \mu_0^2 L^2(x_0) = \cdots = \mu_0^n L^n x_0. \qquad (10.15)$$

By (10.14) and (10.15), we have

$$\frac{x}{n} = \mu_0^n L^n(\frac{x}{n} + \lambda x_0).$$

Because $x_0 \in \overset{\circ}{K}$, we have $\frac{x}{n} + \lambda x_0 \in K$ for large $n : \frac{x}{n} \in K$, and consequently, $x = n \cdot \frac{x}{n} \in K$. However

$$\frac{x}{n} = \mu_0^n L^n(\frac{x}{n} + \lambda x_0),$$

or

$$\frac{x}{n} - \mu_0^n L^n(\lambda x_0) = \mu_0^n L^n(\frac{x}{n}),$$

and consequently

$$\frac{x}{n} - \lambda x_0 = \mu_0^n L^n(\frac{x}{n}).$$

Letting $n \to \infty$,

$$\mu_0^n L^n(\frac{x}{n}) \to -\lambda x_0 \notin K \quad \text{but } \mu_0^n L^n(\frac{x}{n}) \in K, \text{ for } n = 1, 2, ...,$$

a contradiction. Thus, $\lambda = 0 : x = \mu_0 Lx = 0$, that is $x \in \mathrm{Ker}(I - \mu_0 L)$. ∎

Let E, F be two Banach spaces and $f : E \to F$ a map. Suppose the equation $f(x) = 0$ possesses a curve of solutions:

$$\mathcal{C} = \{x(t) : t \in I\},$$

where $I \subset \mathbb{R}$ is an open interval.

Definition 10.40 *A point $x(\tau)$ in C is a bifurcation point of the equation $f(x) = 0$ with respect to C if for each neighborhood U of $x(\tau)$ there is a solution y of $f(x) = 0$, but $y \notin C$.*

Let E be a real Banach space, and consider the bifurcation of the equation

$$f(\lambda, u) = 0, \tag{10.16}$$

where $f(\lambda, u) = u - \lambda Tu + g(\lambda, u)$ satisfying:
(i) $T : E \to E$ is a compact linear operator;
(ii) $g : \mathbb{R} \times E \to E$ is a compact nonlinear operator satisfying $g(\lambda, 0) \equiv 0$ for all $\lambda \in \mathbb{R}$, and near 0,

$$g(\lambda, u) = o(\|u\|)$$

uniformly for any bounded intervals of λ.
The set $\{(\lambda, 0) \mid \lambda \in \mathbb{R}\}$ is a solution set of equation (10.11), called the trivial solution set. To study the bifurcations of the solutions of (10.16) through the line of trivial solutions is to study the structure of the set of nontrivial solutions. A point $(u, 0)$ is a bifurcation point of equation (10.16) with respect to $\mathbb{R} \times \{0\}$ if for each neighborhood U of $(\mu, 0)$ in which there is a nontrivial solution (λ, y) of equation (10.16). For the study of global results of bifurcation problems see Chapter 4.

10.5   Sturm-Liouville Nonlinear Problems

Assume $p \in C^1([0, \pi])$, $p > 0$ on $[0, \pi]$, $q \in C([0, \pi])$, and

$$Lu = -(pu')' + qu.$$

Let $a \in C([0, \pi])$, $a > 0$ on $[0, \pi]$, and $F_1(x, \xi, \eta, \lambda) = o(|\xi^2 + \eta^2|^{1/2})$ near $(0, 0)$ uniformly for $x \in [0, \pi]$, and λ in the bounded interval. Let

$$F(x, \xi, \eta, \lambda) = \lambda a(x) \xi + F_1(x, \xi, \eta, \lambda).$$

Consider the equation

$$\begin{cases} Lu = F(x, u, u', \lambda) & \text{in } (0, \pi), \\ a_0 u(0) + b_0 u'(0) = 0, \\ a_1 u(\pi) + b_1 u'(\pi) = 0, \\ (a_0^2 + b_0^2)(a_1^2 + b_1^2) \neq 0. \end{cases} \qquad (10.17)$$

Suppose $F_1 = 0$ in equation (10.17), then it becomes the Sturm-Liouville linear eigenvalue problem

$$\begin{cases} Lu = \lambda a u & \text{in } (0, \pi), \\ a_0 u(0) + b_0 v'(0) = 0, \\ a_1 u(\pi) + b_1 v'(\pi) = 0, \\ (a_0^2 + b_0^2)(a_1^2 + b_1^2) \neq 0. \end{cases} \qquad (10.18)$$

For the theory of (10.18) see Dieudonne [18, Chapter 9], and Coddington-Levinson [14, Chapters 8, 12]. Let

$$E = \{ u \in C^1([0, \pi]) \mid a_0 u(0) + b_0 u'(0) = 0, \ a_1 u(\pi) + b_1 u'(\pi) = 0,$$
$$(a_0^2 + b_0^2)(a_1^2 + b_1^2) \neq 0 \}.$$

Then E is a Banach space under the norm

$$\|u\|_1 = \sup_{x \in [0, \pi]} |u(x)| + \sup_{x \in [0, \pi]} |u'(x)|.$$

Let S_k^+ consist of $u \in E$ such that u has exactly $k - 1$ zeros in $(0, \pi)$, all zeros are simple, and $u > 0$ near 0.

Theorem 10.41 $S_k^- = -S_k^+$ and $S_k = S_k^+ \cup S_k^-$.

Theorem 10.42 S_k^+ is open in E.

Proof. For $f \in S_k^+ : f(x_i) = 0$, $f'(x_i) \neq 0$ for $i = 1, \cdots, k-1$. Take the closed neighborhood I_{x_i} of x_i, and let $J = [0, \pi] \setminus \bigcup_{i=1}^{k-1} \overset{\circ}{I}_{x_i}$, $I = \bigcup_{i=1}^{k-1} I_{x_i}$. Let $\epsilon_1 = \sup_{x \in J} |f(x)| > 0$, $\epsilon_2 = \sup_{x \in J} |f'(x)| > 0$, $\epsilon = \min(\frac{\epsilon_1}{2}, \frac{\epsilon_2}{2})$, $u \in E$, $\|u - f\|_1 < \epsilon$, then $f - \epsilon < u < f + \epsilon$ on J, $f' - \epsilon < u' < f' + \epsilon$ on I, so u and f have the same sign on J and u' and f' have the same sign on I. Thus, u has exactly one zero in each of I_{x_i} and $u(0) > 0$, so $u \in S_k^+$. ∎

Theorem 10.43 *The eigenvalues of (10.18) are simple, and form an unbounded sequence*

$$\mu_1 < \mu_2 < \cdots, \quad \text{and } \mu_n \to \infty \text{ as } n \to \infty.$$

A nontrivial solution v_n, associated with μ_n, has exactly $n-1$ zeros in $(0, \pi)$ and the zeros of v_n are simple: $v_n \in S_n$. The proper subspace associated with μ_n is of dimension 1. The solution v_n of (10.18) is unique if by normalized $v_n \in S_n^+$, $\|v_n\| = 1$. In $\mathbb{R} \times E$, the solutions of (5.2) are formed by the sequence of lines

$$\{(\mu_n, \alpha v_n) \mid \alpha \in \mathbb{R}\} \subset \{\mu_n\} \times (S_n \cup \{0\}).$$

Proof. See Coddington-Levinson [14, Chapter 8]. ∎

Assume that 0 is not an eigenvalue of equation (10.17), otherwise we may replace L by $L_\epsilon = L + \epsilon a$. Let g be the Green function of L associated with the boundary conditions in equation (10.17). Then equation (10.17) is equivalent to

$$
\begin{aligned}
u(x) &= \int_0^\pi g(x, y) F(y, u(y), u'(y), \lambda) dy \\
&= \lambda \int_0^\pi g(\cdot, y) a(y) u(y) dy + \int_0^\pi g(\cdot, y) F_1(y, u(y), u'(y), \lambda) dy \\
&= \lambda T u + G(\lambda, u),
\end{aligned}
$$

where $Tu = \int_0^\pi g(\cdot, y) a(y) u(y) dy : E \to E$ is a compact linear operator, and $H(\lambda, u) = \int_0^\pi g(\cdot, y) F_1(y, u(y), u'(y), \lambda) dy : \mathbb{R} \times E \to E$ is a compact operator, satisfying $H(\lambda, u) = o(\|u\|_1)$ near 0, uniformly for λ in bounded intervals:

$$
\begin{aligned}
|H(\lambda, u)| &\leq \int_0^\pi |g(\cdot, y)| \left| F_1(y, u(y), u'(y), \lambda) \right| dy \\
&\leq \int_0^\pi |g(\cdot, y)| o(|u(y)|^2 + |u'(y)|^2)^{1/2} dy \\
&\leq o(\|u\|_1) \int_0^\pi |g(\cdot, y)| dy \\
&\leq o(\|u\|_1) \quad \text{for } \lambda \text{ in bounded intervals.}
\end{aligned}
$$

Note that

$$\lambda T u = \int g(\cdot, y) \lambda a(y) u(y) dy = w \quad \text{if and only if} \quad Lw = -(pw')' + qw = \lambda a u.$$

The characteristic values of T are the eigenvalues of L; by Theorem 10.42, they are simple. By the Rabinowitz Theorem 4.22, there is a connected component C_k of $N \cup \{(\mu_k, 0)\}$, where N is the closure of nontrivial solutions of (10.17) in $\mathbb{R} \times E$, containing $(\mu_k, 0)$, and C_k is either unbounded or meets another $(\mu_j, 0)$, $j \neq k$.

Lemma 10.44 *If (λ, u) is a solution of (5.1) such that u admits a double zero in $[0, \pi]$, then $u = 0$.*

Proof. Suppose we have $\tau \in [0, \pi]$, $u(\tau) = u'(\tau) = 0$. Write equation (10.17) as

$$Lu = \lambda a(x) u + q_1(x) u + q_2(x) u', \tag{10.19}$$

where

$$q_1(x) = \frac{F_1(x, u(x), u'(x), \lambda) u(x)}{u(x)^2 + u'(x)^2} \quad x \in [0, \pi],$$

$$q_2(x) = \frac{F_1(x, u(x), u'(x), \lambda) u(x)}{u(x)^2 + u'(x)^2} \quad x \in [0, \pi],$$

$q_1, q_2 \in C([0, \pi])$, $q_1(\tau) = q_2(\tau) = 0$ because $F_1 = o(|\xi|^2 + \eta^2)^{1/2})$. Let $v = u'$, then equation (10.17) can be written as a system of linear equations of degree 1:

$$\begin{cases} Lu = -(pu')' + qu = -(pv)' + qu \\ \quad = \lambda a(x) + q_1(x) u + q_2(x) v \\ v = u'. \end{cases}$$

The solution exists and is unique. $u(\tau) = u'(\tau) = v(\tau) = 0$, so $u = 0$. ∎

Lemma 10.45 *For each integer i, there is a neighborhood U_i of $(\mu_i, 0)$ in $\mathbb{R} \times E$ such that $(\lambda, u) \in U_i \cap N$, $u \neq 0$ implies $u \in S_i$.*

Proof. By contradiction. Suppose there is a sequence (λ_n, u_n) in N such that $(\lambda_n, u_n) \to (\mu_i, 0)$ but $u_n \notin S_i$, $u_n \neq 0$. Then

$$\frac{u_n}{\|u_n\|_1} = \lambda_n L \left(\frac{u_n}{\|u_n\|_1} \right) + \frac{H(\lambda_n, u_n)}{\|u_n\|_1}.$$

L is compact, there is a subsequence of $\{u_n\}$, still denoted by $\{u_n\}$, such that $L \left(\frac{u_n}{\|u_n\|_1} \right) \to w$, and we have

$$\frac{u_n}{\|u_n\|_1} = \lambda_n L \left(\frac{u_n}{\|u_n\|_1} \right) + \frac{H(\lambda_n, u_n)}{\|u_n\|_1} \to \mu_i w.$$

Therefore,

$$L\left(\frac{u_n}{\|u_n\|_1}\right) \to \mu_i L(w),$$

and we obtain

$$w = \mu_i L w, \quad \|w\|_1 \neq 0,$$

so $w \in S_i$. However S_i is open, $\frac{u_n}{\|u_n\|_1} \in S_i$ for large i. Note that $\alpha S_i \subset S_i$ for $\alpha > 0$. We have $u_n = \|u_n\|_1 \frac{u_n}{\|u_n\|_1} \in S_i$ but S_i is a cone, a contradiction. $\blacksquare$

Theorem 10.46 *The component C_k is contained in $(\mathbb{R} \times S_k) \cup \{(\mu_k, 0)\}$ and is unbounded.*

Proof. It suffices to show that

$$C_k \subset (\mathbb{R} \times S_n) \cup \{(\mu_k, 0)\}.$$

In fact, if C_k is bounded, then by Lemma 10.44 C_k meets one point $(\mu_j, 0)$, $j \neq k$. Thus, C_k meets S_j, which contradicts $S_j \cap S_k = \emptyset$. Suppose $C_k \subset (\mathbb{R} \times S_k) \cup \{(u_k, 0)\}$, because C_k is connected, and $C_k \cap U_k \subset (\mathbb{R} \times S_k) \cup \{(u_k, 0)\}$. Then we have $(\bar{\lambda}, \bar{u}) \in C_k \cap (\mathbb{R} \times \partial S_k)$, $(\bar{\lambda}, \bar{u}) \neq (\mu_k, 0)$, a sequence $(\lambda_n, u_n) \to (\bar{\lambda}, \bar{u})$. Now $\bar{u} \in \partial S_k$, so $\bar{u}$ has at least a double zero in $[0, \pi]$. In fact, $\bar{u}$ has $\ell's$ simple zeros, as before u_k had $\ell's$ simple zeros for large n, so $\ell = k$, $\bar{u} \in S_k$, a contradiction. By Lemma 10.43, $\bar{u} \equiv 0$, so $(\bar{\lambda}, \bar{u}) = (\mu_j, 0)$, $j \neq k$. By Lemma 10.45, $u_n \in S_j$ for large n, so $S_k \cap S_j \neq \emptyset$ for $k \neq j$, a contradiction. $\blacksquare$

10.6   Strictly Nonlinear Sturm-Liouville Problems

Consider the equation

$$\begin{cases} -u'' = h(x, u)u & \text{in } (0, \pi), \\ u(0) = u(\pi) = 0, \end{cases} \tag{10.20}$$

where $h : [0, \pi] \times \mathbb{R} \to \mathbb{R}$ is continuous, $h(x, 0) = 0$ for $x \in [0, \pi]$, $h(x, z) \geq 0$ on $[0, \pi] \times \mathbb{R}$, and $h(x, z) \to \infty$ as $|z| \to \infty$ uniformly for $x \in [0, \pi]$. For equation (10.20) there is no bifurcation on the line of trivial solutions. In fact, the Fréchet derivative of $h(x, u)u$ with respect to u is 0 at $u = 0$. This introduces the nonlinear Sturm-Liouville problem:

$$\begin{cases} -u'' + u = \sigma(1 + h(x, u))u & \text{in } (0, \pi), \ \sigma \in \mathbb{R}, \\ u(0) = u(\pi) = 0. \end{cases} \tag{10.21}$$

By Theorem 10.46, for $k \in N$ there is a component C_k of the solution of equation (10.21), which is unbounded, and $C_k \subset (\mathbb{R} \times S_k) \cup \{(\sigma_k, 0)\}$, where σ_k is the k-th eigenvalue of the following linear problem associated with (10.21):

$$\begin{cases} -u'' + u = \sigma u & \text{in } (0, \pi), \ \sigma \in \mathbb{R}, \\ u(0) = u(\pi) = 0. \end{cases} \tag{10.22}$$

We claim that $\sigma_k = k^2 + 1$. In fact $-u'' = (\sigma - 1)u = \lambda u$, for $\lambda = \sigma - 1$, then

$$u(t) = c_1 e^{\sqrt{\lambda}it} + c_2 e^{-\sqrt{\lambda}it},$$

but $u(0) = 0 : c_2 = -c_1$, and consequently $u = 2ic_2 \sin\sqrt{\lambda}t$. Apply $u(\pi) = 0$ to show that $\sqrt{\lambda} = k$. We obtain $\sigma = k^2 + 1$. Note that the least eigenvalue of (10.22) is $\sigma_1 = 2$.

Let $Lu = -(pu')' + qu$, where $p \in C^1([0, \pi])$, $p > 0$ on $[0, \pi)$, and $q \in C([0, \pi])$.

Theorem 10.47 (*Sturm Comparison Theorem*) *Assume* w_1, $w_2 \in C^2([\alpha, \beta])$, *and* $Lw_1 = a_1 w_1$, $Lw_2 = a_2 w_2$ *on* $[\alpha, \beta]$. *If* $a_1(x) > a_2(x)$ *for* $x \in [\alpha, \beta]$, $w_2(\alpha) = w_2(\beta) = 0$, *then there exists* $\gamma \in (\alpha, \beta)$ *such that* $w_1(\gamma) = 0$.

Proof.   We may assume $w_2(x) \neq 0$ for $x \in (\alpha, \beta)$. Suppose by contradiction, $w_1(x) \neq 0$ for $x \in (\alpha, \beta)$. It suffices to prove the case $w_2 > 0$, and $w_1 > 0$ on (α, β). Then

$$\int_\alpha^\beta [(Lw_1)w_2 - (Lw_2)w_1]dx = \int_\alpha^\beta (a_1 - a_2)w_1 w_2 dx. \tag{10.23}$$

Note that

$$(Lw_1)w_2 - (Lw_2)w_1 = -(pw_1')'w_2 + qw_1 w_2 + (pw_2')'w_1 - qw_1 w_2.$$

Therefore,

$$\begin{aligned}
\int_\alpha^\beta & [(Lw_1)w_2 - (Lw_2)w_1]dx \\
&= \int_\alpha^\beta -w_2 d(pw_1') + \int_\alpha^\beta w_1 d(pw_2') \\
&= [-w_2 pw_1']\big|_\alpha^\beta + \int_\alpha^\beta pw_1' w_2' + [w_1 pw_2']\big|_\alpha^\beta - \int_\alpha^\beta pw_1' w_2' \\
&= [pw_2' w_1]\big|_\alpha^\beta \\
&= (pw_2' w_1)(\beta) - (pw_2' w_1)(\alpha) \leq 0.
\end{aligned} \tag{10.24}$$

By (10.23) and (10.24)

$$0 \geq [p\, w_2' w_1]\big|_\alpha^\beta = \int_\alpha^\beta (a_1 - a_2)w_1 w_2 > 0.$$

We have a contradiction.   ∎

Lemma 10.48 *For* $k \in N$, $C_k \subset ([0, \sigma_k] \times S_k) \cup \{(\sigma_k, 0)\}$.

Proof. Let $(\lambda, u) \in C_k : u \in S_k$, $u \neq 0$. Let u be a solution of the equation

$$\begin{cases} -u'' + u = \lambda(1 + h(x, u))u & \text{on } (0, \pi), \\ u(0) = u(\pi) = 0, \end{cases}$$

and v a solution of the equation

$$\begin{cases} v'' + v = \sigma_k v, \\ v(0) = v(\pi) = 0. \end{cases}$$

Suppose $\lambda > \sigma_k$, then $\lambda(1 + h(x, u)) > \sigma_k$. Now $v(0) = v(\pi) = 0$, and v admits $k - 1$ zeros in $(0, \pi)$. By the Sturm Comparison Theorem 10.47, u admits at least k zeros in $(0, \pi)$, contradicting $u \in S_k$. We conclude that $\lambda \leq \sigma_k$. ∎

Lemma 10.49 *For $k \in \mathbb{N}$, there exists $c_k > 0$ such that for $(\lambda, u) \in [1, \sigma_k] \times S_k$, a solution of (10.21), we have*

$$\|u\|_1 \leq c_k.$$

Proof. By contradiction. Assume there is a sequence (λ_n, u_n) of solutions of (10.21) such that $\lambda_n \in [1, \sigma_k]$, $u_n \in S_k$ for $n \in \mathbb{N}$, $\|u_n\|_1 \to \infty$ as $n \to \infty$. Because $u_n \in S_k$, we may write $[0, \pi] = I_{1,n} \cup \cdots \cup I_{k,n}$, and u_n has constant sign in each $I_{i,n}$. We claim that $\|u_n\|_{C^1(I_{j,n})} \to \infty$ as $n \to \infty$ for $j = 1, 2, \cdots, k$. In fact, suppose, for some j, there is a subsequence of $\{u_n\}$, still denoted by $\{u_n\}$, such that $\|u_n\|_{C^1(I_{j,n})} \leq c$ for $n = 1, 2, \cdots$ We claim that $\|u_n\|_{C^1([0,\pi])} \leq c$ for all n. In fact, let $I_{j,n} = [z_j, z_{j+1}]$. Because $-u_n'' = (\lambda_n + \lambda_n h - 1)u$ and $\lambda_n + \lambda_n h - 1 \geq 0$, u_n'' has constant sign on $I_{j,n}$: u_n is convex or concave on $I_{j,n}$, $u_n' = 0$ at only one point $y_j \in I_{j,n}$, and $|u_n'|$ attains a maximum on z_j or z_{j+1}. We claim that $\|u_n\|_{C^1(I_{j-1,n})} \leq c$. It suffices to suppose $u_n \geq 0$ on $I_{j,n}$. We have

$$0 \leq u_n'(x) \leq u_n'(z_j) \leq c \quad \text{for } x \in [y_{j-1}, z_j].$$

Therefore,

$$\begin{aligned} \|u_n\|_{C^0(I_{j-1,n})} &= |u_n(y_{j-1})| \\ &= |u_n(y_{j-1}) - u_n(z_j)| \\ &= |u_n'(x)| \, |y_{j-1} - z_j| \quad \text{for some } x \in [y_{j-1}, z_j] \\ &\leq \pi c. \end{aligned} \tag{10.25}$$

By (10.21), $u_n'' = -(\lambda_n - 1)u_n - \lambda_n h(x, u_n)u_n$, so

$$\left|u_n''(x)\right| \leq c |u_n(x)| \quad \text{for } x \in [z_{j-1}, z_j].$$

We have for $x \in [z_{j-1}, z_j]$

$$|u_n'(x)| = \left| \int_{y_{j-1}}^{x} u_n''(t)dt \right| \leq c\pi \|u_n\|_{C^0(I_{j-1, n})} \leq c$$

or

$$\|u_n'\|_{C^0(I_{j-1,n})} \leq c. \tag{10.26}$$

By (10.25), (10.26), $\|u_n\|_{C^1(I_{j, n})} \leq c$. Similarly $\|u_n\|_{C^1(I_{j+1, n})} \leq c$. After $k-1$ steps, we obtain $\|u_n\|_{C^1([0,\pi])} \leq c$, a contradiction. For each n, choose the length $\ell(I_{j(n), n}) \geq \frac{\pi}{k}$, and $\|u_n\|_{C^1(I_{j(n), n})} \to \infty$ as $n \to \infty$. By (10.26), we have,

$$\|u_n\|_{C^1(I_{j(n), n})} \leq c\|u_n\|_{C^0(I_{j(n),n})}.$$

For $s \in \mathbb{R}^+$, $Y_n(s) = \{x \in I_{j(n),n} \mid |u_n(x)| \geq s\}$,

$$\min_{x \in Y_n(s)} (1 + h(x, u_n(x))) \geq \min_{\substack{x \in [0,\pi] \\ |z| \geq s}} (1 + h(x, z)) = t(s).$$

Because $h(x, z) \to \infty$ as $|z| \to \infty$, we have $t(s) \to \infty$ as $s \to \infty$. On the other hand, if s is large and fixed, $\ell(Y(s)) \leq \frac{\pi}{3k}$ for each n. Otherwise, assume $\ell(Y_n(s)) > \frac{\pi}{3k}$, then the equation

$$\begin{cases} -w'' + w = \lambda w & \text{in } Y_n(s), \\ \quad\quad\quad w = 0 & \text{on } \partial Y_n(s), \end{cases} \tag{10.27}$$

admits an eigenvalue $\lambda_1 = \left[\frac{\pi}{\ell(Y_n(s))} \right]^2 + 1$. Take $t(s) > (3k)^2 + 1$, and we have

$$(3k)^2 > \left(\frac{\pi}{\ell(Y_n(s))} \right)^2,$$

or

$$t(s) > (3k)^2 + 1 > \left(\frac{\pi}{\ell(Y_n(s))} \right)^2 + 1 = \lambda_1.$$

If we apply Lemma 10.49 to compare u_n and a solution w_1 of equation (10.27) associated with λ_1, we have $u_n = 0$ at one point of $Y_n(s)$, a contradiction. Let α_n, β_n be the extrema of $I_{j(n),n}$ and p_n, q_n the extrema of $Y_n(s)$. Let

$$X_n(s) = [\alpha_n, p_n], \quad Z_n(s) = [q_n, \beta_n].$$

One of the two intervals X_n, Z_n is longer than $\frac{\pi}{3k}$, for example $\ell(x_n(s)) \geq \frac{\pi}{3k}$ for $x \in X_n(s)$. If $u_n'' < 0$ on $[\alpha_n, \beta_n]$, then $u_n'(x) \geq u_n'(p_n)$. If $u_n'' > 0$ on $[\alpha_n, \beta_n]$, then $-u_n'(x) \geq -u_n'(p_n)$. We have in any case $|u_n'(x)| \geq |u_n'(p_n)|$,

$$s = |u_n(p_n)| = \int_{\alpha_n}^{p_n} |u_n'(x)| \, dx \geq \frac{\pi}{3k} |u_n'(p_n)|.$$

By (10.21) $u_n''(x) \le c$ for $x \in [\alpha_n, p_n]$, and

$$\left| u_n'(p_n) - u_n'(\alpha_n) \right| = \left| \int_{\alpha_n}^{p_n} u_n''(x)dx \right| \le \int_{\alpha_n}^{p_n} \left| u_n''(x) \right| \le c.$$

Because $|u_n'(\alpha_n)| \to \infty$, we have $|u_n'(p_n)| \to \infty$. However $\|u_n\|_{C^0(I_{j(n),n})} \le \pi |u_n'(\alpha_n)|$, so $|u_n(p_n)| \to \infty$, but $|u_n(p_n)| \le s$, a contradiction. ∎

By the above results, we have

Theorem 10.50 *For $k \in \mathbb{N}$, there is a solution u_k of equation (10.20) in S_k.*

Theorem 10.51 *Theorem 10.50 still holds if we drop the condition $h(x, z) \ge 0$, and $-u''$ is replaced by $Lu = -(pu')' + qu$.*

10.7   Applications of Bifurcation Problems

Consider the problem

$$\begin{cases} -\sum_{i,j=1}^{N} A_{ij(x,u\nabla u)}u_{x_i x_j} + \sum_{i=1}^{N} B_i(x, u, \nabla u)u_{x_i} + C(x, u, \nabla u)u \\ \qquad = \lambda a(x)u + F(x, u, \nabla u, \lambda) \quad \text{in } \Omega \\ u = 0 \qquad \text{on } \partial\Omega \end{cases} \tag{10.28}$$

where $\Omega \subset \mathbb{R}^N$ is a regular bounded domain, A_{ij}, B_i, C, a, $F \in C^1$, then there exists $\beta > 0$ such that

$$\sum_{i,j=1}^{N} A_{ij}(x, u, p)\xi_i\xi_j \ge \beta |\xi|^2 \quad \text{for } x \in \bar{\Omega}, \ u \in \mathbb{R}, \ p, \ \xi \in \mathbb{R}^N,$$

$a(x) > 0$ for $x \in \bar{\Omega}$, and $F(x, u, p, \lambda) = o((|u|^2 + |p|^2)^{1/2})$ near $(u, p) = (0, 0)$ uniformly for $x \in \bar{\Omega}$ and for λ in bounded intervals. Let

$$E = \{u \in C^{1,\alpha}(\Omega) \mid u = 0 \quad \text{on } \partial\Omega\}.$$

Fix $(\lambda, u) \in \mathbb{R} \times E$, let v be the solution of the equation

$$\begin{cases} \mathcal{L}(u)v = \lambda a(x)u - C(x, u, \nabla u)u + F(x, u, \nabla u, \lambda) \quad \text{in } \Omega, \\ v = 0 \quad \text{on } \partial\Omega, \end{cases}$$

where

$$\mathcal{L}u = -\sum_{i,j=1}^{N} A_{ij}(x, u, \nabla u)\frac{\partial^2}{\partial x_i \partial x_j} + \sum_{i=1}^{N} B_i(x, u, \nabla u)\frac{\partial}{\partial x_i}.$$

Set $T(\lambda, u) = v$ and apply the Schauder Estimation to show that T is a compact operator. For $(\lambda, u) \in \mathbb{R} \times E$, let w be the unique solution of the equation

$$\begin{cases} \mathcal{L}(0)w = \lambda a(x)u - C(x, 0, 0)u & \text{in } \Omega, \\ w = 0 & \text{on } \partial\Omega. \end{cases} \tag{10.29}$$

Set $L(\lambda)u = w$. Consider $L(\lambda_1)u = w_1$, $L(\lambda_2)u = w_2$:

$$\mathcal{L}(0)w_1 = \lambda_1 au - Cu \quad \text{in } \Omega, \quad w_1 = 0 \quad \text{on } \partial\Omega, \tag{10.30}$$

$$\mathcal{L}(0)w_2 = \lambda_2 au - Cu \quad \text{in } \Omega, \quad w_2 = 0 \quad \text{on } \partial\Omega. \tag{10.31}$$

By (10.30), (10.31) and the Schauder Estimation

$$\mathcal{L}(0)(w_1 - w_2) = (\lambda_1 - \lambda_2)a\, u,$$

$$\|L(\lambda_1)u - L(\lambda_2)u\|_{C^{1,\alpha}} \leq \|L(\lambda_1)u - L(\lambda_2)u\|_{C^{3,\alpha}}$$

$$= \|w_1 - w_2\|_{C^{3,\alpha}}$$

$$\leq c\,|\lambda_1 - \lambda_2|\,\|u\|_{C^{1,\alpha}}$$

or

$$\|L(\lambda_1) - L(\lambda_2)\| \leq c\,|\lambda_1 - \lambda_2|,$$

and consequently $\lambda \to L(\lambda)$ is Lipschitz-continuous. Moreover, $L(\lambda) : E \to E$ is a compact linear operator. Set

$$H(\lambda, u) = T(\lambda, u) - L(\lambda)u.$$

Then $\|H(\lambda, u)\|_{C^{1,\alpha}} = o(\|u\|_{C^{1,\alpha}})$ near 0 uniformly for λ in bounded intervals. In fact by contradiction, suppose that $\epsilon > 0$ and there is a sequence $(\lambda_n, u_n) \to (\lambda, 0)$ such that

$$\frac{\|H(\lambda_n, u_n)\|_{C^{1,\alpha}}}{\|u_n\|_{C^{1,\alpha}}} > \epsilon \quad \text{for each } n.$$

Set $v_n' = T(\lambda_n, u_n)$, $w_n' = L(\lambda_n)u_n$, $v_n = \frac{v_n'}{\|u_n\|_{1,\alpha}}$, $w_n = \frac{w_n'}{\|u_n\|_{1,\alpha}}$, where v_n and w_n are solutions of the equation

$$\begin{cases} \mathcal{L}(u_n)v_n = [\lambda_n a(x)u_n - C(x, u_n, \nabla u_n)u_n + F(x, u_n, \nabla u_n, \lambda_n)]\|u_n\|_{1,\alpha}^{-1} \\ \qquad \text{in } \Omega, \\ v_n\,|_{\partial\Omega} = 0. \end{cases} \tag{10.32}$$

$$\begin{cases} \mathcal{L}(0)w_n = [\lambda_n a(x)u_n - C(x, 0, 0)u_n]\|u_n\|_{1,\alpha}^{-1} & \text{in } \Omega, \\ w_n\,|_{\partial\Omega} = 0. \end{cases} \tag{10.33}$$

Because $(\lambda_n, u_n) \to (\lambda, 0)$, $\{\lambda_n\}$ and $\{u_n\}$ are bounded in $\mathbb{R}$ and $C^{1,\alpha}$, respectively. Therefore, the C^α-norms of the right hand sides of (10.32) and (10.33) are bounded for each n. By the Schauder Estimation, $\|v_n\|_{C^{2,\alpha}} \le c$, $\|w_n\|_{C^{2,\alpha}} \le c$ for each n. However the injections $i : C^{1,\alpha}(\Omega) \to C^1(\Omega)$, and $i : C^{2,\alpha}(\Omega) \to C^2(\Omega)$ are compact, and there is a subsequence of $\{u_n\}$, still denoted by $\{u_n\}$, $\varphi \in C^1(\Omega)$, v, $w \in C^2(\Omega)$ such that

$$\frac{u_n}{\|u_n\|_{C^{1,\alpha}}} \to \varphi \quad \text{in } C^1(\Omega),$$

$$v_n \to v \quad \text{in } C^2(\Omega),$$

$$w_n \to w \quad \text{in } C^2(\Omega),$$

where v, w are two solutions of the equation in ψ :

$$\begin{cases} \mathcal{L}(0)\psi = \lambda a(x)\psi - C(x,0,0)\psi \quad \text{in } \Omega, \\ \psi\,|_{\partial\Omega} = 0. \end{cases}$$

By the Schauder Estimation $v = w$. Thus, there is a subsequence, still denoted by $\{v_n\}$, such that

$$\frac{\|H(\lambda_n, u_n)\|_{1,\alpha}}{\|u_n\|_{1,\alpha}} = \left\| \frac{v_n'}{\|u_n\|_{1,\alpha}} - \frac{w_n'}{\|u_n\|_{1,\alpha}} \right\|_{1,\alpha}$$

$$= \|v_n - w_n\|_{1,\alpha}$$

$$\to \|v - w\|_{1,\alpha} = 0,$$

a contradiction.

The equation (10.28) is equivalent to

$$u = L(\lambda)u + H(\lambda, u).$$

Take $M > 0$ such that

$$Ma(x) + C(x,0,0) \ge 0 \quad \text{for } x \in \overline{\Omega}.$$

Introduce $v = L_M u$, the solution of the equation

$$\begin{cases} \mathcal{L}(0)v + [Ma(x) + C(x,0,0)]v = a(x)u \quad \text{in } \Omega \\ v = 0 \quad \text{on } \partial\Omega. \end{cases}$$

Consider the convex cone

$$K = P^+ = \{u \in E \mid u > 0 \quad \text{in } \Omega, \ \frac{\partial u}{\partial n} < 0 \quad \text{on } \partial\Omega\},$$

$$P^- = -P^+,$$

$$P = P^+ \cup P^-.$$

P, P^-, P^+ are open in E. L_M is a compact linear operator on E, and is strictly positive by the order defined by K. Let $u \in K$, then by the strong maximum principle

$$\begin{cases} \mathcal{L}(0)v + kv = a(x)u > 0 & \text{in } \Omega, \\ \qquad\qquad\quad v = 0 & \text{on } \partial\Omega. \end{cases}$$

where $k(x) = Ma(x) + C(x, 0, 0)$. That is,

$$\begin{cases} -\sum a_{ij}\dfrac{\partial^2 v}{\partial x_i\, \partial x_j} + \sum b_i \dfrac{\partial v}{\partial x_i} + kv = au > 0 \text{ in } \Omega, \\ v = 0 \text{ on } \partial\Omega. \end{cases}$$

If v attains its nonpositive minimum at $x_0 \in \Omega$, then

$$\begin{cases} v(x_0) \le 0, \\ \nabla v(x_0) = 0, \\ D_{ij}v(x_0) \text{ is nonnegative.} \end{cases}$$

Because (a_{ij}) is positive, we have $-\sum a_{ij}\frac{\partial^2 v}{\partial x_i\, \partial x_j} \le 0$. Thus ,

$$0 \ge -\sum_{i,j} a_{ij}\frac{\partial^2 v}{\partial x_i\, \partial x_j}(x_0) + \sum_i b_i \frac{\partial v}{\partial x_i}(x_0) + kv(x_0) > 0,$$

a contradiction, showing $v > 0$ in Ω. We can also obtain $\frac{\partial v}{\partial n} < 0$. Therefore, $L_M K \subset \overset{\circ}{K} = K$. By the Krein-Rutman Theorem 10.39, L_M admits a simple characteristic value $\widetilde{\mu}_1 > 0$ associated with an eigenvector $v_1 \in P^+$ such that $\|v_1\|_{1,\alpha} = 1$. Note that

$$u = L(\lambda)u \qquad \text{if and only if} \qquad \mathcal{L}(0)u = \lambda a u - C u,$$
$$u = (\lambda + M)L_M u \qquad \text{if and only if} \qquad \mathcal{L}(0)u + [Ma + C]u = (\lambda + M)a u.$$

Therefore, $u = L(\lambda)u$ and $u = (\lambda + M)L_M u$ are equivalent. Let $\mu_1 = \widetilde{\mu}_1 - M$, then L_M is a linear compact operator, so if $0 < |\lambda - \mu_1| < \eta$, then 0 is the only solution of the equation

$$u = (\lambda + M)L_M u$$

that is equivalent to $u = L(\lambda)u$. In fact

$$0 < |\lambda - \mu_1| < \eta \qquad \text{if and only if} \qquad 0 < |(\lambda + M) - (\mu_1 + M)| < \eta,$$

and because $\sigma(L_M)\backslash\{0\}$ is discrete. Therefore, the index $i(I - L(\lambda), 0, 0)$ is defined.

Remark 10.52 *In the equation*

$$\Phi(\lambda, u) = 0$$

where $\Phi(\lambda, u) = u - G(\lambda, u) = u - (\lambda\, Lu + H(\lambda, u))$ if we replace λL by a family $L(\lambda)$ of linear compact operators on E such that $\lambda \to L(\lambda)$ is continuous and the index $i(I - L(\lambda), 0, 0)$ is defined for λ such that $|\lambda - \mu| > 0$ and it is sufficiently small. The index changes sign when λ passes through μ. Then from the proof of the Rabinowitz Theorem, there is a connected component C of $N \cup \{(\mu, 0)\}$, containing $(\mu, 0)$, which is either unbounded or meets the point $(\hat{\mu}, 0)$ where $\hat{\mu} \neq \mu$, a characteristic value. Because $\hat{\mu}_1$ is of odd multiplicity of $u = (\lambda + M)L_M u$, to prove that $u = L(\lambda)u$ changes sign when λ passes through μ_1, it suffices to prove

Theorem 10.53 $i(I - L(\lambda), 0, 0) = i(I - (\lambda + M)L'_M, 0, 0)$.

Proof.   We require

$$\mathrm{d}(I - L(\lambda), B_1(0), 0) = \mathrm{d}(I - (\lambda + M)L_M, B_1(0), 0),$$

for $t \in [0, 1]$, $u \in E$. Let $v = S(t, u)$ be the unique solution of

$$\begin{cases} \mathcal{L}(0)v + t[M(a(x) + C(x, 0, 0)]v = \lambda\, a(x)u - (1 - t)C(x, 0, 0)u \\ \quad + tMa(x)u \ \text{in}\, \Omega \\ v = 0 \quad \text{on}\, \partial\Omega. \end{cases}$$

By the Schauder estimation and because the injection $i : C^{2,\alpha}(\Omega) \to C^{1,\alpha}(\Omega)$ is compact, $S : [0, 1] \times \overline{B_1(0)} \to E$ is compact. $u = S(t, u)$ can be written as

$$\begin{cases} \mathcal{L}(0)u = \lambda\, a(x)u - C(x, 0, 0)u \ \text{in}\, \Omega \\ u = 0 \ \text{on}\, \partial\Omega. \end{cases}$$

Note that

$$u = S(t, u) \ \text{if and only if}\ u = L(\lambda)u \ \text{if and only if}\ u = (\lambda + M)L_M u.$$

Thus for $0 < |\lambda - \mu_1| < \eta$, S does not vanish at $[0, 1] \times \partial B_1(0)$. Because $S(0, \cdot) = L(\lambda)$, $S(1, \cdot) = (\lambda + M)L_M$, by Homotopic Invariance $d(I - S(t, \cdot), B_1(0), 0)$ is constant for $t \in [0, 1]$, and we obtain

$$\mathrm{d}(I - L(\lambda), B_1(0), 0) = \mathrm{d}(I - (\lambda + M)L_M, B_1(0), 0).$$

By the Rabinowitz Theorem 4.22, there is a connected component C of solutions of (10.28) containing $(\mu_1, 0)$, which is either unbounded or meets $(\hat{\mu}, 0)$ such that $\hat{\mu} \neq \mu_1$, a characteristic value of $L(\cdot) : I - L(\hat{\mu})$, is not invertible.   ∎

Let N be the closure of the set of all nontrivial solutions of (10.28) in $\mathbb{R} \times E$.

Lemma 10.54 (*i*) *There is a neighborhood N of $(\mu_1, 0)$ such that $(\lambda, u) \in N \cap N$, $u \neq 0$ implies $u \in P$.*
(*ii*) *Given $\hat{\mu}$ such that $I - L(\hat{\mu})$ is not invertible, $\hat{\mu} \neq \mu$, there is a neighborhood $\hat{N}$ of $(\hat{\mu}, 0)$ such that $(\lambda, \mu) \in \hat{N} \times N$. Then $u \notin P$.*

Proof.

(*i*) By contradiction. Suppose there is a sequence $(\lambda_n, u_n) \in N$, $u_n \neq 0$, $u_n \notin P$, $(\lambda_n, u_n) \to (\mu_1, 0)$. Then

$$\frac{u_n}{\|u_n\|_{1,\alpha}} = L(\lambda_n) \frac{u_n}{\|u_n\|_{1,\alpha}} + \frac{H(\lambda_n, u_n)}{\|u_n\|_{1,\alpha}}$$

$$= [L(\lambda_n) - L(\mu_1)] \frac{u_n}{\|u_n\|_{1,\alpha}} + L(\mu_1) \frac{u_n}{\|u_n\|_{1,\alpha}} + \frac{H(\lambda_n, u_n)}{\|u_n\|_{1,\alpha}}.$$

Now $\|[L(\lambda_n) - L(\mu_1)] \frac{u_n}{\|u_n\|_{1,\alpha}}\| \leq c \, |\lambda_n - \mu_1| \to 0$, and $\frac{H(\lambda_n, u_n)}{\|u_n\|_{1,\alpha}} \to 0$. $L(\mu_1)$ is a compact linear operator, $\{\frac{u_n}{\|u_n\|_{1,\alpha}}\}$ is bounded, then there is a subsequence, still denoted by $\frac{u_n}{\|u_n\|_{1,\alpha}}$, $u \in E$ such that $L(\mu_1) \frac{u_n}{\|u_n\|_{1,\alpha}} \to u$, so $\frac{u_n}{\|u_n\|_{1,\alpha}} \to u$, and consequently $u = L(\mu_1)u$, $\|u\|_{1,\alpha} = 1$ or

$$\begin{cases} u = (\mu_1 + M)L_M u, \\ \|u\|_{1,\alpha} = 1. \end{cases}$$

However, the eigenvector space of L_M in P is 1-dimensional. Therefore, $u = \pm v_1 \in P$, where v_1 is the unique eigenfunction of L_M in P^+, $\|v_1\|_{1,\alpha} = 1$, $u_{n_j} \in P$ for large j because P is a cone.

(*ii*) By contradiction. We have $(\lambda_n, u_n) \in N$ and $u_n \in P$, $(\lambda_n, u_n) \to (u, 0)$, so

$$\frac{u_n}{\|u_n\|_{1,\alpha}} = L(\lambda_n) \frac{u_n}{\|u_n\|_{1,\alpha}} + \frac{H(\lambda_n, u_n)}{\|u_n\|_{1,\alpha}}.$$

As in (*i*)

$$\frac{u_n}{\|u_n\|_{1,\alpha}} \to u, \ \|u\|_{1,\alpha} = 1, \ u \in \bar{P}, \ u = (\hat{\mu} + M)L_M u.$$

Because $\mu_1 \neq \hat{\mu}$, if w is the eigenvector associated with μ_1, then w and u are linearly independent: $\bar{P}$ contains an eigenspace of dimension 2, a contradiction. ∎

Lemma 10.55 *Let $(\lambda, u) \in C$, $(\lambda, u) \neq (\mu, 0)$. If (λ, u) is the limit of a sequence $(\lambda_n, u_n) \in (\mathbb{R} \times P) \cap N$, then $u \in P$.*

Proof.   Suppose $u \notin P$. If $u \in \partial P^+$ then either (a) $\xi \in \Omega$ and $u(\xi) = 0$ exists, or (b) $\eta \in \partial \Omega$ and $\frac{\partial u}{\partial n}(\eta) = 0$ exists. Take $M_1 < 0$ large such that

$$\begin{cases} M_1 a(x) + C(x, 0, 0) > 0 & \text{in } \Omega, \\ \lambda + M_1 > 0. \end{cases}$$

u satisfies

$$\begin{cases} [\mathcal{L}(u) + M_1 a(x) + C(x, u, \nabla u)]u \\ \quad = (\lambda + M_1)a(x)u + F(x, u, \nabla u, \lambda) \text{ in } \Omega \\ u = 0 \text{ on } \partial \Omega \end{cases}$$

(a) $u(\xi) = 0$ for some $\xi \in \Omega$: because $F(x, u, p, \lambda) = o(|u|^2 + |p|^2)^{1/2})$ near $(u, p) = (0, 0)$ uniformly $x \in \bar{\Omega}$, and for λ in bounded intervals. There is a regular neighborhood Ω_0 of ξ, $\Omega_0 \subset \Omega$ such that

$$(\lambda + M_1)a(x)u + F(x, u, \nabla u, \lambda) \geq 0 \qquad \text{for } x \in \Omega_0.$$
$$M_1 a(x) + C(x, u, \nabla u) \geq 0.$$

By the Maximum Principle, $u = 0$ on Ω_0, then by the connectedness $u = 0$ on Ω.

(b) Similarly to (a), $u = 0$ on Ω.

By (a), (b), $(\lambda, u) = (\hat{\mu}, 0)$, $\hat{\mu} \neq \mu_1$ and $I - L(\hat{\mu})$ is not invertible, contradicting Lemma 10.54.   ∎

Theorem 10.56 *the component C of solutions of the equation (10.28) containing $(\mu_1, 0)$ is contained in $(\mathbb{R} \times P) \cup \{(\mu, 0)\}$ and is unbounded.*

Proof.   By Lemma 10.54, it suffices to show that $C \subset (\mathbb{R} \times P) \cup \{(\mu, 0)\}$. Because C is connected and $C \cap N \subset (\mathbb{R} \times P) \cup \{(\mu, 0)\}$. If $C \subset (\mathbb{R} \times P) \cup \{(\mu_1, 0)\}$, then $(\lambda, u) \in (\mathbb{R} \times \partial P) \cap C$ is a limit of a sequence in $(\mathbb{R} \times P) \cap C$, contradicting Lemma 10.55.   ∎

10.8   Bifurcation at Infinity

Let E be a real Banach space, $L : E \to E$ a linear compact operator, and $K : \mathbb{R} \times E \to E$ a compact operator. Consider the equation

$$u = \lambda Lu + K(\lambda, u). \tag{10.34}$$

Definition 10.57 (μ, ∞) *is a bifurcation point for (10.34) if there is a sequence (λ_n, u_n) of solutions of (10.34) such that $\lambda_n \to \mu$, $\|u_n\| \to \infty$.*

Let $w = \frac{u}{\|u\|^2}$, then $\|w\| = \frac{1}{\|u\|}$, $u = \frac{w}{\|w\|^2}$. We can write (10.34) as

$$w = \lambda L w + H(\lambda, w), \qquad (10.35)$$

where

$$H(\lambda, w) = \|w\|^2 K(\lambda, \frac{w}{\|w\|^2}).$$

Therefore, the study of the bifurcation of (μ, ∞) for (10.34) is the study of the bifurcation $(\mu, 0)$ for (10.35).

Lemma 10.58 *If $K : \mathbb{R} \times E \to E$ is compact with $K(\lambda, u) = o(\|u\|)$ near ∞ uniformly for λ in bounded intervals, let*

$$H(\lambda, w) = \begin{cases} \|w\|^2 K(\lambda, \dfrac{w}{\|w\|^2}) & \text{for } w \neq 0, \\[2mm] 0 & \text{for } w = 0. \end{cases}$$

Then H is compact, $H(\lambda, w) = o(\|w\|)$ near 0, uniformly for λ in bounded intervals.

Proof.

(*i*) Let $w = \frac{u}{\|u\|^2}$. Then $u = \frac{w}{\|w\|^2}$, and $\|u\| \to \infty$ if and only if $\|w\| \to 0$. If $\|w\| \to 0$, then $\|u\| \to \infty$, and consequently

$$\frac{1}{\|w\|} H(\lambda, w) = \|w\| K(\lambda, \frac{w}{\|w\|^2})$$

$$= \frac{K(\lambda, u)}{\|u\|} \longrightarrow 0 \text{ uniformly for } \lambda \text{ in bounded intervals.}$$

(*ii*) H is compact: let (λ_n, w_n) be a bounded sequence in $\mathbb{R} \times E$, and $z_n = H(\lambda_n, w_n)$.

(a) If there is a subsequence of $\{w_n\}$, still denoted by $\{w_n\}$, such that $w_n \to 0$, then, for $w_n = \frac{u_n}{\|u_n\|^2}$,

$$z_n = H(\lambda_n, w_n) = \frac{K(\lambda_n, u_n)}{\|u_n\|^2} \to 0 \quad \text{as } n \to \infty.$$

(b) If there are $\delta > 0$ and $N > 0$ such that $\delta \leq \|w_n\| \leq c$ for $n \geq N$, then $\{\|w_n\|^2\}$, $\{\frac{w_n}{\|w_n\|^2}\}$, $\{\lambda_n\}$ are bounded for $n \geq N$. Because K is compact, there is a convergent subsequence of $\{\|w_n\|^2\}$, $\{\frac{w_n}{\|w_n\|^2}\}$, $\{\lambda_n\}$ respectively, still denoted by the original sequences, such that

$$z_n = H(\lambda, w_n) = \|w_n\|^2 K(\lambda_n, \frac{w_n}{\|w_n\|^2}) \quad \text{converges.}$$

By (a) and (b) we conclude that H is compact.   ∎

By the Rabinowitz Theorem 4.22, we obtain

Theorem 10.59 *Let $L : E \to E$ be a linear compact operator, and let $K : \mathbb{R} \times E \to E$ be compact with $K(\lambda, u) = o(\|u\|)$ near ∞ uniformly for λ in bounded intervals. Suppose μ is a characteristic value of L of multiplicity odd. Then there is a connected component D of $N \cup \{(\mu, \infty)\}$ containing (μ, ∞), where N is the set of all solutions of (8.1), and D is unbounded. Moreover, if M is a neighborhood of (μ, ∞) and its projection on $\mathbb{R}$ is bounded and does not contain another characteristic value of L except μ, and if the closure of the projection on E does not contain 0, then we have either:*
(i) $D \backslash M$ is bounded in $\mathbb{R} \times E$; in this case $D \backslash M$ meets the line $\mathbb{R} \times \{0\}$; or
(ii) $D \backslash M$ is unbounded in $\mathbb{R} \times E$; in this case, if $D \backslash M$ has a bounded projection in $\mathbb{R}$, then $D \backslash M$ meets a point $(\hat{\mu}, \infty)$, or $\hat{\mu} \neq \mu$, and $\hat{\mu}$ is a characteristic value of L. Let

$$E = \{u \in C^1([0, \pi]) \mid a_0 u(0) + b_0 u'(0) = 0, a_1 u(\pi) + b_1 u'(\pi) = 0,$$
$$(a_0^2 + b_0^2)(a_1^2 + b_1^2) \neq 0\},$$

under the norm $\|\cdot\|_1$. Let

$$\mathcal{L}u = -(pu')' + qu,$$

where $p \in C^1([0, \pi])$, $p > 0$ on $[0, \pi]$, $q \in C^0([0, \pi])$.

Consider the Sturm-Liouville nonlinear equation.

$$\begin{cases} \mathcal{L}u = F(x, u, u', \lambda) & \text{in } (0, \pi), \\ a_0 u(0) + b_0 u'(0) = 0, \\ a_1 u(\pi) + b_1 u'(\pi) = 0, \\ (a_0^2 + b_0^2)(a_1^2 + b_1^2) \neq 0, \end{cases} \tag{10.36}$$

where F is continuous, and

$$F(x, u, p, \lambda) = \lambda a(x)u + f(x, u, p, \lambda),$$
$$f(x, u, p, \lambda) = o((u^2 + p^2)^{1/2})$$

as $|u| + |p| \to \infty$ uniformly for $x \in [0, \pi]$ and for λ in bounded intervals.
Let $\mu_1 < \mu_2 < \cdots < \mu_k < \cdots$ be the eigenvalues of the linear problem

$$\begin{cases} \mathcal{L}v = \mu a(x)v & \text{in } (0, \pi), \\ a_0 v(0) + b_0 v'(0) = 0, \\ a_1 v(\pi) + b_1 v'(\pi) = 0, \\ (a_0^2 + b_0^2)(a_1^2 + b_1^2) \neq 0. \end{cases} \tag{10.37}$$

Theorem 10.60 *For each $k \in \mathbb{N}$, there is a connected component D_k of $N \cup \{(\mu_k, \infty)\}$ containing (μ_k, ∞), where N is the set of all solutions of equation (10.36), and D_k is unbounded and obtained by the alternative of Theorem 10.59. Moreover, there is a neighborhood O_k of (μ_k, ∞) such that $(\lambda, \mu) \in O_k \cap D_k$ implies $u \in S_k$.*

Proof. If 0 is an eigenvalue of L, then we can replace L by the operator $L_\epsilon = L + \epsilon$ for small ϵ. Therefore, we can assume that 0 is not an eigenvalue of L. Let $g(x, y)$ be the Green function associated with L together with boundary conditions. The equation (10.36) is equivalent to the equation on E :

$$u = \lambda Lu + K(\lambda, u),$$

where

$$Lu(x) = \int_0^\pi g(x, y) a(y) u(y) dy,$$

$$K(\lambda, u)(x) = \int_0^\pi g(x, y) f(y, u(y), u'(y), \lambda) dy.$$

Then $L : E \to E$ is a linear compact operator, and $K : \mathbb{R} \times E \to E$ is a compact operator. We claim that $K(\lambda, u) = o(\|u\|)$ near $u = \infty$ uniformly for λ in bounded intervals. In fact, let $\Lambda \subset \mathbb{R}$ be a bounded interval. For $t \geq 0$, set

$$\varphi(t) = \max_{\substack{x \in [0,\pi] \\ \lambda \in \Lambda \\ |u| + |p| \leq t}} |f(x, u, p, \lambda)|.$$

Given $\epsilon > 0$, choose $M > 0$ such that $\lambda \in \Lambda$, $|u| + |p| \geq M$ implies $|f(x, u, p, \lambda)| \leq \epsilon(|u| + |p|)$. Now

$$\|K(\lambda, u)\|_1 = \max_{x \in [0, \pi]} \left| \int_0^\pi g f dy \right| + \max_{x \in [0, \pi]} \left| \int_0^\pi g_x' f dy \right|,$$

where $g \in C^1([0, x])$, $g \in C^1([x, \pi])$, $\frac{\partial g}{\partial x}(x+) - \frac{\partial g}{\partial x}(x-) = -1$. Then there is $c_1 > 0$ such that

$$\|K(\lambda, u)\|_1 \leq c_1 \int_{\{y \in [0,\pi] \mid |u(y)| + |u'(y)| \leq M\}} |f(y, u(y), u'(y), \lambda)| \, dy$$

$$+ c_1 \int_{\{y \in [0,\pi] \mid |u(y)| + |u'(y)| \geq M\}} |f(y, u(y), u'(y), \lambda)| \, dy$$

$$\leq c_1 \pi [\varphi(M) + \epsilon \|u\|_1] \quad \text{for } \lambda \in \Lambda$$

$$\leq 2 c_1 \pi \epsilon \|u\|_1 \quad \text{for } \lambda \in \Lambda, \ \|u\|_1 \geq \frac{\varphi(M)}{\epsilon}.$$

Applying Theorem 10.59, we obtain D_k, which is unbounded and contains (μ_k, ∞) thus obtaining the alternative of Theorem 10.59. ∎

Theorem 10.61 *The solutions of equation (10.36) near ∞ correspond by the inversion $w = \frac{u}{\|u\|_1^2}$ to solutions near 0 of the equation*

$$\begin{cases} \mathcal{L}w = \lambda\,a(x)w + \bar{f}(x,w,w',\lambda) & \text{in } (0,\pi), \\ \text{boundary condition}, \end{cases}$$

where

$$\bar{f}(x,w,w',\lambda) = \begin{cases} \|w\|_1^2\,f(x,\dfrac{w}{\|w\|_1^2},\dfrac{w'}{\|w\|_1^2},\lambda) & \text{if } w \neq 0, \\ 0 & \text{if } w = 0. \end{cases}$$

In this case there is a neighborhood O_k of (μ_k,∞) such that $(\lambda,u) \in O_k \cap D_k$ implies $u \in S_k$.

Remark 10.62 *The bifurcation component $C_k \subset \mathbb{R} \times S_k$ with respect to the trivial solution line containing $(\mu_k,0)$, in the case of bifurcation asymptotic, in general we do not have the inclusion $D_k \subset (\mathbb{R} \times S_k) \cup \{(u_k,\infty)\}$ as the following example shows:*

Example 10.63 *Consider the equation*

$$\begin{cases} -u'' = \lambda(u+1) & \text{in } (0,\pi), \\ u(0) = u(\pi) = 0. \end{cases}$$

The components D_k in the sense of the space $(\mathbb{R} \times E) \cup \{(\mu_k,\infty)\}$ are as follows:

$$D_1 = \{(0,0)\} \cup \{(\lambda,u_\lambda) \mid \lambda \in (0,1) \cup (1,9)\}$$
$$\cup\{(1,\infty)\} \cup \{(4,u_4 + \alpha\sin 2x),\ \alpha \in \mathbb{R}\},$$
$$D_2 = \{(\lambda,u_\lambda) \mid \lambda \in (1,9)\} \cup \{(4,\ u_4 + \alpha\sin 2x),\ \alpha \in \mathbb{R}\} \cup \{(4,\infty)\},$$
$$D_3 = (D_2\backslash\{(4,\infty)\}) \cup \{(9,\infty)\} \cup (D_4\backslash\{(16,\infty)\}),\ \text{etc.},$$

where $u_\lambda(x) = -1 + \cos\sqrt{\lambda}x + \frac{1-\cos\sqrt{\lambda}\pi}{\sin\sqrt{\lambda}\pi}\sin\sqrt{\lambda}x$ for $\lambda \neq n^2$, $n \in \mathbb{N}$, and $u_4(x) = -1 + \cos 2x$. Because $S_1 \cap S_2 = \emptyset$, $S_1 \cap S_3 = \emptyset$ and $S_2 \cap S_3 = \emptyset$, $D_1 \cap D_2 \cap D_3 \supset \{(\lambda,u_\lambda) \mid \lambda \in (1,9)\}$. It is impossible to have the previous inclusion.

Corollary 10.64 *In equation (10.36), set*

$$f(x,u,p,\lambda) = f_1(x,u,p,\lambda)u + f_2(x,u,p,\lambda)p,$$

where f_1 and f_2 are continuous, then $D_k\backslash O_k$ contains a connected subset in $\mathbb{R} \times (S_k \cup \{0\})$, which is either unbounded or meets the line $\mathbb{R} \times \{0\}$. Let T_k be the maximal connected subset of $D_k \subset \mathbb{R} \times (S_k \cup \{0\})$. If $T_k\backslash O_k$ is bounded and does not meet $\mathbb{R} \times \{0\}$, then for some $(\lambda,1) \in \bar{D}_k$, $u \in \partial S_k$: u admits at least a double zero in $[0,\pi]$. Apply $f = f_1 u + f_2 p$ to obtain $u = 0$.

Example 10.65 *Consider the equation*

$$\begin{cases} -u'' = \lambda(1 + f_1(u))u & \text{in } (0, \pi), \\ u(0) = u(\pi) = 0, \end{cases}$$

where

$$f_1(x) = \begin{cases} -1 & \text{if } |x| \leq 1, \\ 0 & \text{if } |x| \geq 2, \\ x - 2 & \text{if } 1 \leq x \leq 2, \\ -x - 2 & \text{if } -2 \leq x \leq -1. \end{cases}$$

Thus, $\lambda(1 + f_1(u))u = 0$ for $\|u\|_1 \leq 1$. There is no bifurcation point on the line $\mathbb{R} \times \{0\}$, so D_k does not meet $\mathbb{R} \times \{0\}$. This is the first situation in Corollary 10.64. In this case D_k is global and unbounded in $(\mathbb{R} \times S_k) \cup \{(\mu_k, \infty)\}$. Note that $D_k \cap (\mathbb{R} \times \partial S_n) = \emptyset$.

Example 10.66 *Consider the equation*

$$\begin{cases} -u'' = \lambda[1 + (1 + u^2)^{-1}]u & \text{in } (0, \pi). \\ u(0) = u(\pi) = 0. \end{cases} \tag{10.38}$$

By the Sturm Comparison Theorem 10.47, $(\lambda, u) \in T_k$ implies $\lambda \in [\frac{k^2}{2}, k^2]$, so T_k does not contain another point (k'^2, ∞), $k' \neq k$. Thus, $T_k \backslash O_k$ is bounded. This is the second situation in Corollary 10.64. T_k contains a point $(\mu, 0)$, $\mu \in [\frac{k^2}{2}, k^2]$. This point may be $(\frac{k^2}{2}, 0)$ because the linear part of equation (10.37) near $u = 0$ is

$$\begin{cases} -u'' = 2\lambda u & \text{in } (0, \pi), \\ u(0) = u(\pi) = 0. \end{cases}$$

The bifurcation on $\mathbb{R} \times \{0\}$ results in

$$D_k = T_k \subset (\mathbb{R} \times S_k) \cup \{(\frac{k^2}{2}, 0)\} \cup \{k^2, \infty\}.$$

Corollary 10.67 *In equation (10.36),*

$$f(x, u, p, \lambda) = f_1(x, u, p, \lambda)u + f_2(x, u, p, \lambda)p$$
$$f_i(x, 0, 0, \lambda) = 0 \quad i = 1, 2.$$

Then D_k only meets $\mathbb{R} \times \{0\}$ at the point $(\mu_k, 0)$. We have

$$D_k \subset (\mathbb{R} \times S_k) \cup \{(\mu_k, \infty)\} \cup \{(\mu_k, 0)\}.$$

Proof.

(*i*) If D_k does not meet $\mathbb{R} \times \{0\}$, then $D_k \cap (\mathbb{R} \times \partial S_k) = \emptyset$, $D_k \subset (\mathbb{R} \times S_k) \cup \{(\mu_k, \infty)\}$.

(*ii*) If D_k meets $\mathbb{R} \times \{0\}$, then it only meets $(\mu_k, 0)$, because by Theorem 10.60, there is a neighborhood N_i in $\mathbb{R} \times E$ of $(\mu_i, 0)$ such that $(\lambda, u) \in N_i \cap N$, $u \neq 0$ implies $u \in S_i$. ∎

10.9   Existence of Pairs of Positive Solutions

Consider the equation

$$\begin{cases} Lu = \lambda f(x,u) & \text{in } \Omega, \\ u = 0 & \text{on } \partial\Omega, \end{cases} \tag{10.39}$$

where L is a differential operator having a divergence structure defined by

$$Lu = \sum_{i,j=1}^{N} \frac{\partial}{\partial x_i}\left(a_{ij}(x)\frac{\partial u}{\partial x_j}\right) + c(x)u,$$

where a_{ij}, c are regular, $c \geq 0$ on $\bar{\Omega}$, $a_{ij} = a_{ji}$ for i, j, and $-L$ is uniformly elliptic in $\bar{\Omega}$. Then for some $\beta > 0$, $\sum a_{ij}(x)\xi_i\,\xi_j \geq \beta\,|\xi|^2$ for $x \in \bar{\Omega}$, $\xi \in \mathbb{R}^N$. Let $f : \bar{\Omega} \times \mathbb{R}^+ \to \mathbb{R}$ satisfy:

(fi) f is locally Lipschitz on $\bar{\Omega} \times \mathbb{R}^+$, $f(x,0) = 0$ for $x \in \bar{\Omega}$;

(fii) there is $\bar{z} > 0$ such that $f(x,z) < 0$ for $z \geq \bar{z}$;

$(fiii)$ $f(x,z) > 0$ for $z \in U\backslash\{0\}$, U a neighborhood of $z = 0$;

(fiv) $f(x,z) = o(|z|)$ in a neighborhood of $z = 0$, uniformly for $x \in \bar{\Omega}$.

For example, the function $f(z) = z^2 - z^3$ satisfies $(fi) - (fiv)$.

Theorem 10.68 *For some $\underline{\lambda} > 0$ such that $\lambda > \underline{\lambda}$, equation (10.39) possesses at least two distinct solutions $\underline{u}(\lambda)$, $\bar{u}(\lambda)$ satisfying $0 < \bar{u}(\lambda)$, $\underline{u}(\lambda) < \bar{z}$ in Ω.*

Proof.   We divide the proof into steps.

(i) Let u be a solution of (10.39) for $\lambda > 0$ with $u \geq 0$ in Ω, then $0 \leq u(x) < \bar{z}$ for $x \in \Omega$: let $\bar{x} \in \Omega$ with $u(\bar{x}) = \max_{\Omega} u$. Suppose $u(\bar{x}) \geq \bar{z}$, by (f_2), $f(\bar{x}, u(\bar{x})) < 0$. Because f is continuous on $\bar{\Omega} \times \mathbb{R}^+$, there is an open ball $B = B_r(\bar{x}) \subset \Omega$ such that $f(x, u(x)) < 0$ for $x \in B$. In this case $Lu(x) = \lambda f(x, u(x)) < 0$ for $x \in B$. Apply the Maximum Principle 10.7 to show that u is constant on B, so

$$0 > \lambda f(x, u(x)) = Lu(x) = c(x)u(x) \geq 0,$$

a contradiction. Thus, $0 \leq u(x) < \bar{z}$ for $x \in \Omega$.

(ii) $|u|_{W^{2,p}} \leq c_3\,\lambda$, $\|u\|_{C^{1,1-N/p}} \leq c_5\,\lambda$: by the Agmon-Douglas-Nirenberg Estimation 10.3, for the solution $u \geq 0$ of (10.39) and $p > 1$,

$$\begin{aligned} |u|_{W^{2,p}} &\leq c_1\lambda\,|f(\cdot,u)|_{L^p} \\ &\leq c_1'\lambda\,|f(\cdot,u)|_{L^\infty} \\ &\leq c_2\lambda_0 \max_{\substack{x\in\bar{\Omega} \\ 0\leq z\leq\bar{z}}} |f(x,z)| = c_2\lambda. \end{aligned}$$

By the Sobolev Inequality, for $p > N$, $\alpha = 2 - \frac{N}{p} - 1 = 1 - \frac{N}{p}$, $0 < \alpha < 1$, so $W_0^{2,p}(\Omega) \subset C^{1,\alpha}(\Omega)$, and we have

$$\|u\|_{C^{1,1-(N/p)}} = \|u\|_{C^{1,\alpha}} \le c\,|u|_{W^{2,p}} \le c_5 \lambda.$$

(iii) Monotone Scheme: by (fi), there is $K > 0$ such that for $x \in \bar{\Omega}$ and $z_1, z_2 \in [0, \bar{z}]$, $|f(x, z_1) - f(x, z_2)| \le K\,|z_1 - z_2|$. Let $h(x, z) = Kz + f(x, z)$, $h(x, 0) = 0$. For $z_1, z_2 \in [0, \bar{z}]$,

$$[h(x, z_1) - h(x, z_2)](z_1 - z_2) = K(z_1 - z_2)^2 + [f(x, z_1) - f(x, z_2)](z_1 - z_2) \ge 0$$

Thus, $h(x, z)$ is monotone increasing on $[0, \bar{z}]$. Set $L_\lambda = L + \lambda K$. Write (10.39) as

$$\begin{cases} L_\lambda u = \lambda h(x, u) & \text{in} \quad \Omega \\ u = 0 & \text{on } \partial\Omega. \end{cases} \tag{10.40}$$

Set $\bar{u}_0(x) = \bar{z}$ for $x \in \Omega$, then

$$\begin{cases} L\bar{u}_0 = c(x)\bar{u}_0 \ge 0 \ge \lambda f(x, \bar{u}_0) & \text{in } \Omega, \\ \bar{u}_0 \ge 0 & \text{on } \partial\Omega, \end{cases}$$

so $\bar{u}_0$ is a superposition of (10.39). Define $(\bar{u}_j)$ by recurrence as the successful solution of the equation

$$\begin{aligned} L_\lambda \bar{u}_{j+1} &= \lambda h(x, \bar{u}_j) & \text{in } \Omega, \\ \bar{u}_{j+1}(x) &= 0 & \text{on } \partial\Omega. \end{aligned}$$

Then the sequence $\{\bar{u}_j\}$ is monotone decreasing of positive functions. In fact,

(a)

$$\begin{aligned} L_\lambda(\bar{u}_1) &= \lambda h(x, \bar{u}_0) = \lambda K \bar{u}_0 + \lambda f(x, \bar{u}_0), \\ L_\lambda(\bar{u}_0) &= (L + \lambda K)\bar{u}_0 = \lambda K \bar{u}_0 + c(x)\bar{u}_0. \end{aligned}$$

Therefore,

$$\begin{cases} L_\lambda(\bar{u}_1 - \bar{u}_0) = \lambda f(x, \bar{u}_0) - c(x)\bar{u}_0 \le 0 & \text{in} \quad \Omega, \\ \bar{u}_1 - \bar{u}_0 \le 0 & \text{on } \partial\Omega, \end{cases}$$

$$\begin{cases} L_\lambda(\bar{u}_1) = \lambda h(x, \bar{u}_0) \ge 0 & \text{in} \quad \Omega, \\ \bar{u}_1 = 0 & \text{on } \partial\Omega. \end{cases}$$

By the Maximum Principle 10.7, because $L_\lambda(\bar{u}_1, \bar{u}_0) \le 0$, we have $\bar{u}_1 - \bar{u}_0 \le 0$: $\bar{u}_1 \le \bar{u}_0$ in Ω. Moreover, because $L_\lambda \bar{u}_1 \ge 0$, we have $\bar{u}_1 \ge 0$ in Ω. We conclude that $0 \le \bar{u}_1 \le \bar{u}_0$.

(b) Set $0 \leq \bar{u}_j \leq \bar{u}_{j-1} \leq \cdots \leq \bar{u}_0$

$$L_\lambda(\bar{u}_{j+1} - \bar{u}_j) = \lambda[h(x, \bar{u}_j) - h(x, \bar{u}_{j-1})] \leq 0 \quad \text{in } \Omega,$$
$$\bar{u}_{j+1} - \bar{u}_j = 0 \quad \text{on } \partial\Omega.$$

By the Maximum Principle 10.7, $0 \leq \bar{u}_{j+1} \leq \bar{u}_j$, and by recurrence, $0 \leq \cdots \leq \bar{u}_{j+1} \leq \bar{u}_j \leq \cdots \leq \bar{u}_0 = \bar{z}$.

By the Agmon-Douglis-Nirenberg Estimation 10.3,

$$\|\bar{u}_j\|_{C^{1,1-N/p}} \leq c \quad \text{for each } j,$$

and by the Schauder Estimation 10.1,

$$\|\bar{u}_j\|_{C^{2,\alpha}} \leq c.$$

Because the imbedding $i : C^{2,\alpha}(\Omega) \to C^2(\Omega)$ is compact there is a subsequence of $\{\bar{u}_j\}$, still denoted by $\{\bar{u}_j\}$, $\bar{u}_\lambda \in C^2$ such that $\bar{u}_j \to \bar{u}_\lambda$ in C^2. Now $\bar{u}_\lambda$ is a solution of (10.39).

(iv) $\bar{u}_\lambda$ is the unique maximal positive solution of (10.40) if v is a positive solution of (10.40). Then $0 \leq v < \bar{z}$ for $j \in \mathbb{N}$, and

$$\begin{cases} L_\lambda(\bar{u}_{j+1} - v) = \lambda[h(x, \bar{u}_j) - h(x, v)] & \text{in } \Omega, \\ \bar{u}_{j+1} - v = 0 \quad \text{on } \partial\Omega. \end{cases}$$

By the Recurrence and Maximum Principles 10.7. $\blacksquare$

Remark 10.69 *We may replace (fii) and $(fiii)$ by*
(fv) : there is $\bar{z} > 0$ such that $f(x, z) \geq 0$, $f(x, z) \not\equiv 0$ for $z \in [0, \bar{z}]$, and $f(x, z) \leq 0$ for $z > \bar{z}$.
We then obtain more precise results: there is a connected component C of solutions of (10.39) in $\mathbb{R}^+ \times P^+$ such that $\lambda > \mu$, in C there is a pair of distinct solutions $(\lambda, \underline{u}(\lambda))$ and $(\lambda, \bar{u}(\lambda))\mu = \inf\{\lambda \in \mathbb{R}^+ \mid. \text{ Equation } (10.39) \text{ has a solution in } P^+\}$, $\mu \leq \underline{\lambda}$.

10.10   Existence of Pairs of Solutions in S_k

Consider the ordinary differential equation

$$\begin{cases} \mathcal{L}u = \lambda f(x, u) & \text{in } (0, \pi), \\ u(0) = u(\pi) = 1, \end{cases} \tag{10.41}$$

where $Lu = -(pu')' + qu$ with $p > 0$, $p \in C^1([0, \pi])$, $q \geq 0$, and $q \in C([0, \pi])$. Let f satisfy:

$(f1)$ $f(x,z) = g(x,z)z$, where $g : [0,\pi] \times \mathbb{R} \to \mathbb{R}$ is locally Lipschitz-continuous, $g(x,0) = 0$;

$(f2)$ for some $\bar{z} > 0$, $g(x,z) < 0$ for $|z| \geq \bar{z}$;

$(f3)$ there exists $r > 0$ such that $g(x,z) > 0$ for $0 < |z| < r$. Let $E = \{u \in C^1([0,\pi]) \mid u(0) = u(\pi) = 0\}$, and let $\varphi \subset \mathbb{R} \times E$ be the set of all solutions of equation (10.41).

Theorem 10.70 *For $k \in \mathbb{N}$, there exists $\lambda_k \in \mathbb{R}^+$ and a connected component C_k of N, unbounded and included in $\mathbb{R} \times S_k$ such that $\lambda \geq \lambda_k$. In C_k there are two distinct solutions $(\lambda, \underline{u}_k(\lambda))$ and $(\lambda, \bar{u}_k(\lambda))$.*

Proof.

(i) If (λ, u) is a solution of (10.41), $\lambda \geq 0$. By the Maximum Principle $\|u\|_0 < \bar{z}$: let $\bar{g} : [0,\pi] \times \mathbb{R} \to \mathbb{R}$ defined by

$$\bar{g}(x,z) = \begin{cases} g(x,z) & \text{for } |z| \leq \bar{z} \\ g(x,\bar{z}) & \text{for } |z| \geq \bar{z} \\ g(x,-\bar{z}) & \text{for } z \leq -\bar{z}, \end{cases}$$

then $\bar{g}$ is Lipschitz on $[0,\pi] \times \mathbb{R}$. Set $\bar{f}(x,z) = \bar{g}(x,z)z$. Consider the truncated equation

$$\begin{cases} \mathcal{L}u = \lambda \bar{f}(x,u) & \text{in } (0,\pi) \\ u(0) = u(\pi) = 0. \end{cases} \tag{10.42}$$

Then the solutions u of equations (10.41) and (10.42) are the same, and satisfy $\|u\|_0 < \bar{z}$.

(ii) For $k \in \mathbb{N}$, there exists $\lambda_k \in \mathbb{R}^+$ such that for $\lambda \geq \lambda_k$, (10.41) possesses at least two distinct solutions $\underline{u}_k(\lambda)$ and $\bar{u}_k(\lambda)$ in S_k. Approach (10.42) by the equation

$$\begin{cases} \mathcal{L}u = \lambda[\theta u + \bar{f}(x,u)] & \text{in } (0,\pi), \\ u(0) = u(\pi) = 0. \end{cases} \tag{10.43}$$

We know that for $k \in \mathbb{N}$, there is a connected component of solutions of equation (10.43), $C_{k,\theta}$ is unbounded and included in $(\mathbb{R} \times S_k) \cup \{(\mu_k(\theta), 0\}$, where $\mu_k(\theta)$ is the kth eigenvalue of

$$\begin{cases} \mathcal{L}u = \mu\theta u & \text{in } (0,\pi), \\ u(0) = u(\pi) = 0. \end{cases} \tag{10.44}$$

We obtain $\mu_k(\theta) = \frac{\mu_k(1)}{\theta}$, and $\mu_k(\theta) \to \infty$ as $\theta \to 0$. The idea of the proof is a priori estimates of the solutions of equation (10.43), contaminant $C_{k,\theta}$ goes through two distinct points of $\{\lambda\} \times E$ for $\lambda \geq \lambda_k$, as $\theta \to 0$. ∎

Let

$$\rho = \inf_{\substack{x \in [0,\pi] \\ z = \pm \bar{z}}} |g(x,z)|, \qquad \rho > 0.$$

Lemma 10.71 *For each solution (λ, u) of (10.42), $u \neq 0$, $\theta \in [0, \rho)$, we have $\|u\|_0 < \bar{z}$ and $\lambda \geq \underline{\lambda}$ where $\underline{\lambda}$ is a positive constant.*

Proof. By the Maximum Principle $\|u\|_0 < \bar{z}$, for $|z| \geq \bar{z}$ implies $\theta + \bar{z}(x,z) < 0$. Set $M = \rho + \max_{\substack{x \in [0,\pi] \\ |z| \leq \bar{z}}} \bar{g}(x,z)$, we have $\lambda \geq \underline{\lambda} = \frac{\mu_1(1)}{M}$. Otherwise $\lambda[\theta + \bar{g}(x,u)] < \mu_1(1)$.

By the Sturm Comparison Theorem, there is a function $v_1 \neq 0$ that is a solution of

$$\begin{cases} \mathcal{L}v_1 = \mu_1(1)v_1 & \text{in } (0,\pi), \\ v_1(0) = v_1(\pi) = 0, \end{cases}$$

and admits a zero in $(0,\pi)$. This contradicts $v_1 \in S_1$. Assume $a < a_1 < c < b_1 < b$ such that $u(a_1) = u(b_1) = \frac{\alpha}{2}$, and $u(c) = \max_{x \in I} u(x)$. We have

$$u'(x) = -\int_x^c u''(t)dt.$$

For $x \in [a,c]$, we have $\lambda k \frac{\alpha}{2}(c-x) \leq u'(x) \leq \lambda K \beta(c-x)$, so

$$\begin{cases} \dfrac{\lambda k \alpha}{4}(c-a_1)^2 \leq \beta - \dfrac{\alpha}{2}, \\ \dfrac{\lambda K \beta}{2}(c-a_1)^2 \geq \dfrac{\alpha}{2}, \end{cases}$$

or

$$\begin{cases} c - a_1 \leq \dfrac{k_1}{\sqrt{\lambda}} \\ u'(a_1) \geq k_1\sqrt{\lambda}. \end{cases}$$

k, k_i', K_i and K_i', are positive constants independent of λ, and $\theta \in [0, \rho)$, and similarly, $b_1 \cdot c \leq \frac{k_1'}{\sqrt{\lambda}}$ and $|u'(b_1)| \geq K_1'\sqrt{\lambda}$. For $x \in [a, a_1]$, $u'(x) \geq u'(a_1)$, so

$$\frac{\alpha}{2} \geq u'(a_1)(a_1 - a),$$

or

$$a_1 - a \leq \frac{k_2}{\sqrt{\lambda}},$$

we have $b_1 - a_1 \leq \frac{k_2'}{\sqrt{\lambda}}$, and $b - a \leq \frac{k_0}{\sqrt{\lambda}}$. Denote by $I_1, \cdots, I_k$ the k subintervals of $[0,\pi]$, where u is constant sign. It suffices to prove there is a constant $\alpha > 0$ independent of

$\lambda > 0$ and $\theta \in [0, \rho]$ such that for $j = 1, \cdots, k$, $\max_{x \in I_j} |u(x)| > \alpha_1$. In fact, if we replace α above by α_1, and repeat the k subintervals, we obtain $\pi \leq \frac{k_0}{\sqrt{\lambda}} : \lambda \leq \lambda_k$. Suppose J is an interval adjacent to I, for example $J = [d, a]$, $u(d) = u(a) = 0$, and $e \in J$ such that $|u(e)| = \max_{x \in J} |u(x)| = M$. For $x \in (d, a)$, $u(x) < 0$, and u is convex on J. Then $K_1 \sqrt{\lambda} \leq u'(a)$ implies $\frac{K_1^2}{2} \lambda \leq \int_e^a u''(t) u'(t) dt \leq \frac{\lambda K M^2}{2}$, so $M \geq \frac{k_1}{\sqrt{k}}$. By $k - 1$ steps we obtain the existence of α_1. ∎

Lemma 10.72 *Let (λ, u) be a solution of equation (10.42) with $u \in S_k$, and $\|u\|_0 < r$. Then $\lambda \leq \mu_k(\theta)$.*

Proof. If $\lambda > \mu_k(\theta)$, then $\lambda[\theta + \bar{g}(x, u)] > \mu_k(\theta)\theta$ for $x \in [0, \pi]$. This contradicts the Sturm Comparison Theorem, because $u \in S_k$. ∎

We continue to prove Theorem 10.70: the above lemmas assert that $C_{k,\theta}$ cannot go through the indicated area. We claim that for each $\lambda \geq \lambda_k$ there are two solutions of equation (10.42) in S_k, $(\lambda, \underline{u}_k(\lambda, \theta))$ and $(\lambda, \bar{u}_k(\lambda, \theta))$ with

$$0 < \|\underline{u}_k(\lambda, \theta)\|_0 < \alpha, \quad \beta \leq \|\underline{u}_k(\lambda, \theta)\|_0 < \bar{z}.$$

However $\|\underline{u}_k(\lambda, \theta)\|_0$ is also bounded below for small θ. In fact, by the Sturm Comparison Theorem $\|\theta + \bar{g}(x, u)\|_0 \geq \frac{\mu_k(1)}{\lambda}$ for a solution $(\lambda, u) \in \mathbb{R} \times S_k$ of $(10.3)_\theta$, $\theta \leq \frac{1}{2} \frac{\mu_k(1)}{\lambda}$. For example $\|\underline{u}_k(\lambda, \theta)\|_0 > \alpha_0 > 0$. The equation (10.42) asserts that $\underline{u}_k(\lambda, \theta)$ and $\bar{u}_k(\lambda, \theta)$ are bounded in $C^2([0, \pi])$-norm. There is a subsequence convergent in $C^1([0, \pi])$. By the equation, in $C^2([0, \pi])$ towards $\underline{u}_k(\lambda)$ and $\bar{u}_k(\lambda)$ respectively, distinct nonzero solutions of (10.42) belong to $\bar{S}_k$. Now it is impossible for one of these two solutions to be in ∂S_k, otherwise the solution admits at least one double zero in $[0, \pi]$. By the classical unicity theorem of systems of first-order equations they are zero, so $\underline{u}_k(\lambda)$, $\bar{u}_k(\lambda) \in S_k$. Let C_k be the connected component of solutions of (10.43) containing $(\lambda_k, \bar{u}_k(\lambda_k))$. It is impossible to have a bifurcation point for (10.3) on the line $\mathbb{R} \times \{0\}$, so C_k does not meet $\mathbb{R} \times \partial S_k$, contained in $\mathbb{R} \times S_k$. Let

$$\mathcal{E}_k = \{(\lambda, u) \in \mathbb{R} \times E \mid \lambda \geq \lambda_k, \beta \leq \|u\|_0 < \bar{z}\},$$
$$\mathcal{F}_k = \{(\lambda, u) \in \mathbb{R} \times E \mid \underline{\lambda} \leq \lambda \leq \lambda_k, \|u\|_0 < \bar{z}\},$$
$$\cup \mathcal{E}_k = \{(\lambda, u) \in \mathbb{R} \times E \mid \lambda \geq \lambda_k, \|u\|_0 < \alpha\},$$
$$C'_k = \mathcal{E}_k \cap C_k, C_k^2 = \mathcal{F}_k \cap C_k.$$

Suppose C_k^1 is bounded in E_k. Denote by $N_\epsilon(C_k^1)$ the neighborhood of order $\epsilon > 0$ of C_k^1 in E_k. $\overline{N_\epsilon(C_k^2)} \cap N$ is a compact metric space. By Lemma, $G \subset E_k$ is open and bounded and contains C_k^1 such that $\partial G \cap N = \emptyset$. Now this is impossible because for $\theta > 0$ small, $C_{k,\theta}$ is unbounded and passes sufficiently close to $(\lambda_k, \mu_k(\lambda_k))$ transverse to ∂O. Then $(\lambda_\theta, u_{k,\theta}) \in C_{k,\theta} \cap \partial G$. By equation (10.43), $u''_{k,\theta}$ is bounded. There is a subsequence $u_{k,\theta}$ convergent in C^1 and also in C^2 by equation (10.43). For some $\theta_n \to 0$ such that

$$(\lambda_{\theta_n}, u_k, \theta_n) \xrightarrow{\mathbb{R} \times C^2} (\lambda_0, u_{k,0}) \in \partial G \cap N,$$

C_k^1 is unbounded. Similarly C_k^2 is unbounded.

Remark 10.73 *We always have $\lambda_k \to \infty$ as $k \to \infty$. In fact, by the Sturm Comparison Theorem on $u_k(\lambda_k)$ and on an eigenvalue v_k of*

$$\begin{cases} \mathcal{L}v_k = \mu_k(1)v_k & \text{in } (0,\pi), \\ v_k(0) = v_k(\pi) = 0. \end{cases}$$

We obtain $\lambda_k \geq \frac{\mu_k(1)}{M}$, where $M = \max\limits_{\substack{x \in [0,\pi] \\ |z| \leq \bar{z}}} |\bar{g}(x,z)|$, so $\lambda_k \to \infty$, because $\mu_k(1) \to \infty$.

Remark 10.74 *By Lemma 10.72, as $\alpha \to 0$ and $\beta \to r$, we obtain*

$$\lim_{r \to \infty} \|\underline{u}_k(\lambda)\|_0 = 0, \quad \text{and} \quad \liminf_{\lambda \to \infty} \|\bar{u}_k(\lambda)\|_0 \geq r.$$

Notes Most material in this Chapter is adapted from Rabinowitz [38]. For related material see Coddington-Levinson [14], Dieudonne [18], Dugundji [20], and Stein [45].

Bibliography

[1] Adams, R.A., **Sobolev Spaces**, Academic Press, 1975.

[2] Adams, R.A., *Compact Sobolev Imbeddings for unbounded domains with discrete boundaries*, J. Math. Anal. and Appl., 24(1968), 326–333.

[3] Ambrosetti, A. and Rabinowitz, P.H., *Dual variational methods in critical point theory and applications*, J. Funct. Anal., 14(1973), 349–381.

[4] Avez, A., **Differential Calculus**, John Wiley and Sons, 1986.

[5] Bahri, A. and Coron, J.M., *On a nonlinear elliptic equation involving the critical Sobolev exponent: the effect of topology of the domain*, Comm. Pure Appl. Math., 16(1988), 253–294.

[6] Berger, M., **Nonlinearity and Functional Analysis,** Academic Press, 1977.

[7] Brezis, H., **Analyse Fonctionnelle; theorie et application**, Masson, 1983.

[8] Brezis, H., **Operteurs Maximaux Monotones et Semi-groupes de Contractions dans les Espaces de Hilbert**, Notas de Mat. 59, American Elsevier, 1973.

[9] Brezis, H., *Elliptic equations with limiting Sobolev exponents—the impact of topology*, Comm. Pure Appl. Math., 39(1986), S17–S39.

[10] Brezis, H. and Nirenberg, L., *Positive solutions of nonlinear elliptic equations involving critical Sobolev exponents*, Comm. Pure Appl. Math., 35(1983), 437–477.

[11] Browder, F., *Fixed point theory and nonlinear problems*, Bull. Amer. Math. Soc., 9(1983), 1–40.

[12] Cartan, H., **Cours de Calcul Differentiel,** 2nd ed., Hermann, Paris, 1977.

[13] Choquet, G., **Lectures on Analysis** (3 volumes), Benjamin 1969

[14] Coddington, E. and Levinson, N., **Theory of Ordinary Differential Equations**, McGraw-Hill, New York, 1955.

[15] Crandall, M.G. and Rabinowitz, P.H., B*ifurcation, perturbation of simple eigenvalues, and liberalized stability*, Arch. Rational Mech. Anal., 52(1973), 161–180.

[16] Dancer, E.N., *Bifurcation theory in real Banach space*, Jour. London Math. Soc., 23(1971), 699–734.

[17] Deimling, K., **Nonlinear Functional Analysis**, Springer Verlag, 1985.

[18] Dieudonne, J., **Foundation of Modern Analysis**, Academic Press, Inc., New York, 1960.

[19] Dugundji, J., *An extension of Tietze Theorem*, Pacific J. of Math., 1(1951), 353–367.

[20] Dugundji, J., **Topology**, Allyn and Bacon, Inc., Boston, 1966.

[21] Dugundji, J. and Granas, A., **Fixed-Point Theory, Vol. 1**, Polish Scientific Publisher, 1982.

[22] Dunford, N. and Schwartz, J.T., **Linear Operators** (3 volumes), Interscience, 1958.

[23] Friedman, A., **Partial Differential Equations**, Robert E. Krieger Publishing Company, 1976.

[24] Gilbarg, D. and Trudinger, N.S., **Elliptic Partial Differential Equations of Second Order**, 2nd Edition, Springer-Verlag, 1977.

[25] Giusti, E., **Minimal Surfaces and Functions of Bounded Variations**, Lecture Notes, Australian National University, Canberra, 1977.

[26] Hormander, L, **Fourier Integral Operators I**, Acta Math., 127(1971), 79-183.

[27] Krasnoselskii, M., **Positive Solutions of Operator Equations**, P. Noordhoff, Groningen, 1964.

[28] Krasnoselskii, M., **Topological Methods in the Theory of Nonlinear Integral Equations**, Macmillan, 1964.

[29] Larsen, R., **Functional Analysis; an introduction**, Dekker, 1973.

[30] Lions, P.L., *The concentration-compactness principle in the calculus of variations, the locally compact case, Part 1*, Ann. Inst. H. Poincaré, 1(1984), 109–145.

[31] Lions, P.L., *The concentration-compactness principle in the calculus of variations, the locally compact case, Part 2*, Ann. Inst. H. Poincaré, 1(1984), 223–283.

[32] Lions, P.L., *The concentration-compactness principle in the calculus of variations, the limit case, Part 1*, Rev. Mat. Iberoamericana 1.1(1985), 145-201.

[33] Lions, P.L., *The concentration-compactness principle in the calculus of variations, the limit case, Part 2*, Rev. Mat. Iberoamericana 1.2(1985), 45–121.

[34] Mazja, V.G., **Sobolev Spaces**, Springer Verlag. 1985.

[35] Milnor, J., **Topology from the Differentiable Point of View**, The University Press of Virginia, 1965.

[36] Nirenberg, L., **Topics in Nonlinear Functional Analysis**, Courant Institute of Mathematical Sciences, New York University, 1974.

[37] Rabinowitz, P., *Theorie du degre topologique et applications a des problemes aux limites non lineaires*, Univ. Laborat. Anal. Num., Paris, 1975.

[38] Rabinowitz, P.H., **Applications of Topological Degrees**, Université de Paris VI, 1975.

[39] Rabinowitz, P.H., *A global theory for nonlinear eigenvalue problems and applications*, in Contrib. Nonlinear Functional Analysis, Academic Press, 1971, 11–36.

[40] Reed, M. and Simon, B., **Methods of Modern Mathematical Physics; 1. Functional Analysis**, Academic Press, 1972.

[41] Rudin, W., **Functional Analysis**, McGraw-Hill, 1973.

[42] Sattinger, D.H., **Topics in Stability and Bifurcation Theory**, Springer-Verlag, New York, 1973.

[43] Schechter, M., **Principles of Functional Analysis**, Academic Press, New York, 1971.

[44] Schwartz, J.T., **Nonlinear Functional Analysis**, Gordon and Breach, 1969.

[45] Stein, E.M., **Singular Integrals and Differentiability Properties of Functions**, Princeton University Press, Princeton, New Jersey, 1970.

[46] Struwe, M., *A global compactness result for elliptic boundary value problem involving limiting nonlinearities*, Math. Z., 187(1984), 511–517.

[47] Struwe, M., **Variational Methods**, Springer-Verlag, Berlin - Heidelberg - New York, Second edition, 1996.

[48] Talenti, G., *Best constants in Sobolev inequality*, Annali di Mat., 110(1970), 353–372.

[49] Temme, N.M., **Nonlinear Analysis**, Vols 1, 2, Mathematisch Centrum, Amsterdam, 1978.

[50] Torchinsky, A., **Real-Variable Methods in Harmonic Analysis**, Academic Press, Inc., 1986.

[51] Wheeden, R.L., and Zygmund, A., **Measure and Integral**, Marcel Dekker, Inc., New York and Basel, 1977.

[52] Yosida, K., **Functional Analysis**, Springer Verlag, 1965.

[53] Zeidler, E., **Nonlinear Functional Analysis and its Applications I: Fixed-Point Theorems**, Springer Verlag, 1984.

[54] Zeidler, E., **Nonlinear Functional Analysis and its Applications II: Monotone Operators**, Springer Verlag, 1984.

[55] Zeidler, E., **Nonlinear Functional Analysis and its Applications III**, Springer, New York, 1988.

Index

additive property, 309, 351
Agmon-Douglis-Nirenberg theorem, 368
algebraic multiplicity, 186
Ascoli Theorem, 9

Banach fixed point theorem, 19
Banach theorem, 2
Banach-Alaoglu theorem, 1
Banach-Steinhaus theorem, 1
best Sobolev constant, 263
bifurcation equation, 54
bifurcation point, 64, 390, 404
bilinear form, 216
Borsuk Antipole theorem, 337
Borsuk-Ulam theorem, 339
boundary manifold, 373
Brezis-Lieb lemma, 16
Brezis-Nirenberg-Stampacchia theorem, 330
Brouwer fixed point theorem, 24, 316, 323, 338
Brouwer invariant domain theorem, 345

Caratheodory function, 14
Cauchy problem, 227
chain rule, 31, 49
characteristic value, 321
characterization theorem, 116
Charles relation, 37
Clarkson inequalities, 8
classical derivative, 77
classical solution, 78
closed curve theorem of Jordan, 345
coercive, 216, 270
compact homotopy invariance, 351
compact linear operator, 185
compact operatorr, 185
concentration function, 287

concentration-compactness principle, 244, 253
condition (P), 376
constant in component, 351
continuity of function, 309, 351
continuous dependence on a parameter, 21
convex, 203
critical point, 35

de Beppo Levi theorem, 13
deformation theorem, 225, 274
degree, 297
dichotomy, 237
diffeomorphism, 39
Dugundji extension theorem, 6, 383

eigenvalue, 321
Ekeland variational principle, 258
elliptic, 367
essential, 24, 315
excision property, 312, 354
expansive, 200
extension operator theorem, 130
extension theorem, 87

F-derivative, 28
F-differentiable, 27
F-differential, 28
F. Riesz lemma, 1, 186
finite-dimensional pertuurbation of the identity, 348
first density theorem, 89
first trace theorem, 176
formula for changing variables, 122
forth density theorem, 131
forward cone, 258
Fréchet-Kolmogorom theorem, 10

Fredholm alternative theorem, 192
Fredholm operator, 194
Friedrichs theorem, 113
fundamental theorem of calculus, 38

G-derivative, 28
G-differentiable, 28
G-differential, 28
generalized Morse lemma, 60
genus, 219
GL(E,F), 39

Hölder space, 367
Hahn-Banach theorem, 1
Hartman-Stampacchia theorem, 25, 207, 324
heaviside function, 79
Helly theorem, 92, 237
Hilbert cube, 8, 383
homotopy invariance, 308, 309
Hopf maximum principle, 370

implicit function theorem, 42
indefinite, 58
index, 194
index of the critical point, 58
invariance of domain theorem, 42
inverse function theorem, 40, 53
inversion of an isomorphism between Banach spaces, 39
isomorphic, 39
Ize lemma, 69

KKM theorem, 25, 324
Kneser-Hukuhara theorem, 380, 381
Krasnosellski-Perov theorem, 378
Krasnoselskii theorem, 68
Krein-Rutman theorem, 386
Ky Fan minimax inequality, 25, 324

Lagrange multiplier theorem, 45
Lax-Milgram theorem, 218
Lebesgue dominated convergence theorem, 13

Leray-Schauder degree, 350
Leray-Schauder fixed point theorem, 26, 364, 365
Leray-Schauder index theorem, linear case, 358
Leray-Schauder index theorem, nonlinear case, 360
Leray-Schauder theorem, 382
level surface, 45
Lieb Theorem, 267
Lipschitz-continuous function, 181
Ljusternik-Schnirelman category, 223
lower semicontinuous, 4
Lyapunov-Schmidt procedure, 44

matching lemma, 179
maximum principle, 369, 370
Mazur theorem, 4, 383
mean value theorem, 33
method of continuity, 20
Meyers-Serrin theorem, 115
Minty lemma, 201
monotone operator, 200
Morrey theorem, 141
Morse lemma, 55
motion, 258
mountain pass lemma, 234
MPL-condition, 271
multiplicative property of degree, 342

negative-definite, 58
Nemytskii lemma, 14
non-retraction of unit ball on its boundary, 24, 315
nonexpansive operator, 200

open mapping theorem, 1

parallelogram law, 9
partition of unity lemma, 114, 130
Picard-Lindelof theorem, 22
Pohozaev theorem, 265
Poincaré Inequality, 101, 159, 167

Poincaré-Wirtinger inequality, 102, 123, 169

positive-definite, 58, 369

primitive, 38

product rule, 31, 48, 92

projection on a closed convex set, 5, 213

proper, 377

property (H), 66

PS condition, 225, 276

PSc condition, 276

PSc sequence, 276

pseudogradient vector, 224

Rabinowitz theorem, 71

Rademacher theorem, 181

Radon theorem, 8

reflexive, 2

reflexive extension lemma, 125

regularity sequence, 10

regulated function, 37

relatively compact, 9

Rellich-Kondrachov theorem, 147, 159

retraction, 24, 315

Riesz-Fréchet representation theorem, 3, 211

Sard theorem, 35

Schauder estimation, 368

Schauder fixed point theorem, 25, 362, 363

Schwarz theorem, 47, 48

second concentration-compactness lemma, 255

second density theorem, 113

second derivative, 47

second trace theorem, 177

seminorm, 367

separation theorem of Jordan, 344

Sobolev embedding theorem, 158

Sobolev-Gagliardo-Nirenberg theorem, 136

Stampacchia theorem, 216

standard MPL, 271

stationary, 185

step function, 37

Stone-Weierstrass theorem, 6

strictly elliptic, 367

strong maximum principle, 370

strong solution, 78

Sturm comparison theorem, 395

Sturm-Liouville nonlinear equation, 406

sum rule, 30

Taylor's formula with integral remainder, 51

Taylor's formula with Lagrange's remainder, 51

the Heinz integral of degree, 297

third density theorem, 115

third trace theorem, 178

Tietze-Urysohn extension theorem, 6

Trudinger inequality, 152, 166

twice differentiable, 47

uniform convergence theorem, 34

uniformly convex, 9

uniformly elliptic, 367

uniformly equicontinuous, 9

unit sphere essential on the unit ball, 24, 316

Urysohn theorem, 8, 383

Von Neumann minimax theorem, 327

weak derivative, 77

weak solution, 78

weakly compact, 5

Weierstrass theorem, 5

非 線 性 分 析
Nonlinear Analysis

作　　者：	王懷權
發 行 人：	徐遐生
出 版 者：	國立清華大學出版社
社　　長：	周懷樸
地　　址：	300 新竹市光復路二段 101 號
電　　話：	03-5714337　03-5715131 轉 5050
傳　　眞：	03-5744691
網　　址：	http://my.nthu.edu.tw/~thup/
電子信箱：	thup@my.nthu.edu.tw
行政編輯：	陳文芳
執行編輯：	周興
總 經 銷：	全華科技圖書股份有限公司
地　　址：	台北市中山區龍江路 76 巷 20 號 2 樓
電　　話：	02-2507-1300　傳眞：02-2506-2993
網　　址：	www.opentech.com.tw
出版日期：	民國 92 年 12 月初版
定　　價：	平裝本新台幣 1,500 元

ISBN　957-01-4476-9

全華科技圖書 敬上

親愛的書友：

感謝您對全華圖書的支持與愛用，雖然我們很慎重的處理每一本書，但尚有疏漏之處，若您發現本書有任何錯誤的地方，請填寫於勘誤表內並寄回，我們將於再版時修正。您的批評與指教是我們進步的原動力，謝謝您！

勘 誤 表

書 號		書 名		作 者
頁 數	行 數	錯誤或不當之詞句		建議修改之詞句

我有話要說：（其它之批評與建議，如封面、編排、內容、印刷品質等…）

書友服務卡

為加強對您的服務，只要您填妥本卡寄回全華圖書（免貼郵票），即可成為全華書友！（詳情見背面說明）

填寫日期： ／ ／

姓　名／□□□　生　日／____年（西元）____月____日　性　別／□男　□女

地　址／□□□　____縣/市　____鄉鎮/市區　____街/路　____段　____號____樓之____

電　話／(H)____　(O)____　(行動)____　(FAX)____

E-mail／____

教育程度／□1.高中、職　□2.專科　□3.大學　□4.研究所

職　業／□1.工程師　□2.教師　□3.學生　□4.軍　□5.公

□6.其他

服務單位／學校____　科系、部門____

購買圖書／書號____　書名____

您的閱讀嗜好／□A.電子　□B.電機　□C.計算機工程　□D.資訊　□E.機械

□F.汽車　□I.工管　□K.化工　□L.美工　□M.商管

□o.其他

您購買本書的原因／□1.個人需要　□2.幫公司採購　□3.老師推薦

□4.書友特惠活動

您從何處購買本書／□1.網站　□2.書局　□3.書友特惠活動　□4.團購

□5.書展　□6.其他

您對本書的評價／1.非常滿意　2.滿意　3.普通　4.不滿意　5.非常不滿意（請填代號）

□內容　□版面編排　□圖片　□文筆流暢　□封面設計

您希望全華加強那些服務／□1.電子報　□2.定期目錄　□3.書友雜誌

□4.促銷活動　□5.專業展覽通知